Vascular Endothelium in Human Physiology and Pathophysiology

The Endothelial Cell Research Series
A series of significant reviews of basic and clinical research related to the endothelium.
Edited by *Gabor M. Rubanyi*, *Berlex Biosciences, Richmond, California.*

Vascular Endothelium in Human Physiology and Pathophysiology

Edited by

Patrick J.T. Vallance

Centre for Clinical Pharmacology
University College London Medical School
UK

and

David J. Webb

Clinical Pharmacology Unit and Research Centre
University of Edinburgh
UK

hoap

harwood academic publishers
Australia • Canada • France • Germany • India • Japan
Luxembourg • Malaysia • The Netherlands • Russia
Singapore • Switzerland

Amsteldijk 166
1st Floor
1079 LH Amsterdam
The Netherlands

British Library Cataloguing in Publication Data

A catalogue record for this book is available from the British Library.

ISBN: 90-5702-489-6
ISSN: 1384-1270

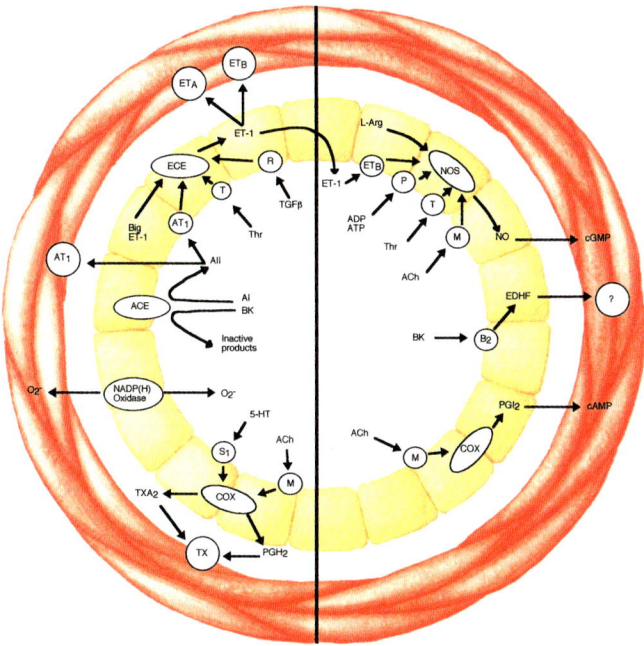

CONTRACTION

RELAXATION

ET_B ET_A
ET-1
L-Arg
ECE R ET-1 ET_B P NOS
T TGFβ ADP ATP T
Big ET-1 AT_1 Thr Thr M
AT_1 AII ACh NO cGMP
ACE AI BK EDHF ?
Inactive products BK B_2
NADP(H) Oxidase O_2⁻
O_2⁻ PGI_2 cAMP
5-HT
S_1 ACh ACh M COX
TXA2 COX M
TX PGH2

CONTENTS

FOREWORD

Once thought to represent an inert lining to the blood vessels, an ever increasing number of crucial functions of the vascular endothelium have emerged in recent years. This monolayer of cells which envelops the circulating blood maintains patency of the vessel through the production of anti-platelet and anti-thrombotic agents. It also regulates vascular tone through the generation of the vasodilator agent, nitric oxide, and the potent vasoconstrictors, endothelin-I and angiotensin II. The endothelium is a signal transducer for vascular effects of circulating hormones as well as a site for metabolism of hormones and blood lipids.

Arising from the discovery of the central role of normal endothelial function in maintaining cardiovascular homeostasis, has emerged the concept that endothelial dysfunction may contribute importantly to cardiovascular disease. Reduced nitric oxide function has been recognised in a wide range of conditions associated with a predisposition to cardiovascular events, including hypertension, hypercholesterolaemia, diabetes, the menopause and heart failure, and in a wide range of vessel types including peripheral resistance vessels, large conduit arteries and the coronary circulation. In some instances the 'fault' seems to be in the pathway generating nitric oxide. However, in others, increased degradation of nitric oxide, perhaps as a result of oxidative stress, may play an important role. In either case the end result is a tendency to increased vascular tone, increased risk of vascular thrombosis, and initiation or potentiation of the atherogenic process. More recently it has become clear that there is a close relationship between the endothelin and nitric oxide systems, and in many situations where the nitric oxide system is underactive, the activity of the endothelin system is enhanced, compounding the adverse effects on vascular structure and function. Indeed, a more 'global' dysfunction of the endothelium, affecting a number of endothelial mediators, may be a feature of many forms of cardiovascular disease.

In recent years, vascular biology, and in particular the biology of the endothelium, has been an exciting and fast moving area of research. The present understanding of the role of the endothelium in cardiovascular health and disease resulted from the rapid application of the results of animal experiments to studies in patients. Endothelial dysfunction is now seen as perhaps one of the earliest markers of vascular disease and an important target for therapeutic intervention. Novel drug approaches have been made possible by the development of new agents targeting the nitric oxide system, including L-arginine and the nitrosothiol nitric oxide donors, the widespread potential for use of anti-oxidant vitamins, and the very rapid clinical development of endothelin receptor antagonists.

Now that the concept of endothelial dysfunction is widely accepted, and the key issues can be addressed by mechanistic studies and clinical trials in patients, it is timely to draw together our understanding of the role of endothelial function in health and cardiovascular disease. This book, involving state-of-the-art reviews by leaders in their respective fields, provides an up-to-date review of the vascular functions of the endothelium and its role in key areas of cardiovascular disease in humans. This book is targeted both at the basic and clinical scientists, in industry or academia, who wish a broad overview of the field and an understanding of the data available from studies in humans. It will also be valuable for physicians who want to understand the role that endothelial dysfunction may play in the patients under their care, and the emerging opportunities for new treatments.

Sir John Vane

CONTRIBUTORS

Bagnall, A.J.
Centre for Genome Research
Roger Land Building
University of Edinburgh
West Mains Road
Edinburgh, EH9 3JQ
UK

Bhagat, K.
Centre for Clinical Pharmacology
The Rayne Institute
University College London
5 University Street
London, WC1E 6JJ
UK

Binggeli, C.
Division of Cardiology
University Hospital Zurich
CH-8091 Zurich
Switzerland

Celermajer, D.S.
Department of Cardiology
Royal Prince Alfred Hospital
Missenden Road
Camperdown
Sydney, NSW 2050
Australia

Chowienczyk, P.J.
Department of Clinical Pharmacology
St Thomas' Hospital, UMDS
Lambeth Palace Road
London, SEI 7EH
UK

Cooke, J.P.
Division of Cardiovascular Medicine
Stanford University School of Medicine
300 Pasteur Drive
Stanford, CA 94305 5406
USA

Corti, R.
Division of Cardiology
University Hospital Zurich
CH-8091 Zurich
Switzerland

Creager, M.A.
Vascular Medicine and Atherosclerosis Unit
Brigham and Women's Hospital and
    Harvard Medical School
Boston, MA 02115
USA

Deanfield, J.E.
Cardio-Thoracic Unit
Great Ormond Hospital for Children
    NHS Trust
Great Ormond Street
London, WC1 3JH
UK

Drexler, H.
Department of Cardiology
Medical University Hannover
Carl-Neuberg-Str. 1
D-30625 Hannover
Germany

Evans, T.W.
Department of Critical Care Medicine
National Heart and Lung Institute
Dovehouse Street
London, SW3 6LY
UK

Félétou, M.
Institut de Recherche Internationales
    Servier
6 place des Pléiades
92415 Courbevoie Cedex
France

Haynes, W.B.
Department of Internal Medicine
University of Iowa
Iowa City, IA 52241
USA

Hingorani, A.D.
Centre for Clinical Pharmacology
The Wolfson Institute for Biomedical
    Research
University College London Medical School
140 Tottenham Court Road
London, W1P 9LN
UK

Kiowski, W.
Division of Cardiology
University Hospital Zurich
CH-8091 Zurich
Switzerland

Lüscher, T.F.
Division of Cardiology
University Hospital Zurich
CH-8091 Zurich
Switzerland

MacAllister, R.J.
Centre for Clinical Pharmacology
The Wolfson Institute for Biomedical
    Research
University College London Medical School
140 Tottenham Court Road
London, W1P 9LN
UK

Mann, G.E.
Vascular Biology Research Centre
School of Biomedical Sciences
King's College London
Campden Hill Road
London, W8 7AH
UK

Nava, E.
Division of Cardiology
University Hospital Zurich
CH-8091 Zurich
Switzerland

Noll, G.
Division of Cardiology
University Hospital Zurich
CH-8091 Zurich
Switzerland

Poston, L.
Department of Obstetrics and Gynaecology
St Thomas' Hospital, UMDS
Lambeth Palace Road
London, SE1 7EH
UK

Radomski, A.S.
Department of Pharmacology
University of Alberta
Edmonton
Alberta, T6G 2H7
Canada

Radomski, M.W.
Department of Pharmacology
University of Alberta
Edmonton
Alberta, T6G 2H7
Canada

Ritter, J.M.
Department of Clinical Pharmacology
St. Thomas' Hospital, UMDS
Lambeth Palace Road
London, SE1 7EH
UK

Sudano, I.
Division of Cardiology
University Hospital Zurich
CH-8091 Zurich
Switzerland

Taddei, S.
Department of Internal Medicine
University of Pisa
Via Roma 67
56100 Pisa
Italy

Vallance, P.J.T.
Centre for Clinical Pharmacology
The Wolfson Institute for Biomedical
    Research
University College London Medical School
140 Tottenham Court Road
London, W1P 9LN
UK

Vanhoutte, P.M.
Institut de Recherche Internationales Servier
6 place des Pléiades
92415 Courbevoie, Cedex
France

Webb, D.J.
Clinical Pharmacology Unit and Research
    Centre
University of Edinburgh
Western General Hospital
Crewe Road
Edinburgh, EH4 2XU
UK

Williams, D.J.
Department of Obstetric and Gynaecology
Imperial College School of Medicine
Chelsea and Westminster Hospital
Fulham Rd
London, SW10 9NH
UK

ACKNOWLEDGEMENTS

None of this would have been possible without the hard work and careful cross checking done by Ann Wemyss. She also managed to keep us all to a time scale in a diplomatic and persuasive manner. In addition we would like to thank Mark Miller for providing the overview figures for each chapter.

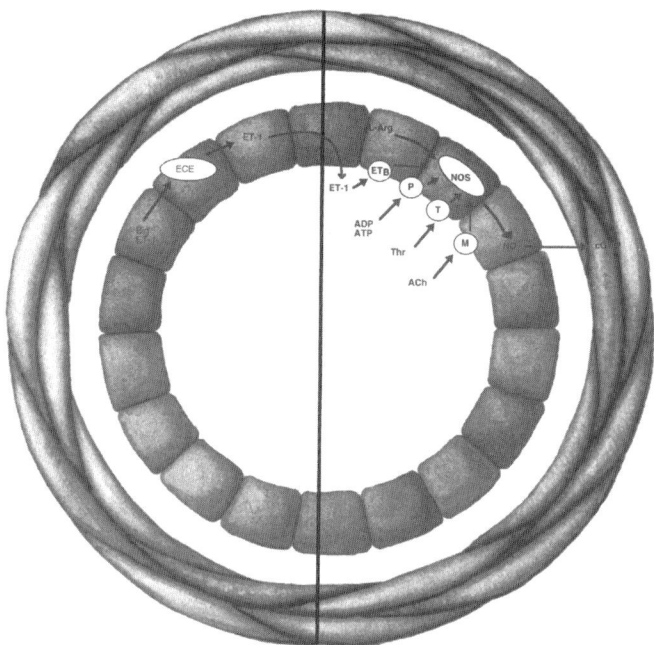

The gaseous mediator nitric oxide (NO) is synthesised in the vascular endothelium from the substrate L-arginine in an enzymatic reaction catalysed by endothelial nitric oxide synthase. Nitric oxide influences the function of circulating cells and the underlying smooth muscle to contribute to the tonic atheroprotective, thromboresistant and vasodilator influence of the endothelium. The endothelial output of nitric oxide can be modulated over the short- and long-term by both chemical and physical signals and might, in part, be genetically-determined. Dysfunction of the endothelial:NO pathway may contribute to a variety of cardiovascular disorders including essential hypertension, atherosclerosis and pre-eclampsia.

1 Endothelial Nitric Oxide

Aroon D. Hingorani and Patrick J. Vallance

Centre for Clinical Pharmacology, University College London, 140 Tottenham Court Road, London, W1P 9LN, UK
Tel/Fax: 0171 209 6351; E-mail: a.hingorani@ucl.ac.uk

INTRODUCTION

In 1980 Furchgott and Zawadzki reported the observation that vascular relaxation induced by acetylcholine was dependent on the presence of the endothelial lining of the blood vessel (Furchgott and Zawadzki, 1980) If the endothelium was removed, the relaxation to acetylcholine was abolished whereas the contractile response to noradrenaline and the dilator response to glyceryl trinitrate were both preserved. Endothelium-dependent relaxation to acetylcholine was shown to be mediated by the release of an unstable, diffusible factor (endothelium-derived relaxant factor; EDRF) which had a biological half-life of a few seconds (Furchgott, 1984; Furchgott, 1990) Subsequently the dilator responses to a variety of other agonists including bradykinin, substance P, adenosine diphosphate, 5-hydroxytryptamine, β-agonists and certain neuropeptides were also shown to be endothelium-dependent and mediated by EDRF (Moncada *et al.*, 1991). The release of EDRF was also shown to attenuate the response to certain endogenous vasoconstrictors (Martin *et al.*, 1986).

For 6 years the nature of EDRF remained speculative until Furchgott and Ignarro suggested independently that it might be a simple inorganic molecule - nitric oxide (NO). This hypothesis was confirmed a year later when Palmer and colleagues and Ignarro showed that vascular endothelial cells synthesise NO and that the known biological effects of EDRF could be reproduced by exogenous administration of NO (Palmer *et al.*, 1987; Ignarro *et al.*, 1987; Radomski *et al.*, 1987a) It is now known that NO is synthesised from L-arginine (Palmer *et al.*, 1988a) in a reaction which utilises molecular oxygen, to generate NO and the co-product L-citrulline (Palmer and Moncada, 1989).

The enzymes responsible for catalysing the conversion of L-arginine to NO (the nitric oxide synthases) and their respective genes have now been identified, the biochemistry of the L-arginine:NO pathway is well established and a variety of drugs are available which modulate the activity of the system. In this chapter we discuss the physiological roles of NO in the endothelium of the human cardiovascular system, describe how it is synthesised and how this synthesis is regulated, and discuss some of the approaches used to study NO biology in humans.

ENDOTHELIUM-DEPENDENT RELAXATION

The experiments of Furchgott and Zawadzki have been replicated in a wide variety of human blood vessels *in vitro* and with a number of agonists. Endothelium-dependent relaxation has been demonstrated in human conduit arteries (O'Neil *et al.*, 1991; Chester *et al.*, 1990; Thom *et al.*, 1987; Greenberg *et al.*, 1987; Toda and Okamura, 1989; Kanamaru *et al.*, 1989; Schoeffter *et al.*, 1988; Luscher and Vanhoutte, 1988; Luscher *et al.*, 1987), small resistance vessels (Woolfson and Poston, 1990; Vila *et al.*, 1991), and veins (Thom *et al.*, 1987; Collier and Vallance, 1990). However, it is clear that the response to agonists varies between vessel types. For example, whilst acetylcholine causes near maximal relaxation of human conduit arteries (Jovanovic *et al.*, 1994; Thom *et al.*, 1987; Schoeffter *et al.*, 1988; Yasue *et al.*, 1990; Yang *et al.*, 1991; Collins *et al.*, 1993) and resistance vessels (Cockcroft *et al.*, 1994a; Chowienczyk *et al.*, 1993; Imaizumi *et al.*, 1992; Linder *et al.*, 1990), human saphenous veins (Yang *et al.*, 1991; Lawrie *et al.*, 1990; Luscher *et al.*, 1990; Thom *et al.*, 1987), hand veins (Collier and Vallance, 1990; Vallance *et al.*, 1989a) and omental venules (Wallerstedt and Bodelsson, 1997) show only a small relaxant response. Furthermore, studies in human and animal vessels show that the response to acetylcholine is complex and comprises several distinct components: endothelium-dependent relaxation (Bruning *et al.*, 1994), endothelium-independent contraction (a direct effect on muscarinic receptors on smooth muscle) (Vallance *et al.*, 1989b; Penny *et al.*, 1995), indirect neurogenic relaxation through stimulation of perivascular nerves (Loke *et al.*, 1994) and, in certain specialised vessels, endothelium-independent relaxation (Brayden and Large, 1986). The relative contribution of each component of the response may vary between vessel types, between vascular beds and between disease states. Therefore, in some instances, the overall response to acetylcholine may not represent an accurate assessment of endothelial dilator function.

Studies in humans, *in vivo* and *in vitro*, confirm that acetylcholine causes vasodilatation of conduit arteries (O'Neil *et al.*, 1991; Chester *et al.*, 1990; Thom *et al.*, 1987; Greenberg *et al.*, 1987; Toda and Okamura, 1989; Kanamaru *et al.*, 1989; Schoeffter *et al.*, 1988; Luscher and Vanhoutte, 1988; Luscher *et al.*, 1987), resistance vessels vessels (Woolfson and Poston, 1990; Hardebo *et al.*, 1987; Vila *et al.*, 1991), and veins (Thom *et al.*, 1987; Collier and Vallance, 1990), but again differences between vessels are apparent. In the resistance beds, infusion of acetylcholine causes a graded vasodilatation and no constrictor element has been reported (Vallance *et al.*, 1989b; Cockcroft *et al.*, 1994b; Chowienczyk *et al.*, 1993; Imaizumi *et al.*, 1992; Linder *et al.*, 1990), whereas in the coronary artery and superficial veins acetylcholine causes a biphasic response with low doses causing vasodilatation and higher doses causing constriction (Collier and Vallance, 1990; Angus *et al.*, 1991) (Figure 1.1). It is assumed that the dilator component is largely endothelium-dependent whereas the constriction is mediated either by a direct action on smooth muscle or by indirect neurogenic mechanisms. In superficial veins it has been demonstrated directly that the dilator component of the response to acetylcholine is endothelium-dependent *in vivo* (Collier and Vallance, 1990).

NITRIC-OXIDE MEDIATED VASODILATION

The generation of NO from L-arginine can be inhibited competitively by certain guanidino-substituted analogues of arginine of which N^G monomethyl-L-arginine (L-NMMA) has been the most widely used (Figure 1.2).

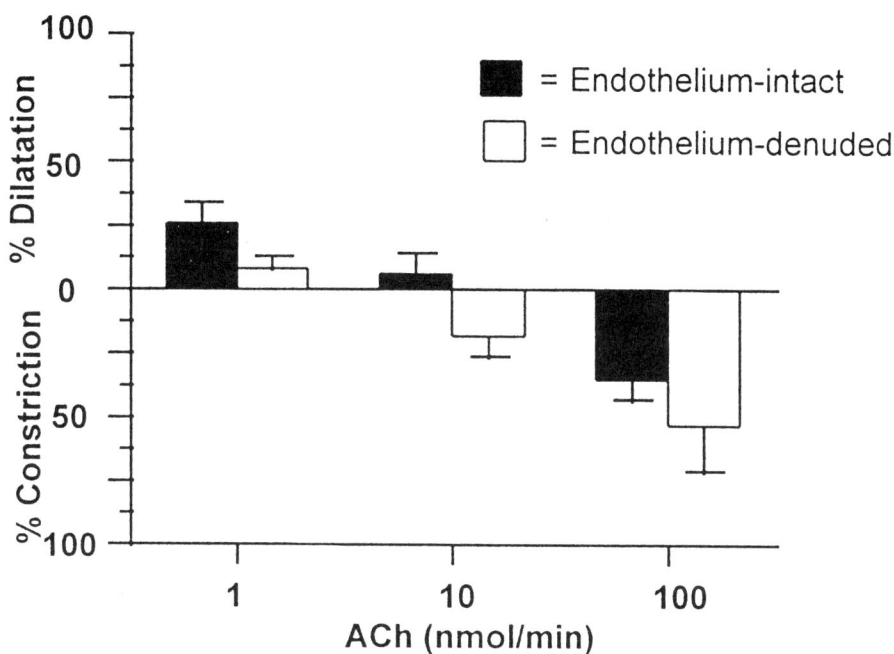

Figure 1.1 Biphasic response to acetylcholine in human hand vein *in vivo.* Acetylcholine was infused into hand veins partially preconstricted with noradrenaline. At low doses, acetylcholine produced vasodilatation, whereas at higher doses it caused vasoconstriction. In some vessels endothelium was removed by infusions of distilled water. In these vessels, the dilator effects of acetylcholine were lost and the constrictor effects enhanced.

Human Vessels *In Vitro*

L-NMMA inhibits the response to "endothelium-dependent" agonists in human vessels *in vitro* (Wallerstedt and Bodelsson, 1997; Raddino *et al.*, 1997; Jovanovic *et al.*, 1994; Vila *et al.*, 1991; Crawley *et al.*, 1990), and blocks the generation of NO by endothelial cells maintained in culture (Palmer *et al.*, 1988b). Its effects are competitive (being overcome by excess L-arginine), stereospecific (the D-enantiomer is without effect) (Radomski *et al.*, 1990; Vallance *et al.*, 1989a; Crawley *et al.*, 1990; Jovanovic *et al.*, 1994; Mollace *et al.*, 1991) and selective (there is no evidence that L-NMMA affects any other enzyme system, ion channel or receptor, in the μM concentration range used [for review see (Vallance, 1996)]). Thus L-NMMA has proved a useful tool to confirm that endothelium-dependent relaxation of human vessels is dependent upon generation of NO. However, the ability of L-NMMA to block the relaxant response to acetylcholine varies from vessel to vessel. For example, L-NMMA completely blocks acetylcholine-induced relaxation in saphenous vein and internal mammary artery but in small resistance arteries a substantial part of the relaxation persists even in the presence of a maximally effective concentration of L-NMMA [see Chapter 4]. Endothelium-dependent but L-NMMA-resistant relaxation has been seen in resistance vessels from animals and man (Urakami Harasawa *et al.*, 1997) and it seems likely that a considerable proportion of

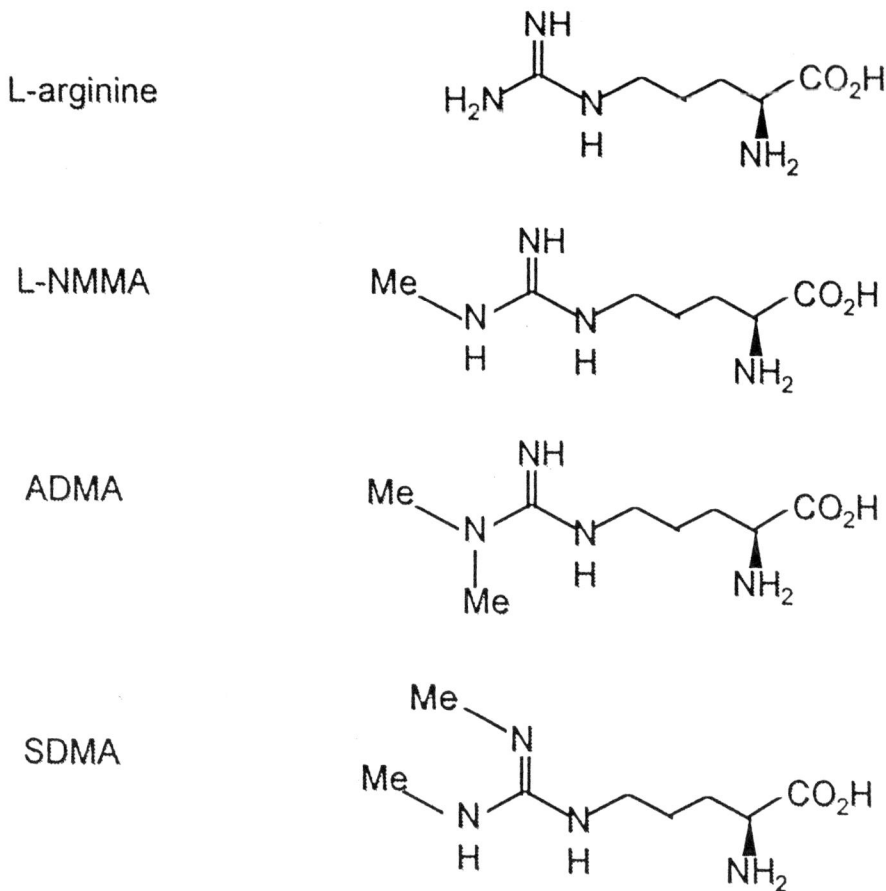

Figure 1.2 **Structures of arginine (substrate for nitric oxide synthesis) and two inhibitors, asymmetric dimethylarginine (ADMA) and N^G-monomethyl-L-arginine (L-NMMA).** L-NMMA and ADMA are both naturally-occuring compounds that compete with arginine and inhibit nitric oxide synthesis. Another endogenous compound, symmetric dimethylarginine (SDMA), in which methyl groups are present on each of the guanidino nitrogen atoms, does not act as an inhibitor of nitric oxide synthesis.

the response to acetylcholine is not mediated by endothelial generation of NO in resistance vessels (Figure 1.3).

Studies *In Vivo*

In studies of the resistance vasculature of the human forearm *in vivo*, L-NMMA inhibits the relaxation to acetylcholine (Vallance *et al.*, 1989b; Chowienczyk *et al.*, 1993), bradykinin (Cockcroft *et al.*, 1994b), vasopressin (Tagawa *et al.*, 1993), substance P (Tagawa *et al.*, 1997), and 5-hydroxytryptamine (Bruning *et al.*, 1993). L-NMMA also inhibits relaxation to ACh in conduit arteries. In superficial hand veins, L-NMMA, but not

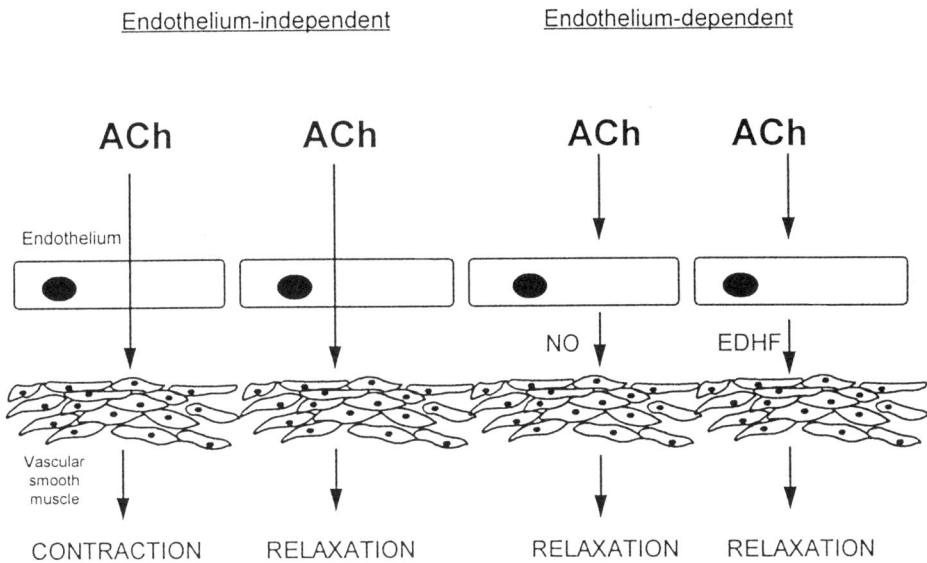

Figure 1.3 Diverse effects of acetylcholine on vascular tone. Acetylcholine may cause vascular contraction or relaxation by endothelium-dependent or endothelium-independent mechanisms. Endothelium dependent relaxation is mediated via the release of nitric oxide or endothelium-derived hyperpolarising factor (EDHF). Endothelium-independent contraction is induced by direct activation of muscarinic cholinoceptors on the vascular smooth muscle cell. In certain specialised vessels (e.g. the lingual artery) acetylcholine may produce endothelium-independent relaxation.

D-NMMA, blocks completely the relaxation to acetylcholine or bradykinin (Vallance *et al.*, 1989a), and inhibits the generation of NO as detected by a porphyrinic microsensor (Vallance *et al.*, 1995). These effects of L-NMMA are reversed by L-arginine but not D-arginine, confirming the specificity of the responses seen. The dilator response to acetylcholine in the coronary artery *in vivo* is also nearly completely abolished by L-NMMA (Quyyumi *et al.*, 1995b; Lefroy *et al.*, 1993). However, similar to the situation *in vitro*, the dilator response to acetylcholine in resistance vessels is only partly inhibited by L-NMMA (Vallance *et al.*, 1989b; Quyyumi *et al.*, 1995a; Lefroy *et al.*, 1993). In the forearm resistance bed, infusion of acetylcholine into the brachial artery causes a large increase in blood flow which can be inhibited by 40–50% by L-NMMA (Vallance *et al.*, 1989b). Again this result is similar to observations in animals and is consistent with the idea that acetylcholine causes vasodilatation through multiple mechanisms in resistance vessels.

The vascular endothelium has receptors for agonists but can also respond to physical stimuli (Davies, 1995; Lansman, 1988). Ion channels that respond to stretch (Lansman *et al.*, 1987) and flow (Olesen *et al.*, 1988; Siegel *et al.*, 1996) have been identified and the endothelium responds to increased shear stress by increasing the generation of NO (Meredith *et al.*, 1996; Kanai *et al.*, 1995; Buga *et al.*, 1991; Lamontagne *et al.*, 1992; Noris *et al.*, 1995). This is responsible for the phenomenon of flow-mediated vasodilatation (Meredith *et al.*, 1996; Lieberman *et al.*, 1996; Kuo *et al.*, 1990; Pohl *et al.*, 1986;

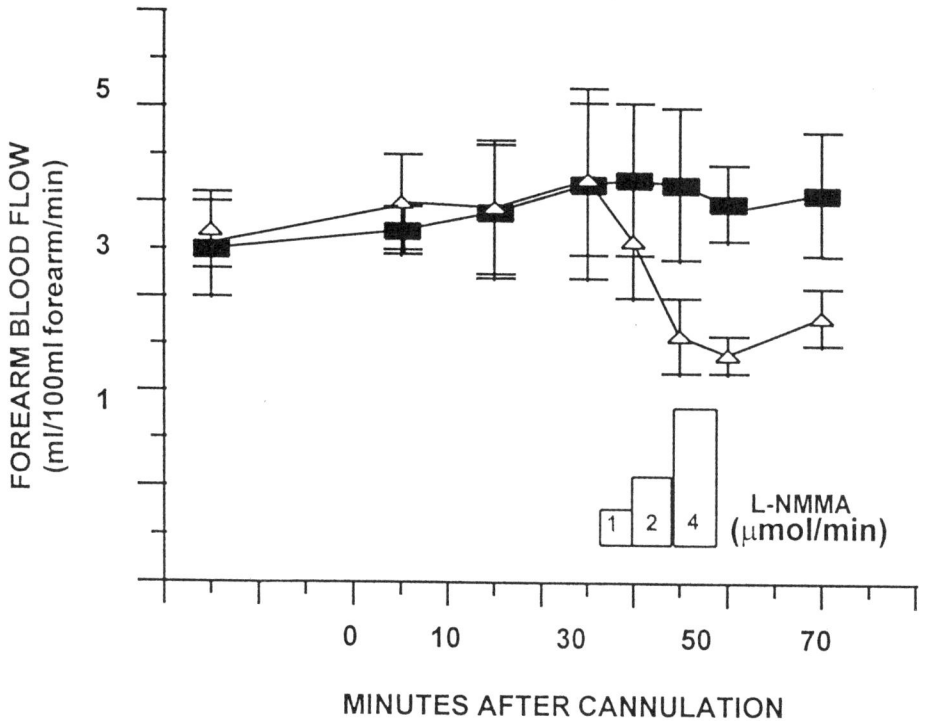

Figure 1.4 **Basal nitric oxide generation in the human forearm.** L-NMMA produces a dose-dependent fall in resting forearm blood flow. L-NMMA (1, 2 and 4mol/min) was infused into the the brachial artery of one arm and blood flow measured in both arms. Local inhibition of nitric oxide synthesis leads to a substantial fall in resting flow and an increase in peripheral resistance in the infused arm (open triangles), with no change in flow in the control arm (closed boxes).

Celermajer *et al.*, 1992), which might be a mechanism for maintaining constant shear stress within the vasculature [see Chapter 10 and (MacAllister and Vallance, 1996)]. Increased flow through the brachial or coronary arteries in humans is associated with vasodilatation of the vessel and this response is abolished by L-NMMA (Meredith *et al.*, 1996; see Chapter 10).

Basal Generation Of Nitric Oxide

In addition to its effects on agonist or shear stress-stimulated dilatation, L-NMMA affects resting vessel tone (Yang *et al.*, 1991). The largest effect is seen in resistance vessels (Vallance *et al.*, 1989b), with human conduit arteries and veins showing little or no increase in basal tone in response to L-NMMA. Vascular resistance nearly doubles when L-NMMA is infused indicating that basal generation of NO provides a major continuous vasodilator influence in the human arterial circulation (Figure 1.4) (Vallance *et al.*, 1989b). This effect has been demonstrated in the forearm arterial bed and in the coronary (Lefroy *et al.*, 1993), renal (Haynes *et al.*, 1997; Dijkhorst Oei and Koomans, 1998) and cerebral (White *et al.*,

1998) circulations of healthy volunteers. Systemic infusion of L-NMMA causes a substantial increase in systemic vascular resistance and elevates blood pressure (Haynes *et al.*, 1993; Stamler *et al.*, 1994).

The mechanisms underlying basal generation of NO remain obscure. It is possible that the increased shear stress on the arterial side of the circulation provides a continuous stimulus for NO generation. However, this would not explain why basal NO generation is seen predominantly in resistance rather than conduit vessels, nor why the basal NO generation persists when arterial vessels are studied *in vitro* under conditions of no flow. The physiological function of basal NO release is also unclear. At the most simplistic level, basal generation of NO might provide a natural counterbalance against the constrictor action of the sympathetic nervous system or the basal generation of endothelin [see Chapter 2]. Like noradrenaline, NO is a rapidly acting mediator and perhaps the two act in concert to allow for central (sympathetic nervous system) and local (endothelial) regulation of vascular tone. Basal, or shear stress activated NO may also contribute to vascular compliance or allow the cardiovascular system to adapt readily to changes in intravascular volume [for discussion see (MacAllister and Vallance, 1996)].

Other Actions Of Endothelial Nitric Oxide

Although vasodilatation is the most obvious and well studied effect of endothelial NO, many other actions have been identified in studies in animals and *in vitro*. Nitric oxide inhibits platelet aggregation (Radomski *et al.*, 1990), prevents adhesion of platelets to the endothelial surface (Radomski *et al.*, 1987b) and induces disaggregation of aggregating platelets [see Chapter 5]. In addition, it inhibits the activation and expression of certain adhesion molecules (Takahashi *et al.*, 1996; Khan *et al.*, 1996; Biffl *et al.*, 1996; De Caterina *et al.*, 1995) and modulates adhesion of white cells (Bath *et al.*, 1991). Smooth muscle cell replication (Garg and Hassid, 1989; Nakaki *et al.*, 1990) and migration (Sarkar *et al.*, 1996) is inhibited by NO and, in experimental models, an intact endothelial NO system is important to limit the degree of intimal hyperplasia that occurs after balloon injury (Lee *et al.*, 1996; Wolf *et al.*, 1995), or the neo-intima formation that occurs during cholesterol feeding (Naruse *et al.*, 1994; Cayatte *et al.*, 1994). *In vivo* transfer of the eNOS gene into balloon injured rat carotid artery restores NO production in this vessel to normal levels and and inhibits neointimal hyperplasia (Janssens *et al.*, 1998; von der Leyen *et al.*, 1995). Whilst many of these effects have also been observed using human cells *in vitro* there are few data on human blood vessels studied *in vitro* or *in vivo*. Infusion of L-NMMA is associated with a decrease in bleeding time but no effects on platelet aggregation have been detected (Simon *et al.*, 1995; Remuzzi *et al.*, 1990). However, inhaled NO (100-884 ppm) has been shown to increase bleeding time and reduce agonist-stimulated platelet aggregation (Gries *et al.*, 1998)

MEASUREMENT OF NITRIC OXIDE IN HUMANS

Nitric oxide has a biological half life of a few seconds in biological solution (Cocks *et al.*, 1985; Griffith *et al.*, 1984) and rapidly degrades in blood to nitrite and thence to nitrate, both of which are largely inactive. Although circulating adducts of NO have been identified (Stamler *et al.*, 1992b; Stamler *et al.*, 1992a; Keaney, Jr. *et al.*, 1993), most data suggest that endothelial NO acts close to its point of synthesis and does not have significant downstream or circulating effects. Because of its short half-life, biochemical assessment of NO generation has proved difficult. Free NO can be measured using a porphyrinic microsensor (Malinski *et al.*, 1993b; Malinski *et al.*, 1993a; Malinski and Taha, 1992; Vallance *et al.*, 1995). Although useful for detecting the kinetics of NO release, it cannot be used as a quantitative measure of NO production because of the instability of the molecule (Baylis and Vallance, 1998). Nitrite is unstable in blood and its concentration in plasma provides only a glimpse of NO in an intermediary state of metabolism. Nitrate is a stable end product of NO and its concentration can readily be assessed in plasma and urine. However, a substantial proportion of the circulating nitrate derives from the diet and the remainder may come from NO-generating cells other than the vascular endothelium (Baylis and Vallance, 1998). An alternative method to quantify NO production involves measuring the conversion of exogenously administered ^{15}N-labelled arginine to ^{15}N-nitrate — an approach which can be used *in vivo* (Hibbs, Jr. *et al.*, 1992; Forte *et al.*, 1997; Macallan *et al.*, 1997). Although this has the advantage that all of the measured nitrate derives from the L-arginine:NO pathway it still does not allow precise identification of the cellular source of the NO, nor does it indicate how much of the NO generated was biologically active.

Overall, measurement of NO metabolites has confirmed the existence of NO generation in humans but each of the methods used has significant drawbacks when assessing the activity of endothelial NO in health or disease. A full critique of biochemical assessment of NO in humans has been published elsewhere (Baylis and Vallance, 1998) and is outside the scope of the present chapter.

NITRIC OXIDE SYNTHASE IN THE VASCULAR ENDOTHELIUM

The biosynthesis of NO from L-arginine and molecular oxygen is mediated by a family of three enzymes — the nitric oxide synthases (NOSs) (Figure 1.5) (Forstermann *et al.*, 1994b; Forstermann *et al.*, 1994a; Forstermann *et al.*, 1993; Knowles and Moncada, 1994). Endothelial NOS (eNOS), like neuronal NOS (nNOS), was originally characterised by: (1) a constitutive activity responsible for generating low levels of NO in the basal state; (2) regulation by intracellular calcium concentration (Mulsch *et al.*, 1989; Mayer *et al.*, 1989); and (3) a restricted pattern of expression which not only includes the endothelium, but also platelets (Sase and Michel, 1995), placenta (Eis *et al.*, 1995; Myatt *et al.*, 1993), myocytes (Feron *et al.*, 1996; Wei *et al.*, 1996) and bronchiolar epithelium (Shaul *et al.*, 1994). This contrasts with the inducible isoform of NOS (iNOS) which requires an inflammatory stimulus for its expression (Hibbs, Jr. *et al.*, 1988; Marletta *et al.*, 1988; de Vera *et al.*, 1996), can be induced by cytokines in a variety of cell types including macrophages (Ravalli *et al.*, 1998; Nicholson *et al.*, 1996), vascular smooth muscle cells (Ravalli *et al.*, 1998; Kolyada *et al.*, 1996), endothelial cells, neutrophils (Evans *et al.*, 1996; Wheeler *et al.*, 1997) and eosinophils (del Pozo *et al.*, 1997), appears functionally independent of calcium (McCall *et al.*, 1991), and synthesises NO at relatively high rates. However, this classification of NOS isoforms according to inducibility and calcium-dependence is now considered over-simplistic since eNOS is subject to transcriptional regulation (Harrison *et al.*, 1996; Wang and Marsden, 1995a), and the apparent calcium-

Protein structure

nNOS — Haem/Arg CaM FMN FAD NADPH

iNOS — Haem/Arg CaM FMN FAD NADPH

eNOS — Haem/Arg CaM FMN FAD NADPH

Key to binding sites

Haem/Arg	Haem / arginine
CaM	Calmodulin
FMN	Flavin mononucleotide
FAD	Flavin adenine dinucleotide
NADPH	Nicotinamide diphosphate hydrogen

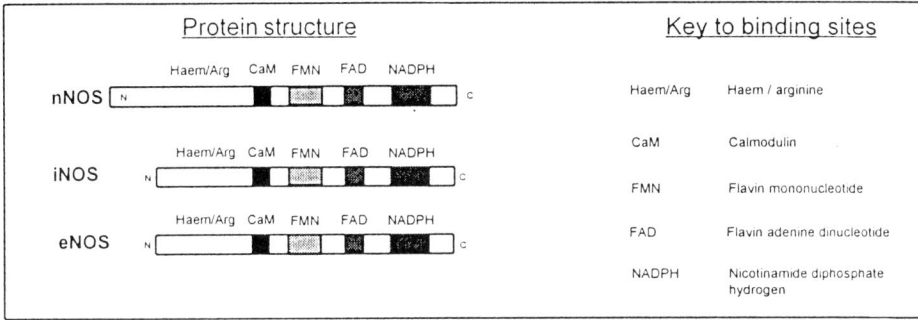

	nNOS	iNOS	eNOS
Gene nomenclature	*NOS 1*	*NOS 2*	*NOS 3*
Chromosomal localisation	12q24.2	17q11.2-q12	7q35-q36
Number of exons	29	26	26
cDNA/kb	8.5-9.5	4.2-4.5	4.3-4.8
M_w of protein/kD	160	130	135
Amino acids	1434	1153	1203

Figure 1.5 Human nitric oxide synthase gene family. Endothelial nitric oxide synthase (eNOS) is one of a family of three isoforms, the other two isoforms being neuronal (nNOS) and inducible (iNOS). Each isoform is a product of a separate gene but all are highly homologous particularly in regions encoding the highly conserved binding sites for calmodulin and enzyme co-factors. The location of these binding sites is illustrated in the upper panel.

independence of iNOS is, in fact, a manifestation of the permanent, tight binding of a Ca^{2+}/ calmodulin complex to this isoform of the enzyme (Cho *et al.*, 1992).

Enzymatic Synthesis Of Endothelial Nitric Oxide

In common with the other isoforms, endothelial nitric oxide synthase has a C-terminal reductase domain (which bears much sequence and functional homology to the cytochrome P450-like haemproteins) and an N-terminal oxidative domain (Wang and Marsden, 1995a). Whilst enzyme monomers can mediate electron exchange, the active form of one eNOS enzyme is a homodimer (Rodriguez Crespo *et al.*, 1996; Venema *et al.*, 1997) catalysing the formation of NO from the amino-acid L-arginine and molecular oxygen in a reaction which involves the five electron oxidation of the terminal guanidino nitrogen of arginine. NADPH is a co-substrate for this reaction, which also involves the co-factors FAD, FMN, enzyme-bound haem and tetrahydrobiopterin. L-arginine is oxidised through an intermediate (N^G-hydroxyl-L-arginine) to L-citrulline — a co-product formed in equimolar quantities with NO. Binding sites for FAD, FMN, calmodulin and NADPH (apparent from consensus sequence motifs in the primary structure of the protein) are located in the C-terminal reductase domain, and are encoded by distinct exons or groups of exons. The binding site for haem/arginine is less well documented but is thought to reside in the N-terminal oxidative end of the protein.

Gene Localisation, Structure And Polymorphism

Each NOS isoform is the product of a separate gene and the structure and chromosomal location of each of these has been ascertained in man (Wang and Marsden, 1995b) (Figure 10.5). At the DNA level, there is a 50–60% sequence homology between isoforms within a species and this is particularly prominent in the regions encoding peptide domains known to subserve a particular biological function (e.g. co-factor binding). Between-species homology for a specific isoform is much greater, with the bovine, murine and human eNOS gene sequences exhibiting 80–90% homology. Despite this evolutionary conservation, human eNOS has been shown to exhibit significant inter-individual sequence variation (polymorphism) (Nadaud *et al.*, 1994; Marsden *et al.*, 1993; Hingorani *et al.*, 1997; Miyamoto *et al.*, 1998; Markus *et al.*, 1998; Wang *et al.*, 1996). A variety of both bi- and multi-allelic polymorphisms have been identified in the promoter and introns of the gene (Figure 1.6). A polymorphism in intron 4 of the gene may be a marker for coronary artery disease and, may also be a marker for (or directly influence) the amount of NO generated by the enzyme (Wang *et al.*, 1997). Thus far, only one genetic variant affecting the amino-acid sequence has been identified. A G→T substitution in exon 7 of the gene predicts a glutamic acid (Glu)→spartic acid (Asp) substitution at residue 298 of the mature protein (Hingorani *et al.*, 1995; Yasue *et al.*, 1995). Whether any of these sequence variants alters transcriptional regulation or activity of the encoded protein is not yet known, but the Glu298Asp polymorphism, which has been identified in both Caucasian (Hingorani *et al.*, 1995) and Japanese subjects (Yasue *et al.*, 1995) (albeit at different allele frequency), has been associated with a variety of cardiovascular disorders including essential hypertension in the Japanese (Miyamoto *et al.*, 1998), coronary artery spasm (Yasue *et al.*, 1995; Yoshimura *et al.*, 1997), coronary atherosclerosis (Hingorani *et al.*, 1997), myocardial infarction (Shimasaki *et al.*, 1998; Hibi *et al.*, 1998) and the *NOS 3* gene has also been linked to pre-eclampsia (Arngrimsson *et al.*, 1998). Comparison with the recently solved crystal structure of murine inducible nitric oxide synthase (iNOS) (Crane *et al.*, 1997), suggests that residue 298 of human eNOS (homologous with residue 308 of murine iNOS) is located in a loop between two β-pleated sheets on the external surface of the protein and is proximal to residues critical for substrate binding e.g. residue Glu 371 in murine iNOS (homologous to Glu 361 in human eNOS).

Expression And Subcellular Localisation

Originally cloned and isolated from vascular endothelium, eNOS is now known to be expressed in a variety of other cell types including platelets (Sase and Michel, 1995), cardiac myocytes (Feron *et al.*, 1996; Wei *et al.*, 1996), the syncytiotrophoblast of the placenta (Buttery *et al.*, 1994) and the hippocampus of the brain (Dinerman *et al.*, 1994). Endothelial NOS is highly membrane-bound (Busconi and Michel, 1993). Recent studies have shown that eNOS can be found in the Golgi apparatus (Sessa *et al.*, 1995) and in small protein-rich invaginations of the plasma membrane called caveolae (Shaul *et al.*, 1996). Within the caveolae of endothelial cells, eNOS is bound to a protein called caveolin-1 — this interaction involves a conserved 20 amino acid region within caveolin-1 and a

A

Exon 7 (894 G/T)   Intron 11 (-30 A/G)   Intron 18 (27 A/C)   Intron 23 (10 G/T)

Transcription start site

17 18 19

1 2 3 4   5 6 7 8 9   10 11 12 13   14 15 16   20 21 22   23   24 25 26

5'   3'

Promoter

Intron 4 VNTR   Intron 13 (CA)n

B

-1468 T/A   -922 G/A   -786 T/C   Transcription start site

5'   3'

AP-1   AP-2   GATA   AP-2 GATA/Sp 1

SHEAR   NF-1

AP-1= activating protein 1
AP-2= activating protein 2
SHEAR= shear response element
◄► ◄ oestrogen response elements

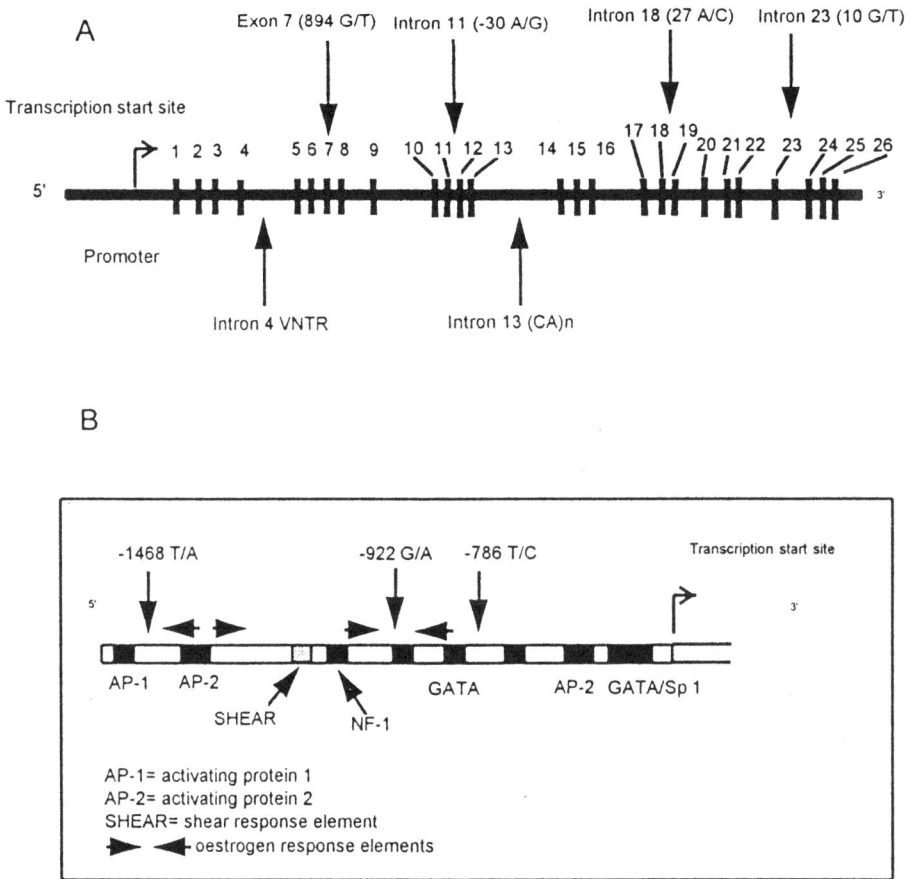

Figure 1.6   Polymorphisms of the human endothelial nitric oxide synthase gene (NOS 3). Panel A illustrates the gene structure intron/exon arrangement and polymorphisms of the human *NOS 3* gene. Single nucleotide bi-allelic polymorphisms are shown above the gene. The variable number tandem repeat (VNTR) polymorphism in intron 4 is also bi-allelic with individuals having 4 or 5 repeats of a 27bp sequence element. The multi-allelic CA repeat polymorphism in intron 13 is highly polymorphic with allele sizes ranging from 18 to 36 CA repeats (Hingorani, 1997).

proposed caveolin-binding sequence in eNOS in the oxidative domain (Michel *et al.*, 1997b; Feron *et al.*, 1996). Additional membrane binding interactions also exist. The presence of glycine as the second amino-acid residue found in eNOS, serves as an acceptor site for myristic acid which is critical for the binding of eNOS to the plasma membrane (Busconi and Michel, 1994; Busconi and Michel, 1993). Artificial site-directed mutagenesis of this residue converts eNOS from a membrane-bound to a cytosolic enzyme (Busconi and Michel, 1993). Palmitoylation of eNOS at two cysteine residues near the N-terminus stabilises this membrane binding (Robinson and Michel, 1995; Robinson *et al.*, 1995). This dual acylation is unique among the NOS isoforms (Michel and Feron, 1997). The functional significance of these processes which affect NOS localisation within the cell is discussed below.

Regulation Of Expression And Activity

The NOS isoforms are regulated at many levels from gene transcription to post-translational modification (Wang and Marsden, 1995a). Sequence analysis of the promoter regions of each enzyme isoform reveals potential sequences for regulation through a variety of transcription factors including, including those induced by TGF-β (NF-1), cAMP (AP-2) and phorbol esters (AP-1) (Wang and Marsden, 1995a). Transcription of the eNOS gene is upregulated by increased shear stress at the endothelial cell surface (Topper *et al.*, 1996; Uematsu *et al.*, 1995; Ranjan *et al.*, 1995), by TGF-β (Inoue *et al.*, 1995) and by oestrogens (Armour and Ralston, 1998; MacRitchie *et al.*, 1997; Kleinert *et al.*, 1998) but the cellular mechanisms and transcription factors mediating these effects have yet to be completely defined. The eNOS gene promoter contains a single shear stress response element (SSRE) (Marsden *et al.*, 1993; Nadaud *et al.*, 1994; Miyahara *et al.*, 1994), comprising a 6 base pair sequence motif originally identified as transducing the shear-induced transcriptional activation of the PDGF-gene and subsequently identified in a variety of other endothelially expressed shear-responsive genes. Whether this sequence underlies shear-induced transcription of eNOS has yet to be formally confirmed. Although the eNOS promoter does not contain the full 22 base pair oestrogen response element (ERE) — a sequence motif mediating the transcriptional response to the oestrogen/oestrogen receptor complex — it does contain a number of so-called "half-sites" comprising half of the full-length ERE consensus sequence which in other genes have been shown to act in concert to mediate a transcriptional response to oestrogen (Miyahara *et al.*, 1994). Again it is not clear whether it is these sequences which confer oestrogen responsiveness (Kleinert *et al.*, 1998). Transcriptional regulation of the eNOS gene by these two stimuli may underlie the increase in NO generation seen physiologically following exercise training (Wilson and Kapoor, 1993) and in pregnancy (Williams *et al.*, 1997).

Post-translational Modification of eNOS

The endothelial NOS is also known to undergo post-translational modification, and this serves as a further potential mechanism for regulating activity. As well as N-terminal myristoylation (Busconi and Michel, 1994; Busconi and Michel, 1993) and palmitoylation (Robinson and Michel, 1995; Robinson *et al.*, 1995) which determine subcellular targeting, the enzyme also undergoes phosphorylation (Michel *et al.*, 1993; Corson *et al.*, 1996). Although myristoylation appears irreversible, palmitoylation/depalmitoylation is a reversible process thought to be subject to regulation (Robinson *et al.*, 1995). Bradykinin, which stimulates NO generation by the endothelium has been shown to modulate this process. Endothelial NOS has the potential to be phosphorylated at both tyrosine (Garcia Cardena *et al.*, 1996) and serine (Corson *et al.*, 1996; Michel *et al.*, 1993) residues. In endothelial cells, phosphorylation of eNOS at serine residues occurs following exposure to agonists and shear stress (Corson *et al.*, 1996; Michel *et al.*, 1993) and in some cases has been associated with subcellular translocation of the protein. Fluid shear stress may induce tyrosine phosphorylation and activation of eNOS independent of Ca^{2+}/calmodulin (Fleming *et al.*, 1997).

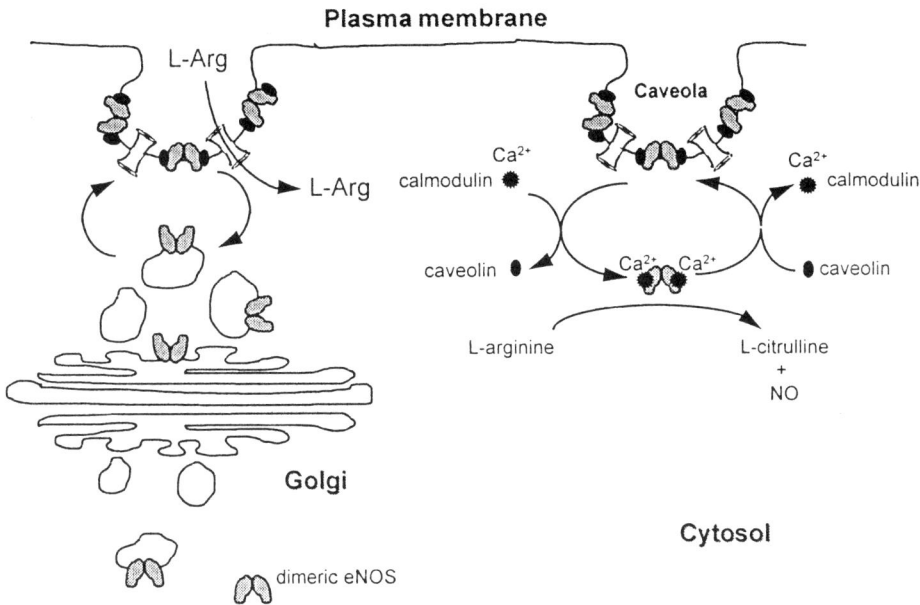

Figure 1.7 Intracellular localisation and activation of eNOS. Active eNOS is a homodimer. Trafficking of the enzyme to subcellular membrane compartments such as the Golgi apparatus and caveolae appears to occur. Caveolae may be sites where arginine transporters and receptors are co-localised in close proximity to membrane bound eNOS. Caveolins are negative allosteric regulators of the enzyme. Caveolin-mediated inhibition of enzyme activity may be removed by activation of calcium-calmodulin by certain agonists e.g. bradykinin. In some cases, agonist-mediated activation is accompanied by enzyme translocation from membrane to cytosol.

Mechanism of Agonist-induced eNOS Activation

The increase in activity of pre-formed eNOS mediated by exposure of endothelial cells to endothelium-dependent agonists has been shown to be dependent on an increased concentration of intracellular calcium and is mediated by the binding to eNOS of the calcium-activated protein calmodulin (Marletta, 1993). This binding and activation appears to be associated with the displacement of caveolin from the eNOS/caveolin protein complex (Michel *et al.*, 1997a). It has been proposed that caveolin and calmodulin act as opposing modulators of eNOS activity, the former acting as an allosteric inhibitor and the latter as an allosteric activator (Michel and Feron, 1997; Michel *et al.*, 1997b). The concentration of eNOS within caveolae has additional potential implications for regulation of enzyme activity. Caveolae may serve as regions of the plasma membrane where there is a high concentration of mechanosensing molecules (e.g. shear stress sensors), molecules which mediate signal transduction, and arginine transporters (McDonald *et al.*, 1997). A microenvironment within caveolae might also modulate eNOS activity by controlling the local concentration of co-factors and substrate. Clearly this has potential implications for changes occuring in disease. Whilst overall expression of eNOS may be unchanged in disease, enzyme activity might be drastically altered by changes in the subcellular localisation of eNOS or the proteins with which it interacts. (Figure 1.7)

Substrate Concentration-Enzyme Activity Relationship

L-arginine circulates in plasma at a concentration of 50–100 µM and enters cells largely but not exclusively through the cationic y^+ amino acid transporter. Within the endothelial cell concentrations of L-arginine of up to 800 µM have been measured (Baydoun *et al.*, 1990). These concentrations are many fold higher than the K_m of the enzyme for L-arginine of (1–4 µM) calculated from the activity of partially purified native or recombinant enzymes (Charles *et al.*, 1996; Bredt and Snyder, 1990). In principle, therefore, there should be no situation in which the activity of the enzyme is substrate-limited. However, several studies in humans *in vivo* have identified bio-logical effects of L-arginine which have been attributed to increased NO generation via the provision of excess substrate (Bode Boger *et al.*, 1994; Mehta *et al.*, 1996; Imaizumi *et al.*, 1992). In some instances the reported effects seem unlikely to be due to this mechanism. For example, the vasodilation and hypotension which occurs when very large doses of L-arginine are infused (sufficient to increase the plasma concen-tration of arginine to 5–10 mM or higher), are also seen with D-arginine and certain other amino acids which are not substrates for NOS (Rhodes *et al.*, 1996; MacAllister *et al.*, 1995). However, particularly in certain disease states such as atherosclerosis, the actions of arginine do appear to be stereospecific and related to NO generation. The mechanism of this effect (termed the "arginine paradox") has not been determined but might relate to: (1) a disease-induced fall in substrate concentration (at least in the microenvironment of the enzyme); (2) alteration of the *Km* of the enzyme; (3) a deficiency of one or more co-factors; or (4) post-translational modifications of the enzyme which affect the binding of substrates or co-factors, or the sub-cellular locali-sation of the enzyme (Boger *et al.*, 1996). Another possibility is that in certain disease states there is an increase in the level of an endogenous enzyme inhibitor which acts at the arginine binding site. The NOS inhibitor L-NMMA, which has been widely used to probe the biology of the L-arginine:NO pathway and a closely related compound, asymmetric dimethylarginine (ADMA), are endogenous naturally occuring compounds.

Endogenous eNOS Inhibitors

Guanidino nitrogens in arginine residues present within proteins are methylated by the action of protein methylase I. Upon hydrolysis of the proteins free L-NMMA and the dimethylarginines ADMA and SDMA are released (Figures 1.2 and 1.8). These compounds are synthesised by human endothelial cells (MacAllister *et al.*, 1996a), circulate in plasma and are excreted in urine (Vallance *et al.*, 1992). L-NMMA and ADMA are both inhibitors of NOS (Calver *et al.*, 1993; MacAllister *et al.*, 1994b) and all three methylated amino acids can prevent arginine entry through the y^+ transporter (MacAllister *et al.*, 1994a). A specific enzyme exists (dimethyl arginine dimethylamino hydrolase; DDAH) for metabo-lising ADMA and L-NMMA and this has been identified in human endothelial cells and blood vessels (MacAllister *et al.*, 1996a). Increased circulating concentrations of ADMA are found in certain disease states (Fickling *et al.*, 1993; Vallance *et al.*, 1992; MacAllister *et al.*, 1996b; Boger *et al.*, 1997) and it is possible that local accumulation of this compound contributes to impaired endothelial NO generation and competes with L-arginine thereby rendering NOS arginine-sensitive *in vivo*. (Figure 1.8)

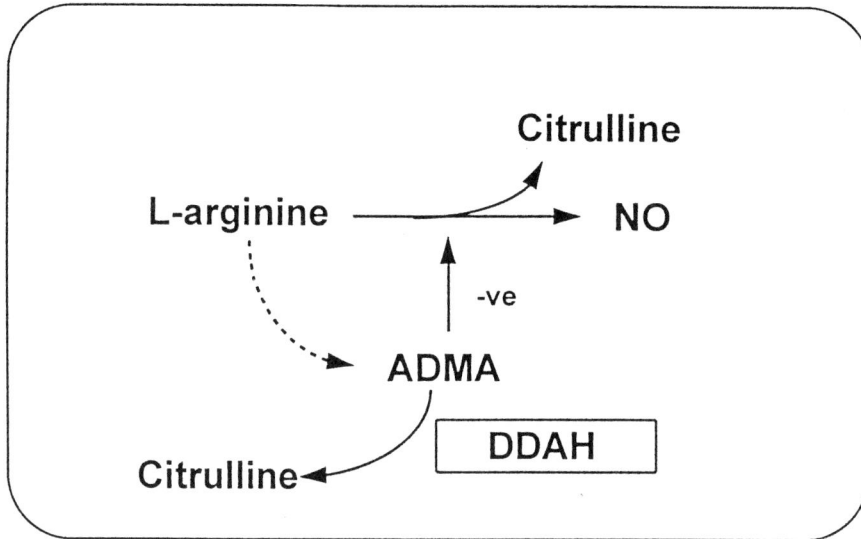

Figure 1.8 Synthesis and metabolism of ADMA. L-arginine is converted to nitric oxide by nitric oxide synthase. However, it can also be converted to the inhibitor ADMA. The precise synthetic route is not clear but probably involves methylation of arginine residues in proteins. Free ADMA is metabolised to citrulline by the action of dimethylarginine dimethylaminohydrolase (DDAH). Citrulline is subsequently recycled to arginine.

MOLECULAR TARGETS FOR ENDOTHELIAL NITRIC OXIDE

Nitric oxide is non-polar, diffuses freely across cell membranes and therefore has a number of potential fates depending on the target cell and chemical entity with which it interacts. It is capable of binding to or reacting with a variety of chemical moieties including reactive oxygen species (e.g. superoxide and hydroxyl radicals) and with metal-containing proteins (e.g. soluble guanylate cyclase, cytochrome c oxidase).

Activation of Soluble Guanylate Cyclase

It is the binding and activation by NO of the haem-containing protein soluble guanylate cyclase (sGC) that explains the elevation in cyclic guanosine monophosphate (cGMP) observed in target cells following exposure to NO, and which accounts for many of its biological effects (Ignarro, 1990a; Furchgott and Jothianandan, 1991). Soluble guanylate cyclase is a heterodimeric protein containing distinct catalytic and haem-binding domains (Hobbs, 1997). NO-induced activation of sGC is thought to require its binding to the haem moiety to form a nitrosyl-haem complex. The conformational change induced in the heterodimer by this binding is thought to expose the catalytic domain of sGC to its substrate GTP, a process which leads to the formation of cGMP (Hobbs, 1997). In the vascular smooth muscle cell (the prototypic target cell), there then follows an activation of cGMP-dependent protein kinases and phosphorylation of intracellular proteins including myosin light chain kinases and calcium-activated potassium channels which completes the

NO-mediated signalling cascade. Although it has been suggested that NO-mediated vasorelaxation can occur in certain systems independently of an elevation in cGMP (Bolotina *et al.*, 1994) these findings require further confirmation and, in most vessels, activation of sGC seems to account fully for the vasorelaxant effects of NO.

Interaction With Cytochrome C Oxidase

Cytochrome c oxidase (a key oxygen-binding protein in the mitochondrial electron transport chain) also appears to be a cellular target of NO in a wide variety of tissues studied (Brown, 1997; Borutaite and Brown, 1996; Brown, 1995; Brown and Cooper, 1994; Torres *et al.*, 1997; Giuffre *et al.*, 1996; Cleeter *et al.*, 1994). NO binds with high affinity to the oxygen binding site of the enzyme raising the apparent Km of cytochrome oxidase for oxygen (Brown, 1997). It has been suggested that this renders mitochondrial respiration sensitive to oxygen concentration over the normal physiological range thereby allowing mitochondrial respiration to be regulated by physiological changes in oxygen concentration. This area is of considerable potential physiological importance (for recent review see (Torres and Wilson, 1996)).

Chemical Reactions of Nitric Oxide

NO released into the bloodstream and tissue fluids also has the potential to participate in a variety of chemical reactions which would be expected to terminate the physiological activity of free NO (Ignarro, 1990b). Potential reactions include combination with oxygen (to form nitrite $[NO_2^-]$), with superoxide anion $[O_2^-]$ (to generate peroxynitrite $[ONOO^-]$) and with -SH groups in peptides, proteins, amino acids and other small molecules to yield nitrosothiols $[R-SNO]$ (Ignarro, 1990b; Darley-Usmar *et al.*, 1995). The reaction of NO with haemoglobin leads rapidly to the formation of NO_2^- and methaemoglobin. Nitrite formed by the oxidation of NO itself undergoes rapid oxidation to nitrate $[NO_3^-]$ (a stable end-product). Under physiological conditions it is likely that the activity of superoxide dismutase is sufficient to prevent the the formation of significant amounts of $ONOO^-$ from the reaction between NO and O_2^- (Beckmann and Koppenol, 1996). However, during inflammation (and other disease states such as diabetes) when the production of both of these species is increased, there may be significant generation of $ONOO^-$ and thence other damaging free radicals which may contribute to the inflammatory process (Bhagat and Vallance, 1996).

   The reversible interaction of NO with haemoglobin at its haem centre and to the sulphydryl (-SH) group of a cysteine at residue 93 in the -globin chain (and the modulating effect of molecular oxygen on this process) has been the subject of much recent interest (Jia *et al.*, 1996). The binding of NO to haemoglobin and its inactivation (see above) would be expected to limit the immediate sphere of action of gaseous NO. However, the recent observation that nitroso-Hb (a potential product of the NO/Hb interaction) might itself act as a vasodilator by releasing NO at a distant site (Jia *et al.*, 1996; Gow and Stamler, 1998; Stamler *et al.*, 1997) requires further verification. For the most part NO can be considered a local rather than a circulating mediator. NO can also react with -SH groups in other circulating substances including, homocysteine and albumin (Keaney *et al.*, 1993). The nitrosothiols which form have also been shown to behave as NO donors *in vitro* and *in vivo*, but it is as yet unclear whether these substances play a major role in normal circulatory physiology.

CONCLUSIONS

The importance of L-arginine:NO pathway in regulation of vascular tone, and systemic blood pressure and in its contribution to the thromboresistant and atheroprotective functions of the blood vessel wall have become clear over the last decade. The identification of this signalling pathway has also fuelled the current interest in the biology of the endothelium as a whole. There are, however, still a number of questions which await answers. What processes are responsible for the basal release of NO from some vessel types but not others? How do the mechanisms mediating physiological release of NO in response to physical stimuli (such as flow) differ from those which subserve agonist-stimulated NO release? Which mechanisms account for the "arginine paradox"? It is hoped that the answers to these questions will be forthcoming and that a better understanding of the mechanisms involved will enhance the potential for the therapeutic manipulation of this important system.

Many of the chapters which follow discuss the evidence that NO-mediated dilation is altered in disease states and that this might contribute to atherogenesis and other cardiovascular disorders. It is increasingly clear that changes in NO-mediated effects may also arise due to individual genetic differences in the activity or expression of eNOS, post-translational modification of the enzyme, endogenous inhibitors, and chemical reactions of NO prior to its effect on target molecules. This book documents some of the functional changes in the pathway seen in disease states.

References

Angus, J.A., Cocks, T.M., McPherson, G.A. and Broughton, A. (1991) The acetylcholine paradox: a constrictor of human small coronary arteries even in the presence of endothelium. *Clin. Exp. Pharmacol. Physiol.*, **18**, 33–36.

Armour, K.E. and Ralston, S.H. (1998) Estrogen upregulates endothelial constitutive nitric oxide synthase expression in human osteoblast-like cells. *Endocrinology*, **139**, 799–802.

Arngrimsson, R., Hayward, C., Nadaud, S., Baldursdottir, A., Walker, J.J., Liston, W.A., Bjarndottir, R.I., Brock, D.J.H., Geirsson, R., Connor, J.M. and Soubrier, F. (1998) Evidence for a familial pregnancy-induced hypertension locus in the eNOS-gene region. *Am. J. Hum. Genet.*, **61** 354–362.

Bath, P.M., Hassall, D.G., Gladwin, A.M., Palmer, R.M. and Martin, J.F. (1991) Nitric oxide and prostacyclin. Divergence of inhibitory effects on monocyte chemotaxis and adhesion to endothelium in vitro. *Arterioscler. Thromb.*, **11**, 254–260.

Baydoun, A.R., Emery, P.W., Pearson, J.D. and Mann, G.E. (1990) Substrate-dependent regulation of intracellular amino acid concentrations in cultured bovine aortic endothelial cells. *Biochem. Biophys. Res. Commun.*, **173**, 940–948.

Baylis, C. and Vallance, P. (1998) Measurement of nitrite and nitrate levels in plasma and urine–what does this measure tell us about the activity of the endogenous nitric oxide system? *Curr. Opin. Nephrol. Hypertens.*, **7**, 59–62.

Beckmann, J.S. and Koppenol, W.H. (1996) Nitric oxide, superoxide, and peroxynitrite: the good, the bad, and the ugly. *Am. J. Physiol.*, **271** C1424–C1437.

Bhagat, K. and Vallance, P. (1996) Inducible nitric oxide synthase in the cardiovascular system. *Heart*, **75**, 218–220.

Biffl, W.L., Moore, E.E., Moore, F.A. and Barnett, C. (1996) Nitric oxide reduces endothelial expression of intercellular adhesion molecule (ICAM)-1. *J. Surg. Res.*, **63**, 328–332.

Bode Boger, S.M., Boger, R.H., Creutzig, A., Tsikas, D., Gutzki, F.M., Alexander, K. and Frolich, J.C. (1994) L-arginine infusion decreases peripheral arterial resistance and inhibits platelet aggregation in healthy subjects. *Clin. Sci. Colch.*, **87**, 303–310.

Boger, R.H., Bode Boger, S.M. and Frolich, J.C. (1996) The L-arginine-nitric oxide pathway: role in atherosclerosis and therapeutic implications. *Atherosclerosis*, **127**, 1–11.

Boger, R.H., Bode-Boger, S.M., Thiele, W., Junker, W., Alexander, K. and rolich, J.C. (1997) Biochemical evidence for impaired nitric oxide synthesis in patients with peripheral arterial occlusive disease. *Circulation*, **95** 2068–2074.

Bolotina, V.M., Najibi, S., Palacino, J.J., Pagano, P.J. and Cohen, R.A. (1994) Nitric oxide directly activates calcium-dependent potassium channels in vascular smooth muscle. *Nature*, **368** 850–853.

Borutaite, V. and Brown, G.C. (1996) Rapid reduction of nitric oxide by mitochondria, and reversible inhibition of mitochondrial respiration by nitric oxide. *Biochem. J.*, **315**, 295–299.

Brayden, J.E. and Large, W.A. (1986) Electrophysiological analysis of neurogenic vasodilatation in the isolated lingual artery of the rabbit. *Br. J. Pharmacol.*, **89**, 163–171.

Bredt, D.S. and Snyder, S.H. (1990) Isolation of nitric oxide synthetase, a calmodulin-requiring enzyme. *Proc. Natl. Acad. Sci. U. S. A.*, **87**, 682–685.

Brown, G.C. (1995) Nitric oxide regulates mitochondrial respiration and cell functions by inhibiting cytochrome oxidase. *FEBS Lett.*, **369**, 136–139.

Brown, G.C. (1997) Nitric oxide inhibition of cytochrome oxidase and mitochondrial respiration: implications for inflammatory, neurodegenerative and ischaemic pathologies. *Mol. Cell Biochem.*, **174**, 189–192.

Brown, G.C. and Cooper, C.E. (1994) Nanomolar concentrations of nitric oxide reversibly inhibit synaptosomal respiration by competing with oxygen at cytochrome oxidase. *FEBS Lett.*, **356**, 295–298.

Bruning, T.A., Chang, P.C., Blauw, G.J., Vermeij, P. and van Zwieten, P.A. (1993) Serotonin-induced vasodilatation in the human forearm is mediated by the "nitric oxide-pathway": no evidence for involvement of the 5-HT3-receptor. *J. Cardiovasc. Pharmacol.*, **22**, 44–51.

Bruning, T.A., Hendriks, M.G., Chang, P.C., Kuypers, E.A. and van Zwieten, P.A. (1994) In vivo characterization of vasodilating muscarinic-receptor subtypes in humans. *Circ. Res.*, **74**, 912–919.

Buga, G.M., Gold, M.E., Fukuto, J.M. and Ignarro, L.J. (1991) Shear stress-induced release of nitric oxide from endothelial cells grown on beads. *Hypertension*, **17**, 187–193.

Busconi, L. and Michel, T. (1993) Endothelial nitric oxide synthase. N-terminal myristoylation determines subcellular localization. *J. Biol. Chem.*, **268**, 8410–8413.

Busconi, L. and Michel, T. (1994) Endothelial nitric oxide synthase membrane targeting. Evidence against involvement of a specific myristate receptor. *J. Biol. Chem.*, **269**, 25016–25020.

Buttery, L.D., McCarthy, A., Springall, D.R., Sullivan, M.H., Elder, M.G., Michel, T. and Polak, J.M. (1994) Endothelial nitric oxide synthase in the human placenta: regional distribution and proposed regulatory role at the feto-maternal interface. *Placenta.*, **15**, 257–265.

Calver, A., Collier, J., Leone, A., Moncada, S. and Vallance, P. (1993) Effect of local intra-arterial asymmetric dimethylarginine (ADMA) on the forearm arteriolar bed of healthy volunteers. *J. Hum. Hypertens.*, **7**, 193–194.

Cayatte, A., Palacino, J., Horten, K. and Cohen, R.A. (1994) Chronic inhibition of nitric oxide production accelerates neointima formation and impairs endothelial function in hypercholesterolaemic rabbits. *Arterioscler. Thromb. Vasc. Biol.*, **14** 753–759.

Celermajer, D.S., Sorensen, K.E., Gooch, V.M., Spiegelhalter, D.J., Miller, O.I., Sullivan, I.D., Lloyd, J.K. and Deanfield, J.E. (1992) Non-invasive detection of endothelial dysfunction in children and adults at risk of atherosclerosis. *Lancet*, **340**, 1111–1115.

Charles, I.G., Scorer, C.A., Moro, M.A., Fernandez, C., Chubb, A., Dawson, J., Foxwell, N., Knowles, R.G. and Baylis, S.A. (1996) Expression of human nitric oxide synthase isozymes. *Methods Enzymol.*, **268**, 449–460.

Chester, A.H., O'Neil, G.S., Moncada, S., Tadjkarimi, S. and Yacoub, M.H. (1990) Low basal and stimulated release of nitric oxide in atherosclerotic epicardial coronary arteries. *Lancet*, **336**, 897–900.

Cho, H.J., Xie, Q.W., Calaycay, J., Mumford, R.A., Swiderek, K.M., Lee, T.D. and Nathan, C. (1992) Calmodulin is a subunit of nitric oxide synthase from macrophages. *J. Exp. Med.*, **176**, 599–604.

Chowienczyk, P.J., Cockcroft, J.R. and Ritter, J.M. (1993) Differential inhibition by NG-monomethyl-L-arginine of vasodilator effects of acetylcholine and methacholine in human forearm vasculature. *Br. J. Pharmacol.*, **110**, 736–738.

Cleeter, M.W., Cooper, J.M., Darley Usmar, V.M., Moncada, S. and Schapira, A.H. (1994) Reversible inhibition of cytochrome c oxidase, the terminal enzyme of the mitochondrial respiratory chain, by nitric oxide. Implications for neurodegenerative diseases. *FEBS Lett.*, **345**, 50–54.

Cockcroft, J.R., Chowienczyk, P.J., Benjamin, N. and Ritter, J.M. (1994a) Preserved endothelium-dependent vasodilatation in patients with essential hypertension [published erratum appears in N Engl J Med 1995 May 25;332(21):1455]. *N. Engl. J. Med.*, **330**, 1036-1040.

Cockcroft, J.R., Chowienczyk, P.J., Brett, S.E. and Ritter, J.M. (1994b) Effect of NG-monomethyl-L-arginine on kinin-induced vasodilation in the human forearm. *Br. J. Clin. Pharmacol.*, **38**, 307–310.

Cocks, T.M., Angus, J.A., Campbell, J.H. and Campbell, G.R. (1985) Release and properties of endothelium-derived relaxing factor (EDRF) from endothelial cells in culture. *J. Cell Physiol.*, **123**, 310-320.

Collier, J. and Vallance, P. (1990) Biphasic response to acetylcholine in human veins in vivo: the role of the endothelium. *Clin. Sci. Colch.*, **78**, 101–104.

Collins, P., Burman, J., Chung, H.I. and Fox, K. (1993) Hemoglobin inhibits endothelium-dependent relaxation to acetylcholine in human coronary arteries in vivo. *Circulation*, **87**, 80–85.

Corson, M.A., James, N.L., Latta, S.E., Nerem, R.M., Berk, B.C. and Harrison, D.G. (1996) Phosphorylation of endothelial nitric oxide synthase in response to fluid shear stress. *Circ. Res.*, **79**, 984–991.

Crane, B.R., Arvai, A.S., Gacchui, R., Wu, C., Ghosh, D.K., Getzoff, E.D., Stuehr, D.J. and Tainer, T.A. (1997) The structure of nitric oxide synthase oxygenase domain and inhibitor complexes. *Science*, **278**, 425–431.

Crawley, D.E., Liu, S.F., Evans, T.W. and Barnes, P.J. (1990) Inhibitory role of endothelium-derived relaxing factor in rat and human pulmonary arteries. *Br. J. Pharmacol.*, **101**, 166–170.

Darley-Usmar, V., Wiseman, H. and Halliwell, B. (1995) Nitric oxide and oxygen radicals: a question of balance. *FEBS Lett.* 131–135.

Davies, P.F. (1995) Flow-mediated endothelial mechanotransduction. *Physiol. Rev.*, **75**, 519–560.

De Caterina, R., Libby, P., Peng, H.B., Thannickal, V.J., Rajavashisth, T.B., Gimbrone, M.A., Jr., Shin, W.S. and Liao, J.K. (1995) Nitric oxide decreases cytokine-induced endothelial activation. Nitric oxide selectively reduces endothelial expression of adhesion molecules and proinflammatory cytokines. *J. Clin. Invest.*, **96**, 60–68.

de Vera, M.E., Shapiro, R.A., Nussler, A.K., Mudgett, J.S., Simmons, R.L., Morris, S.M., Jr., Billiar, T.R. and Geller, D.A. (1996) Transcriptional regulation of human inducible nitric oxide synthase (NOS2) gene by cytokines: initial analysis of the human NOS2 promoter. *Proc. Natl. Acad. Sci. U. S. A.*, **93**, 1054–1059.

del Pozo, V., de Arruda Chaves, E., de Andres, B., Cardaba, B., Lopez Farre, A., Gallardo, S., Cortegano, I., Vidarte, L., Jurado, A., Sastre, J., Palomino, P. and Lahoz, C. (1997) Eosinophils transcribe and translate messenger RNA for inducible nitric oxide synthase. *J. Immunol.*, **158**, 859–864.

Dijkhorst Oei, L.T. and Koomans, H.A. (1998) Effects of a nitric oxide synthesis inhibitor on renal sodium handling and diluting capacity in humans. *Nephrol. Dial. Transplant.*, **13**, 587–593.

Dinerman, J.L., Dawson, T.M., Schell, M.J., Snowman, A. and Snyder, S.H. (1994) Endothelial nitric oxide synthase localized to hippocampal pyramidal cells: implications for synaptic plasticity. *Proc. Natl. Acad. Sci. U. S. A.*, **91**, 4214–4218.

Eis, A.L., Brockman, D.E., Pollock, J.S. and Myatt, L. (1995) Immunohistochemical localization of endothelial nitric oxide synthase in human villous and extravillous trophoblast populations and expression during syncytiotrophoblast formation in vitro. *Placenta.*, **16**, 113–126.

Evans, T.J., Buttery, L.D., Carpenter, A., Springall, D.R., Polak, J.M. and Cohen, J. (1996) Cytokine-treated human neutrophils contain inducible nitric oxide synthase that produces nitration of ingested bacteria. *Proc. Natl. Acad. Sci. U. S. A.*, **93**, 9553–9558.

Feron, O., Belhassen, L., Kobzik, L., Smith, T.W., Kelly, R.A. and Michel, T. (1996) Endothelial nitric oxide synthase targeting to caveolae. Specific interactions with caveolin isoforms in cardiac myocytes and endothelial cells. *J. Biol. Chem.*, **271**, 22810–22814.

Fickling, S.A., Williams, D., Vallance, P., Nussey, S.S. and Whitley, G.S. (1993) Plasma concentrations of endogenous inhibitor of nitric oxide synthesis in normal pregnancy and pre-eclampsia. *Lancet*, **342**, 242–243.

Fleming, I., Bauersachs, J. and Busse, R. (1997) Calcium-dependent and calcium-independent activation of the endothelial NO synthase. *J. Vasc. Res.*, **34**, 165–174.

Forstermann, U., Nakane, M., Tracey, W.R. and Pollock, J.S. (1993) Isoforms of nitric oxide synthase: functions in the cardiovascular system. *Eur. Heart J.*, **14 Suppl I**, 10–15.

Forstermann, U., Closs, E.I., Pollock, J.S., Nakane, M., Schwarz, P., Gath, I. and Kleinert, H. (1994a) Nitric oxide synthase isozymes. Characterization, purification, molecular cloning, and functions. *Hypertension*, **23**, 1121–1131.

Forstermann, U., Pollock, J.S., Tracey, W.R. and Nakane, M. (1994b) Isoforms of nitric-oxide synthase: purification and regulation. *Methods Enzymol.*, **233**, 258–264.

Forte, P., Copland, M., Smith, L.M., Milne, E., Sutherland, J. and Benjamin, N. (1997) Basal nitric oxide synthesis in essential hypertension. *Lancet*, **349**, 837–842.

Furchgott, R.F. (1984) The role of endothelium in the responses of vascular smooth muscle to drugs. *Annu. Rev. Pharmacol. Toxicol.*, **24**, 175–197.

Furchgott, R.F. (1990) The 1989 Ulf von Euler lecture. Studies on endothelium-dependent vasodilation and the endothelium-derived relaxing factor. *Acta Physiol. Scand.*, **139**, 257–270.

Furchgott, R.F. and Jothianandan, D. (1991) Endothelium-dependent and -independent vasodilation involving cyclic GMP: relaxation induced by nitric oxide, carbon monoxide and light. *Blood Vessels*, **28**, 52–61.

Furchgott, R.F. and Zawadzki, J.V. (1980) The obligatory role of endothelial cells in the relaxation of arterial smooth muscle by acetylcholine. *Nature*, **288**, 373–376.

Garcia Cardena, G., Fan, R., Stern, D.F., Liu, J. and Sessa, W.C. (1996) Endothelial nitric oxide synthase is regulated by tyrosine phosphorylation and interacts with caveolin-1. *J. Biol. Chem.*, **271**, 27237–27240.

Garg, U.C. and Hassid, A. (1989) Nitric oxide-generating vasodilators and 8–bromo-cyclic guanosine monophosphate inhibit mitogenesis and proliferation of cultured rat vascular smooth muscle cells. *J. Clin. Invest.*, **83**, 1774-1777.

Giuffre, A., Sarti, P., D'Itri, E., Buse, G., Soulimane, T. and Brunori, M. (1996) On the mechanism of inhibition of cytochrome c oxidase by nitric oxide. *J. Biol. Chem.*, **271**, 33404-33408.

Gow, A.J. and Stamler, J.S. (1998) Reactions between nitric oxide and haemoglobin under physiological conditions. *Nature*, **391**, 169–173.

Greenberg, B., Rhoden, K. and Barnes, P.J. (1987) Endothelium-dependent relaxation of human pulmonary arteries. *Am. J. Physiol.*, **252**, H434–8.

Gries, A., Bode, C., Peter, K., Herr, A., Bohrer, H., Motsch, J. and Martin, E. (1998) Inhaled nitric oxide inhibits human platelet aggregation, P-selectin expression, and fibrinogen binding in vitro and in vivo. *Circulation*, **97**, 1481–1487.

Griffith, T.M., Edwards, D.H., Lewis, M.J., Newby, A.C. and Henderson, A.H. (1984) The nature of endothelium-derived vascular relaxant factor. *Nature*, **308**, 645–647.

Hardebo, J.E., Kahrstrom, J., Owman, C. and Salford, L.G. (1987) Vasomotor effects of neurotransmitters and modulators on isolated human pial veins. *J. Cereb. Blood Flow Metab.*, **7**, 612–618.

Harrison, D.G., Sayegh, H., Ohara, Y., Inoue, N. and Venema, R.C. (1996) Regulation of expression of the endothelial cell nitric oxide synthase. *Clin. Exp. Pharmacol. Physiol.*, **23**, 251–255.

Haynes, W.G., Noon, J.P., Walker, B.R. and Webb, D.J. (1993) Inhibition of nitric oxide synthesis increases blood pressure in healthy humans. *J. Hypertens.*, **11**, 1375–1380.

Haynes, W.G., Hand, M.F., Dockrell, M.E., Eadington, D.W., Lee, M.R., Hussein, Z., Benjamin, N. and Webb, D.J. (1997) Physiological role of nitric oxide in regulation of renal function in humans. *Am. J. Physiol.*, **272**, F364–71.

Hibbs, J.B., Jr., Taintor, R.R., Vavrin, Z. and Rachlin, E.M. (1988) Nitric oxide: a cytotoxic activated macrophage effector molecule [published erratum appears in Biochem Biophys Res Commun 1989 Jan 31;158(2):624]. *Biochem. Biophys. Res. Commun.*, **157**, 87–94.

Hibbs, J.B., Jr., Westenfelder, C., Taintor, R., Vavrin, Z., Kablitz, C., Baranowski, R.L., Ward, J.H., Menlove, R.L., McMurry, M., Kushner, J.P. and Samlowski, W.E. (1992) Evidence for cytokine-inducible nitric oxide synthesis from L-arginine in patients receiving interleukin-2 therapy. *J. Clin. Invest.*, **89** 867–877.

Hibi, K., Ishigami, T., Tamura, K., Mizushima, S., Nyui, N., Fujita, T., Ochiai, H., Kosuge, M., Watanabe, Y., Yoshii, Y., Kihara, M., Kimura, M., Kimura, K., Ishii, M. and Umemura, S. (1998) Endothelial nitric oxide synthase gene polymorphism and acute myocardial infarction. *Hypertension*, **32** 521–526.

Hingorani, A.D., Jia, H., Stevens, P.A., Monteith, M.S. and Brown, M.J. (1995) A common variant in exon 7 of the endothelial constitutive nitric oxide synthase gene. *Clin. Sci.*, **88**, 21P(Abstract)

Hingorani, A.D. (1997) Studies of candidate genes for essential hypertension and coronary artery disease. Cambridge University.

Hingorani, A.D., Liang, C.F., Fatibene, J., Parsons, A., Hopper, R.V., Trutwein, D., Stephens, N.G., O'Shaughnessy, K.M. and Brown, M.J. (1997) A common variant of the endothelial nitric oxide synthase gene is a risk factor for coronary atherosclerosis in the East Anglian region of the UK. *Circulation*, **96**, I–545(Abstract)

Hobbs, A.J. (1997) Soluble guanylate cyclase: the forgotten sibling. *Trends. Pharmacol. Sci.*, **18** 484-490.

Ignarro, L.J., Buga, G.M., Wood, K.S., Byrns, R.E. and Chaudhuri, G. (1987) Endothelium-derived relaxing factor produced and released from artery and vein is nitric oxide. *Proc. Natl. Acad. Sci. U. S. A.*, **84**, 9265–9269.

Ignarro, L.J. (1990a) Haem-dependent activation of guanylate cyclase and cyclic GMP formation by endogenous nitric oxide: a unique transduction mechanism for transcellular signaling. *Pharmacol. Toxicol.*, **67**, 1–7.

Ignarro, L.J. (1990b) Biosynthesis and metabolism of endothelium-derived nitric oxide. *Annu. Rev. Pharmacol. Toxicol.*, **30**, 535–560.

Imaizumi, T., Hirooka, Y., Masaki, H., Harada, S., Momohara, M., Tagawa, T. and Takeshita, A. (1992) Effects of L-arginine on forearm vessels and responses to acetylcholine. *Hypertension*, **20**, 511–517.

Inoue, N., Venema, R.C., Sayegh, H.S., Ohara, Y., Murphy, T.J. and Harrison, D.G. (1995) Molecular regulation of the bovine endothelial cell nitric oxide synthase by transforming growth factor-beta 1. *Arterioscler. Thromb. Vasc. Biol.*, **15**, 1255–1261.

Janssens, S., Flaherty, D., Nong, Z., Varenne, O., van Pelt, N., Haustermans, C., Zoldhelyi, P., Gerard, R. and Collen, D. (1998) Human endothelial nitric oxide synthase gene transfer inhibits vascular smooth muscle cell proliferation and neointima formation after balloon injury in rats. *Circulation*, **97**, 1274–1281.

Jia, L., Bonaventura, C., Bonaventura, J. and Stamler, J.S. (1996) S-nitrosohaemoglobin: a dynamic activity of blood involved in vascular control. *Nature*, **380**, 221–226.

Jovanovic, A., Grbovic, L. and Tulic, I. (1994) Predominant role for nitric oxide in the relaxation induced by acetylcholine in human uterine artery. *Hum. Reprod.*, **9**, 387–393.

Kanai, A.J., Strauss, H.C., Truskey, G.A., Crews, A.L., Grunfeld, S. and Malinski, T. (1995) Shear stress induces ATP-independent transient nitric oxide release from vascular endothelial cells, measured directly with a porphyrinic microsensor. *Circ. Res.*, **77**, 284–293.

Kanamaru, K., Waga, S., Fujimoto, K., Itoh, H. and Kubo, Y. (1989) Endothelium-dependent relaxation of human basilar arteries. *Stroke*, **20**, 1208–1211.

Keaney, J.F., Simon, D.I., Stamler, J.S., Jaraki, O., Scharfstein, J., Vita, J.A. and Loscalzo, J. (1993) NO forms an adduct with serum albumin that has endothelium-derived relaxing factor-like properties. *J. Clin. Invest.*, **91** 1582–1589.

Keaney, J.F., Jr., Simon, D.I., Stamler, J.S., Jaraki, O., Scharfstein, J., Vita, J.A. and Loscalzo, J. (1993) NO forms an adduct with serum albumin that has endothelium-derived relaxing factor-like properties. *J. Clin. Invest.*, **91**, 1582–1589.

Khan, B.V., Harrison, D.G., Olbrych, M.T., Alexander, R.W. and Medford, R.M. (1996) Nitric oxide regulates vascular cell adhesion molecule 1 gene expression and redox-sensitive transcriptional events in human vascular endothelial cells. *Proc. Natl. Acad. Sci. U. S. A.*, **93**, 9114–9119.

Kleinert, H., Wallerath, T., Euchenhofer, C., Ihrig-Biedert, I., Li, H. and Forstermann, U. (1998) Estrogens increase transcription of the human endothelial nitric oxide synthase gene. *Hypertension*, **31** 582–588.

Knowles, R.G. and Moncada, S. (1994) Nitric oxide synthases in mammals. *Biochem. J.*, **298**, 249–258.

Kolyada, A.Y., Savikovsky, N. and Madias, N.E. (1996) Transcriptional regulation of the human iNOS gene in vascular-smooth-muscle cells and macrophages: evidence for tissue specificity. *Biochem. Biophys. Res. Commun.*, **220**, 600–605.

Kuo, L., Davis, M.J. and Chilian, W.M. (1990) Endothelium-dependent, flow-induced dilation of isolated coronary arterioles. *Am. J. Physiol.*, **259**, H1063–70.

Lamontagne, D., Pohl, U. and Busse, R. (1992) Mechanical deformation of vessel wall and shear stress determine the basal release of endothelium-derived relaxing factor in the intact rabbit coronary vascular bed. *Circ. Res.*, **70**, 123–130.

Lansman, J.B., Hallam, T.J. and Rink, T.J. (1987) Single stretch-activated ion channels in vascular endothelial cells as mechanotransducers? *Nature*, **325**, 811–813.

Lansman, J.B. (1988) Endothelial mechanosensors. Going with the flow. *Nature*, **331**, 481–482.

Lawrie, G.M., Weilbacher, D.E. and Henry, P.D. (1990) Endothelium-dependent relaxation in human saphenous vein grafts. Effects of preparation and clinicopathologic correlations. *J. Thorac. Cardiovasc. Surg.*, **100**, 612–620.

Lee, J.S., Adrie, C., Jacob, H.J., Roberts, J.D., Jr., Zapol, W.M. and Bloch, K.D. (1996) Chronic inhalation of nitric oxide inhibits neointimal formation after balloon-induced arterial injury. *Circ. Res.*, **78**, 337–342.

Lefroy, D.C., Crake, T., Uren, N.G., Davies, G.J. and Maseri, A. (1993) Effect of inhibition of nitric oxide synthesis on epicardial coronary artery caliber and coronary blood flow in humans. *Circulation*, **88**, 43–54.

Lieberman, E.H., Gerhard, M.D., Uehata, A., Selwyn, A.P., Ganz, P., Yeung, A.C. and Creager, M.A. (1996) Flow-induced vasodilation of the human brachial artery is impaired in patients <40 years of age with coronary artery disease. *Am. J. Cardiol.*, **78**, 1210–1214.

Linder, L., Kiowski, W., Buhler, F.R. and Luscher, T.F. (1990) Indirect evidence for release of endothelium-derived relaxing factor in human forearm circulation in vivo. Blunted response in essential hypertension. *Circulation*, **81**, 1762–1767.

Loke, K.E., Sobey, C.G., Dusting, G.J. and Woodman, O.L. (1994) Requirement for endothelium-derived nitric oxide in vasodilation produced by stimulation of cholinergic nerves in rat hindquarters. *Br. J. Pharmacol.*, **112**, 630–634.

Luscher, T.F., Cooke, J.P., Houston, D.S., Neves, R.J. and Vanhoutte, P.M. (1987) Endothelium-dependent relaxations in human arteries. *Mayo Clin. Proc.*, **62**, 601–606.

Luscher, T.F., Yang, Z., Tschudi, M., von Segesser, L., Stulz, P., Boulanger, C., Siebenmann, R., Turina, M. and Buhler, F.R. (1990) Interaction between endothelin-1 and endothelium-derived relaxing factor in human arteries and veins. *Circ. Res.*, **66**, 1088–1094.

Luscher, T.F. and Vanhoutte, P.M. (1988) Endothelium-dependent responses in human blood vessels. *Trends. Pharmacol. Sci.*, **9**, 181–184.

Macallan, D.C., Smith, L.M., Ferber, J., Milne, E., Griffin, G.E., Benjamin, N. and McNurlan, M.A. (1997) Measurement of NO synthesis in humans by L-[15N2]arginine: application to the response to vaccination. *Am. J. Physiol.*, **272**, R1888–96.

MacAllister, R.J., Fickling, S.A., Whitley, G.S. and Vallance, P. (1994a) Metabolism of methylarginines by human vasculature; implications for the regulation of nitric oxide synthesis. *Br. J. Pharmacol.*, **112**, 43–48.

MacAllister, R.J., Whitley, G.S. and Vallance, P. (1994b) Effects of guanidino and uremic compounds on nitric oxide pathways. *Kidney Int.*, **45**, 737–742.

MacAllister, R.J., Calver, A.L., Collier, J., Edwards, C.M., Herreros, B., Nussey, S.S. and Vallance, P. (1995) Vascular and hormonal responses to arginine: provision of substrate for nitric oxide or non-specific effect? *Clin. Sci. Colch.*, **89**, 183–190.

MacAllister, R.J., Parry, H., Kimoto, M., Ogawa, T., Russell, R.J., Hodson, H., Whitley, G.S. and Vallance, P. (1996a) Regulation of nitric oxide synthesis by dimethylarginine dimethylaminohydrolase. *Br. J. Pharmacol.*, **119**, 1533–1540.

MacAllister, R.J., Rambausek, M.H., Vallance, P., Williams, D., Hoffmann, K.H. and Ritz, E. (1996b) Concentration of dimethyl-L-arginine in the plasma of patients with end-stage renal failure. *Nephrol. Dial. Transplant.*, **11**, 2449–2452.

MacAllister, R.J. and Vallance, P. (1996) Systemic vascular adaptation to increases in blood volume: the role of the blood-vessel wall [editorial]. *Nephrol. Dial. Transplant.*, **11**, 231–234.

MacRitchie, A.N., Jun, S.S., Chen, Z., German, Z., Yuhanna, I.S., Sherman, T.S. and Shaul, P.W. (1997) Estrogen upregulates endothelial nitric oxide synthase gene expression in fetal pulmonary artery endothelium. *Circ. Res.*, **81**, 355–362.

Malinski, T., Kapturczak, M., Dayharsh, J. and Bohr, D. (1993a) Nitric oxide synthase activity in genetic hypertension. *Biochem. Biophys. Res. Commun.*, **194**, 654–658.

Malinski, T., Radomski, M.W., Taha, Z. and Moncada, S. (1993b) Direct electrochemical measurement of nitric oxide released from human platelets. *Biochem. Biophys. Res. Commun.*, **194**, 960–965.

Malinski, T. and Taha, Z. (1992) Nitric oxide release from a single cell measured in situ by a porphyrinic-based microsensor. *Nature*, **358**, 676–678.

Markus, H.S., Ruigrok, Y., Ali, N. and Powell, J.F. (1998) Endothelial nitric oxide synthase exon 7 polymorphism, ischaemic cerebrovascular disease, and carotid atheroma. *Stroke*, **29** 1908–1911.

Marletta, M.A., Yoon, P.S., Iyengar, R., Leaf, C.D. and Wishnok, J.S. (1988) Macrophage oxidation of L-arginine to nitrite and nitrate: nitric oxide is an intermediate. *Biochemistry*, **27**, 8706–8711.

Marletta, M.A. (1993) Nitric oxide synthase structure and mechanism. *J. Biol. Chem.*, **268**, 12231–12234.

Marsden, P.A., Heng, H.H., Scherer, S.W., Stewart, R.J., Hall, A.V., Shi, X.M., Tsui, L.C. and Schappert, K.T. (1993) Structure and chromosomal localization of the human constitutive endothelial nitric oxide synthase gene. *J. Biol. Chem.*, **268**, 17478–17488.

Martin, W., Furchgott, R.F., Villani, G.M. and Jothianandan, D. (1986) Depression of contractile responses in rat aorta by spontaneously released endothelium-derived relaxing factor. *J. Pharmacol. Exp. Ther.*, **237**, 529–538.

Mayer, B., Schmidt, K., Humbert, P. and Bohme, E. (1989) Biosynthesis of endothelium-derived relaxing factor: a cytosolic enzyme in porcine aortic endothelial cells Ca2+-dependently converts L-arginine into an activator of soluble guanylyl cyclase. *Biochem. Biophys. Res. Commun.*, **164**, 678–685.

McCall, T.B., Feelisch, M., Palmer, R.M. and Moncada, S. (1991) Identification of N-iminoethyl-L-ornithine as an irreversible inhibitor of nitric oxide synthase in phagocytic cells. *Br. J. Pharmacol.*, **102**, 234–238.

McDonald, K.K., Zharikov, S., Block, E.R. and Kilberg, M.S. (1997) A caveolar complex between the cationic amino acid transporter 1 and endothelial nitric-oxide synthase may explain the "arginine paradox". *J. Biol. Chem.*, **272**, 31213–31216.

Mehta, S., Stewart, D.J. and Levy, R.D. (1996) The hypotensive effect of L-arginine is associated with increased expired nitric oxide in humans. *Chest*, **109**, 1550–1555.

Meredith, I.T., Currie, K.E., Anderson, T.J., Roddy, M.A., Ganz, P. and Creager, M.A. (1996) Postischemic vasodilation in human forearm is dependent on endothelium-derived nitric oxide. *Am. J. Physiol.*, **270**, H1435–40.

Michel, J.B., Feron, O., Sacks, D. and Michel, T. (1997a) Reciprocal regulation of endothelial nitric-oxide synthase by Ca2+-calmodulin and caveolin. *J. Biol. Chem.*, **272**, 15583-15586.

Michel, J.B., Feron, O., Sase, K., Prabhakar, P. and Michel, T. (1997b) Caveolin versus calmodulin. Counterbalancing allosteric modulators of endothelial nitric oxide synthase. *J. Biol. Chem.*, **272**, 25907–25912.

Michel, T., Li, G.K. and Busconi, L. (1993) Phosphorylation and subcellular translocation of endothelial nitric oxide synthase. *Proc. Natl. Acad. Sci. U. S. A.*, **90**, 6252–6256.

Michel, T. and Feron, O. (1997) Nitric oxide synthases: which, where, how, and why? *J. Clin. Invest.*, **100**, 2146–2152.

Miyahara, K., Kawamoto, T., Sase, K., Yui, Y., Toda, K., Yang, L., Hattori, R., Aoyama, T., Yamamoto, Y., Doi, Y., Ogoshi, S., Hashimoto, K., Kawai, C., Sasayama, S. and Shizuta, Y. (1994) Cloning and structural organisation of the human endothelial nitric oxide synthase gene. *Eur. J. Biochem.*, **223** 719–726.

Miyamoto, Y., Saito, Y., Kajiyama, N., Yoshimura, M., Shimasaki, Y., Nakayama, N., Kamitani, S., Harada, M., Ishikawa, M., Kuwahara, K., Ogawa, E., Hamanaka, I., Takahashi, N., Kaneshige, T., Teraoka, H., Akamizu, T., Azuma, N., Yoshimasa, Y., Yoshimasa, T., Itoh, H., Masuda, I., Yasue, H. and Nakao, K. (1998) Endothelial nitric oxide synthase gene is positively associated with essential hypertension. *Hypertension*, **32** 3-8.

Mollace, V., Salvemini, D., Anggard, E. and Vane, J. (1991) Nitric oxide from vascular smooth muscle cells: regulation of platelet reactivity and smooth muscle cell guanylate cyclase. *Br. J. Pharmacol.*, **104**, 633–638.

Moncada, S., Palmer, R.M. and Higgs, E.A. (1991) Nitric oxide: physiology, pathophysiology, and pharmacology. *Pharmacol. Rev.*, **43**, 109–142.

Mulsch, A., Bassenge, E. and Busse, R. (1989) Nitric oxide synthesis in endothelial cytosol: evidence for a calcium-dependent and a calcium-independent mechanism. *Naunyn Schmiedebergs Arch. Pharmacol.*, **340**, 767–770.

Myatt, L., Brockman, D.E., Eis, A.L. and Pollock, J.S. (1993) Immunohistochemical localization of nitric oxide synthase in the human placenta. *Placenta.*, **14**, 487–495.

Nadaud, S., Bonnardeaux, A., Lathrop, M. and Soubrier, F. (1994) Gene structure, polymorphism and mapping of the human endothelial nitric oxide synthase gene. *Biochem. Biophys. Res. Commun.*, **198**, 1027–1033.

Nakaki, T., Nakayama, M. and Kato, R. (1990) Inhibition by nitric oxide and nitric oxide-producing vasodilators of DNA synthesis in vascular smooth muscle cells. *Eur. J. Pharmacol.*, **189**, 347–353.

Naruse, K., Shimizu, K., Muramatsu, M., Toki, Y., Miyazaki, Y., Okumura, K., Hashimoto, H. and Ito, T. (1994) Long-term inhibition of NO synthesis promotes atherosclerosis in hypercholesterolaemic rabbit thoracic aorta: PGH2 does not contribute to impaired endothelium-dependent relaxation. *Arterioscler. Thromb. Vasc. Biol.*, **14** 746–752.

Nicholson, S., Bonecini Almeida, M., Lapa, e.R., Nathan, C., Xie, Q.W., Mumford, R., Weidner, J.R., Calaycay, J., Geng, J., Boechat, N. and et al (1996) Inducible nitric oxide synthase in pulmonary alveolar macrophages from patients with tuberculosis. *J. Exp. Med.*, **183**, 2293–2302.

Noris, M., Morigi, M., Donadelli, R., Aiello, S., Foppolo, M., Todeschini, M., Orisio, S., Remuzzi, G. and Remuzzi, A. (1995) Nitric oxide synthesis by cultured endothelial cells is modulated by flow conditions. *Circ. Res.*, **76**, 536–543.

O'Neil, G.S., Chester, A.H., Allen, S.P., Luu, T.N., Tadjkarimi, S., Ridley, P., Khagani, A., Musumeci, F. and Yacoub, M.H. (1991) Endothelial function of human gastroepiploic artery. Implications for its use as a bypass graft [see comments]. *J. Thorac. Cardiovasc. Surg.*, **102**, 561–565.

Olesen, S.P., Clapham, D.E. and Davies, P.F. (1988) Haemodynamic shear stress activates a K+ current in vascular endothelial cells. *Nature*, **331**, 168-170.

Palmer, R.M., Ferrige, A.G. and Moncada, S. (1987) Nitric oxide release accounts for the biological activity of endothelium-derived relaxing factor. *Nature*, **327**, 524–526.

Palmer, R.M., Ashton, D.S. and Moncada, S. (1988a) Vascular endothelial cells synthesize nitric oxide from L-arginine. *Nature*, **333**, 664–666.

Palmer, R.M., Rees, D.D., Ashton, D.S. and Moncada, S. (1988b) L-arginine is the physiological precursor for the formation of nitric oxide in endothelium-dependent relaxation. *Biochem. Biophys. Res. Commun.*, **153**, 1251–1256.

Palmer, R.M. and Moncada, S. (1989) A novel citrulline-forming enzyme implicated in the formation of nitric oxide by vascular endothelial cells. *Biochem. Biophys. Res. Commun.*, **158**, 348-352.

Penny, W.F., Rockman, H., Long, J., Bhargava, V., Carrigan, K., Ibriham, A., Shabetai, R., Ross, J., Jr. and Peterson, K.L. (1995) Heterogeneity of vasomotor response to acetylcholine along the human coronary artery. *J. Am. Coll. Cardiol.*, **25**, 1046–1055.

Pohl, U., Holtz, J., Busse, R. and Bassenge, E. (1986) Crucial role of endothelium in the vasodilator response to increased flow in vivo. *Hypertension*, **8**, 37–44.

Quyyumi, A.A., Dakak, N., Andrews, N.P., Gilligan, D.M., Panza, J.A. and Cannon, R.O. (1995a) Contribution of nitric oxide to metabolic coronary vasodilation in the human heart. *Circulation*, **92**, 320-326.

Quyyumi, A.A., Dakak, N., Andrews, N.P., Husain, S., Arora, S., Gilligan, D.M., Panza, J.A. and Cannon, R.O. (1995b) Nitric oxide activity in the human coronary circulation. Impact of risk factors for coronary atherosclerosis. *J. Clin. Invest.*, **95**, 1747–1755.

Raddino, R., Pela, G., Manca, C., Barbagallo, M., D'Aloia, A., Passeri, M. and Visioli, O. (1997) Mechanism of action of human calcitonin gene-related peptide in rabbit heart and in human mammary arteries. *J. Cardiovasc. Pharmacol.*, **29**, 463–470.

Radomski, M.W., Palmer, R.M. and Moncada, S. (1987a) Comparative pharmacology of endothelium-derived relaxing factor, nitric oxide and prostacyclin in platelets. *Br. J. Pharmacol.*, **92**, 181–187.

Radomski, M.W., Palmer, R.M. and Moncada, S. (1987b) Endogenous nitric oxide inhibits human platelet adhesion to vascular endothelium. *Lancet*, **2**, 1057–1058.

Radomski, M.W., Palmer, R.M. and Moncada, S. (1990) Characterization of the L-arginine:nitric oxide pathway in human platelets. *Br. J. Pharmacol.*, **101**, 325–328.

Ranjan, V., Xiao, Z. and Diamond, S.L. (1995) Constitutive NOS expression in cultured endothelial cells is elevated by fluid shear stress. *Am. J. Physiol.*, **269**, H550–5.

Ravalli, S., Albala, A., Ming, M., Szabolcs, M., Barbone, A., Michler, R.E. and Cannon, P.J. (1998) Inducible nitric oxide synthase expression in smooth muscle cells and macrophages of human transplant coronary artery disease. *Circulation*, **97**, 2338–2345.

Remuzzi, G., Perico, N., Zoja, C., Corna, D., Macconi, D. and Vigano, G. (1990) Role of endothelium-derived nitric oxide in the bleeding tendency of uremia. *J. Clin. Invest.*, **86**, 1768–1771.

Rhodes, P., Barr, C.S. and Struthers, A.D. (1996) Arginine, lysine and ornithine as vasodilators in the forearm of man. *Eur. J. Clin. Invest.*, **26**, 325–331.

Robinson, L.J., Busconi, L. and Michel, T. (1995) Agonist-modulated palmitoylation of endothelial nitric oxide synthase. *J. Biol. Chem.*, **270**, 995–998.

Robinson, L.J. and Michel, T. (1995) Mutagenesis of palmitoylation sites in endothelial nitric oxide synthase identifies a novel motif for dual acylation and subcellular targeting. *Proc. Natl. Acad. Sci. U. S. A.*, **92**, 11776–11780.

Rodriguez Crespo, I., Gerber, N.C. and Ortiz de Montellano, P.R. (1996) Endothelial nitric-oxide synthase. Expression in Escherichia coli, spectroscopic characterization, and role of tetrahydrobiopterin in dimer formation. *J. Biol. Chem.*, **271**, 11462–11467.

Sarkar, R., Meinberg, E.G., Stanley, J.C., Gordon, D. and Webb, R.C. (1996) Nitric oxide reversibly inhibits the migration of cultured vascular smooth muscle cells. *Circ. Res.*, **78**, 225–230.

Sase, K. and Michel, T. (1995) Expression of constitutive endothelial nitric oxide synthase in human blood platelets. *Life Sci.*, **57**, 2049–2055.

Schoeffter, P., Dion, R. and Godfraind, T. (1988) Modulatory role of the vascular endothelium in the contractility of human isolated internal mammary artery. *Br. J. Pharmacol.*, **95**, 531–543.

Sessa, W.C., Garcia Cardena, G., Liu, J., Keh, A., Pollock, J.S., Bradley, J., Thiru, S., Braverman, I.M. and Desai, K.M. (1995) The Golgi association of endothelial nitric oxide synthase is necessary for the efficient synthesis of nitric oxide. *J. Biol. Chem.*, **270**, 17641–17644.

Shaul, P.W., North, A.J., Wu, L.C., Wells, L.B., Brannon, T.S., Lau, K.S., Michel, T., Margraf, L.R. and Star, R.A. (1994) Endothelial nitric oxide synthase is expressed in cultured human bronchiolar epithelium. *J. Clin. Invest.*, **94**, 2231–2236.

Shaul, P.W., Smart, E.J., Robinson, L.J., German, Z., Yuhanna, I.S., Ying, Y., Anderson, R.G. and Michel, T. (1996) Acylation targets endothelial nitric-oxide synthase to plasmalemmal caveolae. *J. Biol. Chem.*, **271**, 6518–6522.

Shimasaki, Y., Yasue, H., Yoshimura, M., Nakayama, M., Kugiyama, K., Ogawa, H., Harada, E., Masuda, T., Koyama, W., Saito, Y., Miyamoto, Y., Ogawa, Y. and Nakao, K. (1998) Association of the missense Glu298Asp variant of the endothelial nitric oxide synthase gene with myocardial infarction. *J. Am. Coll. Cardiol.*, **31**, 1506–1510.

Siegel, G., Malmsten, M., Klussendorf, D., Walter, A., Schnalke, F. and Kauschmann, A. (1996) Blood-flow sensing by anionic biopolymers. *J. Auton. Nerv. Syst.*, **57**, 207–213.

Simon, D.I., Stamler, J.S., Loh, E., Loscalzo, J., Francis, S.A. and Creager, M.A. (1995) Effect of nitric oxide synthase inhibition on bleeding time in humans. *J. Cardiovasc. Pharmacol.*, **26**, 339–342.

Stamler, J.S., Jaraki, O., Osborne, J., Simon, D.I., Keaney, J., Vita, J., Singel, D., Valeri, C.R. and Loscalzo, J. (1992a) Nitric oxide circulates in mammalian plasma primarily as an S-nitroso adduct of serum albumin. *Proc. Natl. Acad. Sci. U. S. A.*, **89**, 7674–7677.

Stamler, J.S., Simon, D.I., Osborne, J.A., Mullins, M.E., Jaraki, O., Michel, T., Singel, D.J. and Loscalzo, J. (1992b) S-nitrosylation of proteins with nitric oxide: synthesis and characterization of biologically active compounds. *Proc. Natl. Acad. Sci. U. S. A.*, **89**, 444–448.

Stamler, J.S., Loh, E., Roddy, M.A., Currie, K.E. and Creager, M.A. (1994) Nitric oxide regulates basal systemic and pulmonary vascular resistance in healthy humans. *Circulation*, **89**, 2035–2040.

Stamler, J.S., Jia, L., Eu, J.P., McMahon, T.J., Demchenko, I.T., Bonaventura, J., Gernert, K. and Piantadosi, C.A. (1997) Blood flow regulation by S-nitrosohemoglobin in the physiological oxygen gradient. *Science*, **276**, 2034–2037.

Tagawa, T., Imaizumi, T., Endo, T., Shiramoto, M., Hirooka, Y., Ando, S. and Takeshita, A. (1993) Vasodilatory effect of arginine vasopressin is mediated by nitric oxide in human forearm vessels. *J. Clin. Invest.*, **92**, 1483–1490.

Tagawa, T., Mohri, M., Tagawa, H., Egashira, K., Shimokawa, H., Kuga, T., Hirooka, Y. and Takeshita, A. (1997) Role of nitric oxide in substance P-induced vasodilation differs between the coronary and forearm circulation in humans. *J. Cardiovasc. Pharmacol.*, **29**, 546–553.

Takahashi, M., Ikeda, U., Masuyama, J., Funayama, H., Kano, S. and Shimada, K. (1996) Nitric oxide attenuates adhesion molecule expression in human endothelial cells. *Cytokine.*, **8**, 817–821.

Thom, S., Hughes, A., Martin, G. and Sever, P.S. (1987) Endothelium-dependent relaxation in isolated human arteries and veins. *Clin. Sci.*, **73**, 547–552.

Toda, N. and Okamura, T. (1989) Endothelium-dependent and -independent responses to vasoactive substances of isolated human coronary arteries. *Am. J. Physiol.*, **257**, H988–95.

Topper, J.N., Cai, J., Falb, D. and Gimbrone, M.A., Jr. (1996) Identification of vascular endothelial genes differentially responsive to fluid mechanical stimuli: cyclooxygenase-2, manganese superoxide dismutase, and endothelial cell nitric oxide synthase are selectively up-regulated by steady laminar shear stress. *Proc. Natl. Acad. Sci. U. S. A.*, **93**, 10417–10422.

Torres, J., Davies, N., Darley Usmar, V.M. and Wilson, M.T. (1997) The inhibition of cytochrome c oxidase by nitric oxide using S-nitrosoglutathione. *J. Inorg. Biochem.*, **66**, 207–212.

Torres, J. and Wilson, M.T. (1996) Interaction of cytochrome-c oxidase with nitric oxide. *Methods Enzymol.*, **269**, 3–11.

Uematsu, M., Ohara, Y., Navas, J.P., Nishida, K., Murphy, T.J., Alexander, R.W., Nerem, R.M. and Harrison, D.G. (1995) Regulation of endothelial cell nitric oxide synthase mRNA expression by shear stress. *Am. J. Physiol.*, **269**, C1371–8.

Urakami Harasawa, L., Shimokawa, H., Nakashima, M., Egashira, K. and Takeshita, A. (1997) Importance of endothelium-derived hyperpolarizing factor in human arteries. *J. Clin. Invest.*, **100**, 2793–2799.

Vallance, P., Collier, J. and Moncada, S. (1989a) Nitric oxide synthesised from L-arginine mediates endothelium dependent dilatation in human veins in vivo. *Cardiovasc. Res.*, **23**, 1053–1057.

Vallance, P., Collier, J. and Moncada, S. (1989b) Effects of endothelium-derived nitric oxide on peripheral arteriolar tone in man. *Lancet*, **2**, 997–1000.

Vallance, P., Leone, A., Calver, A., Collier, J. and Moncada, S. (1992) Accumulation of an endogenous inhibitor of nitric oxide synthesis in chronic renal failure. *Lancet*, **339**, 572–575.

Vallance, P., Patton, S., Bhagat, K., MacAllister, R., Radomski, M., Moncada, S. and Malinski, T. (1995) Direct measurement of nitric oxide in human beings. *Lancet*, **346**, 153–154.

Vallance, P. (1996) Use of L-arginine and its analogs to study nitric oxide pathway in humans. *Methods Enzymol.*, **269**, 453–459.

Venema, R.C., Ju, H., Zou, R., Ryan, J.W. and Venema, V.J. (1997) Subunit interactions of endothelial nitric-oxide synthase. Comparisons to the neuronal and inducible nitric-oxide synthase isoforms. *J. Biol. Chem.*, **272**, 1276–1282.

Vila, J., Esplugues, J.V., Martinez Cuesta, M.A., Martinez Martinez, M.C., Aldasoro, M., Flor, B. and Lluch, S. (1991) NG-monomethyl-L-arginine and NG-nitro-L-arginine inhibit endothelium-dependent relaxations in human isolated omental arteries. *J. Pharm. Pharmacol.*, **43**, 869–870.

von der Leyen, H.E., Gibbons, G.H., Morishita, R., Lewis, N.P., Zhang, L., Nakajima, M., Kaneda, Y., Cooke, J.P. and Dzau, V.J. (1995) Gene therapy inhibiting neointimal vascular lesion: in vivo transfer of endothelial cell nitric oxide synthase gene. *Proc. Natl. Acad. Sci. U.S.A.*, **92**, 1137–1141.

Wallerstedt, S.M. and Bodelsson, M. (1997) Endothelium-dependent relaxation by substance P in human isolated omental arteries and veins: relative contribution of prostanoids, nitric oxide and hyperpolarization. *Br. J. Pharmacol.*, **120**, 25–30.

Wang, X.L., Sim, A.S., Badenhop, R.F., McCredie, R.M. and Wilcken, D.E. (1996) A smoking-dependent risk of coronary artery disease associated with a polymorphism of the endothelial nitric oxide synthase gene. *Nat. Med.*, **2**, 41–45.

Wang, X.L., Mahaney, M.C., Sim, A.S., Wang, J., Blangero, J., Almasy, L., Badenhop, R.B. and Wilcken, D.E. (1997) Genetic contribution of the endothelial constitutive nitric oxide synthase gene to plasma nitric oxide levels. *Arterioscler. Thromb. Vasc. Biol.*, **17**, 3147–3153.

Wang, Y. and Marsden, P.A. (1995a) Nitric oxide synthases: gene structure and regulation. *Adv. Pharmacol.*, **34**, 71–90.

Wang, Y. and Marsden, P.A. (1995b) Nitric oxide synthases: biochemical and molecular regulation. *Curr. Opin. Nephrol. Hypertens.*, **4**, 12–22.

Wei, C., Jiang, S., Lust, J.A., Daly, R.C. and McGregor, C.G. (1996) Genetic expression of endothelial nitric oxide synthase in human atrial myocardium. *Mayo Clin. Proc.*, **71**, 346–350.

Wheeler, M.A., Smith, S.D., Garcia Cardena, G., Nathan, C.F., Weiss, R.M. and Sessa, W.C. (1997) Bacterial infection induces nitric oxide synthase in human neutrophils. *J. Clin. Invest.*, **99**, 110–116.

White, R.P., Deane, C., Vallance, P. and Markus, H.S. (1998) Nitric oxide synthase inhibition in humans reduces cerebral blood flow but not the hyperemic response to hypercapnia. *Stroke*, **29**, 467–472.

Williams, D.J., Vallance, P.J., Neild, G.H., Spencer, J.A. and Imms, F.J. (1997) Nitric oxide-mediated vasodilation in human pregnancy. *Am. J. Physiol.*, **272**, H748–52.

Wilson, J.R. and Kapoor, S. (1993) Contribution of endothelium-derived relaxing factor to exercise-induced vasodilation in humans. *J. Appl. Physiol.*, **75**, 2740–2744.

Wolf, Y.G., Rasmussen, L.M., Sherman, Y., Bundens, W.P. and Hye, R.J. (1995) Nitroglycerin decreases medial smooth muscle cell proliferation after arterial balloon injury. *J. Vasc. Surg.*, **21**, 499–504.

Woolfson, R.G. and Poston, L. (1990) Effect of NG-monomethyl-L-arginine on endothelium-dependent relaxation of human subcutaneous resistance arteries. *Clin. Sci. Colch.*, **79**, 273–278.

Yang, Z.H., von Segesser, L., Bauer, E., Stulz, P., Turina, M. and Luscher, T.F. (1991) Different activation of the endothelial L-arginine and cyclooxygenase pathway in the human internal mammary artery and saphenous vein. *Circ. Res.*, **68**, 52–60.

Yasue, H., Matsuyama, K., Okumura, K., Morikami, Y. and Ogawa, H. (1990) Responses of angiographically normal human coronary arteries to intracoronary injection of acetylcholine by age and segment. Possible role of early coronary atherosclerosis. *Circulation*, **81**, 482–490.

Yasue, H., Yoshimura, M., Sugiyama, S., Sumida, H., Okumura, K., Ogawa, H., Kugiyama, K., Ogawa, Y. and Nakao, K. (1995) Association of a point mutation of the endothelial cell nitric oxide synthase (eNOS) gene with coronary spasm. *Circulation*, **92 (Suppl I)**, I–363(Abstract)

Yoshimura, M., Yasue, H., Nakayama, N., Shimasaki, Y., Kugiyama, K., Ogawa, H., Saito, Y., Miyamoto, Y., Ogawa, Y. and Nakao, K. (1997) Mutations in the endothelial nitric oxide synthase gene and susceptibility to coronary spasm in the Japanese. *Jap. J. Pharm.*, **75**, p.22(Abstract)

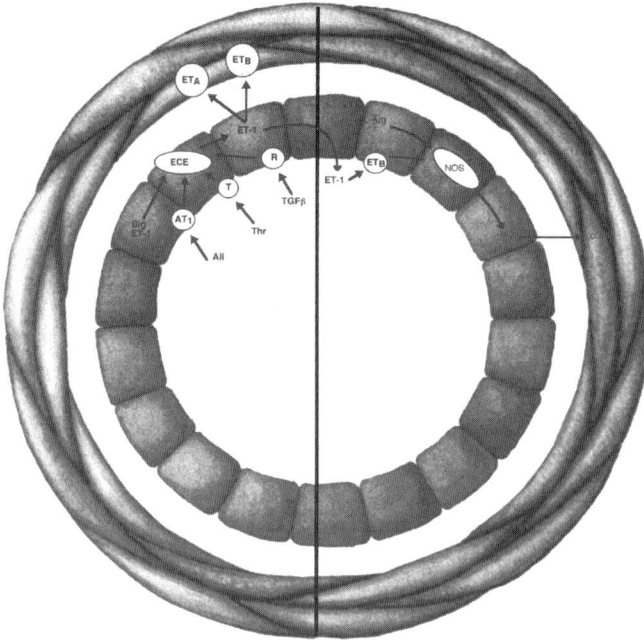

The endothelium plays a key role in the regulation of vascular tone, coagulation, lipid transport and immunological reactivity. Endothelin-1, a member of a family of 21-amino acid peptides, is produced by the endothelium in response to a number of stimuli, including exposure to epinephrine, angiotensin II and hypoxia. Three distinct endothelin isoforms have been identified, termed endothelin-1, endothelin-2 and endothelin-3, of which endothelin-1 is the most potent vasoconstrictor and of most importance functionally. Endothelin-1 produces slow onset and sustained vasoconstriction through its actions on vascular smooth muscle cells. There are two distinct subtypes of endothelin receptor, termed ET_A and ET_B, each the product of a separate gene. ET_A receptors are located on vascular smooth muscle cells and mediate vasoconstriction. The importance of endogenous endothelin-1, acting via the ET_A receptor, in the maintenance of basal vascular tone has been demonstrated through the use of selective endothelin receptor antagonists in both local and systemic studies. ET_B receptors are found on both endothelial and vascular smooth muscle cells where their stimulation mediates vasodilatation and vasoconstriction respectively. Local studies suggest the overall balance of ET_B activation favours vasodilatation. In addition to its potent pressor effects, endothelin-1 exerts mitogenic effects on the cardiovascular system and is intimately linked with the renin-angiotensin, L-arginine/nitric oxide and sympathetic nervous systems. Endothelin-1 has been implicated in the pathogenesis of atherosclerosis, in the increase in peripheral resistance and the structural cardiac and vascular changes seen in hypertension and chronic heart failure, and in a range of pathological conditions affecting the lung, kidney, gut and central nervous system. Following promising results in animal studies, and early clinical results in humans, large scale clinical trials are now underway in hypertension and heart failure to assess the therapeutic potential of endothelin receptor antagonists in humans.

Key words: Endothelin, endothelin converting enzyme, human, cardiovascular, physiology, blood pressure.

2   The Endothelin System: Physiology

Alan J. Bagnall[1] and David J. Webb[2]

[1]Centre for Genome Research, Roger Land Building, The University of Edinburgh, West Maino Road, Edingurgh, EH9 3JQ, UK
[2]Clinical Pharmacology Unit and Research Centre, The University of Edinburgh, Western General Hospital, Edinburgh EH4 2XU, Crewe Road, UK

INTRODUCTION

Endothelin-1 is the most potent vasoconstrictor and pressor agent currently identified and was originally isolated and characterized by Yanagisawa and colleagues (1988) from the culture media of aortic endothelial cells. Subsequently, two further isoforms, termed endothelin-2 and endothelin-3, were identified along with structural homologues isolated from the venom of *Actractaspis engaddensis*, known as the sarafotoxins. Each of the mature isoforms consists of 21 amino acids linked by two constraining intra-chain disulphide bonds. A highly conserved C-terminal sequence is mandatory for biological function of the peptide (Figure 2.1). Although the isoforms are structurally similar, endothelin-1 appears to be the predominant isoform involved in cardiovascular regulation and is the only isoform produced constitutively by endothelial cells (Inoue *et al.*, 1989). The rapid development of selective endothelin receptor antagonists has led to an explosion of research in this field. This work has demonstrated the therapeutic potential for pharmacological manipulation of the endothelin system in a range of cardiovascular conditions.

ENDOTHELIN GENERATION

The human genes for endothelin-1, endothelin-2 and endothelin-3 are located on chromosomes 6, 1 and 20 respectively. Regulation of endothelin-1 synthesis is determined primarily at the level of gene transcription via the influence of promoter regions located upstream (5') of the preproendothelin-1 gene. Of these, a GATA binding site mediates basal levels of gene transcription, whilst AP-1, nuclear factor and a hexonucleotide sequence are thought to be regulated by angiotensin II, transforming growth factor β and acute phase reactants respectively. Further post-transcriptional modulation may occur via selective destabilisation of preproendothelin-1 mRNA via 'suicide motifs' present in the non-translated 3' region of this molecule. These may determine the short (15 minute) half-life of preproendothelin-1 mRNA and thereby prevent excessive endothelin-1 production (Inoue *et al.*, 1989). Factors known to promote endothelin-1 production include thrombin, insulin, cyclosporine, epinephrine, angiotensin II, cortisol, inflammatory mediators, hypoxia and vascular shear stress. Endothelin production is inhibited by nitric oxide, nitric oxide donor drugs and dilator prostanoids via an increase in cellular cGMP, and natriuretic peptides via an increase in cAMP levels (reviewed by

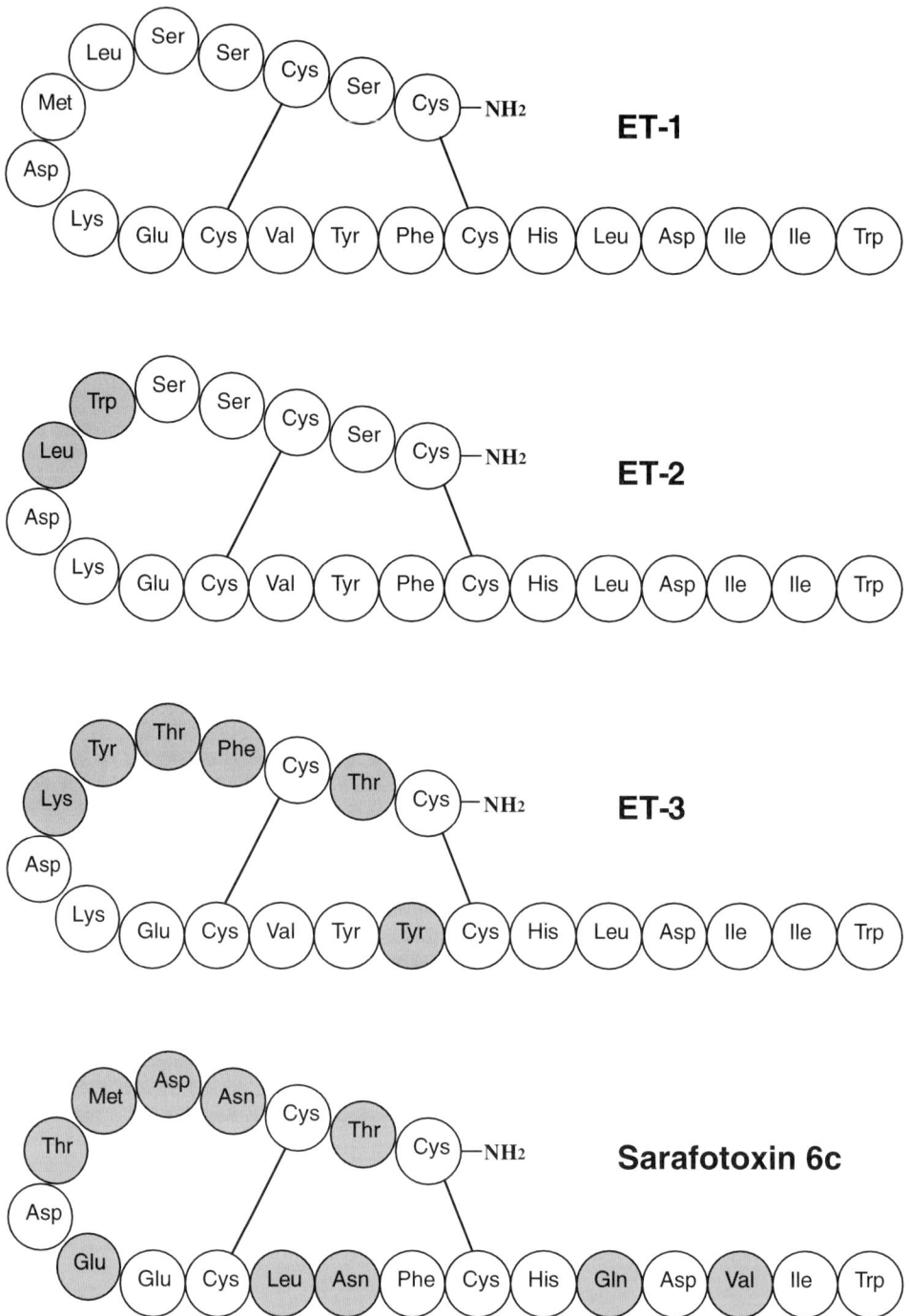

Figure 2.1 Structure of the endothelin family of peptides, and the related snake venom peptide sarafotoxin S6c. Shaded amino acids indicate differences from endothelin-1.

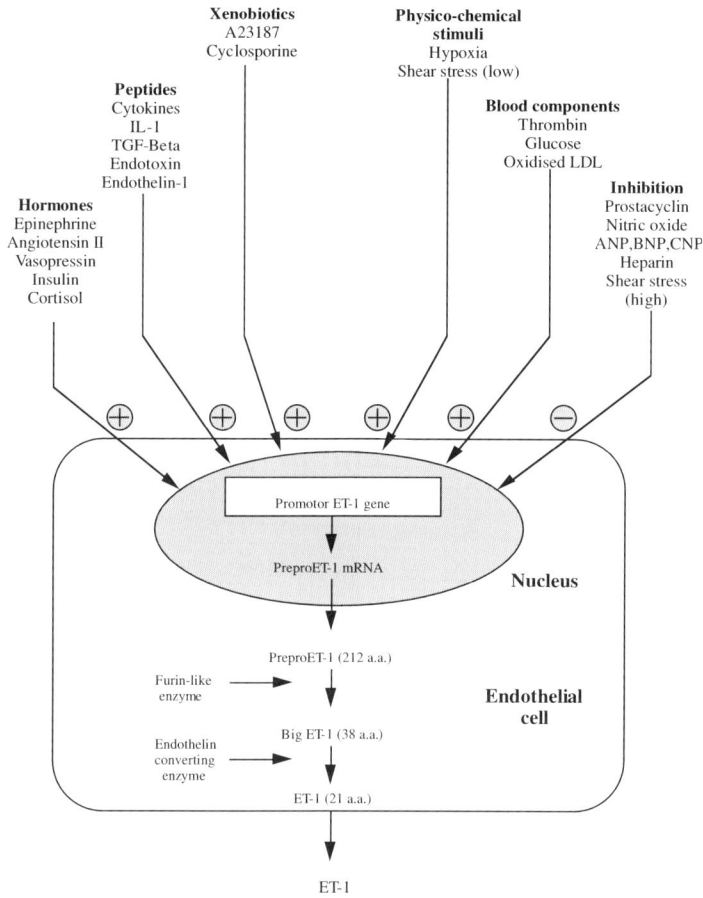

Figure 2.2 Factors that alter endothelin-1 (ET-1) synthesis and the pathway for endothelin-1 generation. IL-1 = interleukin-1; TGFβ = transforming growth factor β; LDL = low density lipoprotein; ANP, BNP, CNP = atrial, brain and c-type natriuretic peptides. See text for details.

Gray and Webb, 1995). The factors affecting endothelin generation and the synthetic pathways involved are illustrated in Figure 2.2.

The mature endothelin-1 peptide is generated by enzymatic cleavage of the initial preproendothelin-1 gene product. A short hydrophobic secretory sequence is first removed to produce proendothelin-1. This is then cleaved at dibasic amino acid pairs by the endopeptidase furin, generating the 39 amino acid peptide big endothelin-1 (Yanagisawa *et al.*, 1988). Subsequent production of mature endothelin-1 by a proteolytic cleavage between Trp^{21} and Val^{22} is catalyzed by the membrane bound metalloprotease endothelin-converting enzyme-1 (ECE-1). Although additional ECE isoforms have been identified in animals, a human ECE-2 and ECE-3 have yet to be identified (Emoto *et al.*, 1995). ECE gene knockout studies suggest that ECE-1 is the major functional ECE for all three endothelin isoforms *in vivo* (Yanagisawa *et al.*, 1996). Endothelin-1 was initially considered to be produced *de novo* in response to the factors described earlier. However, secretory vesicles containing both mature endothelin-1 and ECE have been identified in endothelial cells (Turner et al., 1996). Recently, a further endothelin peptide has been identified in humans formed by the cleavage of big endothelin-1 at the Tyr^{31} and Gly^{32} bonds by a human chymase enzyme expressed in mast cells. This product has been termed endothelin-1_{1-31} though its role in vivo has yet to be determined (Nakano et al., 1997).

ENDOTHELIN CONVERTING ENZYME

ECE-1 was first isolated and purified by Ohnaka and colleagues (1990) from aortic endothelial cells. It is inhibited by the combined ECE and neutral endopeptidase (NEP) inhibitor phosphoramidon, but not by selective NEP inhibitors such as thiorphan or kelatorphan. Structurally, ECE-1 exists as a transmembrane 758 amino acid dimer, linked by a single disulphide bridge. A short (1–56) N-terminal intracellular region is followed by a 21 amino acid transmembrane region. A zinc-binding and catalytic site (595–599) is essential for enzymatic activity. ECE-1 belongs to a family of neutral metalloprotease enzymes which includes NEP and the human Kell blood group protein (Xu et al., 1994). However, ECE is unique amongst this group in that it recognises a relatively long C-terminal portion of big endothelin-1 (residues His^{27} to Gly^{34}) in addition to the cleavage site between residues 21 and 22 (Takayanagi et al., 1998).

The ECE-1 gene is located on chromosome 1 at the p36 band (Valdenaire et al., 1995). cDNA cloning studies have demonstrated differential gene splicing leading to the production of two isoforms of ECE-1, termed ECE-1a and ECE-1b, which differ in structure only at the N-terminus. ECE-1a is responsible for generation of the majority of functional endothelin-1 from big endothelin-1. ECE-1a is expressed by endothelial cells and is located intracellularly, the enzymatically active C-terminal segment facing the intra-luminal region of the Golgi apparatus. A generator role for ECE-1a is further suggested by the presence of characteristic promoter regions for this gene, indicating that it is a constitutively expressed 'housekeeping' gene.

In contrast, ECE-1b spans the plasma membrane of effector cells, such as vascular smooth muscle cells, converting extracellular big endothelin-1 to endothelin-1. A 'responder/regulator' role for ECE-1b to extracellular big endothelin-1 is suggested by its promoter region containing potential receptor sites for transcription factors, allowing modulation of activity. Transfection of preproendothelin-1 and ECE-1b genes into cultured cells demonstrates that ECE-1b expressed at the cell surface is relatively inefficient at proteolysis of exogenous big endothelin-1, with only around 10% converted to endothelin-1. In contrast, between 50–90% of the endothelin peptides secreted were in the mature endothelin-1 form (Xu et al., 1994). This suggests that endogenously generated endothelin-1 secreted abluminally is the most functionally important source and confirms a predominantly autocrine/paracrine mechanism of action for endothelin-1. Such a theory is supported by the low (< 5pM) concentrations of endothelin-1 in the plasma, concentrations probably insufficient to activate endothelin receptors. Concentrations of angiotensin II and atrial natriuretic peptide in plasma are normally up to ten times greater than those of circulating endothelin-1. Also, endothelin-1 has a half-life of less than five minutes in plasma, with clearance occurring mainly in the lungs and kidneys (Dupuis et al., 1996). It is likely that much higher concentrations of endothelin-1 occur at the junctions between endothelial and vascular smooth muscle cells and that at least some of the plasma endothelin-

1 represents overspill from this site. One might conclude, therefore, that plasma levels of endothelin-1 in pathological states represent an unreliable index of vascular endothelin activity (Goddard and Webb, 1999). Similarly, urinary concentrations of endothelin-1 may reflect local renal endothelin activity better than they reflect systemic changes.

ENDOTHELIN RECEPTORS

The isoforms of endothelin exert their physiological effects in a receptor-mediated fashion. Pharmacological analysis of this process suggested the existence of at least two endothelin receptor types in humans, termed ET_A and ET_B receptors, and this has been confirmed by molecular characterisation studies (Arai *et al.*, 1990, Sakurai *et al.*, 1990, Sakamoto *et al.*, 1991). ET_A receptors are located on vascular smooth muscle cells (Arai *et al.*, 1990, Hori *et al.*, 1992) and, when activated, produce a sustained vasoconstriction that is of slow onset. In contrast, ET_B receptors are located on both endothelial (Hosada *et al.*, 1991, Molenaar *et al.*, 1993) and vascular smooth muscle cells (Davenport *et al.*, 1993). Activation of ET_B receptors on endothelial cells causes vasodilatation (Takayanagi *et al.*, 1991) through the release of dilator mediators acting on smooth muscle cells, whilst activation of ET_B receptors on smooth muscle cells produces vasoconstriction directly (Williams *et al.*, 1991, Moreland *et al.*, 1992, Sumner *et al.*, 1992). There is some pharmacological evidence to suggest subdivision within the ET_A and ET_B receptor types, derived from comparison of binding affinities, agonist/antagonist potencies and cell signalling following agonist binding. At a molecular level, however, there is currently no evidence to support this approach. The agonist binding characteristics of the endothelin receptors are shown in Table 2.1.

Endothelin Receptor Genes

The human EDNR*A* gene is located on chromosome 4 and consists of eight exons and seven introns spanning 40 kilobases (Hosada *et al.*, 1992). The human EDNR*B* gene on chromosome 13 is somewhat smaller in size, spanning 24 kilobases and consisting of seven

Table 2.1 Properties of the endothelin receptor subtypes.

	ET_A	ET_B	
Agonist Potency	ET-1>ET-2>>ET-3	ET-1=ET-2=ET-3	
Tissue	Vascular smooth muscle	Endothelium	Vascular smooth muscle
		All vessels	Resistance and Capacitance vessels
Action	Vasoconstriction	Vasodilatation	Vasoconstriction
Selective Agonists	None	ET-3 (600–fold specific) Sarafotoxin 6c (30000–fold specific)	
Antagonists	BQ-123	BQ-788	

exons and six introns (Arai *et al.*, 1993). Analysis of cDNA clones predicts ET_A and ET_B receptors consisting of 427 and 442 amino acids respectively and a sequence homology of ~58%. In a manner analogous to the preproendothelin-1 genes, both the ET_A and ET_B receptor genes have promoter regions which control gene transcription levels in response to factors such as nuclear factor-1, RNA polymerase II transcription factor and acute phase reactant regulatory elements. The receptor genes also encode regions for the post-translational modification of the receptors, altering tertiary structure, membrane anchorage sites and linkage to intracellular effector mechanisms (Elshourbagy *et al.*, 1993).

Distribution of Endothelin Receptor Genes

Autoradiography using labelled endothelin-1 and gene probes for the detection of endothelin receptor mRNA has allowed the qualitative and quantitative assessment of both the tissue distribution and the receptor subtype expressed in various tissues and cell lines. Such analysis has demonstrated the localisation of ET_A receptors predominantly to the vascular smooth muscle cells of large and medium-sized arteries, the highest densities being found in the aorta. In addition, renal arterioles, bronchial smooth muscle and glandular tissues, such as those of the pituitary and adrenal glands, have been shown to preferentially express ET_A receptors (Arai *et al.*, 1990). In contrast, the liver and endothelial cells are devoid of ET_A receptors (Hosada *et al.*, 1991). ET_B receptors are found on both smooth muscle and vascular endothelial cells, particularly in the brain, lungs, liver, kidney, bowel and adrenal glands. Human kidneys express ET_B receptors as the dominant receptor subtype with the collecting ducts exhibiting the highest density of ET_B receptors (Hori *et al.*, 1992). Human cardiac tissue expresses both ET_A and ET_B receptors throughout the A-V node, His-bundle, myocardium and endocardium (Molenaar *et al.*, 1993). Differential receptor subtype expression is thought to account for the wide variety of responses to the endothelins seen in different cell types and tissues.

The pattern of receptor expression in different tissues may reflect local differences in the transcription factors acting upon promoter regions to alter receptor mRNA expression. Alternatively, up- or down-regulation of receptor numbers could be mediated by local environmental factors such as endothelin-1 exposure or insulin concentrations, either causing receptor internalisation or directly altering gene transcription. It has been proposed that alterations in the spectrum or density of receptor expression by such mechanisms may be responsible for certain pathological states. For example, Wang and colleagues (1996) have used reverse transcription-polymerase chain reaction to quantitatively examine mRNA expression in the rat carotid artery balloon angioplasty model. They demonstrated a two-fold increase in preproendothelin-1 and ECE mRNA expression and a 30-fold increase in both ET_A and ET_B receptor mRNA expression post-angioplasty. Neointimal lesions also showed increased endothelin-1 immunoreactivity, suggesting a role for altered ET_A and ET_B receptor expression, in addition to altered endothelin-1 expression, in the pathogenesis of angioplasty-induced neointima formation. The precise role of changes in endothelin receptor expression in humans, however, will require similar mRNA expression studies to be performed on the relevant tissues.

Structure of Endothelin Receptors

The ET_A and ET_B receptors are part of the rhodopsin G-protein-coupled superfamily of receptors. These share a structure consisting of an extracellular N-terminal region, seven helical transmembrane loops connected by hydrophilic domains, and a 60 amino acid intracellular carboxy terminal region. The hydrophobic transmembrane domains and the interconnecting cytoplasmic loops are highly conserved, in contrast to the N-terminal region, which shows only 4% sequence homology between the ET_A and ET_B receptors (Elshourbagy *et al.*, 1993). Investigation into the structural determinants of receptor function has been achieved using the techniques of site directed mutagenesis of specific amino acid residues and the formation of receptor chimeras.

N-terminus

The amino acid residues which lie nearest the first transmembrane region seem to be essential for endothelin-1 binding. In particular, the Asp^{75} and Pro^{93} in this region of the ET_B receptor are thought to be responsible for the high stability of the complex formed between endothelin-1 and the ET_B receptor, compared to the ET_A receptor (Takasuka *et al.*, 1994).

Transmembrane domains

The transmembrane regions I, II, III, and VII are the major determinants of ligand binding. Ligand selectivity appears localised to transmembrane regions IV, V and VI and their intervening loops. The boundary region between the first extracellular loop and the second transmembrane domain regulates BQ-123 binding, whist the Lys^{140} in this area seems to be necessary for the binding of endothelin-1 to the ET_A receptor, possibly via the mediation of conformational changes (Adachi *et al.*, 1994).

Cytoplasmic loops

The cytoplasmic loops are known to mediate receptor/G-protein coupling in other rhodopsin-like G protein receptors. Substitution of the C-terminal end of the third cytoplasmic domain with the corresponding section of the β_2-adrenoceptor results in no changes in the affinity of the ET_A receptor for endothelin-1, but receptor/ligand binding fails to elicit the subsequent rise in intracellular calcium ($[Ca^{2+}]_i$) that normally follows receptor activation (Adachi *et al.*, 1993).

C-terminus

The C-terminal amino acids are thought to be responsible for anchorage of the endothelin receptor to the plasma membrane. They may also mediate elements of signal transduction, as evidenced by the loss of $[Ca^{2+}]_i$ increases on ligand binding following removal of C-terminal amino acids (Adachi *et al.*, 1993).

Signal Transduction

Binding of endothelin-1 to the ET_A receptor on vascular smooth muscle cells initiates a complex cascade of events resulting in a biphasic rise in $[Ca^{2+}]_i$, and, ultimately, cellular contraction and mitogenesis. Typically, the contractions induced by endothelin-1 develop

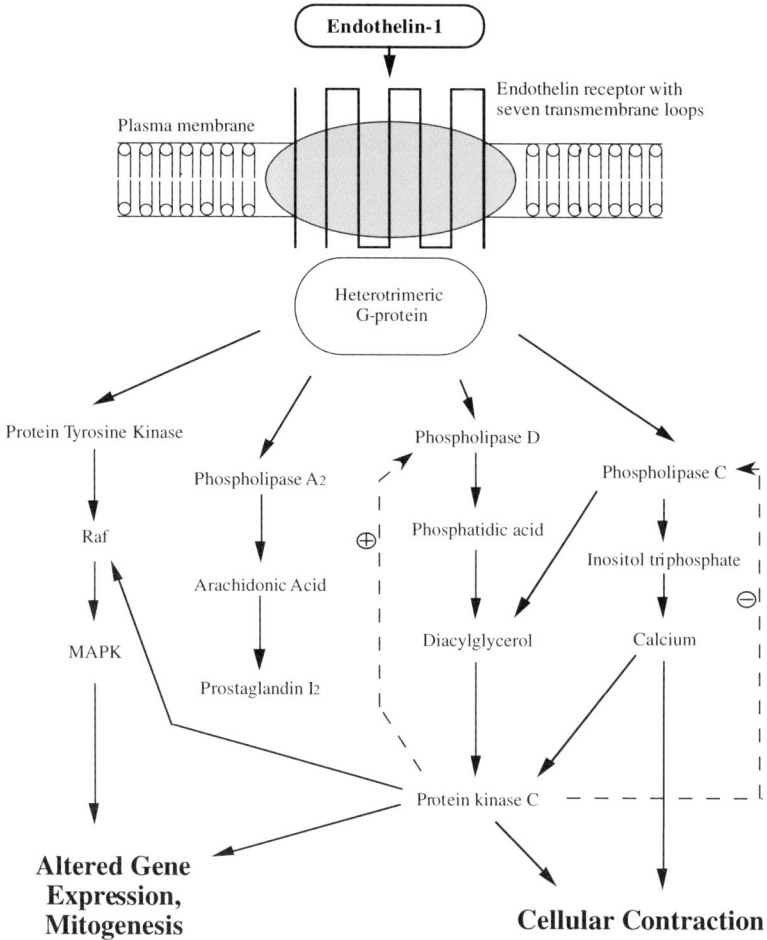

Figure 2.3 Signal transduction pathways of the endothelin receptors. MAPK = mitogen activated protein kinase. Adapted from Decker & Brock, 1998 with kind permission of the authors. See text for details.

slowly but are sustained and resistant to agonist removal. These effects are mediated by the interaction of the endothelin receptor with specific guanine nucleotide regulatory proteins (G-proteins) in the cell membrane (Bitar *et al.*, 1992). G-proteins are heterotrimeric structures involved in receptor signal transduction for a wide variety of cellular processes. Evidence from G-protein inhibitors such as *Bordatella pertussis* toxin and *Vibrio cholera* toxin suggest that the endothelin receptor is able to link with a variety of different G-protein subtypes. These include both inhibitory and stimulatory G-proteins which are able to activate an array of intracellular effector mechanisms, resulting in a wide variety of biological responses. These second messenger systems are summarized below and illustrated in Figure 2.3. The precise signalling mechanisms utilised by the ET_A and ET_B receptors may exhibit subtle differences, although a consensus opinion in this area is awaited (Douglas *et al.*, 1997).

Phospholipase C

G-protein stimulated activation of phospholipase C (PLC) causes hydrolysis of phosphatidylinositol 4,5-bisphosphonate (PIP_2) to produce inositol 1,4,5-triphosphate (IP_3) and sn1,2-diacylglycerol (DAG). The rise in IP_3 concentration stimulates release of intracellular Ca^{2+} stores from the sarcoplasmic reticulum via both ryanodine- and IP_3-sensitive Ca^{2+} channels and is responsible for the rapid initial rise in $[Ca^{2+}]_i$ seen following endothelin-1 binding (Marsden *et al.*, 1989). DAG production initiates activation of protein kinase C (PKC) allowing sensitisation of the cellular contractile elements to changes in $[Ca^{2+}]_i$ (see below). The precise G-protein connecting the endothelin-receptor to phospholipase C is currently unknown.

Calcium

As already indicated, endothelin receptor activation results in a rapid initial rise in $[Ca^{2+}]_i$ followed by a plateau phase dependent on the presence of extracellular Ca^{2+}. However, although usual, a rise in cellular IP_3 via PLC activation is not mandatory to cause a rise in $[Ca^{2+}]_i$. Endothelin-3 and SX6c acting on ET_B receptors appear to cause an IP_3-independent rise in $[Ca^{2+}]_i$ in certain cell types in which both the initial and plateau phases of Ca^{2+} mobilisation are dependent on movement of Ca^{2+} into the cell (Little *et al.*, 1992).

Extracellular Ca^{2+} entry appears to be necessary for the sustained increase in $[Ca^{2+}]_i$ that permits prolonged cellular contraction in response to endothelin-1. Both voltage operated and receptor operated calcium channels are involved in this process. L-type calcium channel blockers (dihydropyridine class) attenuate the sustained rise in $[Ca^{2+}]_i$ by antagonism of voltage operated calcium channels. Prolonged cellular contraction may also be prevented by nickel ions, which selectively block receptor operated channels. The T-type calcium channel is a putative candidate for the receptor operated channel (Stasch *et al.*, 1989). The opening of voltage dependent L-type calcium channels is dependent upon prolonged membrane depolarisation. The precise mechanism for achieving and maintaining such a period of membrane depolarisation remains controversial but is likely to involve calcium-activated chloride channels with resultant efflux of chloride ions (James *et al.*, 1994), and/or an influx of cations through non-selective cation channels. Other candidates for the mediators of the increased $[Na^+]_i$ that produces membrane depolarisation include ATP-sensitive K^+ channels, the Na^+/K^+-ATPase, or activation of the Na^+/H^+ antiporter (Meyer-Lehnart *et al.*, 1989, Danthaluri and Brock, 1990, Miyoshi *et al.*, 1992).

Protein kinase C

PKC exists in multiple isoforms and is one of the key regulatory enzymes involved in the endothelin-1 signalling process. Activation of the endothelin receptor stimulates a rise in DAG levels which, along with phosphatidylserine and Ca^{2+}, results in the activation and translocation of PKC from the cytosol to membranes within vascular smooth muscle cells. Activated PKC then catalyses the phosphorylation of heavy and light myosin chains along with caldesmons, which thereby increases the sensitivity of the contractile elements to $[Ca^{2+}]_i$ (Nishimura *et al.*, 1992). Specific PKC inhibitors block the sustained contractile responses to endothelin-1 and reduce Ca^{2+} sensitivity. PKC also modulates the hydrolysis of PIP_2 following activation of endothelin receptors by exerting an inhibitory effect on PLC and may form part of a negative feedback loop controlling PLD and arachidonic acid metabolism. The mitogenic effects of endothelin-1 appear to be related to both PKC- and tyrosine kinase-dependent mechanisms

Phospholipase D

The rise in intracellular DAG levels in response to endothelin-1 occurs in a biphasic manner, the initial rise resulting from PIP_2 hydrolysis by PLC. The prolonged secondary rise in DAG levels is thought to originate from phosphatidylcholine hydrolysis mediated by PLD and lasts for up to 20 minutes. Activation of PLD occurs via PKC-dependent and PKC-independent mechanisms (Liu *et al.*, 1992). The phosphatidic acid (PA) produced enhances Ca^{2+} influx and further enhances PLC activity (Shukla *et al.*, 1991). It is proposed that PLD activation may contribute to the mitogenic effects of endothelin-1 (Boarder, 1994).

Phospholipase A_2

Phospholipase A2 (PLA_2) catalyses the generation of arachidonic acid metabolites including the leukotrienes, thromboxane A_2 (TxA_2) and prostacyclin (PGI_2) from membrane lipids. Inhibitors of the arachidonic acid metabolism cascade have been shown to inhibit many of the actions of endothelin-1, and PGI_2 production by endothelial cells can be stimulated by endothelin-1. Linkage of endothelin-receptors to PLA_2 may be direct via a G-protein, or occur in response to endothelin-1-activated changes in $[Ca^{2+}]_i$ (Barnett *et al.*, 1994).

Tyrosine kinases

Endothelin-1-induced tyrosine phosphorylation of cellular proteins modulates activation of PLD and in part regulates its mitogenic effects. Stimulation of ET_A receptors results in phosphorylation of tyrosine residues on cytosolic proteins ranging in size from 45–225 kDa by members of the tyrosine kinase family known as Src, focal adhesion kinase (FAK) and Janus kinase (Jak). Although the precise mechanisms are as yet unknown, it seems likely that activation of transcription factors by these kinases in response to endothelin-1 stimulation is involved in the mitogenic response.

Nuclear/mitogenic signalling mechanisms

Alone, endothelin-1 is a weak mitogen, as measured by thymidine incorporation or expression of the proto-oncogenes c-*jun* and c-*fos*. However, when combined with other growth factors such as platelet derived growth factor (PDGF) and epidermal growth factor (EGF) or co-mitogens such as insulin, it acts synergistically to produce marked mitogenic effects. As indicated earlier, a precise mechanistic model is still awaited, but both phorbol ester-sensitive PKCs and tyrosine kinase signal transduction mechanisms are involved. PKCs, activated by the binding of endothelin to its receptor, catalyse the c-Raf-1 cascade leading to activation of mitogen activated protein kinases (MAPKs).

MAPKs are known to be critically important in the transduction of mitogenic signals in a number of other growth factor systems. Where the mitogenic response pathways of endothelin-1 differ from those of other growth factors, however, is in the proposed coupling of the activated ET_A receptor to the c-Raf-1 cascade via PKC. Endothelin-1 activation of the tyrosine kinases Src, FAK and JAK occurs in a non-PKC dependent manner. Their activation may result in the switching on of transcription factors of the signal transducers and activators of transcription family (STAT) and subsequent trans-location of STATs to the nucleus where they interact with promoter regions to initiate gene transcription (reviewed by Decker and Brock, 1998). The study of the mitogenic responses to the endothelins is currently in its infancy when compared to that of other aspects of endothelin physiology. However, the similarities in the effector mechanisms utilised in response to endothelin-1 and other growth factors provide interesting parallels. ET_A and ET_B receptor activation is also known to increase the expression of cellular adhesion molecules, induce chemotactic factors such as $TNF\alpha$, $IL-1\beta$, IL-6, and IL-8, and alter matrix synthesis (reviewed by Douglas and Ohlstein, 1997). Thus, in addition to short-term actions upon cellular contractility, the endothelin isoforms also influence cellular growth and differentiation, a process critically linked with the proposed role of endothelin-1 as a mediator of vascular hypertrophy and cardiac remodelling seen in conditions such as hypertension and chronic cardiac failure.

Nitric oxide and cGMP

The ET_B receptor present on endothelial cells is responsible for vasodilatation in response to activation by endothelin-1, endothelin-3 and the selective ET_B receptor agonist SX6c. This effect is produced by ET_B receptor-stimulated release of the vasodilating factors nitric oxide (NO) and PGI_2. Activation of ET_B receptor-coupled G-proteins leads to an increase in PLC activity with generation of IP_3 and DAG. This releases Ca^{2+} from intracellular stores (see above) with a resultant increase in the activity of a constitutive Ca^{2+}/calmodulin dependent NO synthase enzyme. Liberated NO diffuses to the smooth muscle cell layer and initiates production of cGMP by soluble guanylate cyclase. Increasing levels of cGMP activate mechanisms that ultimately decrease $[Ca^{2+}]_i$ leading to cellular relaxation (Lincoln, 1989). Inhibition of NO synthase by L-N^G-monomethyl arginine (L-NMMA) has been shown to inhibit the vasodilation in response to ET_B receptor stimulation. Further mechanisms of NO production have been proposed. These include a PKC-dependent pathway and direct production of cGMP via activation of a pertussis toxin-insensitive G-protein by endothelin-1 on the kidney epithelial cell line LLC-PK1 (Ozaki *et al.*, 1994). NO also exerts an inhibitory effect on endothelin-1 stimulated receptor signalling. In Chinese hamster ovary (CHO) cells transfected with ET_A receptors it has been shown that NO can displace endothelin-1 from ET_A receptors and disrupt calcium mobilisation (Aramori *et al.*, 1992).

Adenylate cyclase

Endothelin-1 may directly inhibit the accumulation of cAMP induced by forskolin, cholera toxin and isoproterenol in endothelial cells (Ladoux *et al.*, 1991). It is at present unclear whether the effects of endothelin-1 on cyclic nucleotide levels are mediated directly via adenylate cyclase or indirectly as a result of PLC activation.

Formation and Internalisation of Receptor Complexes

Early studies of endothelin-1-induced responses demonstrate a slow onset of action followed by a period of sustained vasoconstriction that resists agonist removal by prolonged washout. This characteristic pattern of vasoconstriction to endothelin-1 is a product of both the ligand/receptor binding relationship and the signal transduction mechanisms thereby induced. Endothelin-1 is a potent vasoconstrictor agent. Saturation of endothelin receptors occurs within one minute of exposure to endothelin-1 at 37°C (Marsault *et al.*, 1991). Once bound, the binding of endothelin to its receptor appears to be a 'pseudo-irreversible' process, with a dissociation half-life in excess of 100 hours (reviewed by Douglas and Ohlstein, 1997). Rapid receptor complex internalisation by a clathrin-mediated endocytic pathway follows, and, once internalised, degradation of the endothelin/endothelin-receptor complex is thought to occur within lysosomes where the local acidic environment favours dissociation. Re-cycling of receptors back to the cell surface then occurs, allowing the cell to regulate the degree of endothelin-1 stimulation. Degradation of receptor complexes by neutral endopeptidases located within the plasma membrane may also occur prior to internalisation (Loffler *et al.*, 1991). Prior to internalisation, however, rates of dissociation of receptor and ligand are dependent upon the duration of exposure, with studies using radio-labelled endothelin-1 showing a progressive decline in the proportion of dissociable endothelin-1 over 60 minutes (Wilkes and Boarder, 1991). Dissociable endothelin-1 is thought to represent the presence of surface bound/non-internalised receptor complexes. There does appear to be some heterogeneity amongst the different receptors and endothelin isoforms in the rate and extent of receptor internalisation. For example, in anterior pituitary cells, the rate of receptor endocytosis is higher for endothelin-1 bound to ET_A receptors than for endothelin-3 bound to the same receptor (Stojilkovic *et al.*, 1992). Overall, the delayed action of endothelin-1 appears to result from the time required to activate the receptor-coupled second messenger systems outlined earlier, as opposed to delays in ligand/receptor binding, whilst the prolonged duration of action may be due to continued receptor signalling following internalisation.

Receptor Downregulation

Repeated exposure of smooth muscle cells to endothelin results in the downregulation of receptors without a change in their affinity. This is manifest as a progressive decrease in tissue or cellular responsiveness (Hirata *et al.*, 1988). This phenomenon occurs in response to both exogenous and endogenously produced endothelins and receptor downregulation is agonist selective. Pre-treatment of cells with the ET_B receptor-selective agonist SX6c results in the loss of further ET_B receptor-agonist-mediated responses, but the response to ET_A receptor agonists is preserved (Henry, 1993). Interestingly, stimulation with endothelin-3 of cells having no ET_B receptor binding sites results in a loss of further responsiveness to endothelin-3 but maintenance of endothelin-1-stimulated responses (Hiley *et al.*, 1992). This suggests that endothelin-3 may selectively desensitise the ET_A receptor to endothelin-3 but not endothelin-1. Alternatively, it has been suggested that endothelin-3 may mediate its effects via a non-ET_A and non-ET_B receptor (Hiley *et al.*, 1992), although no molecular evidence for such a receptor has yet emerged.

Endothelin Receptor Subtypes

Two further subtypes of endothelin receptor have been isolated in *Xenopus laevis* but not from mammalian species. The first of these receptor subtypes, termed ET_{AX}, displays agonist binding characteristics identical to the ET_A receptor but is insensitive to the selective ET_A receptor antagonist BQ-123 (Kumar *et al.*, 1994). The second, isolated only from *Xenopus laevis* melanocytes, has a greater binding affinity for endothelin-3 than endothelin-1 and may represent a third morphological receptor subtype called ET_C (Karne *et al.*, 1993). The *Xenopus* ET_C receptor shares a 50% amino acid sequence homology with the human ET_A and ET_B receptors and complementary DNAs for each of these receptor subtypes have now been isolated and characterised. Analysis of human genomic DNA with cDNA probes of low stringency has, as yet, failed to identify any further homologous sequences. Any putative endothelin-3-selective ET_C receptor would, therefore, be likely to have a structure that differs widely from the other endothelin receptors (Sakamoto *et al.*, 1991).

Sub-division of the ET_B receptor type has also been proposed. This stems from the observation that activation of ET_B receptors may produce either vasodilatation or vasoconstriction. Vasodilatation in response to ET_B receptor stimulation is mediated by ET_B receptors located on vascular endothelial cells through the production of nitric oxide or dilator prostanoids (Takayanagi *et al.*, 1991). In contrast, vasoconstrictor responses are mediated directly by ET_B receptors located on vascular smooth muscle cells (Williams *et al.*, 1991, Moreland *et al.*, 1992, Sumner *et al.*, 1992). However, despite reports that production of two receptor subtypes from a single gene is possible by varying transcription initiation sites or post-translational modification (Shyamala *et al.*, 1994), it is more likely that these differing responses are due to differences in signal transduction pathways in the effector cells.

Pharmacological differences in the responses of ET_A receptors in various tissues to antagonists such as BQ-123 (a selective ET_A receptor antagonist) have also been demonstrated (Sudjarwo *et al.*, 1994). This has led to suggestions that subtypes of the ET_A receptor may also exist though, again, validation by molecular studies has not yet been forthcoming.

DEVELOPMENTAL BIOLOGY

Targetted gene knockout studies have provided further clues to the role of the endothelin isoforms. Studies originally conceived to examine the physiological effects of deletion of the genes for endothelin-1, ECE and the ET_A and ET_B receptors in the adult animal have produced unexpected results. Mouse embryonic stem cells manipulated to carry a mutant endothelin-1$^-$allele underwent homologous recombination to produce endothelin-1$^{-/-}$ mice. The offspring were characterised by lethal abnormalities of the craniofacial and pharyngeal pouch structures. These structures are derived from the neural crest ectomesenchymal cells indicating that endothelin-1 is crucially involved in normal ontogeny of the pharyngeal arches (Kurihara *et al.*, 1994). Targetted disruption of the ECE-1 gene produces identical phenotypic abnormalities of the craniofacial structures, as does deletion of the ET_A receptor gene. Deletion of the ET_B receptor or endothelin-3 gene produces mice characterised by aganglionic megacolon and coat colour spotting (Hosada *et al.*, 1994). The ET_B receptor/endothelin-3 interaction therefore appears to be critical for the normal development of epidermal melanocytes and enteric neurones. Additionally, ECE-1 knockout mice include the phenotype of endothelin-3 knockout mice, suggesting that ECE-1 is functionally responsible for the conversion of big endothelin-3 to endothelin-3 (Baynash *et al.*, 1994). Two human conditions share phenotypical similarities to endothelin-knockout mice models. The first, the Pierre-Robin and Treacher-Collins syndromes share the same craniofacial abnormalities as endothelin-1 or ET_A receptor-knockout mice, indicating that the human condition might occur secondary to ET_A receptor/endothelin-1 anomalies (Ong, 1996). Secondly, abnormalities of preproendothelin-3 or ET_B receptor genes have been documented to occur in the neurocristopathies associated clinically with Hirschsprung's disease (Puffenberger *et al.*, 1994) and the Waardenburg-Shah syndrome (Attie *et al.*, 1995) — both of which are varieties of aganglionic megacolon. This has obvious implications for the clinical use of ET_A receptor antagonists and ECE-inhibitors, rendering them unsuitable for use during pregnancy or at the time of conception, these abnormalities having been detected in teratogenicity studies with endothelin antagonists in animals.

It is interesting to note that endothelin-1-knockout mice are hypertensive — at odds with the predicted role of endothelin-1 as a mediator of vasoconstriction and contributing to basal vascular tone. The explanation for this is thought to lie in the sympathetic-adrenergic overactivity that is induced by hypoxia secondary to the severe craniofacial abnormalities, or due to disruption of the central control of cardiorespiratory function, in which endothelin-1 is thought to play a role (Kurihara *et al.*, 1994).

EFFECTS OF ENDOTHELIN ADMINISTRATION

The endothelins act in a predominantly autocrine and paracrine fashion. Studies involving the administration of the endothelin isoforms, either by bolus or constant infusion are, therefore, unlikely to reproduce the *in vivo* physiological effects of endothelin action. Indeed, such studies may be frankly misleading, despite the presence of increased circulating concentrations of endothelin-1 having been documented in a number of pathological conditions. For example, increased plasma concentrations of endothelin-1 in any given disease state may result in downregulation of endothelin receptors with a resultant *reduction* in the magnitude of responses to exogenous endothelin-1, in contrast to the predicted outcome of higher endothelin-1 concentrations producing an amplification of the normal physiological response. Here, studies with antagonists are more revealing. However, agonist studies are helpful in defining target organs and the receptor subtype involved in the response, and these data will, therefore, be reviewed below.

Effects of Endothelin on Blood Pressure and the Heart

Administration of endothelin-1 in humans by constant infusion produces a dose-dependent increase in blood pressure that is of prolonged duration and accompanied by sodium retention (Vierhapper *et al.*, 1990). The sustained pressor effect is also seen with bolus endothelin-1 administration, despite rapid clearance of the peptide from the blood. Pretreatment with either cyclosporine, nifedipine, or the cyclo-oxygenase inhibitor indomethacin has no effect on the blood pressure response to endothelin-1 (Vierhapper *et al.*, 1992). In cardiac studies, infusion of endothelin-1 at 8 pmol/Kg/min to healthy human subjects produced a reduction in coronary blood flow of ~30%, with coronary vascular resistance increasing by ~100% (Pernow *et al.*, 1996). Mean arterial pressure increased during infusion of endothelin-1 in this study by ~10 mmHg. Big endothelin-1 was shown to be as potent as endothelin-1 in producing these effects, probably reflecting local cardiac, or systemic, conversion of big endothelin-1 to endothelin-1 rather than a direct action of big endothelin-1 itself. Endothelin-1 exerts a negatively inotropic effect *in vivo*, impairs diastolic filling of both the right and left ventricles, decreases cardiac output and reduces heart rate (Kiely *et al.*, 1997), probably through a baroreceptor-mediated reflex. The impairment of diastolic function was also found at doses insufficient to alter systemic or pulmonary blood pressure. Sarafotoxin-containing bites from *Actractaspis* have been shown to cause myocardial infarction (Tony and Bhat, 1995), and coronary vasospasm has been demonstrated *in vitro* in response to SX6c (Bax *et al.*, 1994).

Effects of Endothelin on the Renal Circulation

The human renal circulation is extremely sensitive to the vasoconstrictor effects of endothelin-1 and has a high capacity for the conversion of circulating big endothelin-1 to endothelin-1 (Ahlborg *et al.*, 1994). Endothelin-1 administered to healthy human subjects at 4 pmol/Kg/min produces an ~25% decrease in renal plasma flow (RPF) and an ~45% increase in renal vascular resistance, the effects lasting up to three hours (Weitzberg *et al.*, 1991). Prolonged low dose infusions of endothelin-1 (0.4 pmol/Kg/min for six hours) generate ~45% decrease in RPF and an ~80% increase in filtration fraction, indicating a more pronounced vasoconstriction of the efferent arteriole *in vivo* (Jilma *et al.*, 1997). Renal vasoconstriction and sodium retention may be reversed by nifedipine, although the increase in filtration fraction remains unaltered (Kaasjager *et al.*, 1995). Nifedipine has a mainly preglomerular site of action and does not, therefore, reverse the significant vasoconstrictor effect of endothelin-1 on the efferent arteriole. *In vitro* studies have also demonstrated vasoconstriction of the arcuate and inter-lobar arteries in response to endothelin-1 (Edwards *et al.*, 1990). Although there are marked interspecies variations in the receptor subtype mediating these effects, the localisation of the ET_A receptor subtype to the vasculature in the human kidney makes this the most favoured candidate (Karet *et al.*, 1993). Indeed, in clinical studies, endothelin-3 infusion (selective for the ET_B receptor) failed to alter renal haemodynamics and electrolyte excretion (Kaasjager *et al.*, 1997).

In vivo, endothelin-1 administration consistently produces sodium retention in humans, even at doses insufficient to affect RPF or glomerular filtration rate (GFR) (Rabelink *et al.*, 1994). In the study by Kiely and colleagues cited earlier, low, medium and high dose endothelin-1 infusions produced a decrease in plasma renin activity, but no significant effect on aldosterone levels.

Effects of Endothelin on the Pulmonary Circulation

Wagner and colleagues (1992) have examined the effects of relatively low dose endothelin-1 infusion (4 pmol/Kg/min) on the pulmonary circulation. No changes in pulmonary haemodynamics were observed despite endothelin-1-induced increases in systolic blood pressure, total and splanchnic vascular resistance and decreases in hepatic blood flow. An absence of effect on the pulmonary vascular tree following intra-pulmonary endothelin-1 infusion was also found in patients with borderline pulmonary hypertension and chronic hypoxaemia, although the difference between arterial and venous oxygen concentrations was markedly increased during endothelin-1 infusion (reviewed by Holm, 1997). In contrast, increases in total pulmonary vascular resistance and mean pulmonary artery pressure have been noted with endothelin-1 infusion in other studies, measured using Doppler echocardiographic techniques (Kiely et al., 1997). The pulmonary circulation does appear responsible for the short half-life of endothelin-1, removing 50% of infused endothelin-1. Concentrations of circulating endothelin-1 are known to be increased in pulmonary hypertension and correlate with the severity of the disease. This has been shown to be at least partially due to reduced pulmonary clearance of endothelin-1 (Dupuis et al., 1998) but may also reflect increased local production (Cacoub et al., 1997).

Effects of Endothelin on Forearm Blood Flow

Many studies have examined the effects of giving endothelin isopeptides to animals. However, because of the marked and sustained vasoconstriction produced by endothelin-1, particularly in the renal and coronary vascular bed, there have been fewer studies in humans. The potential risks associated with administering systemic doses of endothelins to humans have prompted a number of investigators to utilise the technique of forearm plethysmography coupled with brachial artery administration of locally active doses of drug. This technique allows precise assessment of drug effects on vascular smooth muscle *in vivo*, without the confounding influences of drug effects on other organs or activation of neurohumoral reflexes. Changes in blood flow may, therefore, be attributed solely to the drug infused. Blood flow in the opposite arm can also be measured to provide a contemporaneous control during such studies and increase their power (Webb, 1995).

Endothelin-1 infusion into the brachial artery in healthy humans produces a slow onset of vasoconstriction that is dose-dependent. The effect lasts for up to two hours after discontinuation of the infusion (Clarke et al., 1989). The infusion of ET_B receptor agonists such as endothelin-3 or SX6c also produces a reduction in forearm blood flow, albeit to a lesser extent, suggesting that ET_B receptors are able to mediate at least part of the vasoconstrictor effect in human resistance vessels. A similar response is seen in forearm capacitance vessels (Haynes et al., 1995). However, the relevance of this vasoconstrictor response to the physiological role of ET_B receptor-mediated vascular responses is uncertain. The forearm vasoconstriction to endothelin-1 may be overcome by co-infusion of Ca^{2+} antagonists but is not modulated by NO (Kiowski et al., 1991).

Bolus administration of either endothelin-1, endothelin-3 or SX6c produces transient vasodilatation before causing sustained vasoconstriction. Vasodilatation is more pronounced and sustained with endothelin-3 and SX6c, indicating that it is likely to be mediated via the endothelial ET_B receptor (Inoue et al., 1989a, Ohlstein et al., 1990). A similar response is found in humans (Haynes et al., 1995b). However, this is seen only following high doses on bolus administration and is, therefore, likely to be a pharmacological rather than a physiological phenomenon.

Effects of Endothelins on Hand Veins

Human dorsal hand veins are an attractive experimental model because they possess no intrinsic tone and their lack of response to big endothelin-1 (Haynes et al., 1995a) suggests that they do not constitutively express ECE. Stimulation with endothelin-1 produces venoconstriction that is modulated by endogenous endothelial PGI_2 activity but not nitric oxide synthesis (Haynes and Webb, 1993a). Sarafotoxin S6c is also able to produce dorsal hand vein constriction, indicating that ET_B receptors may contribute to venoconstriction. The venoconstrictor response is dependent on the endothelin-1-stimulated closure of K^+_{ATP} channels causing membrane depolarisation, an effect reversed by the K^+_{ATP} channel opener cromakalim (Haynes and Webb, 1993b). Dihydropyridine Ca^{2+} channels are also responsible for signal transduction in hand veins but to a lesser extent.

Effects of Endothelin on the Skin Microcirculation

Injection of endothelin-1, but not endothelin-3, into the skin produces vasoconstriction of the microcirculation (Wenzel *et al.*, 1994). Co-injection of ET_A receptor antagonists reverses the effects of endothelin-1 and elicit vasodilatation when given alone, indicating that endothelin-1 may contribute to basal tone in the microcirculation of the skin. A flare reaction is produced surrounding the central area of vasoconstriction at the site of injection of endothelin-1. This flare results from the endothelin-1-stimulated release of nitric oxide, possibly from polimodal nociceptor fibres located in the dermis, causing local vasodilatation around the injection site (Wenzel *et al.*, 1998). The authors suggest that endothelin-1 may, therefore, be involved in the pathogenesis of neurogenic inflamation.

Effects of Endothelin on the Adrenals

Systemic infusion of endothelin-1 has no effect on basal aldosterone concentrations in healthy men but does cause a rise in serum K^+ concentration. However, there is selective augmentation of aldosterone secretion in response to exogenous ACTH if endothelin-1 is administered concomitantly (Vierhapper *et al.*, 1995). In contrast, there is no augmentation of the ACTH-induced secretion of cortisol, corticosterone or 18-OH-corticosterone, suggesting that endothelin-1 exerts its effects in the later stages of steroid production.

Effects of Endothelin on the Pituitary

The pituitary gland contains an abundance of ET_A receptors (Kanyicska and Freeman, 1993) and the release of endothelin-3 from rat pituitary cells has been demonstrated (Matsumoto *et al.*, 1989). In humans, endothelin-1 infusion augments the corticotrophin releasing hormone-induced rise in plasma ACTH levels and possibly also the responses to luteinising hormone and follicle stimulating hormone induced by their respective releasing hormones. In contrast, endothelin-1 inhibits the rise in prolactin, growth hormone and thyroid stimulating hormone seen following stimulation by their releasing hormones. Basal hormone levels are unaffected by endothelin-1 administration (Vierhapper *et al.*, 1993). The plasma concentrations of endothelin-1 achieved during these experiments were greater than physiological but similar to those documented in pathophysiological states. The effects of endothelin-1 upon the posterior pituitary are complex. In the rat, the paraventricular and supraoptic nuclei of the hypothalamus and the axons of those neurons ending in the posterior pituitary exhibit endothelin-containing secretory vesicles (Yoshizawa *et al.*, 1990). These vesicles become depleted during water deprivation. In man, upright tilt results in a parallel rise in the concentrations of endothelin-1 and anti diuretic hormone (ADH). This response is absent in patients with diabetes insipidus (Kaufmann *et al.*, 1991). Further support for the theory that endothelin-1 regulates the release of ADH comes from the observation that infusion of endothelin-1 in dogs results in a rise in plasma ADH concentrations (Nakamoto *et al.*, 1989). Activation of the hypothalamic-pituitary axis by endothelin-1 may, therefore, be clinically relevant.

CARDIOVASCULAR PHYSIOLOGY OF THE ENDOTHELINS

As already indicated, agonist studies are potentially poor predictors of the physiological role of the endothelin peptides and the physiology of autocrine and paracrine systems is best examined using antagonists. Studies with endothelin receptor antagonists and ECE inhibitors have provided important insights into the role of the endothelin system in the maintenance of basal vascular tone and also indicated important interactions with the sympathetic nervous system and the L-arginine/nitric oxide system.

Local Cardiovascular Effects

One of the first available antagonists of the endothelin system was phosphoramidon, an inhibitor of ECE-1 and neutral endopeptidase (NEP), and this was utilised to study the *in vivo* role of ECE. Brachial artery administration of big endothelin-1 in healthy subjects caused a dose-dependent forearm vasoconstriction that could be blocked completely by phosphoramidon (30 nmol/min), suggesting that the effects of the precursor are mediated through conversion to the mature peptide by ECE (Haynes and Webb, 1994). The blockade of constriction to big endothelin-1 by phosphoramidon is unlikely to have been due to inhibition of endothelin receptor binding as vasoconstriction to endothelin-1 was un-affected by this dose of phosphoramidon and because conversion of infused big endothelin-1 to endothelin-1 and its C-terminal fragment was confirmed in plasma samples taken from the veins draining the infused forearm (Plumpton *et al.*, 1995). Big endothelin-1 conversion in the forearm presumably occurs via vascular, probably endothelial, ECE situated within the forearm resistance vessels since circulating blood exhibits little ECE activity (Watanabe *et al.*, 1991). The difference in potency between big endothelin-1 and endothelin-1, and the ratio of concentrations of C-terminal fragment to big endothelin-1 in venous blood, both indicate that local ECE converts ~10% of luminally presented big endothelin-1 to endothelin-1, consistent with ~10% conversion of exogenous big endothelin-1 by cells expressing the ECE-1 gene (Xu *et al.*, 1994). Big endothelin-1 does not cause venoconstriction in hand veins (Haynes *et al.*, 1995a), even though these vessels respond to endothelin-1 (Clarke *et al.*, 1989), suggesting that ECE activity may not be present in all vessel types.

Administration of phosphoramidon (30 nmol/min) alone results in a slowly progressive vasodilatation, consistent with a role for endothelin-1 in maintenance of basal vascular tone (Haynes and Webb, 1994) (Figure 2.4). Although phosphoramidon also inhibits NEP, this latter action is unlikely to explain the vasodilatation because potent and selective inhibitors of NEP, such as thiorphan and candoxatril, cause slowly progressive forearm *vasoconstriction* (Ferro *et al.*, 1998). This effect of NEP inhibitors is likely to be caused by accumulation of endothelin-1, because it is a substrate for metabolism by NEP and the vasoconstriction is blocked by local infusion of the ET_A receptor antagonist BQ-123, but not by systemic ACE inhibition. This observation may also account for the increase in plasma endothelin by NEP inhibitors in clinical trials (Ando *et al.*, 1995) and their failure to lower blood pressure in hypertensive subjects (reviewed by Ferro *et al.*, 1998).

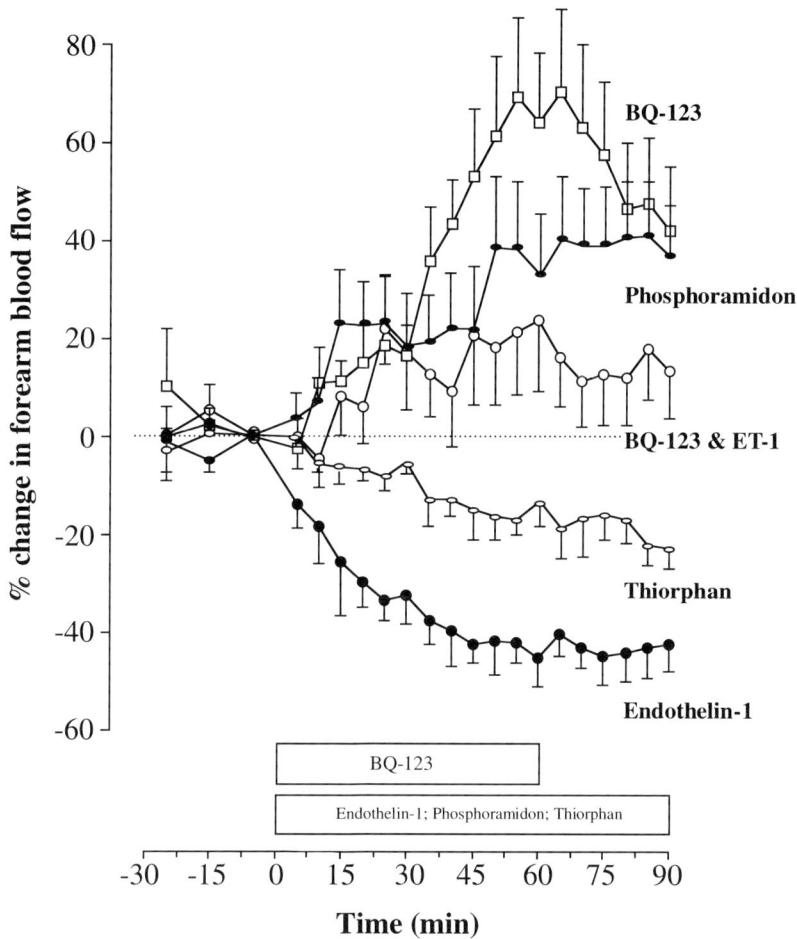

Figure 2.4   Forearm vasoconstriction to brachial artery infusion of endothelin-1 (5 pmol/min; closed circles) is abolished by the co-infusion of the ET_A antagonist BQ-123 (100 nmol/min; open circles). Infusion of BQ-123 (100 nmol/min; open squares) or the ECE/NEP inhibitor phosphoramidon (30 nmol/min; closed ovals) alone produce progressive forearm vasodilatation whereas the selective NEP inhibitor thiorphan (30 nmol/min; open ovals) causes progressive vasoconstriction. Adapted from Haynes & Webb, 1994, with kind permission of the *Lancet*. See text for details.

Using the forearm blood flow model, Haynes and Webb (1994) provided further evidence supporting a role for endothelin-1 in the maintenance of basal vascular tone. Using the selective cyclic pentapeptide BQ-123, they demonstrated inhibition of endothelin-1-induced vasoconstriction with local ET_A receptor antagonism and BQ-123 caused vasodilatation when given alone. The vasodilatation was of slow onset but persisted for up to one hour after the infusion was discontinued. These results have since been confirmed by others (Berrazueta *et al.*, 1997, Verhaar *et al.*, 1998) and suggest that the contribution of endogenous endothelin-1 to the maintenance of basal vascular tone is largely mediated via the ET_A receptor.

Systemic Cardiovascular Effects

The relevance of the observations with endothelin antagonists in the forearm model are confirmed by the effects of systemically administered TAK-044, a combined ET_A/ET_B receptor antagonist. TAK-044 causes a reduction of both blood pressure and peripheral vascular resistance when infused into normotensive individuals (Haynes *et al.*, 1996), with a dose of 1000 mg over 15 minutes causing a 4% and 18% reduction of systolic and diastolic blood pressures respectively, along with a 26% decrease in peripheral vascular resistance, thus indicating that its main effect is on resistance vessels. The summation of the physiological effects of the endothelin isoforms acting on both ET_A and ET_B receptors in human resistance vessels *in vivo* is, therefore, one of vasoconstriction. Bosentan, a combined ET_A/ET_B receptor antagonist, has also been shown to exert a modest hypotensive effect in healthy subjects (Weber *et al.*, 1996). Studies with bosentan given over four weeks to patients with essential hypertension have also shown a hypotensive effect, comparable to that of enalapril, indicating that the endothelin system contributes importantly to blood pressure in hypertensive individuals (Krum *et al.*, 1998).

Role of the ET_B Receptor

Whether endothelial ET_B receptor-mediated vasodilatation or vascular smooth muscle ET_B receptor-mediated vasoconstriction is the predominant *in vivo* response to basal endogenous endothelin-1 is an important question relevant to the development of ET_A receptor-selective or combined ET_A/ET_B receptor antagonists. Indeed, the potentially beneficial effects of NO release (with its vascular growth and platelet aggregation inhibiting properties) and of vasodilatation produced by endogenous ET_B receptor stimulation might be lost with the use of truly non-selective antagonists or ECE-inhibitors. Verhaar and colleagues (1998) have recently addressed this question in the human forearm. In agreement with previous studies, they demonstrated an increase in forearm blood flow following administration of the selective ET_A receptor antagonist BQ-123. In addition, this vasodilatatory response was shown to be largely mediated by, presumably increased, generation of NO (see Figure 2.5). Inhibition of vascular ET_B receptors with the selective ET_B antagonist BQ-788 produced vasoconstriction and, when co-infused with the ET_A receptor antagonist BQ-123, BQ-788 attenuated the vasodilator response (see Figure 2.6). Thus, the summation of effects of endogenous activation of ET_B receptors is one of vasodilatation via endothelin-stimulated generation of NO. Dilator PGI_2 generation following ET_B receptor activation appears to contribute little to this response because pre-treatment with aspirin has no influence on the response. Interestingly, many cardiovascular diseases have shown to cause dysfunction of the endothelium, with a resultant decrease in the production of NO or an increase in its breakdown (reviewed by Ferro and Webb, 1997). In such pathological states, the NO mediated vasodilator response to endogenous ET_B receptor activation may be lost, allowing unopposed vasoconstriction to predominate. The preservation of ET_B receptor mediated effects may be particularly important in the renal circulation. As indicated earlier, the ET_A receptor appears to be responsible for the mediation of vasoconstrictor responses to endothelin-1 in the human kidney. In contrast, animal studies suggest that ET_B receptors located on the proximal tubules are important in the control of diuresis and natriuresis (King *et al.*, 1989) and may mediate renal vasodilatation. Indeed, rescued ET_B receptor knockout mice are hypertensive

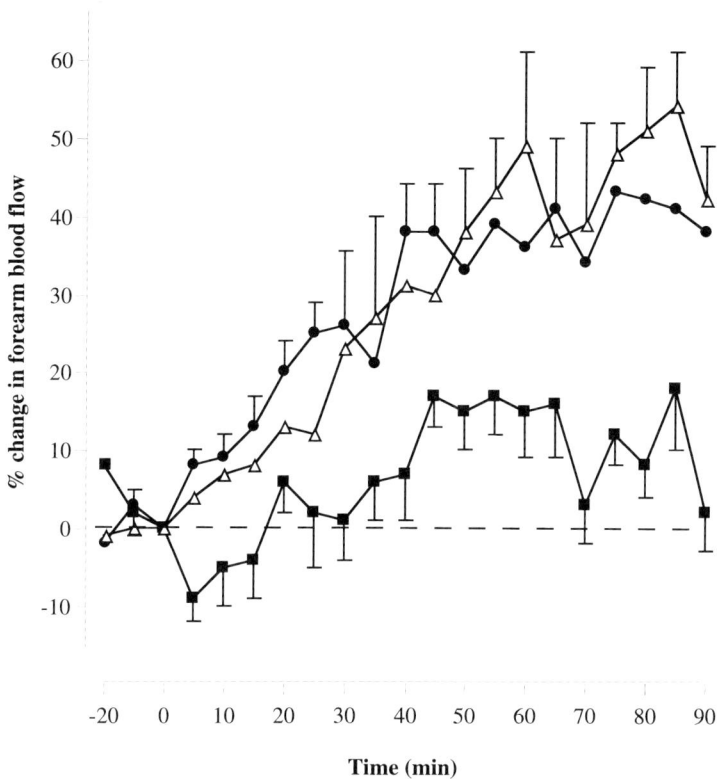

Figure 2.5 Slow onset of forearm vasodilatation in response to brachial artery infusion of BQ-123 (100 nmol/min; closed circles) during coinfusion of saline. Inhibition of prostanoid generation during BQ-123 infusion (100 nmol/min) had no significant effect (open triangles). Vasodilator effects of BQ-123 (100nmol/min) were attenuated by inhibition of NO (closed squares). Adapted from Verhaar *et al.*, 1998, with kind permission of *Circulation.* See text for details.

secondary to renal retention of sodium and chronic treatment of rats with selective ET_B receptor antagonists results in hypertension (reviewed by Webb *et al.*, 1998). Taken together, these findings suggest an important protective role for the ET_B receptor in the renal responses to endothelin-1.

Interactions Between the Endothelin System and the Nitric Oxide, Renin-Angiotensin and Sympathetic Nervous Systems

It is well recognised that endothelin-1 release is inhibited by endothelium-derived nitric oxide in the short term and that prolonged exposure to NO stimulates upregulation of ET_A receptors. Inhibition of the NO synthetic pathway by L-NMMA results in the potentiation of the vasoconstrictor responses to endothelin-1 (Lerman *et al.*, 1992) and an increase in circulating endothelin-1 levels (Ahlborg *et al.*, 1997). Recent studies in pigs have confirmed a close interaction between the NO and endothelin systems. A high cholesterol diet over a 10 week period resulted in enhanced coronary vasoconstriction in response to

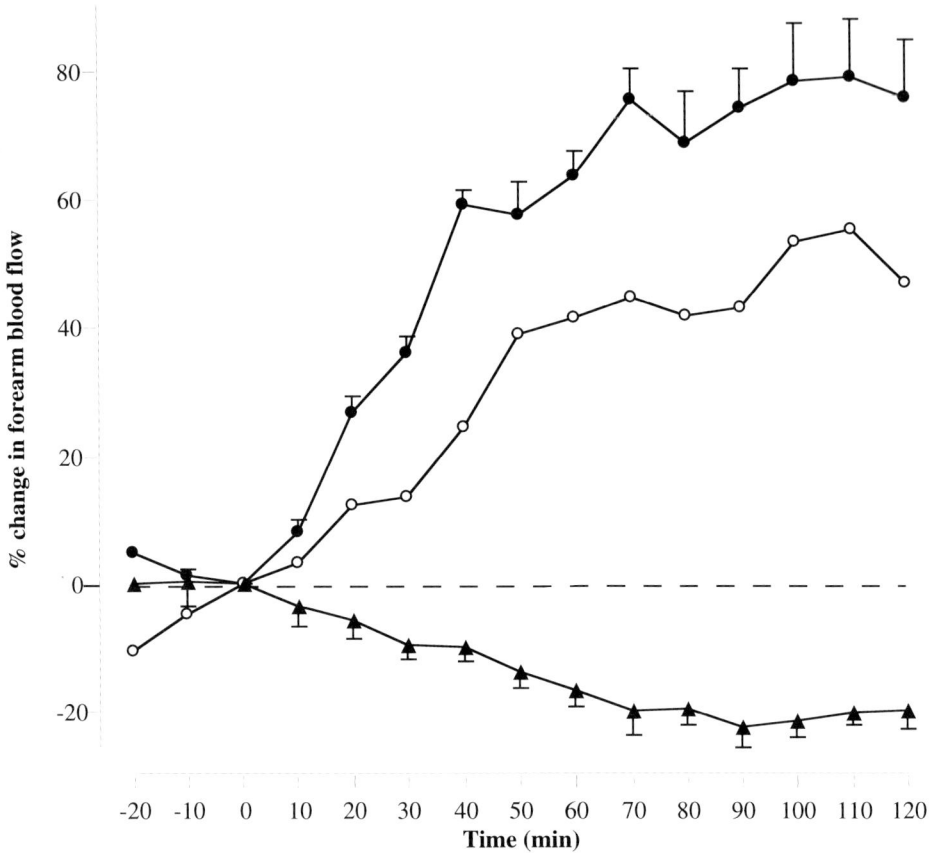

Figure 2.6 Slow onset forearm vasodilatation in response to brachial artery infusion of BQ-123 (10 nmol/min; closed circles) alone, was attenuated by co-infusion of BQ-788 (1 nmol/min; open circles). Infusion of BQ-788 alone caused a small but significant vasoconstriction (closed triangles). Adapted from Verhaar *et al.*, 1998, with kind permission of *Circulation*. See text for details.

pathophysiological doses of endothelin-1, whilst basal NO activity was reduced, suggesting an overall disturbance of vascular reactivity in this experimental model (Mathew *et al.*, 1997). In addition, the important role of the endothelin ET_B receptor in modulating responses to endothelin-1 in resistance vessels in man has already been described (Verhaar *et al.*, 1998).

In vitro experiments with human blood vessels have shown a potentiation of norepinephrine-stimulated vascular contractions when endothelin-1 is co-administered (Yang *et al.*, 1990) and sympathetically mediated venoconstriction of capacitance vessels may be potentiated by endothelin-1 in hypertensive patients (Haynes *et al.*, 1994). Norepinephrine infusions in rats increase the expression of endothelin-1 mRNA in ventricular myocytes and cause an increase in cardiac mass, an effect that may be blocked with the combined ET_A/ET_B receptor antagonist, bosentan (Kaddoura *et al.*, 1996).

Similarly, angiotensin II stimulates both increased expression of endothelin-1 mRNA (Imai *et al.*, 1992) and increased release of endothelin-1 from endothelial and vascular

Table 2.2 Cardiovascular and related conditions that have been associated with subtypes of endothelial dysfunction.

Disease State	NO pathway	Sympathetic pathway	ET pathway
Hypertension	+	v	v
Heart failure	+	v	v
Atherosclerotic vascular disease	+	v	v
Myocardial infarction	+	v	v
Unstable angina	+	v	v
Variant angina	+	v	v
Restenosis after angioplasty	+	?	v
Ischaemic stroke	+	v	v
Subarachnoid haemorrhage	+	v	v
Migraine	+	v	?
Raynaud's disease	+	v	v
Hyperhomocysteinaemia	+	?	?
Chronic renal failure	+	?	v
Diabetes mellitus	+	v	v
Ageing	+	v	v
Primary pulmonary hypertension	+	v	v

^{v} represents an association between these disease states and endothelial dysfunction affecting the particular pathway.

smooth muscle cells (Dohi *et al.*, 1992, Sung *et al.*, 1994). Angiotensin II-induced vaso-constriction may be potentiated by endothelin-1 administration (Yoshida *et al.*, 1992), whilst the pressor and mitogenic effects of prolonged angiotensin II infusion are abolished by concomitant infusion of ET_A receptor antagonists (Moreau *et al.*, 1997).

The close interactions between the endothelin, renin-angiotensin and sympathetic nervous systems suggest important synergistic effects leading to the vascular hypertrophy and increased peripheral vascular resistance seen in hypertension and chronic heart failure. Endothelin antagonists may, therefore, be of use in the treatment of these conditions to prevent the complications associated with overactivity of the renin-angiotensin and sympathetic nervous systems and/or decreased activity of the NO system.

The vascular responses to NO and the sympathetic nervous system are characteristically of rapid onset and are relatively short-lived. It may, therefore, be that NO and the sympathetic nervous system provide minute-by-minute, whilst endogenous endothelin-1 secretion controls long-term, maintenance of vascular tone. Endothelial dysfunction affecting the L-arginine/NO system, allowing unopposed activity of the endothelin system, is a proposed model for a range of pathological conditions. A number of these conditions are also associated with altered sympathetic activity. Examples of conditions proposed to be associated with endothelial dysfunction affecting the L-arginine/NO, endothelin and sympathetic nervous systems, indicating their potential interactions, are given in Table 2.2.

PATHOPHYSIOLOGY OF ENDOTHELIN

The normal function of the endothelin system and the balance with other local and hormonal cardiovascular regulatory mechanisms may become disrupted by a number of pathophysiological mechanisms (see Table 2.3). Many cardiovascular diseases have been

Table 2.3 Proposed mechanisms of endothelial system dysfunction in various pathological conditions.

Pathological mechanism	Proposed disease model
Decreased production	Hirschsprung's Disease
Decreased clearance	Renal Failure
Increased sensitivity	Cyclosporin-A toxicity
Increased production	Chronic Heart Failure
Decreased NO production	Hypertension
	Chronic Heart Failure
	Hypercholesterolaemia

shown to be associated with increased circulating concentrations of plasma endothelin-1, though whether these represent primary or secondary phenomena is undetermined. Changes in gene expression for the endothelin precursors, endothelin converting enzymes or endothelin receptors may cause congenital conditions such as Hirschsprung's disease and the Waardenburg-Shah syndromes. More subtle genetic mutations may result in changes in receptor expression, affinity or selectivity, perhaps contributing to hypertension in some patients. Enhanced receptor number or affinity is thought to mediate the hypertension and renal dysfunction during cyclosporine treatment, and renal dysfunction may reduce peptide clearance, thereby increasing circulating endothelin concentrations. Enhanced production of endothelin-1, in response to a number of factors including vasoactive hormones and tissue hypoxia, may further augment the neurohormonal activation seen in chronic heart failure and enhance the effects of other mediators. Indeed, the combined ET_A/ET_B receptor antagonist bosentan has been shown to produce favourable haemodynamic responses in the short term in patients with chronic heart failure in whom ACE inhibitors had been withheld (Kiowski *et al.*, 1995). In a similar context, the selective ET_A receptor antagonist BQ-123 elicits short term improvements in cardiac index, and produces decreases in both systemic vascular resistance and mean arterial pressure, when administered to small groups of patients with stable chronic heart failure also taking ACE inhibitors (Cowburn *et al.*, 1998).

The actions and concentrations of the endothelins would also be potentiated in conditions characterised by dysfunction of the L-arginine/NO system, allowing uncompensated changes in vascular tone and vascular remodelling to occur. This has been suggested as a mechanism for Raynaud's disease and cerebral vasospasm, amongst others. For a review of the pathophysiology of endothelial dysfunction, readers are referred to the appropriate chapters later in this book.

SUMMARY

Endothelin-1 is a 21-amino acid peptide produced by the vascular endothelium that has potent and long-lasting vasoconstrictor effects. Responses to endothelin-1 are mediated by two subtypes of endothelin receptor, ET_A and ET_B, acting in an autocrine and paracrine fashion. Clinical studies demonstrate a central role for endothelin-1 in the regulation of vascular tone in concert with the L-arginine/NO and sympathetic nervous systems. In addition to its effects on vessel tone, endothelin-1 exerts mitogenic effects which may contribute to the cardiac and vascular hypertrophy or remodelling central to the pathology of hypertension and chronic cardiac failure. Endothelin-1 has also been implicated in a wide variety of pathological conditions including renal, pulmonary, gastrointestinal and neurological disorders (Ferro and Webb, 1997). The use of both peptide and non-peptide receptor antagonists in experimental models of cardiovascular disease have provided valuable insights into the potential therapeutic uses of these agents. Large scale clinical trials are now underway in several different cardiovascular diseases, including hypertension, heart failure and sub-arachnoid haemorrhage, to further assess their efficacy and safety.

REFERENCES

Adachi, M., Furuichi, Y. and Miyamoto, C. (1994) Identification of a ligand binding site of the human endothelin-A receptor and specific regions required for ligand selectivity. *European Journal of Biochemistry*, **220**, 37–43.

Adachi, M., Hashido, K., Trzeciak, A., Watanabe, T., Furuichi, Y. and Miyamoto, C. (1993) Functional domains of human endothelin receptor. *Journal of Cardiovascular Pharmacology*, **22**, S121–S124.

Ahlborg, G. and Lundberg, J.M. (1997) Nitric oxide-endothelin-1 interaction in humans. *Journal of Applied Physiology*, **82**, 1593–1600.

Ahlborg, G., Ottosson-Seeberger, A., Hemsen, A. and Lundberg, J.M. (1994) Big-ET-1 infusion in man causes renal ET-1 release, renal and splanchnic vasoconstriction, and increased mean arterial blood pressure. *Cardiovascular Research*, **28**, 1559–1563.

Ando, S., Rahman, M.A., Butler, G.C., Senn, B.L. and Floras, J.S. (1995) Comparison of candoxatril and atrial natriuretic factor in healthy men: effects on hemodynamics, sympathetic activity, heart rate variability and endothelin. *Hypertension*, **26**, 1160–1166.

Arai, H., Hori, S., Aramori, I., Ohkubu, H. and Nakanishi, S. (1990) Cloning and expression of a cDNA encoding an endothelin receptor. *Nature*, **348**, 730–732.

Arai, H., Nakao, K., Takayo, K., Hosoda, K., Ogawa, Y., Nakanishi, S., *et al.* (1993) The human endothelin-B receptor gene: structural organisation and chromosomal assignment. *Journal of Biological Chemistry*, **268**, 3463–3470.

Aramori, I. and Nakanishi, S. (1992) Coupling of two endothelin receptor subtypes to different signal transduction in transfected Chinese hamster ovary cells. *Journal of Biological Chemistry*, **267**, 12468–12474.

Attie, T., Pelet, A., Edery, P., Eng, C., Mulligan, L., Amiel, J., *et al.* (1995) Mutation of the endothelin-receptor B gene in the Waardenburg-Hirschsprung disease. *Human Molecular Genetics*, **4**, 2407–2409.

Barnett, R.L., Ruffini, L., Hart, D., Mancuso, P. and Nord, E.P. (1994) Mechanism of endothelin activation of phospholipase A2 in rat renal medullary interstitial cells. *American Journal of Physiology*, **266**, F46–F56.

Bax, W.A., Aghai, Z., Van Tricht, C.L.J., Wassenaar, C. and Saxena, P.R. (1994) Different endothelin receptors involved in endothelin-1 and sarafotoxin S6B-induced contractions of the human isolated coronary artery. *British Journal of Pharmacology*, **113**, 1471–1479.

Baynash, A.G., Hosoda, K., Giaid, A., Richardson, J.A., Emoto, N., Hammer, R.E., *et al.* (1994) Interaction of endothelin-3 with endothelin-B receptor is essential for development of epidermal melanocyte and enteric neurons. *Cell*, **79**, 1277–1285.

Berrazueta, J.R., Bhagat, K., Vallance, P. and MacAllister, R.J. (1997) Dose- and time-dependency of the dilator effects of the endothelin antagonist, BQ-123, in the human forearm. *British Journal of Clinical Pharmacology*, **44**, 569–571.

Bitar, K.M., Stein, S. and Omann, G.M. (1992) Specific G-proteins mediate endothelin-induced contraction. *Life Sciences*, **50**, 2119–2124.

Boarder, M.R. (1994) A role for phospholipase D in control of mitogenesis. *Trends in Pharmacological Science*, **15**, 57–62.

Cacoub, P., Dorent, R., Nataf, P., Carayon, A., Riquet, M., Noe, E., *et al.* (1997) Endothelin-1 in the lungs of patients with pulmonary hypertension. *Cardiovascular Research*, **33**, 196–200.

Clarke, J.G., Benjamin, N., Larkin, S.W., Webb, D.J., Keogh, B.E., Davies, G.J., *et al.* (1989) Endothelin is a potent and long-lasting vasoconstrictor in man. *American Journal of Physiology*, **1989**, H2033–H2035.

Cowburn, P.J., Cleland, J.G.F. and McArthur, J.D. (1998) Short-term haemodynamic effects of BQ-123, a selective endothelin ET-A receptor antagonist, in chronic cardiac failure. *Lancet*, **352**, 201–202.

Danthaluri, N.R. and Brock, T.A. (1990) Endothelin-receptor coupling mechanisms in vascular smooth muscle: a role for protein kinase C. *Journal of Pharmacology and Experimental Therapeutics*, **254**, 393–399.

Davenport, A.P., O'Reilly, G., Molenaar, P., Maguire, J.J., Kuc, R.E., Sharkey, A., *et al.* (1993) Human endothelin receptors characterised using reverse transcriptase-polymerase chain reaction, in-situ hybridisation and subtype selective ligands. BQ123 and BQ3020: evidence for expression of ET-B receptors in human vascular smooth muscle. *Journal of Cardiovascular Pharmacolgy*, **22 Supplement 8**, S22–S25.

Decker, E.R. and Brock, T.A. (1998) Endothelin receptor-signalling mechanisms in vascular smooth muscle. In *Endothelin Molecular Biology, Physiology and Pathology*, (Ed, Highsmith, R.F.) Humana Press, Totowa, New Jersey, pp. 93–120.

Dohi, Y.A., Hahn, W.A., Boulanger, C.M., Buhler, F.R. and Luscher, T.F. (1992) Endothelin stimulated by angiotensin II augments contractility of spontaneously hypertensive rat resistance arteries. *Hypertension*, **19**, 131–137.

Douglas, S.A. and Ohlstein, E.H. (1997) Signal transduction mechanisms mediating the vascular actions of endothelin. *Journal of Vascular Research*, **34**, 152–164.

Dupuis, J., Cernacek, P., Tardif, J.C., Stewart, D.J., Gosselin, G., Dydra, I., *et al.* (1998) Reduced pulmonary clearance of endothelin-1 in pulmonary hypertension. *American Heart Journal*, **135**, 614–620.

Dupuis, J., Stewart, D.J., Cernacek, P. and Gosselin, G. (1996) Human pulmonary circulation is an important site for both clearance and production of endothelin-1. *Circulation*, **94**, 1578–1584.

Edwards, R.N., Trizna, W. and Ohlstein, E.H. (1990) Renal microvascular effects of endothelin. *American Journal of Physiology*, **259**, F217–F221.

Elshourbagy, N.A., Korman, D.R., Wu, H.L., Sylvester, D.R. and Lee, J.A. (1993) Molecular characterisation and regulation of the human endothelin receptors. *Journal of Biological Chemistry*, **268**, 3873–3879.

Emoto, N. and Yanagisawa, M. (1995) Endothelin converting enzyme-2 is a membrane bound, phosphoramidon-sensitive metalloprotease with acidic pH optimum. *Journal of Biological Chemistry*, **270**, 15262–15268.

Ferro, C.J., Spratt, J.C., Haynes, W.G. and Webb, D.J. (1998) Inhibition of neutral endopeptidase causes vasoconstriction of human resistance vessels in vivo. *Circulation*, **97**, 2323–2330.

Ferro, C.J. and Webb, D.J. (1997) Endothelial dysfunction and hypertension. *Drugs*, **53**, 30–41.

Goddard, J. and Webb, D.J. (1999) Plasma endothelin concentrations in hypertension. *Journal of Cardiovascular Pharmacology*, In Press.

Gray, G. (1995) Generation of Endothelin. In *Molecular Biology and Pharmacology of the Endothelins*, (Eds. Webb, D.J. and Gray, G.) R.G. Landes Company, Georgetown, Texas, pp. 1–173.

Haynes, W.G., Ferro, C.J., O'Kane, K.P.J., Somerville, D., Lomax, C.C. and Webb, D.J. (1996) Systemic endothelin receptor blockade decreases peripheral vascular resistance and blood pressure in humans. *Circulation*, **93**, 1860–1870.

Haynes, W.G., Hand, M.F., Johnstone, H.A., Padfield, P.L. and Webb, D.J. (1994) Direct and sympathetically mediated venoconstriction in essential hypertension. Enhanced responses to endothelin-1. *Journal of Clinical Investigation*, **94**, 1359–1364.

Haynes, W.G., Moffat, S. and Webb, D.J. (1995a) An investigation into the direct and indirect venoconstrictor effects of endothelin-1 and big endothelin-1 in man. *British Journal of Clinical Pharmacology*, **40**, 307–311.

Haynes, W.G., Strachan, F.E. and Webb, D.J. (1995b) Endothelin ET-A and ET-B receptors cause vasoconstriction of human resistance and capacitance vessels in vivo. *Circulation*, **92**, 357–363.

Haynes, W.G. and Webb, D.J. (1993a) Endothelium-dependent modulation of responses to endothelin-1 in human veins. *Clinical Science*, **84**, 427–433.

Haynes, W.G. and Webb, D.J. (1993b) Venoconstriction to endothelin-1 in humans: role of calcium and potassium channels. *American Journal of Physiology*, **265**, H1676–H1681.

Haynes, W.G. and Webb, D.J. (1994) Contribution of endogenous generation of endothelin-1 to basal vascular tone. *Lancet*, **344**, 852–854.

Henry, P.J. (1993) Endothelin-1 induced contraction in rat isolated trachea: involvement of ET-A and ET-B receptors and multiple signal transduction systems. *British Journal of Pharmacology*, **110**, 435–441.

Hiley, C.R., McStay, M.K.G. and Bottrill, F.E. (1992) Cross-desensitisation studies with endothelin isopeptides in the rat isolated superior mesenteric arterial bed. *Journal of Vascular Research*, **29**, 135.

Hirata, Y., Yoshimi, H., Takaichi, S., Yanagisawa, M. and Masaki, T. (1988) Binding and receptor down regulation of a novel vasoconstrictor endothelin in cultured rat vascular smooth muscle cells. *FEBS Letters*, **239**, 13–17.

Holm, P. (1997) Endothelin in the pulmonary circulation with special reference to hypoxic pulmonary vasoconstriction. *Scandinavian Cardiovascular Journal*, **46**, 1–40.

Hori, S., Komatsu, Y., Shigemoto, R., Mizuno, N. and Nakanishi, S. (1992) Distinct tissue distribution and cellular location of two messenger ribonucleic acids encoding different subtypes of rat endothelin receptors. *Endocrinology*, **130**, 1885–1895.

Hosada, K., Hammer, R.E., Richardson, J.A., Baynash, A.G., Cheung, J.C., Giaid, A., *et al.* (1994) Targetted and natural (piebald-lethal) mutations of endothelin-B receptor gene produce megacolon associated with spotted coat colour in mice. *Cell*, **79**, 1267–1276.

Hosada, K., Nakao, K., Arai, H., Suga, S., Ogawa, Y., Mukoyama, M., *et al.* (1991) Cloning and expression of human endothelin-1 receptor cDNA. *FEBS Letters*, **287**, 23–26.

Hosada, K., Nakao, K., Tamura, N., Arai, H., Ogawa, Y., Suga, S.I., *et al.* (1992) Organisation, structure, chromosomal assignment and expression of the gene encoding the human endothelin-A receptor. *Journal of Biological Chemistry*, **267**, 18797–18804.

Imai, T., Hirata, Y., Emori, T., Yanagisawa, M., Masaki, T. and Marumo, F. (1992) Induction of endothelin-1 gene by angiotensin and vasopressin in endothelial cells. *Hypertension*, **19**, 753–757.

Inoue, A., Yanagisawa, M. and Kimura, S. (1989a) The human endothelin family: three structurally and pharmacologically distinct isopeptides predicted by three separate genes. *Proceedings of the National Academy of Science USA*, **86**, 2863–2867.

Inoue, A., Yanagisawa, M., Takuwa, Y., Mitsui, Y., Kobayashi, M. and Masaki, T. (1989b) The human preproendothelin-1 gene: complete nucleotide sequence and regulation of expression. *Journal of Biological Chemistry*, **264**, 14954–14959.

James, A.F., Xie, L.-H., Fujitani, Y., Hayashi, S. and Horie, M. (1994) Inhibition of the cardiac protein kinase A-dependent chloride conductance by endothelin-1. *Nature*, **370**, 297–300.

Jilma, B., Szalay, E., Dirnberger, E., Eichler, H.-G., Stohlawetz, P., Schwarzinger, I., *et al.* (1997) Effects of endothelin-1 on circulating adhesion molecules in man. *European Journal of Clinical Investigation*, **27**, 850–856.

Kaasjager, K.A.H., Shaw, S., Koomans, H.A. and Rabelink, T.J. (1997) Role of endothelin receptor subtypes in the systemic and renal responses to endothelin-1 in humans. *Journal of the American Society of Nephrology*, **8**, 32–39.

Kaasjager, K.A.H., Van Rijn, H.J.M., Koomans, H.A. and Rabelink, T.J. (1995) Interactions of nifedipine with the renovascular effects of endothelin in humans. *Journal of Pharmacology and Experimental Therapeutics*, **275**, 306–311.

Kaddoura, S., Firth, J.D., Boheler, K.R., Sugden, P.H. and Poole-Wilson, P.A. (1996) Endothelin-1 is involved in norepinephrine-induced ventricular hypertrophy in vivo: acute effects of bosentan, an orally active, mixed endothelin ET(A) and ET(B) receptor antagonist. *Circulation*, **93**, 2068–2079.

Kanyicska, B. and Freeman, M.E. (1993) Characterisation of endothelin receptors in the anterior pituitary gland. *American Journal of Physiology*, **265**, E601–E608.

Karet, F., Kuc, R. and Davenport, A. (1993) Novel ligands BQ123 and BQ3020 characterise endothelin receptor subtypes ET-A and ET-B in human kidney. *Kidney International*, **44**, 36–42.

Karne, S., Jayawickreme, C.K. and Lerner, M.R. (1993) Cloning and characterisation of an endothelin-3 specific receptor (ETC receptor) from Xenopus laevis dermal melanophores. *Journal of Biological Chemistry*, **268**, 19126–19133.

Kaufmann, E., Oribe, E. and Oliver, J.A. (1991) Plasma endothelin during upright tilt: relevance for orthostatic hypotension? *Lancet*, **338**, 1542–1545.

Kiely, D.G., Cargill, R.I., Struthers, A.D. and Lipworth, B.J. (1997) Cardiopulmonary effects of endothelin-1 in man. *Cardiovascular Research*, **33**, 378–386.

King, A.J., Brenner, B.M. and Anderson, S. (1989) Endothelin: a potent renal and systemic vasoconstrictor peptide. *American Journal of Physiology*, **256**, F1051–F1058.

Kiowski, W., Luscher, T.F. and Linder, L. (1991) Endothelin-1–induced vasoconstriction in humans: reversal by calcium channel blockade but not by nitrovasodilators or endothelium-derived relaxing factor. *Circulation*, **83**, 469–475.

Kiowski, W., Sutsch, G., Hunziker, P., Muller, P., Kim, J., Oechslin, E., *et al.* (1995) Evidence for endothelin-1–mediated vasoconstriction in severe heart failure. *Lancet*, **346**, 732–736.

Krum, H., Viskoper, R.J., Lacourciere, M.D., Budde, M. and Charlon, V. (1998) The effect of an endothelin receptor antagonist, bosentan, on blood pressure in patients with essential hypertension. *New England Journal of Medicine*, **338**, 784–790.

Kumar, C., Mwangi, V., Nuthulaganti, P., Wu, H.L., Pullen, M., Brun, K., *et al.* (1994) Cloning and characterisation of a novel endothelin receptor from Xenopus heart. *Journal of Biological Chemistry*, **269**, 13414–13420.

Kurihara, Y., Kurihara, H., Suzuki, H., Kodama, T., Maemura, K. and Nagai, R. (1994) Elevated blood pressure and craniofacial abnormalities in mice deficient in endothelin-1. *Nature*, **368**, 703–710.

Ladoux, A. and Frelin, C. (1991) Endothelins inhibit adenylate cyclase in brain capillary cells. *Biochemical and Biophysical Research Communications*, **180**, 169–173.

Lerman, A., Sandok, E.K., Hildebrand, F.L. and Burnett, J.C. Jr. (1992) Inhibition of endothelium-derived relaxing factor enhances endothelin-mediated vasoconstriction. *Circulation*, **85**, 1894–1898.

Lincoln, T.M. (1989) Cyclic GMP and mechanisms of vasodilatation. *Pharmacological Therapeutics*, **41**, 479–502.

Little, P.J., Neylon, C.B., Tkachuk, V.A. and Bobik, A. (1992) Endothelin-1 and endothelin-3 stimulate calcium mobilisation by different mechanisms in vascular smooth muscle. *Biochemical and Biophysical Research Communications*, **183**, 694–700.

Liu, Y., Geisbuhler, B. and Jones, A.W. (1992) Activation of multiple mechanisms including phospholipase D by endothelin-1 in rat aorta. *American Journal of Physiology*, **262**, C941–C949.

Loffler, B.M., Kalina, B. and Kunze, H. (1991) Partial characterisation and sub-cellular distribution patterns of endothelin-1, -2 and -3 binding sites in human liver. *Biochemical and Biophysical Research Communications*, **181**, 840–845.

Marsault, R., Vigne, P., Breittmayer, J.P. and Frelin, C. (1991) Kinetics of vasoconstrictor actions of endothelins. *American Journal of Physiology*, **261**, C986–C993.

Marsden, P.A., Danthaluri, N.R., Brenner, B.M., Ballermann, B.J. and Brock, T.A. (1989) Endothelin action on vascular smooth muscle involves inositol triphosphate and calcium mobilisation. *Biochemical and Biophysical Research Communications*, **158**, 86–93.

Mathew, V., Cannan, C.R., Miller, V.M., Barber, D.A., Hasdai, D., Schwartz, R.S., *et al.* (1997) Enhanced endothelin-mediated coronary vasoconstriction and attenuated basal nitric oxide activity in experimental hypercholesterolemia. *Circulation*, **96**, 1930–1936.

Matsumoto, H., Suzuki, N., Onda, H. and Fujimo, M. (1989) Abundance of endothelin-3 in rat intestine, pituitary gland and brain. *Biochemical and Biophysical Research Communications*, **164**, 74–80.

Meyer-Lehnart, H., Wanning, C., Predel, H.G., Backer, A. and Kramer, H.J. (1989) Effects of endothelin on sodium transport mechanisms: potential role in cellular calcium mobilisation. *Biochemical and Biophysical Research Communications*, **163**, 458–465.

Miyoshi, Y., Nakayi, Y., Wakatsuki, T., Nomura, M., Saito, K., Nakaya, Y., *et al.* (1992) Endothelin blocks ATP-sensitive potassium channels and depolarises smooth muscle cells of porcine coronary artery. *Circulation Research*, **70**, 612–616.

Molenaar, P., O' Reilly, G., Sharkey, A., Kuc, R.E., Harding, D.P., Plumpton, C., *et al.* (1993) Characterisation and localisation of endothelin receptor subtypes in the human atrioventricular conducting system and myocardium. *Circulation*, **72**, 526–538.

Moreau, P., D'Uscio, L.V., Shaw, S., Takase, H., Barton, M. and Luscher, T.F. (1997) Angiotensin II increases tissue endothelin and induces vascular hypertrophy: reversal by ET(A)-receptor antagonist. *Circulation*, **96**, 1593–1597.

Moreland, S., McMullen, D.M., Delaney, C.L., Lee, V.G. and Hunt, J.T. (1992) Venous smooth muscle contains vasoconstrictor ET-B-like receptors. *Biochemical and Biophysical Research Communications*, **184**, 100–106.

Nakamoto, H., Suzuki, H., Murakami, M., Ohishi, A., Fukuda, K., Hori, S., *et al.* (1989) Effects of endothelin on systemic and renal hemodynamics and neuroendocrine hormones in conscious dogs. *Clinical Science*, **77**, 567–572.

Nakano, A., Kishi, F., Minanmi, K., Wakabayashi, H., Yutaka, N. and Kido, H. (1997) Selective conversion of big endothelins to tracheal smooth muscle-constricting 31 amino acid length endothelins by chymase from human mast cells. *Journal of Immunology*, **159**, 1987–1992.

Nishimura, J., Moreland, S., Ahn, H.Y., Kawase, T., Moreland, R.S. and Van Breemen, C. (1992) Endothelin increases myofilament calcium sensitivity in alpha-toxin-permeabilized rabbit mesenteric artery. *Circulation Research*, **71**, 951–959.

Ohlstein, E.H., Vickery, L., Sauermelch, C. and Willette, R.N. (1990) Vasodilatation induced by endothelin: role of EDRF and prostanoids in rat hindquarters. *American Journal of Physiology*, **259**, H1835–H1841.

Ohnaka, K., Takayanagi, R., Yamauchi, T., Okazaki, H., Ohashi, M., Umeda, F., *et al.* (1990) Identification and characterisation of endothelin converting activity in cultured bovine endothelial cells. *Biochemical and Biophysical Research Communications*, **168**, 1128–1136.

Ong, A.C.M. (1996) Surprising new roles for endothelins. *British Medical Journal*, **312**, 195–196.

Ozaki, S., Ihara, M., Saeki, T., Fukami, T., Ishikawa, K. and Yano, M. (1994) Endothelin ET-B receptors couple to two distinct signalling pathways in porcine kidney epithelial LLC-PK 1 cells. *Journal of Pharmacology and Experimental Therapeutics*, **270**, 1035–1040.

Pernow, J., Kaijser, L., Lundberg, J.M. and Ahlborg, G. (1996) Comparable potent coronary constrictor effects of endothelin-1 and big endothelin-1 in humans. *Circulation*, **94**, 2077–2082.

Plumpton, C., Haynes, W.G., Webb, D.J. and Davenport, A.P. (1995) Phosphoramidon inhibition of the in vivo conversion of big endothelin to endothelin-1 in the human forearm. *British Journal of Pharmacology*, **116**, 1821–1828.

Puffenberger, E.G., Hosada, K., Washington, S.S., Nakao, K., deWit, D., Yanagisawa, M., *et al.* (1994) A missense mutation of the endothelin receptor B gene in multigenic Hirschsprung's disease. *Cell*, **79**, 1257–1266.

Rabelink, T.J., Kaasjager, K.A.H., Boer, P., Stroes, E.G., Braam, B. and Koomans, H.A. (1994) Effects of endothelin-1 on renal function in humans: implications for physiology and pathophysiology. *Kidney International*, **46**, 376–381.

Sakamoto, A., Yanagisawa, M., Sakurai, T., Takuwa, Y., Yanagisawa, H. and Masaki, T. (1991) Cloning and functional expression of human cDNA for the ET-B endothelin receptor. *Biochemical and Biophysical Research Communications*, **178**, 656–663.

Sakurai, T., Yanagisawa, M., Takuwa, Y., Miyazaki, H., Kimura, S., Goto, K., *et al.* (1990) Cloning of a cDNA encoding a non-isopeptide selective subtype of the endothelin receptor. *Nature*, **348**, 732–735.

Shukla, S.D. and Halkenda, S.P. (1991) Phospholipase D in cell signalling and its relationship to phospholipase C. *Life Sciences*, **48**, 851–866.

Shyamala, V., Moulthrop, D.H., Stratton-Thomas, J. and Tekamp-Olsen, P. (1994) Two distinct human endothelin B receptors generated by alternative splicing from a single gene. *Cellular and Molecular Biological Research*, **40**, 285–296.

Stasch, J.P. and Kazda, S. (1989) Endothelin-1 induced vascular contractions: interactions with drugs affecting the calcium channel. *Journal of Cardiovascular Pharmacology*, **13**, S63–S66.

Stojilkovic, S.S., Balla, T., Fukuda, S., Merelli, F., Krsmanovic, L.Z. and Catt, K.J. (1992) Endothelin-A receptors mediate the signalling and secretory actions of endothelins in pituitary gonadotrophs. *Endocrinology*, **130**, 469–474.

Sudjarwo, S.A., Hori, M., Tanaka, T., Matsuda, Y., Okada, T. and Karaki, H. (1994) Subtypes of endothelin ET-A and ET-B receptors mediating venous smooth muscle contraction. *Biochemical and Biophysical Research Communications*, **200**, 627–633.

Sumner, M.J., Cannon, T.R., Mundin, J.W., White, D.G. and Watts, I.S. (1992) Endothelin ET-A and ET-B receptors mediate vascular smooth muscle contraction. *British Journal of Pharmacology*, **107**, 858–860.

Sung, C.P., Arleth, A.J., Storer, B.L. and Ohlstein, E.H. (1994) Angiotensin type 1 receptors mediate smooth muscle proliferation and endothelin biosynthesis in rat vascular smooth muscle. *Journal of Pharmacology and Experimental Therapeutics*, **271**, 429–437.

Takasuka, T., Sakurai, T., Goto, K., Furuichi, Y. and Watanabe, T. (1994) Human endothelin receptor ET-B: amino acid requirements for superstable complex formation with its ligand. *Journal of Biological Chemistry*, **269**, 7509–7513.

Takayanagi, R., Kitazumi, K., Takasaki, C., Ohnaka, K., Aimoto, S., Tasaka, K., *et al.* (1991) Presence of a non-selective type of endothelin receptor on vascular endothelium and its linkage to vasodilatation. *FEBS Letters*, **282**, 103–106.

Takayanagi, R., Ohnaka, K., Liu, W., Ito, T. and Nawata, H. (1998) Molecular biology of endothelin-converting enzyme. In *Endothelin Molecular Biology, Physiology and Pathology*, (Ed, Highsmith, R.F.) Humana Press, Totowa, New Jersey, pp. 75–92.

Tony, J.C. and Bhat, R. (1995) Acute myocardial infarction following snake bite. *Tropical Doctor*, **25**, 137.

Turner, A.J. and Murphy, L.J. (1996) Molecular pharmacology of endothelin converting enzyme. *Biochemical Pharmacology.*, **51**, 91–102.

Valdenaire, O., Rohrbachere, E. and Mattei, M.G. (1995) Organisation of the gene encoding the human endothelin converting enzyme (ECE-1). *Journal of Biological Chemistry*, **270**, 29794–29798.

Verhaar, M.C., Strachan, F.E., Newby, D.E., Cruden, N.L., Koomans, H.A., Rabelink, T.J., *et al.* (1998) Endothelin-A receptor antagonist-mediated vasodilatation is attenuated by inhibition of nitric oxide synthesis and by endothelin-B receptor blockade. *Circulation*, **97**, 752–756.

Vierhapper, H., Hollenstein, U., Roden, M. and Nowotny, P. (1993) Effect of endothelin-1 in man - impact on basal and stimulated concentrations of LH, FSH, TSH, GH, ACTH, and PRL. *Metabolism*, **42**, 902–906.

Vierhapper, H., Nowotny, P. and Waldhausl, W. (1995) Effect of endothelin-1 in man - impact on basal and ACTH-stimulated concentrations of aldosterone. *Journal of Clinical Endocrinology and Metabolism*, **80**, 948–951.

Vierhapper, H., Wagner, O.F., Nowotny, P. and Waldhausl, W. (1992) Effect of endothelin-1 in man: pretreatment with nifedipine, with indomethacin and with cyclosporine A. *European Journal of Clinical Investigation*, **22**, 55–59.

Vierhapper, H., Wagner, P., Nowotny, P. and Walhausl, W. (1990) Effect of endothelin in man. *Circulation*, **81**, 1415–1418.

Wagner, O.F., Vierhapper, H., Gasic, S., Nowotny, P. and Waldhausl, W. (1992) Regional effects and clearance of endothelin-1 across pulmonary and splanchnic circulation. *European Journal of Clinical Investigation*, **22**, 277–282.

Wang, X., Douglas, S.A. and Ohlstein, E.H. (1996) The use of quantitative RT-PCR to demonstrate the increased expression of endothelin-related mRNAs following angioplasty- induced neointima formation in the rat. *Circulation Research*, **78**, 322–328.

Watanabe, Y., Naruse, M., Monzen, C., Naruse, K., Ohsumi, K., Horiuchi, J., *et al.* (1991) Is big endothelin converted to endothelin-1 in circulating blood? *Journal of Cardiovascular Pharmacology*, **17 Supplement 7**, S503–S505.

Webb, D.J. (1995) The pharmacology of human blood vessels in vivo. *Journal of Vascular Research*, **32**, 2–15.

Webb, D.J., Monge, J.C., Rabelink, T.J. and Yanagisawa, M. (1998) Endothelin: new discoveries and rapid progress in the clinic. *TiPS*, **19**, 5–8.

Weber, C., Schmitt, R., Bimboeck, H., Hopfgarter, G., Van Marle, S.P., Peeters, P.A.M., *et al.* (1996) Pharmacokinetics and pharmacodynamics of the endothelin-receptor antagonist bosentan in healthy human subjects. *Clinical Pharmacology and Therapeutics*, **60**, 124–137.

Weitzberg, E., Ahlborg, G. and Lundberg, J.M. (1991) Long-lasting vasoconstriction and efficient regional extraction of endothelin-1 in human splanchnic and renal tissues. *Biochemical and Biophysical Research Communications*, **180**, 1298–1303.

Wenzel, R.R., Noll, G. and Luscher, T.F. (1994) Endothelin receptor antagonists inhibit endothelin in human skin microcirculation. *Hypertension*, **23**, 581–586.

Wenzel, R.R., Zbinden, S., Noll, G., Meier, B. and Luscher, T. (1998) Endothelin-1 induces vasodilation in human skin by nociceptor fibres and release of nitric oxide. *British Journal of Clinical Pharmacology*, **45**, 441–446.

Williams, D.L., Jones, K.L., Pettibone, D.J., Lis, E.V. and Clineschmidt, B.V. (1991) Sarafotoxin S6c: an agonist which distinguishes between endothelin receptor subtypes. *Biochemical and Biophysical Research Communications*, **175**, 556–561.

Xu, D., Emoto, N., Giaid, A., Slaughter, C., Kaw, S., DeWit, D., *et al.* (1994) ECE-1: a membrane bound metalloprotease that catalyses the proteolytic activation of big endothelin-1. *Cell*, **78**, 473–485.

Yanagisawa, H. and Yanagisawa, M. (1996) Endothelin and the differentiation of the neural crest: analysis by gene targeting. *Folia Endocrinologica Japonica*, **72**, 739.

Yanagisawa, M., Kurihara, H., Kimura, S., Tomobe, Y., Kobayashi, M., Mitsui, Y., *et al.* (1988) A novel potent vasoconstrictor peptide produced by vascular endothelial cells. *Nature*, **332**, 411–415.

Yang, Z., Richard, V. and von Segesser, L. (1990) Threshold concentrations of endothelin-1 potentiate contractions to norepinephrine and serotonin in human arteries: a new mechanism for vasospasm? *Circulation*, **82**, 188–195.

Yoshida, K., Yasujima, M., Kohzuki, M., Kanazawa, M., Yoshinaga, K. and Abe, K. (1992) Endothelin-1 augments response to angiotensin II infusion in rats. *Hypertension*, **20**, 292–297.

Yoshizawa, T., Shinmi, O., Giaid, A., Yanagisawa, M., Gibson, S.J., Kimura, S., *et al.* (1990) Endothelin: a novel peptide in the posterior pituitary system. *Science*, **247**, 462–464.

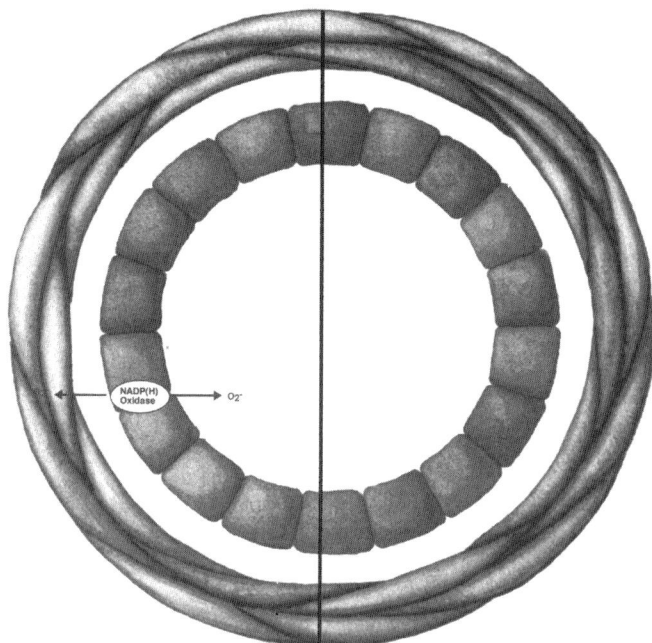

C
O
N
T
R
A
C
T
I
O
N

R
E
L
A
X
A
T
I
O
N

NADP(H) Oxidase → O_2^-

Endothelial cells not only mediate relaxation, but may also produce vasoconstrictor substances in response to a number of agents and physical stimuli. Of relevance is the activity of an endothelial cyclooxygenase pathway which can produce thromboxane A_2, prostaglandin H_2 and oxygen free radicals, mainly superoxide anions. In agreement with experimental evidence, cyclooxygenase-dependent endothelium-derived contracting factors were at first identified as responsible for impaired endothelium-dependent vasodilation in patients with essential hypertension. This alteration does not seem to be related to the increase in blood pressure, since it was not seen in patients with hypertension secondary to primary aldosteronism or renovascular disease, who also have endothelial dysfunction. In effect, production of cyclooxygenase-dependent endothelium-derived contracting factors is a phenomenon which is specific to aging, with essential hypertension merely causing an acceleration and enhancement of this alteration. It is worth noting that both in aging and hypertension appearance of cyclooxygenase-derived contracting factors is associated with a parallel decrease in nitric oxide availability, suggesting that the contracting factor could be an oxygen free radical. In line with this possibility, in essential hypertension both indomethacin, a cyclooxygenase inhibitor, and vitamin C, an antioxidant, increase the vasodilation to acetylcholine by restoring nitric oxide availability. Moreover the magnitude of effect of these two substances is similar and not additive, consistent with the idea that cyclooxygenase might be a source of oxidative stress in human hypertension. Other clinical conditions characterized by production of cyclooxygenase-dependent endothelium-derived contracting factors are acute estrogen deprivation and heart failure. Thus in normotensive women, ovariectomy and acute estrogen deprivation causes endothelial dysfunction resulting from production of cyclooxygenase-dependent vasoconstrictor substances. Finally, in patients with heart failure, indomethacin increases forearm vasodilation to acetylcholine, demonstrating that cyclooxygenase-dependent endothelium-derived contracting factors can play a major role in determining endothelial dysfunction in this pathological condition.

Key words: cyclooxygenase, acetylcholine, nitric oxide, oxidative stress, aging, essential hypertension.

3 Cyclooxygenase-Dependent Endothelium-Derived Contracting Factors

Stefano Taddei, Agostino Virdis, Lorenzo Ghiadoni and Antonio Salvetto

Department of Internal Medicine, University of Pisa, Via Roma, 67, 56100 Pisa, Italy
Tel: +39-50-551110; Fax: +39-50-502617; E-mail: s.taddei@int.med.unipi.it

INTRODUCTION

Soon after the initial discovery by Furchgott and Zawadzki (1980) of the obligatory role of endothelial cells in relaxations of rabbit isolated arteries to acetylcholine, De Mey and Vanhoutte (1982; 1983) found that the endothelium can also induce contractions of isolated canine arteries and veins. A large number of agents and physical stimuli produce such contractions and, depending on the agent, stimulus, and anatomic origin of the blood vessel, several different endothelium-derived contracting factors have been described, including cyclooxygenase-dependent endothelium derived contracting factors (EDCFs), endothelin and angiotensin II (Lüscher, 1990). Although relaxing factors play an important physiological role in circulatory regulation, existing experimental evidence supports the concept that contracting factors may become important regulators of vascular tone and structure in aging or under pathological conditions such as hypertension, diabetes, vasospasm and reperfusion injury (Katušic, 1991; Lüscher, 1991) .

Among the different pathways leading to endothelium-dependent contractions, it became apparent that cyclooxygenase could play a primary role. Arachidonic acid induces endothelium-dependent contractions in arteries and veins which can be inhibited by cyclooxygenase blockers (Miller, 1985; Katušic, 1998).

Moreover cyclooxygenase-dependent endothelium-derived contracting factors can be also induced by acetylcholine and the calcium ionophore A23187 in different vessels in experimental models of hypertension or diabetes (Konishi, 1983; Lüscher, 1986; Katušic, 1988; Tesfamarian, 1989).

So far in animal vessels, two kinds of mediators of cyclooxygenase-dependent EDCFs have been identified including oxygen free radicals (mainly superoxide anions), generated by the hydroperoxidase activity of the enzyme, and prostanoids such as thromboxane A_2 or prostaglandin H_2 (Figure 3.1) (Lüscher, 1990). It is relevant to observe that while prostanoids act exclusively as direct vasoconstrictors, oxygen free radicals can either directly constrict vascular smooth muscle, possibly by acting on prostaglandin H_2 receptors, or indirectly since they enhance NO breakdown (Figure 3.1).

This brief review will focus on current knowledge of cyclooxygenase-dependent EDCFs and their role in the control of vascular tone in humans.

ENDOTHELIUM

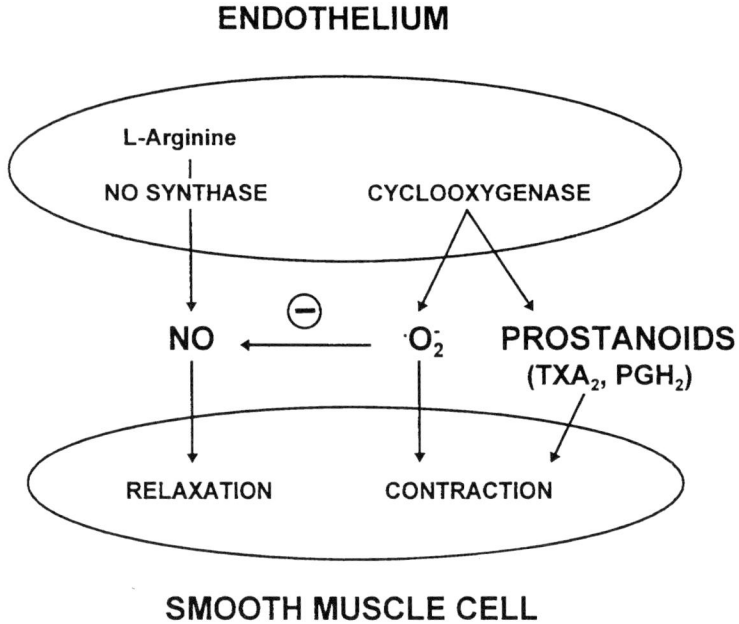

Figure 3.1 Schematic diagram showing endothelium-derived nitric oxide (NO) and cyclooxygenase-dependent contracting factors in the vessel wall. In certain conditions, including advancing age or essential hypertension, endothelial stimulation can activate not only L-arginine-NO pathway, but also cyclooxygenase to produce and secrete prostanoids such as thromboxane A_2 (TX A_2) or prostaglandin H_2 (PG H_2) and oxygen free radicals which cause vasoconstriction. Oxygen free radicals are also potent NO breakdown inductors.

IDENTIFICATION OF CYCLOOXYGENASE-DEPENDENT EDCFS IN HUMANS

Studies in humans have been conducted in the peripheral vasculature by using the technique of forearm blood flow measurement. Intra-arterial infusion of agonists and antagonists at concentrations which are inactive systemically allows adequate local vascular stimulation to be obtained without systemic effects or stimulation of neurohormonal reflexes. In this way, the increase or decrease in forearm blood flow, measured by strain gauge plethysmography, is an index of local vasodilation and vasoconstriction.

The possible production of cyclooxygenase-dependent EDCFs in humans was first tested in genetic and non-genetic models of secondary hypertension. Thus the response to acetylcholine is impaired in the forearm circulation of patients with essential hypertension as compared to normotensive controls (Linder, 1990; Panza, 1990; Taddei, 1993). In these patients, but not in healthy controls, infusion of intrabrachial indomethacin, a cyclooxygenase inhibitor, increases the response to acetylcholine (Taddei, 1993). This finding clearly indicates the production of cyclooxygenase-dependent products which contribute to the pathogenesis of endothelial dysfunction in human primary hypertension. These effects of indomethacin are obtained with an infusion rate of 50 μg/100 ml forearm tissue/min, which should provide a local plasma drug concentration of around 10^{-5} M. Although this is the concentration of indomethacin often employed in animal experiments, it should be recognised that at these doses the drug is no longer selective.

ESSENTIAL HYPERTENSIVE PATIENTS

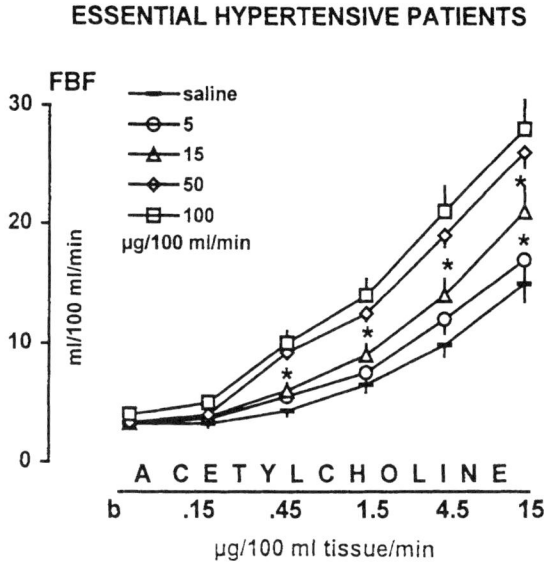

Figure 3.2 Acetylcholine-induced increase in forearm blood flow (FBF) in the presence of saline (0.2 ml/min) or indomethacin at 5, 15, 50 and 100 μg/100 ml forearm tissue/min in essential hypertensive patients (n = 6 each). Data are shown as means ± SD and expressed as absolute values. * denotes a significant difference between infusion in control conditions and in the presence of different infusion rates of indomethacin (p < 0.05 or less) (adapted from Taddei, 1998).

However when we tested different indomethacin infusion rates to titrate the lowest and highest effective dose, we observed that an indomethacin infusion rate of 5 μg/100 forearm tissue/min, 10 times lower than the standard dose, did not change the vascular response to acetylcholine (Figure 3.2) (Taddei, 1998). Moreover the infusion rate of 50 μg/100 ml forearm tissue/min maximally potentiates the vasodilation to acetylcholine (Figure 3.2). It is therefore crucial to employ this compound at an adequate local concentration to be able to demonstrate the possible presence of cyclooxygenase EDCFs, at least in essential hypertensive patients, since lower concentrations such as those obtained with systemic administration would be devoid of any effectiveness in blocking EDCF production.

Patients with hypertension secondary to primary aldosteronism or renovascular disease are also characterized by suppressed endothelium-dependent vasodilation (Taddei, 1993). However in these models of secondary hypertension, indomethacin did not improve the response to acetylcholine, indicating that EDCFs play no role in determining endothelial dysfunction in human secondary hypertension. Taken together these results are in agreement with experimental findings, demonstrating that cyclooxygenase-dependent EDCFs are produced in the presence of genetic hypertension, but not in the presence of secondary hypertension. The major relevance of this information is that EDCF production is probably not a consequence of the increase in blood pressure but is likely to be pathogenically related to essential hypertension.

Despite its marked effect on vasodilation to acetylcholine in the forearm circulation of essential hypertensive patients, intrabrachial infusion of indomethacin does not change local basal flow. This finding indicates that cyclooxygenase-dependent EDCFs are not tonically produced and therefore do not participate in the control of basal tone. Only selective receptor-mediated endothelial activation can induce the production of these substances.

ARE CYCLOOXYGENASE-DEPENDENT EDCFS CHARACTERISTIC OF ESSENTIAL HYPERTENSION?

The finding that indomethacin does not improve the blunted vasodilation to acetylcholine in patients with secondary hypertension seems to exclude the possibility that an increase in blood pressure *per se* could be the main mechanism leading to production of such substances. Moreover the possibility that EDCFs could be related to the pathogenesis of essential hypertension seems to be excluded by the results obtained in young normotensive offspring of essential hypertensive patients. These subjects showed an impaired response to acetylcholine as compared to matched offspring of normotensive subjects, while vasodilation to sodium nitroprusside was similar in the two groups (Taddei, 1996a). Thus a genetic predisposition to develop hypertension is associated with impaired endothelium-dependent vasodilation, suggesting that the alteration is a primary defect and is not dependent upon high blood pressure. However, in contrast to older essential hypertensive patients, the alteration seen in the normotensive offspring was not reversed by indomethacin infusion, but was sensitive to administration of L-arginine, the substrate for NO-synthase (Taddei, 1996a). This might indicate that a primary defect in the L-arginine-NO pathway, and not production of cyclooxygenase-dependent EDCFs, is responsible for the endothelial dysfunction present in individuals with a familial predisposition to develop hypertension. Thus EDCFs cannot universally participate in endothelial dysfunction in essential hypertension.

Therefore, to identify the significance of cyclooxygenase-dependent EDCF production in human cardiovascular physiopathology it is crucial to consider the impact of increasing age on endothelium-dependent vasodilation. There is evidence that aging is one of the main determinants of endothelial dysfunction in human vessels. The effect of aging can be detected both in the micro and macrocirculation of forearm (Taddei, 1995a; Gerhard, 1996) and coronary vessels (Vita, 1990; Yasue, 1990; Egashira, 1990; Zeiher, 1993) and is so strong that in the forearm microcirculation its negative effect can be detected even in the presence of essential hypertension (Taddei, 1995a). During exploration of the mechanisms responsible for age-related endothelial dysfunction, it was observed that in normotensive subjects the principle mechanism responsible for this alteration is a primary defect in the L-arginine-NO pathway. In contrast, only in old subjects, around after 60 years, did EDCF production starts to be detected and become relevant (Taddei, 1997a). In these older subjects, production of cyclooxygenase-dependent factors is associated with a further and parallel impairment in the L-arginine-NO pathway (Taddei, 1997a). If we consider essential hypertensive patients, the mechanisms involved in age-related endothelial dysfunction are similar to those found in healthy individuals, but characterized by an earlier onset. Therefore the production of cyclooxygenase-dependent EDCF starts in an age range of 31–45 years, and in patients older than 45 years the potentiating effect of indomethacin is augmented in parallel with increasing age (Taddei, 1997a). Taken together, this series of results supports the possibility that cyclooxygenase-dependent EDCF production is a phenomenon characteristic of aging, with essential hypertension merely causing earlier onset of this endothelial alteration.

ESSENTIAL HYPERTENSIVE PATIENTS

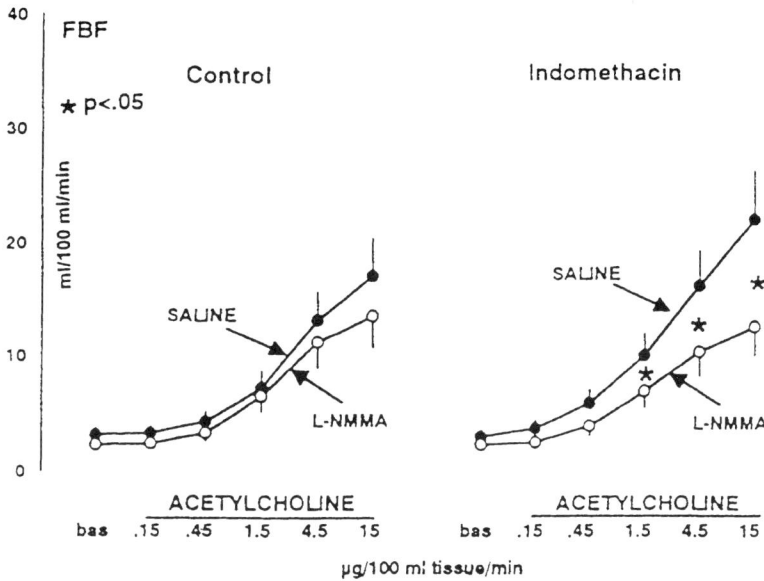

Figure 3.3 Acetylcholine-induced increase in forearm blood flow (FBF) in the absence (left) and presence (right) of indomethacin (50 µg/100 ml forearm tissue/min) under control conditions (saline at 0.2 ml/min) and in the presence of N^G-monomethyl-L-arginine (L-NMMA, 100 µg/100 ml forearm tissue/min) in essential hypertensive patients (n = 7). Data are shown as means ± SD and expressed as absolute values. * denotes a significant difference between infusion with and without L-NMMA (p < 0.05 or less) (adapted from Taddei, 1997a).

WHAT IS THE NATURE OF CYCLOOXYGENASE-DEPENDENT EDCFS?

Experimental evidence suggests that cyclooxygenase-dependent EDCFs could be prostanoids such as thromboxane A_2 or prostaglandin H_2 or oxygen free radicals (mainly superoxide anions). It is significant that in essential hypertensive patients, EDCF production is not the only endothelial alteration identified; rather it seems to be associated with a simultaneous impairment in the L-arginine-NO pathway (Panza, 1993a; Panza, 1993b; Taddei, 1994; Taddei, 1995b). Thus vasodilation to acetylcholine is resistant to L-arginine or L-NMMA induced facilitation or inhibition, respectively, suggesting the presence of a decrease in NO availability in patients with essential hypertension. In these patients, intrabrachial indomethacin not only increases the vasodilating response to acetylcholine but also restores the facilitating and inhibiting effect of L-arginine and L-NMMA (Figure 3.3) respectively, indicating that cyclooxygenase activity produces substances which reduce NO availability (Taddei, 1997b). These substances are probably oxygen free radicals since intrabrachial infusion of the scavenger vitamin C not only increases the response to acetylcholine, but also restores the inhibiting effect of L-NMMA (Taddei, 1998). Moreover, when simultaneously tested in the same study population, vitamin C and

ESSENTIAL HYPERTENSIVE PATIENTS

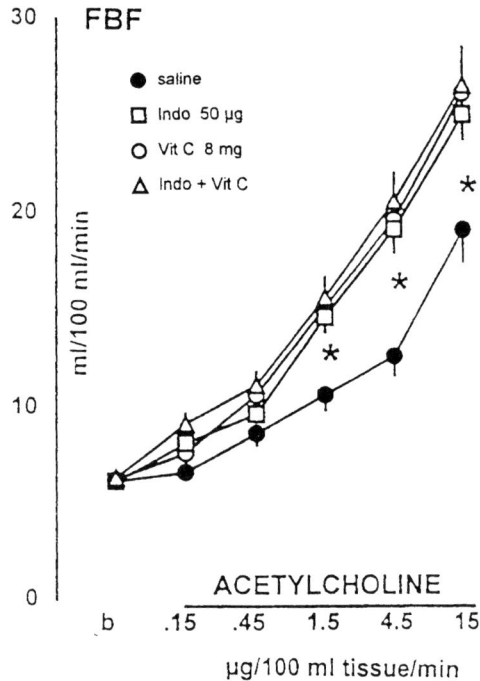

Figure 3.4 Acetylcholine-induced increase in forearm blood flow (FBF) in the presence of saline (0.2 ml/min); indomethacin (50 µg/100 ml forearm tissue/min); vitamin C (8 mg/100 ml forearm tissue/min) and simultaneous indomethacin and vitamin C (Δ) in essential hypertensive patients (n = 7). Data are shown as means ± SD and expressed as absolute values. *denotes a significant difference between infusion in control conditions and in the presence of vitamin C, indomethacin or vitamin C plus indomethacin (p < 0.05 or less) (adapted from Taddei, 1998).

indomethacin potentiate the response to acetylcholine to a similar degree (Figure 3.4) and infusing both compounds together does not cause a further additional improvement in endothelium-dependent vasodilation (Figure 3.4) (Taddei, 1998). Finally, preliminary observations in aged (older than 60 years) normotensive subjects indicate that vitamin C can improve vasodilation to acetylcholine, with an effect similar to that observed with indomethacin.

Taken together these findings seem to indicate that, in ageing and essential hypertension, EDCFs are likely to be oxygen free radicals produced by cyclooxygenase activity. However, whether these oxidative substances are directly produced by cyclooxygenase or indirectly by thromboxane A_2 or prostaglandin H_2 activity is still to be established. Future experiments with specific inhibitors of thromboxane synthase or selective antagonists for thromboxane A_2 or prostaglandin H_2 receptors, would be necessary to determine the contribution of these substances.

Until now, cyclooxygenase-dependent EDCF production has been identified only during stimulation of the endothelium by acetylcholine, and there are no data for other specific receptor operated endothelial agonists such as bradykinin, substance P, serotonin or endothelial responses activated by physical stimuli such as flow increase in large arteries (see chapter 10).

Furthermore, cyclooxygenase-derived vasoconstrictor factors are not produced in baseline conditions and do not modulate tonic NO release. Thus intrabrachial L-NMMA injection causes a dose-dependent vasoconstriction, which is related to basal NO production (Vallance, 1989). In essential hypertension the vascular response to L-NMMA is reduced, indicating a decrease in basal NO release (Calver, 1992; Taddei, 1995b). However, when L-NMMA is tested in the presence of indomethacin in the forearm circulation of essential hypertensive patients, cyclooxygenase inhibition does not improve the blunted vasoconstrictor response to the NO-synthase inhibitor, demonstrating that EDCF production is not responsible for the impaired basal release of NO (Taddei, 1997b). Vitamin C is also devoid of effect on vasoconstrictor response to L-NMMA in essential hypertension, making it unlikely that oxidative stress contributes to this impairment (Taddei, 1998).

Thus although patients with essential hypertension are characterized by a dysfunction in basal NO-mediated dilatation and receptor-operated endothelium-dependent relaxation, the mechanisms responsible for these defects are profoundly different, with the involvement of cyclooxygenase activity only in the latter.

OTHER CLINICAL CONDITIONS CHARACTERIZED BY CYCLOOXYGENASE-DEPENDENT EDCFs: ACUTE ESTROGEN DEFICENCY AND HEART FAILURE

Acute endogenous estrogen deprivation is a clinical condition characteristic of women who undergo ovariectomy. Estrogen can protect the vessel wall not only by improving lipid profile (Bush, 1987), but also by acting on endothelial responses (Giscard, 1988; Williams, 1990). Thus the menopause is characterized by endothelial dysfunction in both normotensive and hypertensive females (Celermajer, 1994; Taddei, 1996b).

Normotensive women who undergo ovariectomy and hysterectomy provide a good model to evaluate the role of acute estrogen in modulating endothelial responses. In these subjects ovariectomy was associated with a decrease in response to acetylcholine, but not to sodium nitroprusside (Pinto, 1997). This phenomenon was detected within one month of surgery and normal endothelium-dependent vasodilation was restored by estrogen replacement therapy (ERT) (Pinto, 1997). These results confirm that estrogen can preserve endothelial function. In line with this interpretation, prior to ovariectomy indomethacin did not change the response to acetylcholine while L-NMMA was able to blunt it indicating that in these subjects, cyclooxygenase-dependent EDCFs are not produced and NO availability is maintained, as expected in a population of normotensive women aged around 45 years. After ovariectomy, indomethacin-induced facilitation of the response to acetylcholine was observed, while the inhibiting effect of L-NMMA disappeared, suggesting that acute endogenous estrogen deprivation leads to EDCF production and a parallel drastic decrease in NO availability. When endothelial responses were tested again after ERT, vasodilation to acetylcholine was sensitive to L-NMMA and no longer affected by indomethacin. EDCFs produced after estrogen deprivation are probably oxygen free radicals since vitamin C, while ineffective on the vasodilation to acetylcholine before ovariectomy, significantly increased the response to the muscarinic agonist after ovariectomy and was again ineffective after ERT administration. Taken together and in line with experimental evidence (Ghin, 1992), these results suggest that estrogen protects endothelial function by inhibiting the production of cyclooxygenase-dependent EDCFs, which are very likely to be oxygen free radicals.

Finally, in patients with congestive heart failure (New York Heart Association functional class II–III), systemic cyclooxygenase inhibition with oral indomethacin (50 mg) induced a slight (+39%), but statistically significant, increase in the response to acetylcholine (Katz, 1993). However no information is available on the characterization of the cyclooxygenase-dependent vasoconstrictor substances released in response to acetylcholine in patients with heart failure.

CONCLUSIONS

Endothelial cells are capable of producing contracting factors including cyclooxygenase derivatives. While production of these substances is characteristic of aging, essential, but not secondary, hypertension, causes an acceleration and enhancement of the change. Cyclooxygenase-dependent EDCFs are also produced in acute estrogen deprivation and in heart failure. At least in the peripheral circulation, the nature of these substances seems to be mainly related to oxidative stress, which reduces NO availability. Since in physiological conditions NO has a protective effect on the vessel wall by inducing constant vasodilation, inhibition of platelet aggregation, smooth muscle cell proliferation, monocyte adhesion and endothelin-1 production, it is conceivable that in pathological conditions production of cyclooxygenase-dependent EDCFs can lead to impaired NO effectiveness, thereby diminishing the protective role of endothelial cells (Taddei, 1997c). However, the mechanisms that regulate the balance between relaxing and contracting factors and the process through which the endothelium loses its protective function by becoming a source of substances such as cyclooxygenase-dependent EDCFs with vasoconstrictor, proaggregatory and promitogenic activity remain to be determined.

REFERENCES

Bush, T.L., Barret-Connor, E., Cowan, L.D., Criqui, M.H., Wallace, R.B., Suchindram, C.M., *et al.* (1987) Cardiovascular mortality and noncontraceptive estrogen use in women: results from the Lipid Research Clinics Program Follow-up Study. *Circulation*, **75**, 1102–1109.

Calver, A., Collier, J., Moncada, S. and Vallance, P. (1992) Effect of local intra-arterial N^G-monomethyl-L-arginine in patients with hypertension: the nitric oxide dilator mechanism appears abnormal. *J. Hypertens.* **10**, 1025–31.

Celermajer, D.S., Sorensen, K.E., Spiegelhalter, D.J., Georgakopoulos, D., Robinson, J. and Deanfield, J.E. (1994) Aging is associated with endothelial dysfunction in healthy men years before the age-related decline in women. *J. Am. Coll. Cardiol.*, **24**, 471–476.

DeMey, J.G. and Vanhoutte, P.M. (1982) Heterogeneous behaviour of the canine arterial and venous wall: importance of the endothelium. *Circ. Res.*, **51**, 439–447.

DeMey, J.G. and Vanhoutte, P.M. (1983) Anoxia and endothelium-dependent reactivity of the canine phemoral artery. *J. Physiol.* (London), **335**, 65–74.

Egashira, K., Inou, T., Hirooka, Y., Kai, H., Sugimachi, M., Suzuki, S., Kuga, T., Urabe, Y. and Takeshita, A. (1993) Effects of age on endothelium-dependent vasodilation of resistance coronary artery by acetylcholine in humans. *Circulation*, **88**, 77–81.

Furchgott, R.F. and Zawadzki J.V. (1980) The obligatory role of endothelial cells in the relaxation of arterial smooth muscle by acetylcholine. *Nature* **288**:373–376.

Gerhard, M., Roddy, M.A., Creager, S.J. and Creager, MA. (1996) Aging progressively impairs endothelium dependent vasodilation in forearm resistance vessels of humans. *Hypertension*, **27**, 849–853.

Ghin, J.H., Azhar, S. and Hoffman, B.B. (1992) Inactivation of endothelium derived relaxing factor by oxidized lipoproteins. *J. Clin. Invest.*, **89**, 10–18.

Giscard, V., Miller, V. and Vanhoutte, P.M. (1988) Effect of 17beta-estradiol on endothelium-dependent responses in the rabbit. *J. Pharmacol. Exp. Ther.*, **244**, 19–22.

Katušic S.Z., Sheperd, J.T. and Vanhoutte, P.M. (1988) Endothelium-dependent contractions to calcium ionophore A23187, rachidonic acid, and acetylcholine in canine basilar arteries. *Stroke*, **19**, 476–479.

Katušic, S.Z. and Sheperd, J.T. (1991) Endothelium-derived vasoactive factors: II. Endothelium-dependent contraction. *Hypertension*, **[suppl III]**, III-86–III-92.

Katz, S.D., Schwarz, M., Yuen, J. and LeJemtel, T.H. (1993) Impaired acetylcholine-mediated vasodilation in patients with congestive heart failure. Role of endothelium-derived vasodilating and vasoconstricting factors. *Circulation*, **88**, 55–61

Konishi, M. and Su, C. (1983) Role of endothelium in dilator responses of spontaneously hypertensive rat arteries. *Hypertension*, **5**, 881–886.

Linder, L., Kiowski, W., Buhler, F.R. and Luscher, T.F. (1990) Indirect evidence for the release of endothelium-derived relaxing factor in the human forearm circulation in vivo: Blunted response in essential hypertension. *Circulation*, **81**, 1762–1767

Lüscher, T.F. and Vanhoutte, P.M. (1986) Endothelium-dependent contractions to acetylcholine in the aorta of spontaneously hypertensive rats. *Hypertension*, **8**, 344–348.

Lüscher, T.F. and Vanhoutte, P.M. (1990) *The endothelium: modulator of cardiovascular function*. Boca Raton, pp. 1–215, Fla, CRC Press.

Lüscher, T.F., Boulanger, C.M., Dohi, Y. and Yang, Z. (1991) Endothelium-derived contracting factors. *Hypertension*, **19**, 117–130.

Miller, V.M. and Vanhout, P.M. (1985) Endothelium-dependent contractions to arachidonic acid are mediated by products of cyclooxygenase. *Am. J. Physiol.*, **248**, H432–H437.

Panza, J.A., Quyyumi, A.A., Brush, J.E. Jr and Epstein, S.E. (1990) Abnormal endothelium dependent vascular relaxation in patients with essential hypertension. *N. Engl. J. Med.*, **323**, 22–27.

Panza, J.A., Casino, P.R., Badar, D.M. and Quyyumi, A.A. (1993a) Effect of increased availability of endothelium-derived nitric oxide precursor on endothelium-dependent vascular relaxation in normal subjects and in patients with essential hypertension. *Circulation*, **87**, 1475–1481.

Panza, J.A., Casino, P.R., Kilcoyne, C.M. and Quyyumi, A.A. (1993b) Role of endothelium-derived nitric oxide in the abnormal endothelium-dependent vascular relaxation of patients with essential hypertension. *Circulation*, **87**, 1468–1474.

Pinto, S., Virdis, A., Ghiadoni, L., Bernini, G.P., Lombardo, M., Petraglia, F., Gennazzani, A.R., Taddei, S. and Salvetti A. (1997) Endogenous estrogen and acetylcholine-induced vasodilation in normotensive women. *Hypertension*, **29[part 2]**, 268–273.

Taddei, S., Virdis, A., Mattei, P. and Salvetti, A. (1993) Vasodilation to acetylcholine in primary and secondary forms of human hypertension. *Hypertension* **21**, 929–33.

Taddei, S., Mattei, P., Virdis, A., Sudano, I., Ghiadoni, L. and Salvetti, A. (1994) Effect of potassium on vasodilation to acetylcholine in essential hypertension. *Hypertension*, **23**, 485–490.

Taddei, S., Virdis, A., Mattei, P., Ghiadoni, L., Gennari, A., Basile-Fasolo, C., Sudano, I. and Salvetti, A. (1995a) Aging and endothelial function in normotensive subjects and essential hypertensive patients. *Circulation*, **91,**, 1981–1987.

Taddei, S., Virdis, A., Mattei, P., Natali, A., Ferrannini, E. and Salvetti, A. (1995b). Effect of insulin on acetylcholine-induced vasodilation in normotensive subjects and patients with essential hypertension. *Circulation*, **92**, 2911–2918.

Taddei, S., Virdis, A., Mattei, P., Ghiadoni, L., Sudano, I. and Salvetti A. (1996a) Defective L-Arginine-Nitric Oxide in offspring of essential hypertensive patients. *Circulation*, **94**, 1298–1303.

Taddei, S., Virdis, A., Ghiadoni, L., Mattei, P., Bernini, G.P., Pinto, S. and Salvetti, A. (1996b) Menopause is associated with endothelial dysfunction in normotensive and essential hypertensive humans. *Hypertension*, **28**, 576–582.

Taddei, S., Virdis, A., Mattei, P., Ghiadoni, L., Basile-Fasolo, C., Sudano, I. and Salvetti, A. (1997a) Hypertension causes premature aging of endothelial function in humans. *Hypertension*, **29**, 736–743

Taddei, S., Virdis, A., Ghiadoni, L., Magagna, A. and Salvetti, A. (1997b) Cyclooxygenase inhibition restores nitric oxide activity in essential hypertension. *Hypertension*, **29[part 2]**, 274–279.

Taddei, S., Virdis, A., Ghiadoni, L. and Salvetti, A. (1997c) Hypertension and endothelial dysfunction. *Cardiovasc. Risk Factors*, **7**, 76–87.

Taddei, S., Virdis, A., Ghiadoni, L., Magagna, A. and Salvetti, A. (1998) Vitamin C improves endothelium-dependent vasodilation by restoring nitric oxide activity in essential hypertension. *Circulation*, **97**, 2222–2229.

Tesfamariam, B., Jakubowski, J.A. and Cohen, R.A. (1989) Contraction of diabetic rabbit aorta caused by endothelium-derived PGH2-TxA2. *Am. J. Physiol.*, **257**, H1327–H1333.

Vallance, P., Collier, J. and Moncada, S. (1989) Effects of endothelium-derived nitric oxide on peripheral arteriolar tone in man. *Lancet*, **ii**, 997–1000.

Vita, J.A., Treasure, C.B., Nabel, E.G., Mclenacham, J.M., Fish, R.D., Yeung, A.C., Vekshtein, V.I., Selwyn, A.P. and Ganz, P. (1990) Coronary vasomotor response to acetylcholine relates to risk factors for coronary artery disease. *Circulation*, **81**, 491–497.

Williams, J.K., Adams, M.R. and Klopfenstein, H.B. (1990) Estrogen modulates responses of atherosclerotic coronary arteries. *Circulation*, **81**, 1680–1687.

Yasue, H., Matsuyama, K., Matsuyama, K., Okumura, K., Morikami, Y. and Ogawa, H. (1990) Responses of angiographically normal human coronary arteries to intracoronary injection of acetylcholine by age and segment. Possible role of early coronary atherosclerosis. *Circulation*, **81**, 482–490.

Zeiher, A.M., Drexter, H., Saurbier, B. and Just, H. (1993) Endothelium-mediated coronary blood flow modulation in humans. Effects of age, atherosclerosis, hypercholesterolemia, and hypertension. *J. Clin. Invest.*, **92**, 652–62.

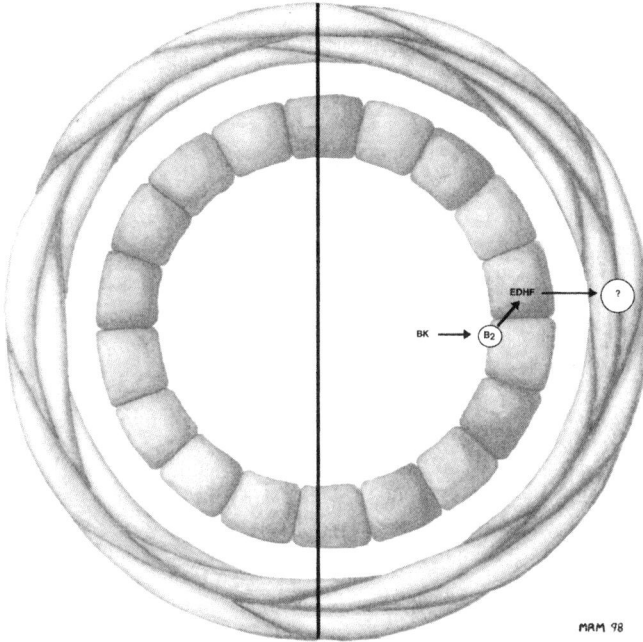

C O N T R A C T I O N

R E L A X A T I O N

EDHF

BK → (B2)

?

MRM 98

Endothelium-dependent relaxations cannot be fully explained by the release of either NO or/and prostacyclin. Another unidentified substance(s) which hyperpolarises the underlying vascular smooth muscle cells, termed endothelium-derived hyperpolarizing factor (EDHF), may contribute to endothelium-dependent relaxations, especially in small arteries. In human blood vessels, endothelium-dependent hyperpolarisations are observed in those blood vessels which exhibit endothelium-dependent relaxations that are partially or totally resistant to inhibitors of NO synthase and cyclooxygenase. The existence of EDHF as a diffusable substance has been demonstrated with isolated animal blood vessels but also with cultured human umbilical vein endothelial cells. Hyperpolarisations caused by EDHF are insensitive to glibenclamide and iberiotoxin but are inhibited by tetraethylammonium or the combination of charybdotoxin plus apamin or apamin plus ciclazindol. This indicates that EDHF does not activate ATP-sensitive or large conductance calcium-activated potassium channels but rather an undetermined population of potassium channels which may contain a subunit of small conductance calcium-activated potassium channels. The identification of EDHF is unresolved and the suggestion that it could be a cytochrome P450 metabolite of arachidonic acid is still controversial. The identification of EDHF and/or the discovery of specific inhibitors of its synthesis and its action will allow a better understanding of its physiological and pathophysiological role(s).

Key words: EDHF, endothelium, smooth muscle, potassium channels, membrane potential, vasodilatation.

4 Endothelium-Derived Hyperpolarizing Factor

Michel Félétou and Paul M. Vanhoutte

Institut de Recherches Internationale Servier, 6 Place des Pleides, 92415 Courbevoie, Cedex, France

INTRODUCTION

Endothelial cells synthesise and release vasoactive mediators in response to various neurohumoral substances (e.g. acetylcholine, adenosine trisphosphate, bradykinin, thrombin, etc.) and physical stimuli (e.g. shear stress). Nitric oxide (NO) produced from the L-arginine by endothelial NO synthase and prostacyclin produced from arachidonic acid by cyclooxygenase have been identified as endothelium-derived vasodilators. However, not all endothelium-dependent relaxations can be fully explained by the release of either NO and/or prostacyclin. Another unidentified substance(s) which hyperpolarizes the underlying vascular smooth muscle cells, termed endothelium-derived hyperpolarizing factor (EDHF), may contribute to endothelium-dependent relaxations (for review see Furchgott and Vanhoutte, 1989, Félétou and Vanhoutte, 1996, Mombouli and Vanhoutte, 1997).

EXISTENCE OF AN ALTERNATIVE PATHWAY

Besides prostacyclin and NO [the endothelium-derived relaxing factor discovered by Furchgott and Zawadzki (1980)], the existence of a third mediator was postulated by DeMey *et al.* in 1982. In 1984, Bolton *et al.* demonstrated that a muscarinic agonist elicited endothelium-dependent hyperpolarization of vascular smooth muscle cells. This phenomenon has been confirmed in various blood vessels from different species (Félétou and Vanhoutte, 1985, 1988; Komori and Suzuki, 1987; Taylor *et al.*, 1988; Chen *et al.*, 1988; Keef and Bowen, 1989; Brayden, 1990). When L-arginine analogues became available as specific inhibitors of the production of NO (Rees *et al.*, 1989), it soon appeared that endothelium-dependent relaxations and hyperpolarizations could be partially or totally resistant to inhibitors of cyclooxygenase and NO synthase (Bény and Brunet, 1988a & b; Huang *et al.*, 1988; Richard *et al.*, 1990; Cowan and Cohen, 1991; Mugge *et al.*, 1991; Hasunuma *et al.*, 1991; Illiano *et al.*, 1992; Nagao *et al.*, 1992a; Nagao and Vanhoutte, 1992b; Suzuki *et al.*, 1992), suggesting the existence of an additional endothelial mechanism. Furthermore, endothelium-dependent responses, which are resistant to inhibitors of NO synthase and cyclooxygenase, are observed without an increase in intracellular levels of cyclic nucleotides (cyclic GMP and cyclic AMP) in the smooth muscle cells (Taylor *et al.*, 1988; Cowan and Cohen, 1991; Mombouli *et al.*, 1992; Zygmunt *et al.*, 1994; Garcia-Pascual *et al.*, 1995). Under these conditions, the hyperpolarization of the cell membrane of vascular smooth muscle and the resulting reduction in Ca^{2+} entry explains the endothelium-dependent relaxations (Nagao and Vanhoutte, 1991, 1992a; Garland and McPherson, 1992). Indeed, hyperpolarization of smooth muscle cells reduces the open probability of voltage-dependent calcium channels, thereby decreasing calcium influx and lowering intracellular calcium levels (Nelson *et al.*, 1990). In addition, the hyperpolarization may reduce the increase in intracellular phosphatidylinositol turnover caused by agonist-induced receptor activation (Itoh *et al.*, 1992).

NO can activate various potassium conductance: ATP-sensitive potassium channels (K-ATP) via a cyclic GMP-dependent mechanism (Miyoshi *et al.*, 1994) and large conductance calcium-activated potassium channels (BK_{Ca}) either via a cyclic GMP-dependent mechanism (Robertson *et al.*, 1993; Taniguchi *et al.*, 1993; Archer *et al.*, 1994) or even directly (Bolotina *et al.*, 1994; Shin *et al.*, 1997). Hence NO could therefore produce hyperpolarization of the smooth muscle cells. Indeed, in blood vessels from species such as rat, guinea-pig and rabbit, endothelium-dependent hyperpolarizations resistant to inhibitors of NO synthase and cyclooxygenase and hyperpolarizations to endothelium-derived or exogenous NO can be observed in the same vascular tissue (Tare *et al.*, 1990; Murphy and Brayden, 1995a; Parkington *et al.*, 1995; Plane *et al.*, 1995; Corriu *et al.*, 1996b). Likewise, prostacyclin and its stable analogue, iloprost, also can hyperpolarize vascular smooth muscle cells (Siegel *et al.*, 1987; Parkington *et al.*, 1993; Corriu *et al.*, 1996b). Thus incomplete inhibition of the NO-synthase and/or cyclooxygenase could account for endothelium-dependent hyperpolarization (Cohen *et al.*, 1997). However, in blood vessels such as the carotid artery of the guinea-pig the combination of NO scavengers and inhibitors of nitric oxide synthase (in the presence of inhibitor of cyclooxygenase) does not inhibit endothelium-dependent hyperpolarizations (Figure 4.1). Finally, in canine and porcine blood vessels hyperpolarization cannot be produced by nitrovasodilators or exogenous NO (Bény and Brunet, 1988a; Komori *et al.*, 1988; Pacicca *et al.*, 1992) indicating that in most species a third pathway has to be involved in order to fully explain endothelium-dependent relaxations.

Mechanism of Endothelium-Dependent Hyperpolarization

Endothelium-dependent hyperpolarization involves the opening of a potassium conductance. The amplitude of the hyperpolarization is inversely related to the extracellular concentration of K^+ ions, and it disappears in K^+ concentrations higher than 25 mM (Chen and Suzuki, 1989; Nagao and Vanhoutte, 1992a; Corriu *et al.*, 1996a). Furthermore, endothelium-dependent hyperpolarizations are associated with an increase in rubidium uptake (Taylor *et al.*, 1988). Finally, non-selective potassium channel inhibitors, such as tetraethylammonium (TEA) or tetrabutylammonium prevent the hyperpolarization (Chen *et al.*, 1991; Nagao and Vanhoutte, 1992a; Van de Voorde *et al.*, 1992). The exact nature of the potassium conductance involved is still unknown and may depend on the species. In arteries of rats, rabbits, bovines and pigs, responses attributed to EDHF are blocked by apamin suggesting, in these species, the involvement of calcium-dependent potassium channels of small conductance (SK_{Ca}; Adeagbo and Triggle, 1993; Hecker *et al.*, 1994; Garcia-Pascual *et al.*, 1995; Murphy and Brayden, 1995b). In rat mesenteric and guinea-pig carotid arteries, the combination of charybdotoxin (an inhibitor of both BK_{Ca} and voltage-sensitive potassium channels) plus apamin is required to abolish the hyperpolarization resistant to inhibitors of NO synthase and cyclooxygenase (Figure 4.2) while each blocker individually is either ineffective (charybdotoxin) or only partially effective (apamin) (Garland and Plane, 1996; Corriu *et al.*, 1996b; Chataigneau *et al.*, 1998a). The combination of iberiotoxin (a specific inhibitor of BK_{Ca}) plus apamin did not mimic the effects

Guinea-pig carotid artery with endothelium
(presence of L-NA : 100 µM + indomethacin : 5 µM)

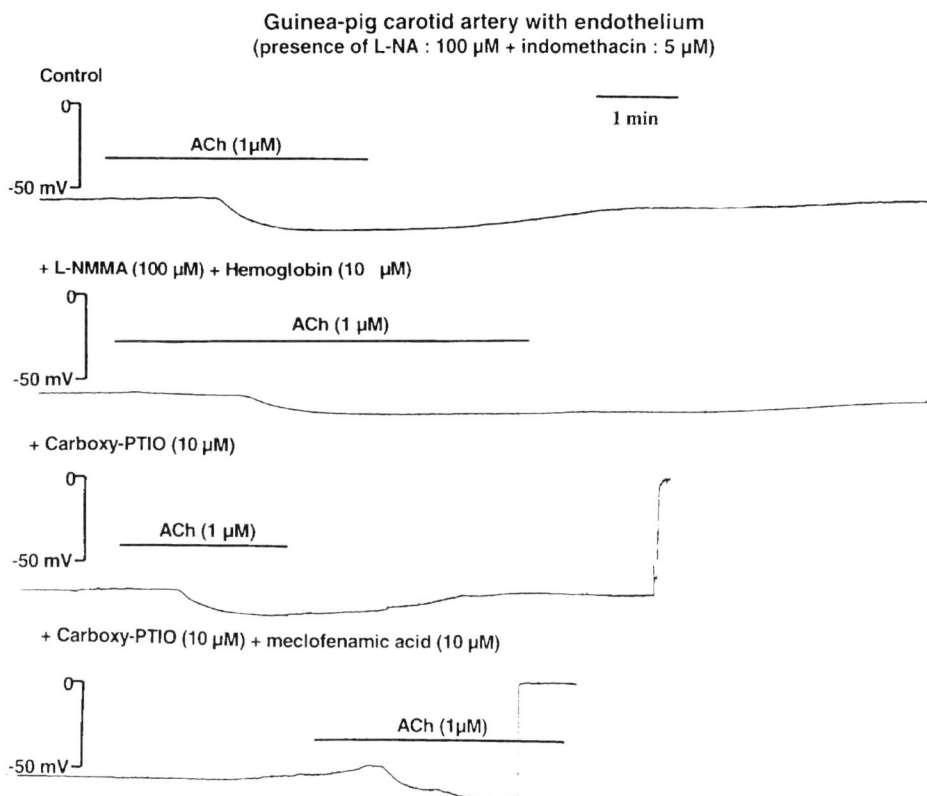

Figure 4.1 Membrane potential recording in guinea-pig isolated carotid artery with endothelium.
* First trace from the top: Acetylcholine (ACh, 1 µM) induced hyperpolarization in the presence of an inhibitor of nitric oxide synthase L-nitro-arginine (L-NA, 100 µM) and an inhibitor of cyclooxygenase (indomethacin, 5 µM). The hyperpolarization was not observed in preparations without endothelium (not shown).
* Second trace from the top: Acetylcholine in the presence of L-NA plus indomethacin plus a second inhibitor of nitric oxide synthase L-nitromonomethyl-arginine (L-NMMA, 100 µM) plus a scavenger of nitric oxide (haemoglobin, 10 µM) still induced endothelium-dependent hyperpolarization.
* Third trace from the top: Acetylcholine in the presence of L-NA plus indomethacin plus a different scavenger of nitric oxide 2-(4-carboxy-phenyl)-4,4,5,5-tetramethylimidazoline-1oxyl-3-oxide (carboxy-PTIO, 10 µM) still induced endothelium-dependent hyperpolarization.
* Fourth trace from the top: Acetylcholine in presence of L-NA plus indomethacin plus carboxy-PTIO plus a second inhibitor of cyclooxygenase (meclofenamic acid, 10 µM) still induced endothelium-dependent hyperpolarization.

of charybdotoxin and apamin (Zygmunt and Högestätt, 1996; Chataigneau *et al.*, 1998a) indicating that BK_{Ca} channels are not involved in the endothelium-dependent hyperpolarization of the guinea-pig carotid and rat mesenteric artery. In contrast, in the rabbit carotid artery relaxations attributed to EDHF are sensitive to charybdotoxin alone (Cowan *et al.*, 1993). A unifying hypothesis once proposed by Zygmunt *et al.* (1997a) suggested that the potassium channel(s) involved in the endothelium-dependent hyperpolarization was either a "classical" type of SK_{Ca} or a channel similar to one of the apamin-insensitive variants

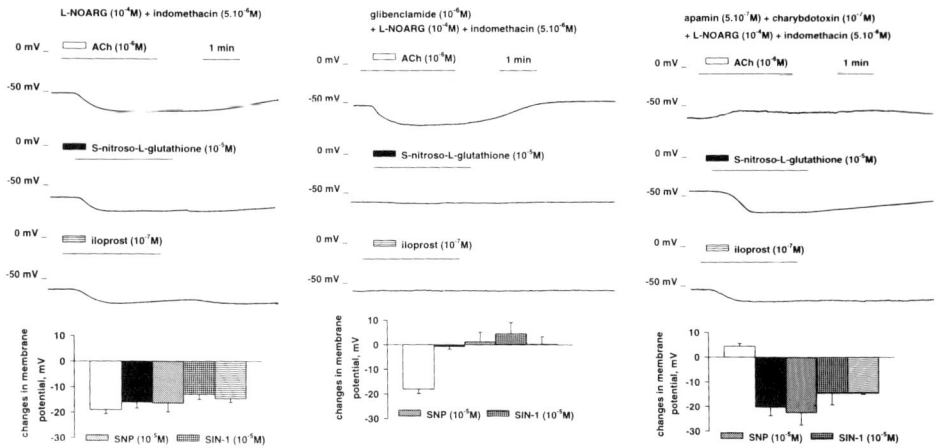

Figure 4.2   Effect of acetylcholine (ACh, 1 µM, upper traces), S-nitroso-L-glutathione (a nitrovasodilator, 10 µM, middle traces) and iloprost (a stable analogue of prostacyclin, 0.1 µM, lower traces) in the guinea-pig isolated carotid artery with endothelium

Left panel: control experiments in presence of L-nitro-arginine (L-NOARG, 100 µM) and indomethacin (5µM).
Middle panel: treatment with glibenclamide (1 µM).
Right panel: treatment with the combination of charybdotoxin (0.1 µM) plus apamin (0.5 µM).
Bar graphs summarised the changes in membrane potential obtained in the various experimental conditions. The effects of S-nitroso-L-glutathione were mimicked by two other nitrovasodilators: sodium nitroprusside (SNP, 10 µM) and SIN-1 (10 µM).
Modified from Corriu *et al.* (1996b).

of SK$_{Ca}$, some of which are sensitive to charybdotoxin alone (Van Renterghem and Lazdunski, 1992; Köhler *et al.*, 1996). It has been assumed that the targets of potassium channel blockers were situated on the vascular smooth muscle cells because these toxins inhibited EDHF-mediated responses without affecting the relaxations or the hyperpolarizations produced by endothelial nitric oxide or prostacyclin e.g. being selective (Corriu *et al.*, 1996b; Zygmunt and Höggestätt, 1996; Petersson *et al.*, 1997; Chataigneau *et al.*, 1998a; Zygmunt *et al.*, 1998). However, calcium-activated potassium channels are also expressed in endothelial cells (Marchenko and Sage, 1996). In the endothelial cells of the guinea-pig coronary artery, the combination of the two toxins does not affect the acetycholine-induced increase in intracellular free calcium concentration (Yamanaka *et al.*, 1998). However, in the hepatic artery of the rat and in the aortic valve of the rabbit, the combination of charybdotoxin plus apamin inhibits the hyperpolarization of the endothelial cells produced by acetylcholine (Edwards *et al.*, 1998; Ohashi *et al.*, 1999). Thus, charybdotoxin and apamin can act on the endothelial cells and this endothelial effect could be responsible for the inhibition of the EDHF-mediated responses. Therefore, the vascular smooth muscle population of potassium channel activated during the endothelium-dependent hyperpolarization is still unknown.

The sustained membrane hyperpolarizations provoked by prostacyclin (and iloprost) are blocked by glibenclamide, suggesting the involvement of ATP-dependent potassium channels (Parkington *et al.*, 1993, 1995; Corriu *et al.*, 1996b; Figure 4.2). Likewise, in blood vessels where NO causes hyperpolarization, as recorded with intracellular microelectrodes, the response is also blocked by glibenclamide (Miyoshi *et al.*, 1994; Plane *et al.*, 1995; Murphy and Brayden, 1995a; Parkington *et al.*, 1995; Corriu *et al.*, 1996b) but not by the combination of charybdotoxin plus apamin (Corriu *et al.*, 1996b, Figure 4.2). By contrast, glibenclamide does not inhibit endothelium-dependent hyperpolarization resistant to inhibitors of NO synthase and cyclooxygenase confirming the existence of a third pathway (Chen *et al.*, 1991; McPherson and Angus, 1991; Fujii *et al.*, 1992; Nakashima *et al.*, 1992; Nagao and Vanhoutte, 1992a; Plane *et al.*, 1995; Corriu *et al.*, 1996b; Figure 4.2).

Nature of EDHF

The term EDHF implicitly refers to a factor released by the endothelial cells. However, theoretically, endothelium-dependent hyperpolarization could involve electrical coupling through myo-endothelial junctions. Indeed, substances which produce endothelium-dependent hyperpolarization of vascular smooth muscle cells, also hyperpolarize endothelial cells, with the same time course (Busse *et al.*, 1988; Davies *et al.*, 1988; Brunet and Beny, 1989; Colden-Stanfield *et al.*, 1990; Chen and Cheung, 1992). However, if electrical coupling from smooth muscle to endothelial cells exists, electrical propagation in the reverse direction does not seem to occur (Marchenko and Sage, 1993; Bény and Gribi, 1989; Bény and Pacicca, 1994). Dye studies demonstrate that couplings between endothelial and smooth muscle cells do not necessarily occur (Beny, 1990). Even in tissues where heterocellular dye coupling between endothelial and smooth muscle cells is observed, hyperpolarization of endothelial cells is not propagated to the smooth muscle cells (Beny *et al.*, 1997). Furthermore, halothane or heptanol, agents which uncouple cells linked by gap junctions, do not inhibit endothelium-dependent hyperpolarizations (Bény and Pacicca, 1994; Bény and Chabaud, 1996; Zygmunt and Högestätt, 1996).

However, more specific blockers of gap junctions, such as 18-glycyrrhetinic acid and Gap27 (a peptide which possesses a conserved sequence homology with a portion of connexin) inhibit EDHF-like responses in rabbit and guinea-pig arteries (Chaytor *et al.*, 1998; Yamamoto *et al.*, 1998; Taylor *et al.*, 1998; Yamamoto *et al.*, 1999). At present, this mechanism needs to be further explored to better understand its potential contribution to EDHF-mediated responses. Gap junctions could be the site of myo-endothelial electrical cell coupling but could also be the site of transfer of low molecular weight compounds. Furthermore, the role of gap junctions between the innermost intimal vascular smooth muscle cells and the deeper layers has also to be clarified.

Thus, endothelium-dependent hyperpolarizations resistant to inhibitors of NO synthase and cyclooxygenase inhibition have been attributed to the release of a yet unidentified diffusable substance (EDHF), the existence of which has been demonstrated under bioassay conditions whereby the source of EDHF was either native vascular segments or cultured endothelial cells (Félétou and Vanhoutte, 1988; Kauser *et al.*, 1989; Chen *et al.*, 1991; Mombouli *et al.*, 1996; Popp *et al.*, 1996a; Harder *et al.*, 1996; Fukuta *et al.*, 1996). The technical difficulties in demonstrating the diffusable nature of EDHF can be explained by a very short half-life of the substance, its preferential abluminal release (Kauser *et al.*, 1992) the simultaneous release of a hypothetical endothelium-derived depolarizing factor (Mombouli *et al.*, 1996; Corriu *et al.*, 1996b) or a combination of these. The release of EDHF requires an increase in endothelial intracellular calcium and the subsequent acti-

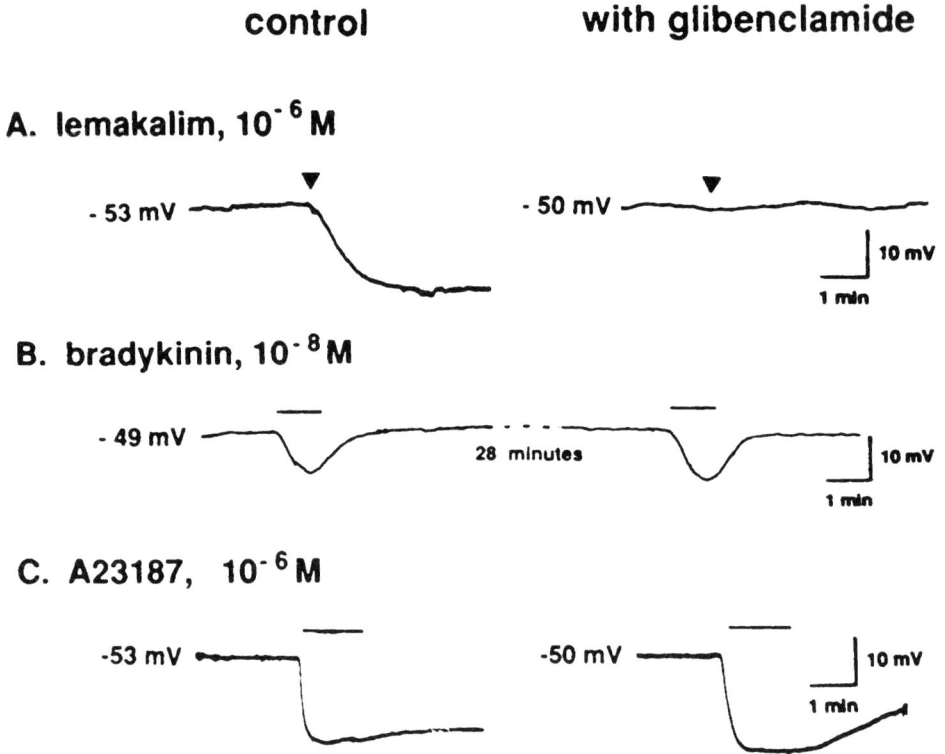

Figure 4.3 Membrane potential of vascular smooth muscle cells in human isolated coronary artery with endothelium in presence of L-nitro-arginine (100 µM) and indomethacin (10 µM) and perindoprilat (1 µM). Effects of glibenclamide on the changes in membrane potential produced by lemakalim (1 µM, A), bradykinin (10 nM, B) an calcium ionophore A 23187 (1 µM, C) on membrane potential. Tissues in A and C were obtained from the heart of a 49 year old alcoholic cardiomyopathy and B from a 62 year old idiopathic cardiomyopathy. From Nakashima *et al.* (1993, with the permission of the *American Heart Association*).

vation of calmodulin (Chen and Suzuki, 1990, Nagao *et al.*, 1992b, Illiano *et al.*, 1992). The calcium and calmodulin dependency of responses mediated by EDHF is similar to that observed with endothelial NO-dependent relaxation although the former appears to be more sensitive to calmodulin blockers than the latter (Bredt and Snyder, 1990; Illiano *et al.*, 1992).

It has been suggested that EDHF may be a short-lived metabolite of arachidonic acid produced through the cytochrome P450 monooxygenase pathway (Rubanyi and Vanhoutte, 1987, Komori and Vanhoutte, 1990). Inhibitors of this pathway inhibit endothelium-dependent vasodilator responses resistant to inhibitors of NO synthase and cyclooxygenase in the perfused heart and kidney of the rat and in isolated porcine and bovine coronary arteries (Bauersachs *et al.*, 1994; Hecker *et al.*, 1994; Fulton *et al.*, 1992 and, 1995). Some metabolites of arachidonic acid formed by the cytochrome P450 activate K^+ channels in vascular smooth muscle cells (Gebremedhin *et al.*, 1992, Hu and Kim, 1993). Muscarinic agonists induce not only endothelium-dependent relaxation and hyperpolarization of bovine coronary arterial smooth muscle but also the release of epoxyeicosatrienoic acids from bovine coronary arterial endothelial cells. These responses are inhibited by SKF 525A and miconazole (Campbell *et al.*, 1996). The cytochrome P450 metabolites, produced by the endothelial cells, increase the open-state probability of calcium-activated potassium channels sensitive to TEA or charybdotoxin, and induce hyperpolarization of coronary arterial smooth muscle cells. Taken in conjonction these observations support the hypothesis that epoxyeicosatrienoic acids could be EDHF. However, cytochrome P450 inhibitors studied at high concentrations, are notoriously unspecific and can inhibit hyperpolarizations induced by potassium channel openers such as levcromakalim (Graier *et al.*, 1995a; Eckman *et al.*, 1995; Edwards *et al.*, 1996; Vanhoutte and Félétou, 1996). In blood vessels of rats, guinea-pigs, dogs and pigs, chemically unrelated inhibitors of cytochrome P450 do not inhibit the EDHF responses or produce a non specific inhibition (Corriu *et al.*, 1996a & c; Graier *et al.*, 1996; Zygmunt *et al.*, 1996; Fukao *et al.*, 1997; Chataigneau *et al.*, 1998a).

Randall *et al.* (1996) postulated that anandamide is EDHF. This arachidonic acid derivative is an endogenous ligand for the cannabinoid CB_1 receptor (Devane *et al.*, 1992; Di Marzo *et al.*, 1994). Anandamide induces dilatation which mimics responses to EDHF in the isolated and perfused mesenteric and coronary arterial bed of the rat (Randall *et al.*, 1997; Randall and Kendall, 1997 and 1998). However, in the kidney dilatation caused by anandamide is due to the release of NO (Deutsch *et al.*, 1997). In isolated blood vessels from various species (pig, guinea-pig, rat) anandamide does not produce hyperpolarization or if it does so the underlying mechanism differs from EDHF-mediated responses. Indeed some of these responses to anandamide are endothelium-dependent (Zygmunt *et al.*, 1997b, Chataigneau *et al.* 1998b). Finally CB1 receptor antagonists do not inhibit endothelium-dependent hyperpolarization. These observations do not support the proposal that an endogenous cannabinoid is the major mediator of endothelium-dependent hyperpolarizations (Campbell *et al.*, 1997; Chataigneau *et al.*, 1997b, 1998b; Plane *et al.*, 1997; Zygmunt *et al.*, 1997; White and Hiley, 1997,).

In the rat hepatic artery, potassium ion could be EDHF (Edwards *et al.*, 1998). Indeed, endothelial cells hyperpolarize in response to neurohumoral substances which produce the release of vasoactive substances such as NO and EDHF. Potassium ions, flowing through the opening of endothelial small and intermediate conductance potassium channels (sensitive to charybdotoxin and apamin), may accumulate in the intercellular space. This rise in potassium concentration could hyperpolarize the smooth muscle by activating the inward rectifying potassium channels (sensitive to low concentrations of barium) and the sodium/potassium pump (sensitive to ouabain). However, in other blood vessels the addition of potassium does not necessarily produce hyperpolarization, possibly because the inward rectifying potassium channel is expressed poorly in certain vascular smooth muscle cells especially in those large blood vessels. Furthermore, in the guinea-pig carotid and porcine coronary arteries, the endothelium-dependent hyperpolarization is not affected by the combination of ouabain and barium (Quignard *et al.*, 1999). Further studies are required to verify the pertinence of this proposal and to determine the vascular beds where it applies (Vanhoutte, 1998).

Molecules such as carbon monoxide, hydroxyl radicals, hydrogen peroxide all are putative EDHF as they are produced by the endothelial cells and induce hyperpolarisation of the smooth muscle cells. However, the evidence conforting the role of these molecules as EDHF is either weak or inexistent (For review see Mombouli and Vanhoutte, 1997).

It has been recently suggested that different pathways could be involved in the EDHF responses. Thus, in the rabbit femoral artery, the relaxation attributed to EDHF could involve the activation of both heme-oxygenase- and cannabinoid CB_1 receptor antagonist-sensitive pathways (Rowe *et al.*, 1998). This proposition should be confirmed and explored in other vascular beds. At present, from the data available it is not possible to conclude that EDHF has been identified with certainty.

ENDOTHELIUM-DERIVED HYPERPOLARIZING FACTOR IN HUMANS

Studies which have specifically addressed EDHF-mediated responses in human blood vessels are scarce. The earliest electrophysiological evidence of an endothelium-dependent hyperpolarization in human blood vessels was obtained in isolated coronary arteries (Nakashima *et al.*, 1993). Bradykinin and the calcium ionophore A23187 induced a transient hyperpolarization which was resistant to inhibitors of nitric oxide synthase and cyclooxygenase, indicating the involvement of EDHF (Figure 4.3). This hyperpolarization was associated with a corresponding endothelium-dependent relaxation resistant to the same inhibitors. These earlier results in the human coronary artery have been confirmed (He, 1997a and b). Similarly, endothelium-dependent hyperpolarizations, associated with endothelium-dependent relaxation resistant to inhibitors of nitric oxide synthase and cyclooxygenase have been observed in human pial arteries in response to substance P (Petersson *et al.*, 1995) and in gastroepiploic arteries of different sizes in response to bradykinin and to a lesser extent to acetylcholine (Urakami-Harasawa *et al.*, 1997). Acetylcholine also induces endothelium-dependent hyperpolarizations in the human saphenous vein in presence of nitric oxide synthase and cyclooxygenase inhibitors (Yang and He, 1997).

Evidence for the presence of a functional role of EDHF comes from isolated human arteries in which endothelium-dependent relaxations resistant to inhibitors of nitric oxide synthase and cyclooxygenase were observed. This phenomenon has been reported in coronary (Nakashima *et al.*, 1993; Stork and Cocks, 1994; Kemp and Cocks, 1997; He, 1997a and b), subcutaneous (Woolfson and Poston, 1991; Deng *et al.*, 1995; Van de Voorde *et al.*, 1997), omental (Pascoal and Umans, 1996; Ohlmann *et al.*, 1997; Wallerstedt and Bodelsson, 1997), renal (Kessler *et al.*, 1996) and radial arteries (Hamilton *et al.*, 1997). By contrast, EDHF-dependent responses are small or inexistent in the internal thoracic artery (Hamilton *et al.*, 1997) and in the basilar artery (Hatake *et al.*, 1990).

As in animal arteries (Mügge *et al.*, 1991; Nagao *et al.*, 1992a; Shimokawa *et al.*, 1996), the contribution of the EDHF response is significantly greater in small than in large human arteries (Woolfson and Poston, 1990; Urakami-Harasawa *et al.*, 1997).

It has not been possible yet to evaluate, in the intact human, the involvement of EDHF in the vasodilator responses to various stimuli as specific inhibitors of its production or its action are not available. An EDHF mechanism is often suggested to explain vasodilatation resistant to inhibitors of nitric oxide synthase. However, numerous other interpretations are possible. First of all, most of the human studies do not involve the administration of an inhibitor of cyclooxygenase. Furthermore, complete blockade of nitric oxide synthase is difficult to obtain, and/or the non-endothelial effect of the vasodilators such as a direct effect on the smooth muscle cells or an inhibitory effect on the sympathetic nerve endings cannot be excluded easily. Therefore, the exact role of EDHF in the control of human blood vessel tone is still unknown.

Mechanism of Endothelium-Dependent Hyperpolarisation in Human Vascular Tissue

The EDHF-mediated responses (hyperpolarizations or/and relaxations resistant to inhibitors of nitric oxide synthase and cyclooxygenase) can be attributed to the activation of potassium channels as they are blocked by elevated potassium concentrations. These responses are independent of K-ATP activation (Figure 4.3) and are minimally or not at all affected by charybdotoxin, iberiotoxin or apamin individually but are blocked either by non-specific potassium channel inhibitors (tetrabutylammonium or tetraethylammonium) or by the combination of charybdotoxin plus apamin (Nakashima *et al.*, 1993; Kessler *et al.*, 1996; Pascoal and Umans, 1996; Ohlmann *et al.*, 1997; Petersson *et al.*, 1997; Urakami-Harasawa *et al.*, 1997; Wallerstedt and Bodelsonn, 1997). In small arteries dissected from gluteal fat, the partial inhibition observed with tolbutamide, a K-ATP inhibitor, is probably due to a non specific effect of the high concentration of the compound (Deng *et al.*, 1995). In the human hand vein, vasodilatations to substance P are partially resistant to inhibitors of nitric oxide synthase and cyclooxygenase but are partially inhibited by the sulfonylurea glyburide. Again, this effect may not be related to an inhibition of the EDHF pathway. Indeed, in the human vein the effect of substance P involves the activation of the cyclooxygenase pathway without apparent involvement of NO synthase. The effects of glyburide and acetylsalicylic acid are not additive suggesting that K-ATP could have been activated by the products of cyclooxygenase (Strobel *et al.*, 1996). Although the potassium conductance involved in the endothelium-dependent hyperpolarization in human blood vessels is not identified with precision, these limited data available are consistent with observations already made in other species.

Nature of EDHF in Human Arteries

Human endothelial cells hyperpolarize in response to mediators which induce endothelium-dependent responses (Groschner *et al.*, 1994; Ochi *et al.*, 1995). Since no data seem available concerning electrical coupling through myo-endothelial junctions in human blood vessels this phenomenon cannot be ruled out. However, a diffusable hyperpolarizing factor released from cultured human umbilical vein endothelial cells has been detected in the presence of combined treatment with inhibitors of NO-synthase and cyclooxygenase (Popp *et al.*, 1996b), demonstrating the production of an EDHF by human endothelial cells.

In intact human arteries, the effect of endogenous NO (or nitrovasodilators) on cell membrane potential is unknown. In cultured human pulmonary vascular smooth muscle cells, NO activates calcium-dependent potassium channels through a cyclic GMP-dependent mechanism (Peng *et al.*, 1996). This may indicate that NO provokes hyperpolarization of vascular smooth muscle in the pulmonary circulation. An incomplete inhibition of the NO-synthase by the inhibitors could explain the endothelium-dependent hyperpolarisation. However, this interpretation is unlikely at least in coronary arteries. In the presence of the combination of inhibitors of NO-synthase and cyclooxygenase plus oxyhemoglobin which scavanges NO, a component of the bradykinin-induced endothelium-dependent

A **B**

endothelium

without   bradykinin
           10^{-7} M   bradykinin
                                          10^{-7} M

- 47 mV ～～～～～～～   - 48 mV ―――――――

                    ⌐ 10 mV   ⌐ 10 mV
                    └ 1 min   └ 1 min

with
           10^{-7} M 10^{-7} M

- 45 mV ～～～＼／＼～   - 48 mV ―――――――

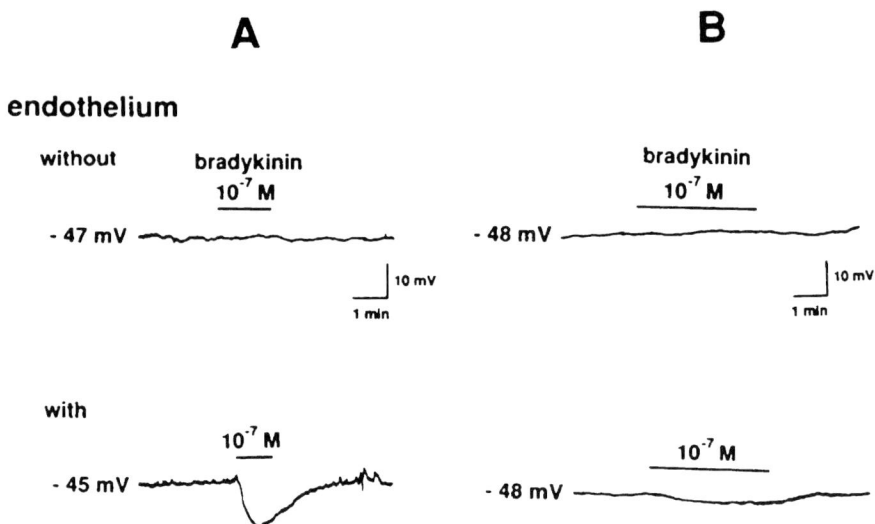

Figure 4.4 Membrane potential of vascular smooth muscle cells in human isolated coronary artery with (bottom traces) and without endothelium (top traces) in presence of L-nitro-arginine (100 μM) and indomethacin (10 μM). Effect of bradykinin in rings with and without endothelium. Tissues were obtained from the heart of a 15 month old (A, congenital heart disease) and from a 64 year old (B, ischemic heart disease). From Nakashima *et al.* (1996, by permission of Harwood Academic Publishers)

relaxation is still observed. This component is abolished by elevation of the extracellular potassium concentration (Kemp and Cocks, 1997).

The involvement of a cytochrome P450 metabolite of arachidonic acid has been ruled out in human coronary and omental arteries as quinacrine, proadifen and clotrimazole have no effect on the EDHF responses (Ohlmann *et al.*, 1997; Urakami-Harasawa *et al.*, 1997; Wallerstedt and Bodelsonn, 1997). In renal arteries the inhibitory effect of two anaesthetics, etomidate and thiopental, has been considered to be an indication of the involvement of cytochrome P450 (Kessler *et al.*, 1996). In cultured endothelial cells from the human umbilical vein a transferable β-naphtoflavone-inducible hyperpolarizing factor is synthesised (Popp *et al.*, 1996b). However, even without considering the non-specific effects of cytochrome P450 inhibitors already mentioned, activation of cytochrome P450 in human endothelial cells appears to be a more general requirement for increasing the intracellular calcium concentration and thus the release of endothelial derived factors such as NO and EDHF (Graier *et al.*, 1995b). This fundamental endothelial function of cytochrome P450 may confuse the issue when interpreting results of studies investigating the effects of inhibitors of cytochrome P450 on EDHF-mediated responses.

CONCLUSION

Besides the release of NO and prostacyclin, a third endothelial pathway involving hyper-polarization of the underlying vascular smooth muscle cells, contributes to the

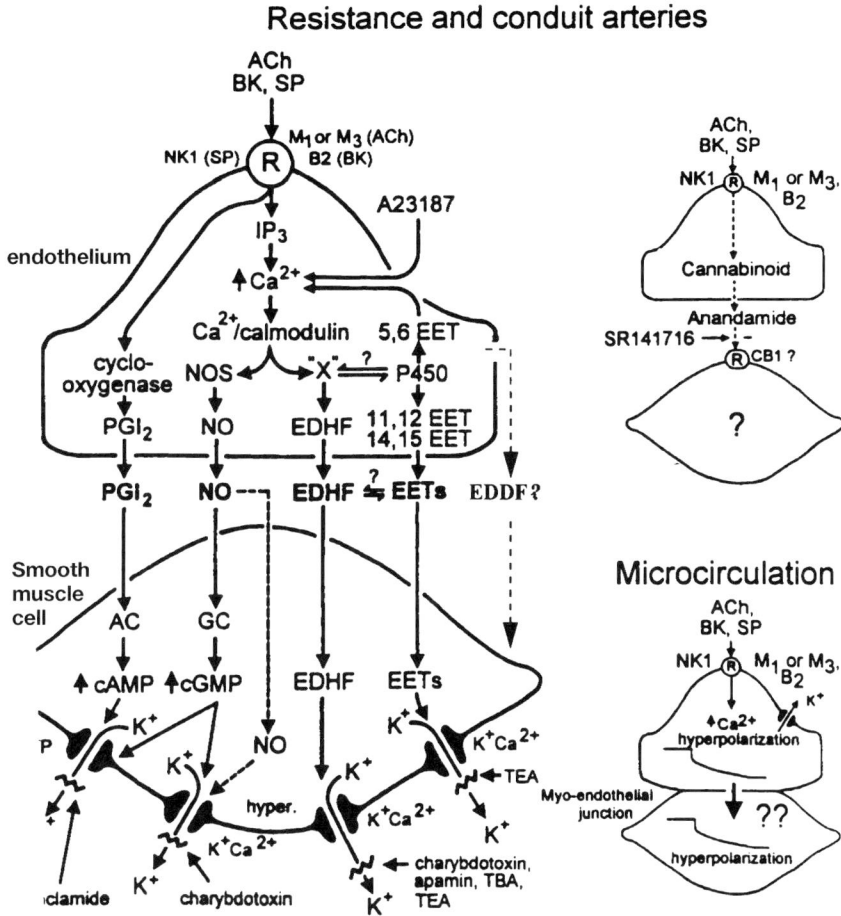

Figure 4.5 Conclusion: Existence of Multiple EDHF(s)?
M_3, B_2, NK_1: M_3 muscarinic, B_2 bradykinin and NK_1 neurokinin receptor subtypes respectively; R: receptor; A23187: calcium ionophore; NOS: nitric oxide synthase; COX: cyclooxygenase; cyt P450: cytochrome P450 monooxygenase; NO: nitric oxide; PGI_2: prostacyclin; EDHF endothelium-derived hyperpolarizing factor; EDDF: endothelium-derived depolarizing factor; 5,6 EET: 5,6-epoxy-eicosatrienoic acid; 11,12 EET: 11,12-epoxy-eicosatrienoic acid; 14,15 EET: 14,15-epoxy-eicosatrienoic acid; GC: guanylate cyclase, cGMP: cyclic guanosine monophosphate; cAMP: cyclic adenosinesine monophosphate; ATP: adenosine trisphosphate; IP_3: inositol trisphosphate; TEA: tetraethyl ammonium; TBA: tetrabutyl ammonium, SR 141716: an antagonist of cannabinoid CB1 receptor. The term EDHF should be restricted to describe phenomena in which endothelium-dependent hyperpolarizations are insensitive to glibenclamide and resistant to inhibitors of nitric oxide synthase and cyclooxygenase. Modified from Vanhoutte and Félétou (1996).

endothelium-dependent relaxation of various human blood vessels. Studies with animal blood vessels show that this phenomenon is due to a diffusible factor, termed endothelium-derived hyperpolarizing factor (EDHF), activating potassium channels. The identity of EDHF and the exact nature of the potassium channel population activated by it remain to be determined. In the absence of selective inhibitors of the synthesis or the action of

EDHF, its role cannot be evaluated fully in humans. The impairment of endothelium-dependent hyperpolarization by ageing (Figure 4.4) or hypercholesterolemia and conversely the enhancement of the responses attributed to EDHF by therapeutic agents such as converting enzyme inhibitors or possibly estrogens, may indicate a pathophysiological role of this third endothelial pathway (Nakashima *et al.*, 1993; Urakami-Harasawa *et al.*, 1997; Tagawa *et al.*, 1997), (Figure 4.5).

REFERENCES

Adeagbo, A.S.O. and Triggle, C.R. (1993) Varying extracellular [K+]: A functional approach to separating EDHF-and EDNO-related mechanisms in perfused rat mesenteric arterial bed. *J. Cardiovascular Pharmacol.*, **21**, 423–429.

Archer, S.L., Huang, J.M.C., Hampl, V., Nelson, D.P., Shultz, P.J. and Weir, E.K. (1994) Nitric oxide and cGMP cause vasorelaxation by activation of a charybdotoxin-sensitive K channel by cGMP-dependent protein kinase. *Proc. Natl. Acad. Sci. USA*, **91**, 7583–7587.

Bauersachs, J., Hecker, M., Busse, R. (1994) Display of the characteristics of endothelium-derived hyperpolarizing factor by a cytochrome P450–derived arachidonic acid metabolite in the coronary microcirculation. *Br. J. Pharmacol.*, **113**, 1548–1553.

Beny, J-L. (1990) Endothelial and smooth muscle cells hyperpolarized by bradykinin are not dye coupled. *Am. J. Physiol.*, **258**, H836–H841.

Beny, J-L., Brunet, P.C. (1988a). Neither nitric oxide nor nitroglycerin accounts for all the characteristics of endothelially mediated vasodilatation of pig coronary arteries. *Blood Vessels*, **25**, 308–311.

Beny, J-L., Brunet, P.C. (1988b) Electrophysiological and mechanical effects of substance P and acetylcholine on rabbit aorta. *J. Physiol.(London)*, **398**, 277–289.

Beny J-L., Chabaud, F. (1996) Kinins and endothelium-dependent hyperpolarization in porcine coronary arteries. In *Endothelium-Derived Hyperpolarizing Factor*, edited by P.M. Vanhoutte, pp. 41–51. Amsterdam: Harwood Academic Publishers.

Beny, J-L., Gribi, F. (1989) Dye and electrical coupling of endothelial cells in situ. *Tissue and Cell*, **21**, 797–802.

Beny, J-L., Paccica, C. (1994) Bidirectional electrical communication between smooth muscle and endothelial cells in the pig coronary artery. *Am. J. Physiol.*, **266**, H1465–1472.

Beny, J-L., Zhu, P.L., Haeflinger, I.O. (1997) Lack of bradykinin-induced smooth muscle hyperpolarization despite heterocellular dye coupling and endothelial cell hyperpolarization in porcine ciliary artery. *J. Vasc. Res.*, **34**, 344–350.

Bolotina, V.M., Najibi, S., Palacino, J.J., Pagano, P.J. Cohen, R.A. (1994) Nitric oxide directly activates calcium-dependent potassium channels in vascular smooth muscle cells. *Nature*, **368**, 850–853.

Bolton, T.B., Lang, R.J. and Takewaki, T. (1984) Mechanism of action of noradrenaline and carbachol on smooth muscle of guinea-pig anterior mesenteric artery. *J. Physiol.*, **351**, 549–572.

Brayden, J.E. (1990) Membrane hyperpolarization is a mechanism of endothelium-dependent cerebral vasodilation. *Am. J. Physiol.*, **259**, H668–H673.

Bredt, D.S., Snyder, S.H. (1990) Isolation of nitric oxide synthase, a calmodulin-requiring enzyme. *Proc. Natl. Acad. Sci. USA*, **87**, 682–685.

Brunet, P.C., Beny, J-L. (1989) Substance P and bradykinin hyperpolarize pig coronary artery endothelial cells in primary culture. *Blood Vessels*, **26**, 228–234.

Busse, R., Fichtner, H., Luckhoff, A., Kohlhardt, M. (1988) Hyperpolarization and increased free calcium in acetylcholine-stimulated endothelial cells. *Am. J. Physiol.*, **255**, H965–H969.

Campbell, W.B., Gebremedhin, D., Pratt, P.F., Harder, D.R. (1996) Identification of epoxyeicosatrienoic acids as endothelium-derived hyperpolarizing. *Circ. Res.*, **78**, 415–423.

Chataigneau T., Félétou M., Duhault J., Vanhoutte P. M. (1998a) Epoxyeicosatrienoic acids, potassium channnel blockers and endothelium-dependent hyperpolarization in the guinea-pig carotid artery. *Br. J. Pharmacol.*, **123**, 574–580.

Chataigneau T., Félétou, M., Thollon, C., Villeneuve, N., Vilaine, J-P., Duhault, J., Vanhoutte., P.M. (1998b) Cannabinoid CB₁ receptor and endothelium-dependent hyperpolarization in guinea-pig carotid, rat mesenteric and porcine coronary arteries. *Br. J. Pharmacol.*, **123**,968–974.

Chataigneau, T., Thollon, C., Iliou, J-P, Villeneuve, N., Feletou, M, Vilaine, J-P, Duhault, J. & Vanhoutte, P. M. (1997) Cannabinoid CB1 receptors and endothelium hyperpolarization in guinea-pig carotid, rat mesenteric and porcine coronary arteries. *J. Vasc. Res.,*, **34**, 11.

Chaytor, A.Y., Evens, W.H. and Griffith, T.M. (1998) Central role of heterocellular gap junction communication in endothelium-dependent relaxations of rabbit arteries. *J. Physiol. (London)*, **508**, 561–73.

Chen, G., Cheung, D.W. (1992) Characterization of acetylcholine-induced membrane hyperpolarization in endothelial cells. *Circ. Res.*, **70**, 257–263.

Chen, G., Suzuki, H. (1989) Some electrical properties of the endothelium-dependent hyperpolarization recorded from rat arterial smooth muscle cells. *J. Physiol.*, **410**, 91–106.

Chen, G., Suzuki, H. (1990) Calcium dependency of the endothelium-dependent hyperpolarization in smooth muscle cells of the rabbit carotid artery. *J. Physiol.*, **421**, 521–534.

Chen, G., Suzuki, H. and Weston, A.H. (1988) Acetylcholine releases endothelium-derived hyperpolarizing factor and EDRF from rat blood vessels. *Br. J. Pharmacol.*, **95**, 1165–1174.

Chen, G., Yamamoto, Y., Miwa, K. and Suzuki, H. (1991) Hyperpolarization of arterial smooth muscle induced by endothelial humoral substances. *Am. J. Physiol.*, **260**, H1888–H1892.

Cohen, R.A., Plane, F., Najibi, S., Huk, I., Malinski, T. And Garland, C.J. (1997) Nitric oxide is the mediator of both endothelium-dependent relaxation and hyperpolarization of the rabbit carotid artery. *Proc. Natl. Acad. Sci. USA*, **94**, 4193–4198

Colden-Stanfield, M., Schilling, W.P., Possani, L.D., Kunz, D.L. (1990) Bradykinin-induced potassium current in cultured bovine aortic endothelial cells. *J. Memb. Biol.*, **116**, 227–238.

Corriu, C., Félétou, M., Canet, E. and Vanhoutte, P.M. (1996a) Inhibitors of the cytochrome P450-monooxygenase and endothelium-dependent hyperpolarizations in the guinea-pig isolated carotid artery. *Br. J. Pharmacol.*, **117**, 607–610

Corriu, C., Félétou, M., Canet, E. and Vanhoutte P.M. (1996b) Endothelium-derived factors and hyperpolarizations of the isolated carotid artery of the guinea-pig. *Br. J. Pharmacol.*, **119**, 959–964.

Corriu, C., Félétou, M., Canet, E., Vanhoutte, P.M. (1996c) Inhibitors of the P-450 monooxygenase pathway do not prevent endothelium-dependent hyperpolarizations in the carotid artery of the guinea-pig. In *Endothelium-Derived Hyperpolarizing Factor*, edited by P.M. Vanhoutte, pp. 91–95. Amsterdam: Harwood Academic Publishers.

Cowan, C.L. and Cohen, R,A. (1991) Two mechanisms mediate relaxation by bradykinin of pig coronary artery: NO-dependent and independent responses. *Am. J. Physiol.*, **261**, H830–H835.

Cowan, C.L., Palacino, J.J., Najibi, S. and Cohen, R.A. (1993) Potassium channel-mediated relaxation to acetylcholine in rabbit arteries. *J. Pharmacol. Exp. Therap.*, **266**, 1482–1489.

Davies, P.F., Oleson, S.P., Clapham, D.E., Morel, E.M, Schoen, F.J. (1988) Endothelial communication: state of the art lecture. *Hypertension*, **11**, 563–572.

Deng, L-Y., Li, J-S., Schiffrin, E.L. (1995) Endothelium-dependent relaxation of small arteries from essential hypertensive patients: mechanisms and comparison with normotensive subjects and with responses of vessels from spontaneously hypertensive rats. *Clin. Sci.*, **88**, 611–622.

De Mey, J.G., Claeys, M. and Vanhoutte, P.M. (1982) Endothelium-dependent inhibitory effects of acetylcholine, adenosine triphosphate, thrombin and arachidonic acid in the canine femoral artery. *J. Pharmacol. Exp. Ther.*, **222**, 166–173.

Deutsch, D.G., Goligorsky, M.S., Schmid, P.C., Krebsbach, R.J., Schmid, H.H.O., Das, S.K., Dey, S.K., Arreaza, G., Thorup, C., Stefano, G. and Moore, L.C. (1997) Production and physiological actions of anandamide in the vasculature of the rat kidney. *J. Clin. Invest.*, **100**, 1538–1546.

Devane, W.A., Hanus, L., Breuer, A., Pertwee, R.G., Stevenson, L.A., Griffin, G., Gibson, D., Mandelbaum, A., Etinger, A. and Mechoulam, R. (1992). Isolation and structure of a brain constituent that binds to the cannabinoid receptor. *Science*, **258**, 1946–1949.

Di Marzo, V., Fontana, A., Cadas, H., Schinelli, S.,Cimino, G., Schwartz, J.C. and Piomelli, D. (1994). Formation and inactivation of endogenous cannabinoid anandamide in central neurons. *Nature*, **372**, 686–691.

Eckman, D.M., Hopkins, N.O. and Keff, K.D. (1995) Effects of inhibitors of cytochrome P450 patway on relaxation and hyperpolarization induced with acetylcholine and lemakalim. *Circulation*, **92**, i–751.

Edwards, G., Gora, K.A., Gardener, M.J. Garland, C.J. and Weston, A.H. (1998) K^+ is am endothelium-derived hyperpolarizing factors in rat arteries. *Nature*, **396**, 269–272.

Edwards, G., Zygmunt, P.M., Högestätt, E.D. and Weston, A.H. (1996). Effects of cytochrome P450 inhibitors on potassium currents in mechanical activity in rat portal vein. *Br. J. Pharmacol.,*, **119**, 691–701.

Félétou, M. and Vanhoutte, P.M. (1985) Endothelium-derived relaxing factor(s) hyperpolarize(s) coronary smooth muscle. *The Physiologist*, **48** :325.

Félétou, M., Vanhoutte, P.M. (1988) Endothelium-dependent hyperpolarisation of canine coronary smooth muscle. *Br. J. Pharmacol.*, **93**, 515–524.

Félétou, M. and Vanhoutte, P.M. (1996) Endothelium-derived hyperpolarizing factor. *Clin. Exp. Pharmacol. Physiol.*, **23**, 1082–1090.

Fujii, K., ominaga, M., Ohmori, S., Kobayashi, K., Koga, T., Takata, Y. and Fujishima, M. (1992) Decreased endothelium-dependent hyperpolarization to acetylcholine in smooth muscle of the mesenteric artery of spontaneously hypertensive rats. *Circ. Res.*, **70**, 660–669.

Fukao, M., Hattori, Y., Kanno, M., Sakuma, I. and Kitabatake, A. (1997). Evidence against a role of cytochrome P450–derived arachidonic acid metabolites in endothelium-dependent hyperpolarization by acetylcholine in rat isolated mesenteric artery. *Br. J. Pharmacol.*, **120**, 439–446.

Fukuta, H., Miwa, K., Hozumi, T., Yamamoto, Y. and Suzuki, H. (1996) Reduction by EDHF of the intracellular calcium concentration in vascular smooth muscle. In *Endothelium-Derived Hyperpolarizing Factor*, edited by P.M. Vanhoutte, pp. 143–153. Amsterdam: Harwood Academic Publishers.

Fulton, D, McGiff, J.C. and Quilley, J. (1992) Contribution of NO and cytochrome P_{450} to the vasodilator effect of bradykinin in the rat kidney. *Br. J. Pharmacol.*, **107**, 722–725.

Fulton, D., Mahboudi, K., Mcgiff, J.C. and Quilley, J. (1995) Cytochrome P450–dependent effects of bradykinin in the rat heart. *Br. J. Pharmacol.*, **114**, 99–102.

Furchgott, R.F. and Vanhoutte, P.M. (1989) Endothelium-derived relaxing and contracting factors. *FASEB J.*, **3**, 2007–2018.

Furchgott, R.F. and Zawadzki, J.V. (1980) The obligatory role of the endothelial cells in the relaxation of arterial smooth muscle by acetylcholine. *Nature*, **288**, 373–376.

Garcia-Pascual, A., Labadia, A., Jimenez, E. and Costa, G. (1995) Endothelium-dependent relaxation to acetyl-choline in bovine oviductal arteries : mediation by nitric oxide and changes in apamin-sensitive K+ conductance. *Br. J. Pharmacol.* **115**, 1221–1230.

Garland, C.J. and Mcpherson, G.A. (1992). Evidence that nitric oxide does not mediate the hyperpolarization and relaxation to acetylcholine in the rat small mesentery artery. *Br. J. Pharmacol.*, **105**, 429–435.

Garland, C.J. and Plane F. (1996) Relative importance of endothelium-derived hyperpolarizing factor for the relaxation of vascular smooth muscle in different arterial beds. In: *Endothelium-Derived Hyperpolarizing Factor*, Volume 1 (P.M. Vanhoutte, ed.), Harwood Academic Publishers, Amsterdam, pp.173–179.

Gebremedhin, D., Ma, Y.H., Falck, J.R., Roman, R.J., VanRollins, M. and Harder, D.R. (1992) Mechanism of action of cerebral epoxyeicosatrienoic acids on cerebral arterial smooth muscle. *Am. J. Physiol.*, **263**,H519–H525.

Graier, W.F., Holzmann, S., Hoebel, B.G. and Kukovetz, W.R. (1995a) L-N$^{\Omega}$-nitro-arginine resistant vessel relaxation is mediated via a pertussis toxin sensitive pathway but not via cytochrome P450 mono-oxygenase in bovine coronary arteries. *Circulation*, **92**, 751.

Graier, W.F., Holzmann, S., Hoebel, B.G., Kukovetz, W.R. and Kostner, G.M. (1996). Mechanisms of L-N^{G} nitroarginine/indomethacin-resistant relaxation in bovine and porcine coronary arteries. *Br. J. Pharmacol.*, **119**, 1177–1186.

Graier, W.F., Simecek, S. and Sturek, M. (1995b) Cytochrome P450 mono-oxygenase-regulated signalling of Ca^{2+} entry in human and bovine endothelial cells. *J. Physiol. (London)*, **482**, 259–274.

Groschner, K., Graier, W.F. and Kukovetz, W.R. (1994) Histamine induces K^{+}, Ca^{2+} and Cl^{-} currents in human vascular endothelial cells – Role of ionic currents in stimulation of nitric oxide biosynthesis. *Cir. Res.*, **75**, 304–314.

Hamilton, C.A., McIntyre, M., Williams, R., Berg, G., Reid, J.L. and Dominiczak, A.F. (1997) Vasorelaxation in response to bradykinin in human veins and arteries in vitro: Effect of ACE inhibitors. *J. Vasc. Res.*, **34**, S20.

Harder, D.R., Campbell, W.B., Gebremedhin, D. and Pratt, P.F. (1996) Biossay of a cytochrome P450–dependent endothelial-derived hyperpolarizing factor from bovine coronary arteries. In *Endothelium-Derived Hyper-polarizing Factor*, edited by P.M. Vanhoutte, pp. 73–81. Amsterdam: Harwood Academic Publishers.

Hasunuma, K., Yamaguchi, T., Rodman, D., O'Brien, R. and McMurtry, I. (1991). Effects of inhibitors of EDRF and EDHF on vasoreactivity of perfused rat lungs. *Am. J. Physiol.* **260**, L97–L104.

Hatake, K., Kakishita, E., Wakabayashi, I., Sakiyama, N. and Hishida, S. (1990) Effect of aging on endothelium-dependent vascular relaxation of isolated human basilar artery to thrombin and bradykinin. *Stroke*, **21**, 1039–1043.

He, G.W. (1997a) Coronary endothelial function in open heart surgery. *Clin. Exp. Pharmacol. Physiol.*, **24**, 955–957.

He, G.W. (1997b) Hyperkalemia exposure impairs EDHF-mediated endothelial function in the human coronary artery. *Ann. Thorac. Surg.*, **63**, 84–87.

Hecker, M., Bara, A.T., Bauersachs, J. and Busse, R. (1994) Characterization of endothelium-derived hyperpolarizing factor as a cytochrome P_{450}-derived arachidonic acid metabolite in mammals. *J. Physiol.*, **481**, 407–414.

Hu S, Kim HS. (1993) Activation of K^+ channel in vascular smooth muscles by cytochrome P450 metabolites of arachidonic acid. *Eur. J. Pharmacol.*, **230**, 215–221.

Huang, A.H., Busse, R. and Bassenge, E. (1988) Endothelium-dependent hyperpolarization of smooth muscle cells in rabbit femoral arteries is not mediated by EDRF (nitric oxide). *Naunyn-Schmiedeberg's Arch. Pharmacol.*, **338**, 438–442.

Illiano, S.C., Nagao, T. and Vanhoutte, P.M. (1992) Calmidazolium, a calmodulin inhibitor, inhibits endothelium-dependent relaxations resistant to nitro-L-arginine in the canine coronary artery. *Br. J. Pharmacol.*, **107**, 387–392.

Itoh, T., Seki, N., Suzuki, S., Ito, S., Kajikuri, J. and Kuriyama, H. (1992) Membrane hyperpolarization inhibits agonist-induced synthesis of inositol 1,4,5-trisphosphate in rabbit mesenteric artery. *J.Physiol.*, **451**, 307–328.

Kauser, K. and Rubanyi, G.M. (1992) Bradykinin-induced, nitro-L-arginine-insensitive endothelium-dependent relaxation of porcine coronary artery is not mediated by bioassayable substances. *J. Cardiovasc. Pharmacol.*, **20** (Suppl. 12), S101–104.

Kauser, K., Stekiel, W.J., Rubanyi, G.M. and Harder, D.R. (1989) Mechanism of action of EDRF on pressurized arteries: Effect on K+ conductance. *Circ. Res.*, **65**, 199–204.

Keef, K.D. and Bowen, S.M. (1989) Effect of ACh on electrical and mechanical activity in guinea pig coronary arteries. *Am. J. Physiol.*, **257**, H1096–H1103.

Kemp, B.K. and Cocks, T.M. (1997) Evidence that mechanisms dependent and independent of nitric oxide mediate endothelium-dependent relaxation to bradykinin in human small resisistance-like coronary arteries. *Br. J. Pharmacol.*, **120**, 757–762.

Kessler, P., Lischke, V. and Hecker, M. (1996) Etomidate and thiopental inhibit the release of endothelium-derived-hyperpolarizing factor in the human renal artery. *Anesthesiology*, **84**, 1485–1488.

Köhler, M., Hirschberg, B., Bond, C.T., Kinzie, J.M., Marrion, N.V., Maylie, J. and Adelman, J.P. (1996) Small-conductance, calcium-activated potassium channels from mammalian brain. *Science*, **273**, 1709–1714.

Komori, K., Lorenz, R.R. and Vanhoutte, P.M. (1988) Nitric oxide, ACh and electrical and mechanical properties of canine arterial smooth muscle. *Am. J. Physiol.*, **255**, H207–H212.

Komori, K. and Suzuki, H. (1987) Electrical responses of smooth muscle cells during cholinergic vasodilation in the rabbit saphenous artery. *Circ Res.*, **61**, 586–593.

Komori, K. and Vanhoutte PM. (1990) Endothelium-Derived Hyperpolarizing Factor. *Blood Vessels.*, **27** 238–245.

Marchenko, S.M. and Sage, S.O. (1993) Electrical properties of resting and acetylcholine-stimulated endothelium in intact rat aorta. *J. Physiol.*, **462**, 735–751.

Marchenko, S.M. and Sage S.O. (1996) Calcium-activated potassium channels in the endothelium – of intact rat aorta. *J. Physiol. (London)*, **492**, 53–60.

McPherson, G.A. and Angus, J.A. (1991) Evidence that acetylcholine-mediated hyperpolarization of the rat small mesenteric artery does not involve the K^+ channel opened by cromakalim. *Br. J. Pharmacol.*, **103**, 1184–1190

Miyoshi, H., Nakaya, Y. and Moritoki, H. (1994) Nonendothelial-derived nitric oxide activates the ATP-sensitive K channel of vascular smooth muscle cells. *FEBS*, **345**, 47–49.

Mombouli, J-V, Bissiriou, I. and Vanhoutte, P.M. (1996) Biossay of Endothelium-derived hyperpolarizing factor: is endothelium-derived depolarizing factor a confounding element? In *Endothelium-Derived Hyperpolarizing Factor*, edited by P.M. Vanhoutte, pp. 51–57. Amsterdam: Harwood Academic Publishers.

Mombouli, J.V., Illiano, S., Nagao, T. and Vanhoutte P.M. (1992) The potentiation of bradykinin-induced relaxations by perindoprilat in canine coronary arteries involves both nitric oxide and endothelium-derived hyperpolarizing factor. *Circ. Res.*, **71**, 137–144.

Mombouli, J-V. and Vanhoutte, P.M. (1997) Endothelium-derived hyperpolarizing factor(s): updating the unknown. *Trends Pharmacol. Sci.*, **18**, 252–256.

Mügge, A., Lopez, J.A.G., Piegors, D.J., Breese, K.R. and Heistad, D.D. (1991) Acetylcholine-induced vasodilatation in rabbit hindlimb *in vivo* is not inhibited by analogues of L-arginine. *Am. J. Physiol.*, **260**, H242–H247.

Murphy, M.E. and Brayden, J.E. (1995a) Nitric oxide hyperpolarization of rabbit mesenteric arteries via ATP-sensitive potassium channels. *J. Physiol.*, **486**, 47–58.

Murphy, M.E. and Brayden, J.E. (1995b) Apamin-sensitive K^+ channels mediate an endothelium-dependent hyperpolarization in rabbit mesenteric arteries. *J. Physiol.*, **489**, 723–734.

Nagao, T., Illiano, S.C. and Vanhoutte, P.M. (1992a) Heterogeneous distribution of endothelium-dependent relaxations resistant to N^G-nitro-L-arginine in rats. *Am. J. Physiol.*, **263**, H1090–H1094.

Nagao, T., Illiano, S.C. and Vanhoutte, P.M. (1992b) Calmodulin antagonists inhibit endothelium-dependent hyperpolarization in the canine coronary artery. *Br. J. Pharmacol.*, **107**, 382–386.

Nagao, T. and Vanhoutte, P.M. (1991). Hyperpolarization contributes to endothelium-dependent relaxations to acetylcholine in femoral veins of rats. *Am. J. Physiol.*, **261**,H1034–H1037.

Nagao, T. and Vanhoutte, P.M. (1992a) Hyperpolarization as a mechanism for endothelium-dependent relaxations in the porcine coronary artery. *J. Physiol.*, **445**, 355–367.

Nagao, T. and Vanhoutte, P.M. (1992b) Characterization of endothelium-dependent relaxations resistant to nitro-L-arginine in the porcine coronary artery. *Br. J. Pharmacol.*, **107**, 1102–1107.

Nakashima, M., Akata, T. and Kuriyama, H. (1992) Effects on the rabbit coronary artery of LP-805, a new type of releaser of endothelium-derived relaxaing factor and a K+ channel opener. *Cir. Res.*, **71**, 859–869.

Nakashima, M., Mombouli, J-V., Taylor, A.A. and Vanhoutte, P.M. (1993) Endothelium-dependent hyperpolarization caused by bradykinin in human coronary arteries. *J. Clin. Invest.*, **92**, 2867–2871.

Nakashima, M., Mombouli, J.V., Taylor, A.A. and Vanhoutte, P.M. (1996) Endothelium-dependent hyperpolarization in the human coronary artery. In *Endothelium-Derived Hyperpolarizing Factor*, edited by P.M. Vanhoutte, pp. 279–285. Amsterdam: Harwood Academic Publishers.

Nelson, M.T., Patlak, J.B., Worley, J.F. and Standen, N.B. (1990) Calcium channels, potassium channels, and voltage dependence of arterial smooth muscle tone. *Am J. Physiol.*, **259**, C3–C18.

Ochi, R., Yumoto, K., Watanabe, M. and Yamaguchi, H. (1995) Regulation of calcium signalling by K+ and Cl- currents in endothelial cells. *Heart Vessels*, **S9**, 80–82.

Ohashi, M. Satoh. K. and Itoh, T. (1999) Acetylcholine-induced membrane potential changes in endothelial cells of rabbit aortic valve. *Br. J. Pharmacol.*, **126**, 19–26.

Ohlmann, P., Martinez, M.C., Schneider, F., Stoclet, J.C. and Andriantsitohaina, R. (1997) Characterization of endothelium-derived relaxaing factors released by bradykinin in human resistance arteries. *Br. J. Pharmacol.*, **121**, 657–664.

Pacicca, C., von der Weid, P. and Beny, J.L. (1992) Effect of nitro-L-arginine on endothelium-dependent hyperpolarizations and relaxations of pig coronary arteries. *J. Physiol.*, **457**, 247–256.

Parkington, H.C., Tare, M., Tonta, M.A. and Coleman, H.A. (1993) Stretch revealed three components in the hyperpolarization of guinea-pig coronary artery in response to acetylcholine. *J. Physiol.*, **465**, 459–476.

Parkington, H.C., Tonta, M., Coleman, H. and Tare, M. (1995). Role of membrane potential in endothelium-dependent relaxation of guinea-pig coronary arterial smooth muscle. *J. Physiol.*, **484**, 469–480.

Pascoal, I.F. and Umans, J.G. (1996) Effect of pregnancy on mechanisms of relaxation in human omental microvessels. *Hypertension*, **28**, 183–187.

Peng, W., Hoidal, J.R. and Farrukh, I.S. (1996) Regulation of Ca^{2+}-activated K^+ channels in pulmonary vascular smooth muscle cells – Role of nitric oxide. *J. Appl. Physiol.*, **81**, 1264–1272.

Petersson, J., Zygmunt, P.M., Brandt, L. and Högestätt, E.D. (1995) Substance P-induced relaxation and hyperpolarization in human cerebral arteries. *Br. J. Pharmacol.*, **115**, 889–894.

Petersson, J., Zygmunt, P.M. and Högestätt, E.D. (1997) Characterization of the potassium channels involved in EDHF-mediated relaxation in cerebral arteries. *Br. J. Pharmacol.*, **120**, 1344–1350.

Plane, F., Holland, M., Waldron, G.J., Garland, C.J. and Boyle, J.P. (1997) Evidence that anandamide and EDHF act via different mechanisms in the rat isolated mesenteric arteries. *Br. J. Pharmacol.*, **121**, 1509–1511.

Plane, F., Pearson, T. and Garland, C.J. (1995) Multiple pathways underlying endothelium-dependent relaxation in the rabbit isolated femoral artery. *Br. J. Pharmacol.*, **115**, 31–38.

Popp, R., Bauersachs, J., Sauer, E., Hecker, M., Fleming, I. and Busse, R. (1996a) The cytochrome P450 monooxygenase pathway and nitric oxide-independent relaxations. In *Endothelium-Derived Hyperpolarizing Factor*, edited by P.M. Vanhoutte, pp. 65–73. Amsterdam: Harwood Academic Publishers.

Popp, R., Bauersachs, J., Sauer, E., Hecker, M., Fleming, I. and Busse, R. (1996b) A transferable, β-naphtoflavone-inducible, hyperpolarizing factor is synthesized by native and cultured porcine coronary endothelial cells. *J. Physiol (London)*, **497**, 699–709.

Quignard, J-F., Félétou, M., Thollon, C., Vilaine, J-P., Duhault, J. and Vanhoutte, P.M. (1999) Potassium ions and endothelium-derived hyperpolarizing factor in guinea-pig carotid and porcine coronary arteries. *Br. J. Pharmacol.*, in press

Randall, M.D., Alexander, S.P.H., Bennett, T., Boyd, E.A., Fry, J.R., Gardiner, S.M., Kemp, P.A., Mcculloch, A.I. and Kendall, D.A. (1996). An endogenous cannabinoid as an endothelium-derived vasorelaxant. *Biochem. Biophys. Res. Commun.*, **229**, 114–120.

Randall, M.D., Mcculloch, A.I. and Kendall, D.A. (1997). Comparative pharmacology of endothelium-derived hyperpolarizing factor and anandamide in rat isolated mesentery. *Eur. J. Pharmacol.*, **333**, 191–197.

Randall, M.D. and Kendall, D.A. (1997). Involvement of a cannabinoid in endothelium-derived hyperpolarizing factor-mediated coronary vasorelaxation. *Eur J. Pharmacol.*, **335**, 205–209.

Randall, M.D. and Kendall, D.A. (1998). Evidence for the involvement of potassium channels in anandamide-induced and EDHF-mediated vasorelaxations in rat isolated mesentery. *Br. J. Pharmacol.* In press.

Rees, D.D., Palmer, R.M.J., Hodson, H.F. and Moncada, S. (1989) A specific inhibitor of nitric oxide formation from L-arginine attenuates endothelium-dependent relaxation. *Br. J. Pharmacol.*, **96**, 418–424.

Richard, V., Tanner, F.C., Tschudi, M.R. and Lüscher, T.F. (1990) Different activation of L-arginine pathway by bradykinin, serotonin, and clonidine in coronary arteries. *Am. J. Physiol.*, **259**, H1433–H1439.

Robertson, B.E., Schubert, R., Hescheler, J. and Nelson, M.T. (1993). cGMP-dependent protein kinase activates Ca-activated K channels in cerebral artery smooth muscle cells. *Am. J. Physiol.*, **265**, C299–303.

Rowe, D.T.D., Garland, C.J. and Plane, F. (1998) Multiple pathways underlie NO-independent relaxation to the calcium ionophore A23187 in the rabbit isolated femoral arteries. *Br. J. Pharmacol.* In press.

Rubanyi, G.M. and Vanhoutte, P.M. (1987) Nature of endothelium-derived relaxing factor: Are there two relaxing mediators? *Circ Res.*, **61**(suppl II), II61–II67.

Shimokawa, H., Yasutake, H., Fujii, K., Owada, M.K., Nakaike, R., Fukumoto, Y., Takayanagani, T., Nagao, T., Egashira, K., Fujishima, M. and Takeshita, A. (1996) The importance of the hyperpolarizing mechanism increases as the vessel size decrease in endothelium-dependent relaxations in rat mesenteric circulation. *J. Cardiovasc. Pharmacol.*, **28**, 703–711.

Shin, J.H., Chung, S., Park, E.J., Uhm, D.Y. and Suh C.K. (1997) Nitric oxide directly activates calcium-activated potassium channels from rat brain reconstituted into planar lipid bilayer. *Febs Lett.*, **415**, 299–302.

Siegel, G., Stock, G., Schnalke, F. and Litza, B. (1987). Electrical and mechanical effects of prostacyclin in canine carotid artery. In: *Prostacyclin and its stable analogue iloprost*, edited by Gryglewski R.J., and Stock, G., pp. 143–149. Berlin Heidelberg: Springer-Verlag.

Stork, A.P. and Cocks, T.M. (1994) Pharmacological reactivity of human epicardial coronary arteries - Characterization of relaxation responses to endothelium-derived relaxing factor. *Br. J. Pharmacol.*, **113**, 1099–1104.

Strobel, W.M., Lüscher, T.F., Simper, D., Linder, L. and Haefeli, W.E. (1996) Substance P in human hand veins *in vivo*: tolerance efficacy, potency, and mechanism of venodilator action. *Clin. Pharmacol. Ther.*, **60**, 435–443.

Suzuki, H., Chen, G., Yamamoto, Y. and Miwa, K. (1992) Nitroarginine-sensitive and insensitive components of the endothelium-dependent relaxation in the guinea-pig carotid artery. *Jpn.J.Physiol.*, **42**, 335–347.

Tagawa, H., Shimokawa, H., Tagawa, T., Kuroiwa-Matsumoto, M., Hirooka, Y. and Takeshita, A. (1997) Short-term estrogen augments both nitric oxide-mediated and non-nitric oxide-mediated endothelium-dependent vasodilation in postmenopausal women. *J. Cardiovasc. Pharmacol.*, **30**, 481–488.

Taniguchi, J., Furukawa, K.I. and Shigekawa, M. (1993) Maxi K^+ channels are stimulated by cyclic guanosine monophosphate-dependent protein kinase in canine coronary artery smooth muscle cells. *Pflügers Arch. Eur. J. Physiol.*, **423**, 167–172.

Tare, M., Parkington, H.C., Coleman, H.A., Neild, T.O. and Dusting, G.J. (1990) Hyperpolarization and relaxation of arterial smooth muscle caused by nitric oxide derived from the endothelium. *Nature*, **346**, 69–71.

Taylor. H.J., Chaytor, A.T., Evans, W.H. and Griffith, T.M. (1998) Inhibition of the gap junctional component of endothelium-dependent relaxations in rabbit iliac artery by 18β-glycyrrhetinic acid. *Br. J. Pharmacol.*, **125**, 1–3.

Taylor, S.G., Southerton, J.S., Weston, A.H. and Baker, J.R.J. (1988) Endothelium-dependent effects of acetylcholine in rat aorta: a comparison with sodium nitroprusside and cromakalim. *Br. J. Pharmacol.*, **94**, 853–863.

Urakami-Harasawa, L., Shimokawa, H., Nakashima, M., Egashira, K. and Takeshita, A. (1997) Importance of endothelium-derived hyperpolarizing factor in human arteries. *J. Clin. Invest.*, **100**, 2793–2799.

Van de Voorde, J., Depypere, H. and Vanheel, B. (1997) The influence of pregnancy on endothelium-derived nitric oxide mediated relaxations in isolated human resistance vessels. *Fund. Clin. Pharmacol.*, **1**, 371–377.

Van de Voorde, J., Vanheel, B. and Leusen, I. (1992) Endothelium-dependent relaxation and hyperpolarization in aorta from control and renal hypertensive rats. *Circ. Res.*, **70**, 1–8.

Vanhoutte, P.M. (1998) An old-timer makes a come-back. *Nature*, **396**, 213–216.

Vanhoutte, P.M. and Félétou, M. (1996) Conclusion: Existence of Multiple EDHF(s)? In *Endothelium-Derived Hyperpolarizing Factor*, edited by P.M. Vanhoutte, pp. 303–307. Amsterdam: Harwood Academic Publishers.

Van Renterghem, C. and Lazdunski, M. (1992) A small conductance charybdotoxin sensitive, apamin resistant Ca^{2+} activated K^+ channel in aortic smooth muscle cells (A7r5 line and primary cultures) *Pflügers Arch.*, **420**, 417–423.

Wallerstedt, S.M. and Bodelsson, M. (1997) Endothelium-dependent relaxations by substance P in human isolated omental arteries and veins: relative contribution of prostanoids, nitric oxide and hyperpolarization. *Br. J. Pharmacol.*, **120**, 25–30.

White, R. and Hiley, C.R. (1997) A comparison of EDHF-mediated responses and anandamide-induced relaxations in the rat isolated mesenteric artery. *Br. J. Pharmacol.*, **122**, 1573–1584.

Woolfson, R.G. and Poston, L. (1990) Effect of NG-monomethyl-L-arginine on endothelium-dependent relaxation of human subcutaneous resistance arteries. *Clin. Sci. London.*, **79**, 273–278.

Woolfson, R.G. and Poston, L. (1991) Effect of ouabain on endothelium-dependent relaxation of human resistance arteries. *Hypertension*, **71**, 619–625.

Yamamoto, Y., Fukuta, H., Nakahira, Y. and Suzuki, H. (1998) Blockade by 18β-glycyrrhetinic acid of intercellular electrical coupling in guinea-pig arterioles. *J. Physiol. (London)*, **511**, 501–508.

Yamamoto, Y. Imaeda, K. and Suzuki, H. (1999) Endothelium-dependent hyperpolarization and intercellular electrical coupling in guinea-pig mesenteric arterioles. *J. Physiol. (London)*, **514**, 505–513.

Yamanaka, A., Ishikawa, T. and Goto, K. (1998) Characterization of endothelium-dependent relaxation independent of NO and postaglandins in guinea pig coronary artery. *J. Pharmacol. Exp. Ther.*, **285**(2), 480–489.

Yang, J-A. and He. G.W. (1997) Surgical preparation abolishes endothelium-derived hyperpolarizing factor-mediated hyperpolarization in the human saphenous vein. *Ann. Thorac. Surg.*, **63**, 429–433.

Zygmunt, P.M., Grundemar, L. and Högestätt, E.D. (1994). Endothelium-dependent relaxation resistant to N -nitro-L-arginine in the rat hepatic artery and aorta. *Acta Physiol. Scand.*, **152**, 107–114.

Zygmunt, P.M., Edwards, G., Weston; A.H., Davis, S.C. and Högestätt, E.D. (1996). Effects of cytochrome P450 inhibitors on EDHF-mediated relaxation in the rat hepatic artery. *Br. J. Pharmacol.*, **118**, 1147–1152.

Zygmunt, P.M., Edwards, G., Weston, A.H., Larsson, B. and Högestätt, E.D. (1997a). Involvement of voltage-dependent potassium channels in the EDHF-mediated relaxation of rat hepatic artery. *Br. J. Pharmacol.*, **121**, 141–149.

Zygmunt, P.M. and Högestätt, E.D. (1996) Endothelium-dependent hyperpolarization and relaxation in the hepatic artery of the rat. In *Endothelium-Derived Hyperpolarizing Factor*, edited by P.M. Vanhoutte, pp., 191–203. Amsterdam: Harwood Academic Publishers.

Zygmunt, P.M., Högestätt, E.D., Waldeck, K., Edwards, G., Kirkup A. J. and Weston, A.H. (1997b) Studies on the effects of anandamide in rat hepatic artery. *Br. J. Pharmacol.*, **122**, 1679–1686.

Zygmunt, P.M., Plane, F., Paulsson, M., Garland, C.J. and Högestätt, E.D. (1998) Interactions betweem endothelium-derived relaxing factors in the rat hepatic artery: focus on regulation of EDHF. *Br. J. Pharmacol.*, **124**, 992–1000.

C
O
N
T
R
A
C
T
I
O
N

R
E
L
A
X
A
T
I
O
N

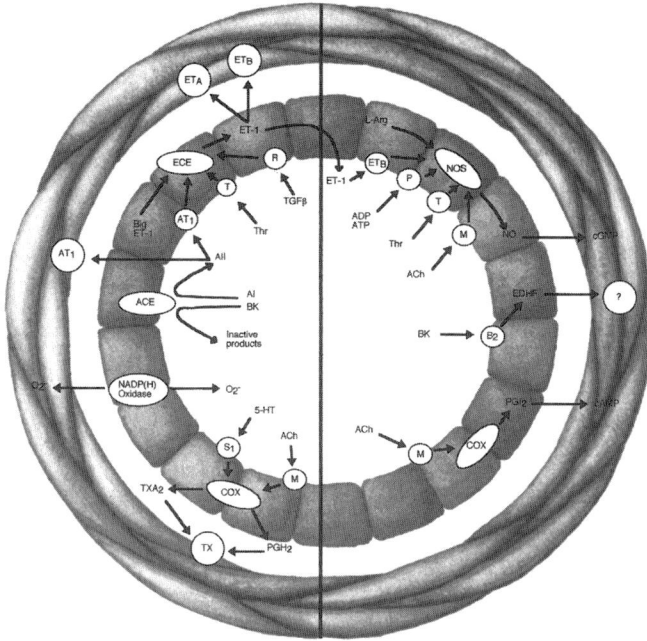

Vascular haemostasis defined as the ability of the vascular system to preserve blood fluidity and vascular integrity is maintained by the reciprocal interactions between the endothelium and blood cells. In physiology, platelets remain in a close vicinity of the endothelial lining, an arrangement supported by central position of erythrocytes in the blood stream. An accidental breach of the endothelial integrity triggers off homeostatic reactions aimed at blood arresting and vessel wall repair. Platelets adhere to the place of injury and form haemostatic aggregate. Platelet mediators such as adenosine diphosphate, thromboxane A_2 and matrix metalloproteinase-2 mediate platelet aggregation. The balancing effects of the inhibitor mediators including nitric oxide and prostacyclin prevent excessive haemostatic reactions (thrombosis). Leukocytes participate in homeostatic endothelial responses, an effect also regulated by the endothelial cells. Thus, the endothelial lining is the "maestro" of the vasculature playing a key role in regulation of vascular haemostasis.

Key words: Platelets, leukocytes, erythrocytes, haemostasis, thrombosis, nitric oxide

5 Regulation of Blood Cell Function by Endothelial Cells

Marek W. Radomski and Anna S. Radomski

Division of Research, Lacer, SA, 08025 Barcelona, Spain and Department of Pharmacology, University of Alberta, Edmonton, AB T6G 2H7, Canada

INTRODUCTION

Regulation of blood cell behaviour is a major endothelial function that ensures blood fluidity. The endothelium actively supports the fluid state of flowing blood and prevents activation of circulating cells. This ability must be carefully balanced against the homeostatic property of blood that is meant to protect the vascular system from loss of blood due to an accidental breach of vascular integrity. These diverse, yet co-ordinated, reactions are often referred to as vascular haemostasis. Thrombosis is a pathological extension of haemostasis that takes place when the regulatory mechanisms are inadequate.

The importance of the endothelial lining for preservation of blood fluidity is best exemplified by the fact that artificial vessel prostheses as well as pathologically altered endothelial cells are highly thrombogenic. In this chapter we will discuss the mechanisms responsible for non-thrombogenic properties of vascular endothelium.

REGULATION OF PLATELET FUNCTION BY ENDOTHELIUM

Platelets

Blood platelets are small (approximately 2 μm in diameter) anucleate elements formed by fragmentation of megakaryocytes. Non-activated (resting) platelets are discoid in shape and contain numerous granules in the cytoplasm. The granules are composed of various activating and proliferating agents whose function is crucial to platelet reactions (White 1988). Rheological studies have shown that pulsatile blood flow and the shear rate are the major determinants of platelet behaviour in vivo (Slack, Cui and Turitto, 1993). The shear stress is responsible for the tendency of suspended particles (blood elements) to move towards the centre of the flowing stream. Erythrocytes are larger and more numerous than platelets and tend to occupy the axial position in the blood stream forcing less numerous and smaller platelets to remain in a close proximity of the endothelial lining.

The endothelial proximity of platelets results in generation of high shear forces acting at the interface between platelet and the vessel wall. This assures, on one hand, prompt and effective platelet recruitment to the place of accidental injury during the haemostatic process, on the other hand, it allows tight endothelial control over the process of platelet activation.

Correspondence: Marek W. Radomski, Departemnt of Pharmacology, University of Alberta, 9–50 MSB, Edmonton T6G 2H7, Canada.

Platelet Activation

The biological signal for initiation of platelet activation is delivered by the exposure of adhesive components of the subendothelium that are normally concealed from the blood by the endothelium.

Platelet adhesion

Platelets make contact with adhesive proteins through specific adhesion receptors. In vivo, under conditions of shear stress, binding of v. Willebrand factor to its receptors on platelets and the counter receptors in the vessel wall, serves as a mechanism to anchor platelets to the subendothelium (Ginsberg, Loftus and Plow, 1988). Following the initial contact phase, platelets change their discoid shape to ameboid-like structures and spread on the endothelium.

Platelet aggregation

The biological role of aggregation is to reinforce the platelet adhesion monolayer with a structure based on web-like interactions between the adjacent platelets. The aggregate, thus formed, is firm enough to withstand disintegrating stimuli brought about by blood flow and shear forces (Figure 5.1).

   The formation of aggregates requires dramatic rearrangements of platelet structure and cytoskeleton and may be brought about by soluble activator agonists including thrombin, adrenaline, serotonin and ADP. These factors trigger a biochemical cascade of events that ultimately leads to the activation of the platelet integrin receptor IIb/IIIa and this allows binding of fibrinogen to the receptors of adjacent platelets. The binding of fibrinogen results in further reinforcement of the existing platelet plug (Radomski and Salas, 1995).

Platelet aggregation pathways

Dramatic reorganisation of platelet structure and formation of effective haemostatic plug is brought about by activation of at least three metabolic pathways that amplify aggregation and plug formation.

   The first pathway is mediated by the biosynthesis from arachidonic acid of thromboxane A_2 (Hamberg, Svensson and Samuelsson, 1975). The generation of this eicosanoid is inhibited by aspirin and aspirin-like drugs (Ferreira, Moncada and Vane, 1973), and this property of aspirin is now widely used to decrease excessive platelet activation that is often associated with vascular disorders including ischaemic heart disease and stroke (Patrono and Renda, 1997).

   The second pathway of aggregation is mediated by the release from platelet granules of ADP that activates platelets via specific receptors (Born, 1985; Puri and Colman, 1997). The pharmacological modulators of ADP-induced platelet aggregation include ticlopidine and dipyridamole (Schafer, 1996).

   We have recently investigated non-thromboxane, non-ADP mechanisms of platelet aggregation and found that the release from platelets of matrix metalloproteinase-2 enzyme (MMP-2) may mediate some of these reactions (Sawicki *et al.*, 1997). Platelet activation leads to the translocation of MMP-2 from the cytosol to the extracellular space. During

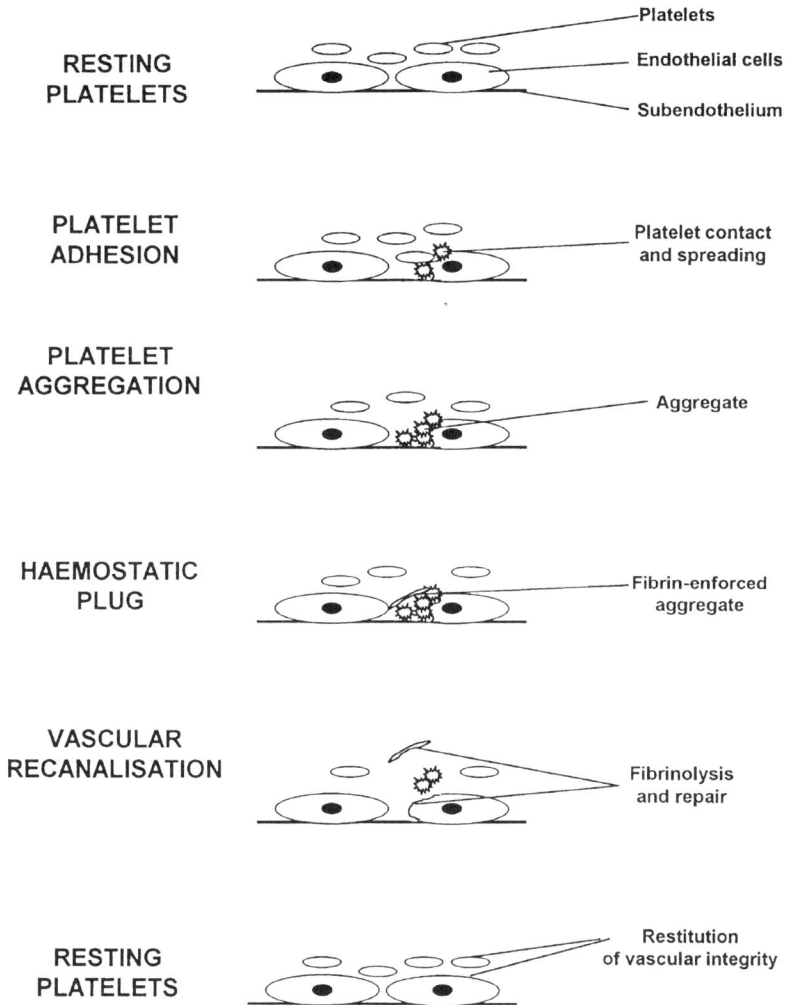

RESTING PLATELETS
Platelets
Endothelial cells
Subendothelium

PLATELET ADHESION
Platelet contact and spreading

PLATELET AGGREGATION
Aggregate

HAEMOSTATIC PLUG
Fibrin-enforced aggregate

VASCULAR RECANALISATION
Fibrinolysis and repair

RESTING PLATELETS
Restitution of vascular integrity

Figure 5.1   The physiology of platelet haemostasis.

the early stages of aggregation MMP-2 remains in close association with the platelet plasma membrane and this association is likely to stimulate platelet aggregation (Sawicki *et al.*, 1998) (Figure 5.2). Endogenous and pharmacological inhibitors of metalloproteinases inhibit MMP-2-induced platelet aggregation.

Physiological Control of Platelet Activation

An inherent capacity of platelets to aggregate and "seal the rents" in the vascular system is controlled by the systems that inhibit platelet activation. Under physiological conditions the balance between the activator and inhibitor systems of platelet activation process is

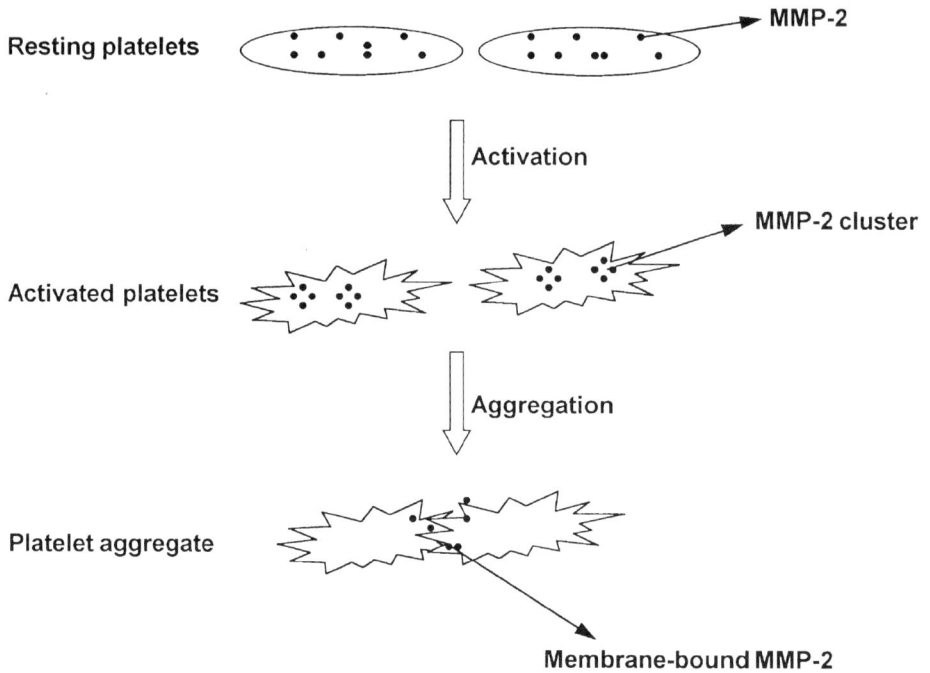

Figure 5.2   Activation-induced translocation of platelet MMP-2.

tightly maintained, preserving the fluid state of blood, thus vascular haemostasis and homeostasis. An imbalance between these groups of mediators may result in a disturbance of haemostasis exemplified by bleeding diathesis or thrombosis.

The endothelium is a major contributor to the inhibitor systems that control platelet activation.

Eicosanoids

In 1976 Vane's group reported that the endothelial cells synthesise prostacyclin, a potent but short-living inhibitor of platelet aggregation and stimulator of platelet disaggregation (Moncada *et al.*, 1976). In fact, prostacyclin can be considered as a biological opponent of thromboxane A_2 on platelets and the vessel wall, as prostacyclin causes vasodilatation and inhibition of aggregation, and these actions are antagonized by thromboxane A_2 (Bunting, Moncada and Vane, 1983).

Prostacyclin binds to its specific receptors present on platelets that are linked to the adenylate cyclase. Stimulation of prostacyclin receptors leads to increased accumulation of the intracellular cAMP and down-regulation of all pathways involved in amplification of platelet aggregation (Bunting, Moncada and Vane, 1983). Prostacyclin exerts little influence on the process of platelet adhesion to subendothelial components of the vessel wall (Radomski, Palmer and Moncada, 1987a).

Prostacyclin acts as a paracrine inhibitor of platelet activation. It is released close to the endothelial surface in response to stimulation with various vasoactive mediators including angiotensins and bradykinin (Nowak *et al.*, 1981). Platelets themselves lack the capacity to synthesise prostacyclin, however, they may contribute to the endothelial synthesis of this eicosanoid by generating and releasing the arachidonic acid cyclic endoperoxides that may be taken up by the endothelial cells for prostacyclin synthesis (Bunting, Moncada and Vane, 1983).

Some of lipoxygenase metabolites of polyunsaturated fatty acids including 12-hydroperoxy and 13-hydroxy derivatives of arachidonic and linoleic acids, respectively, have been shown to inhibit the process of platelet activation (Aharony, Smith and Silver, 1981; Buchanan and Brister, 1991). The precise mode of action of these metabolites on platelets is not known.

Nitric oxide

Nitric oxide (NO) accounts for the vasodilator activity of endothelium-derived relaxing factor (EDRF), a non-prostaglandin vasorelaxant substance first described in the endothelial cells by Furchgott and Zawadzki (1980).

In 1987 we found that cultured endothelial cells released NO to inhibit platelet adhesion, aggregation and cause dis-aggregation of pre-formed platelet aggregates (Radomski, Palmer and Moncada, 1987a-d). Nitric oxide is a gaseous mediator synthesised from L-arginine by the family of isoformic enzymes termed NO synthases (NOS). To date three isoforms eNOS, iNOS and nNOS have been identified (Radomski, 1995). Although cDNAs for the respective proteins are found almost in all mammalian cells, under physiological conditions, eNOS is a major NOS isoform expressed in the endothelial cells (Radomski, 1995). In contrast, inflammation and cell damage are often associated with the expression of iNOS (Radomski, 1995).

Nitric oxide is generated and released from the endothelial cells both under basal and agonist-stimulated conditions. Shear stress and pulsatile flow are major stimuli that cause release of NO under basal conditions (Cooke *et al.*, 1991). These physical forces are also responsible for the rheological arrangement of platelets in the flowing blood close to the surface of endothelial cells (see above). It is, therefore, hardly surprising that the endothelial NO regulates both adhesion and aggregation of platelets in vivo acting locally as a paracrine mediator (Yao *et al.*, 1992).

In 1990, we also discovered that NO can exert an autocrine influence on platelet function (Radomski, Palmer and Moncada, 1990). This is due to the intraplatelet NOS that is regulated by platelet activation generating amounts of NO sufficient for down-regulation of platelet recruitment to the site of the endothelial injury as well as platelet aggregation (Radomski, Palmer and Moncada, 1990; Bode-Boger *et al.*, 1998; Freedman *et al.*, 1997).

Ecto-nucleotidase ATP-diphosphohydrolases (ADP-ases)

These enzymes act as biological opponents of ADP, as they metabolise ADP to AMP and adenosine (Marcus and Safier, 1993). Adenosine can per se inhibit platelet activation contributing to the overall platelet-inhibitory action of ADP-ases.

Figure 5.3 Platelet haemostasis is regulated as a tightrope balance between the activator and inhibitor mediator systems.

PHYSIOLOGICAL MODULATION OF PLATELET BEHAVIOUR BY THE ENDOTHELIUM: THE INTERACTIONS BETWEEN MEDIATORS

Vascular haemostasis is mediated by the interactions between mediators generated in the endothelium-blood cell microenvironment. Physiological platelet haemostasis is best viewed as a tightrope balance between mediators. Both interactions between the activator and inhibitor mediator systems (Figure 5.3) and the interactions within the groups of mediators (Figure 5.4 and 5) are likely to occur.

Interactions between the Activator and Inhibitor Mediator Systems

System interactions

The best example of the interactions between the activator and inhibitor systems controlling platelet activation is the generation and release of NO and prostacyclin by platelet-aggregating agents such as thrombin (Yang *et al.*, 1994). The biological significance of this generation is to down-regulate the extent of the activator response.

Biological opponents

Interestingly, some of the antagonistic pairs of mediators share the same biosynthetic pathway. Prostacyclin acts as a biological opponent of thromboxane A_2 on aggregation and vessel wall reactivity and both eicosanoids are synthesised by cyclooxygenase via the intermediate stage of cyclic endoperoxides. At this stage the pathways separate and prostacyclin and thromboxane are generated by prostacyclin synthase and thromboxane synthase, respectively (Hamberg, Svensson and Samuelsson, 1975; Moncada *et al.*, 1976). This is why inhibition of this enzyme by cyclooxygenase inhibitors such as aspirin and aspirin-like drugs decrease the levels of both eicosanoids. It has been proposed, however,

Activator Mediators

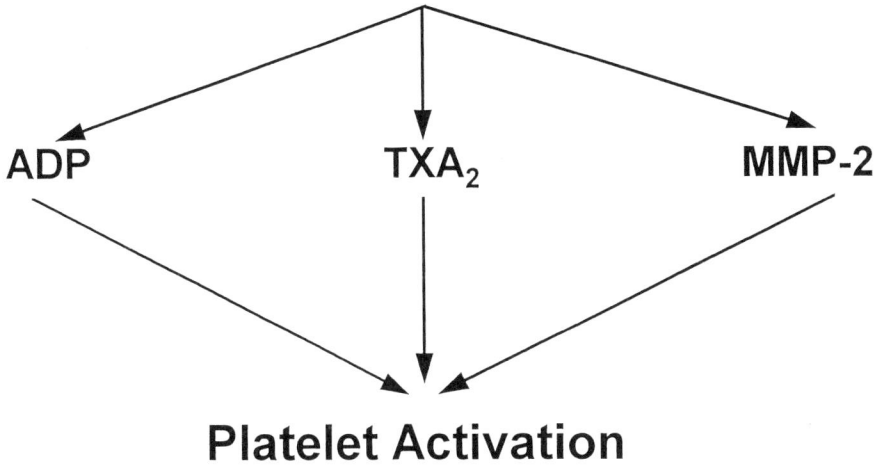

ADP   TXA$_2$ MMP-2

Platelet Activation

Figure 5.4   The activator pathways of platelet aggregation.

Inhibitor Mediators

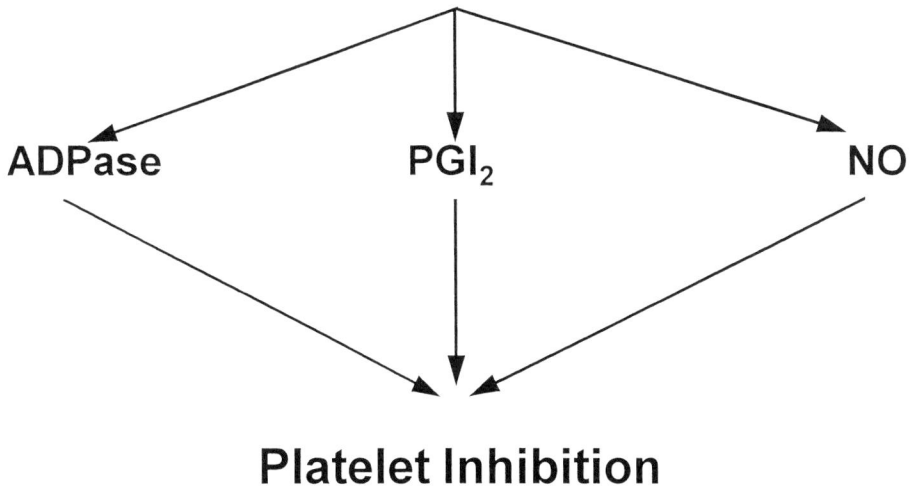

ADPase   PGI$_2$ NO

Platelet Inhibition

Figure 5.5   The inhibitor pathways of platelet aggregation.

that lower doses of aspirin may more selectively inhibit generation of thromboxane A_2 in platelets without affecting generation of prostacyclin in the endothelial cells (Moncada, 1982). The importance of prostacyclin as a biological opponent of thromboxane is further emphasised by the fact that the antithrombotic activity of thromboxane synthase inhibitors may be prostacyclin-dependent (De Clerck *et al.*, 1990).

Peroxynitrite is a potent oxidant that is generated from NO and superoxide as a result of rapid non-enzymatic reaction (Beckman and Tsai, 1994). Peroxynitrite opposes the platelet inhibitor and vessel wall relaxant effects of NO (Moro *et al.*, 1994; Villa *et al.*, 1994). In this context it is worth to emphasise that superoxide dismutase that scavenges superoxide inhibits platelet aggregation (Salvemini *et al.*, 1989).

Synergistic interactions

Synergy of two or more pharmacological agents takes place when their combined effect exceeds the sum of individual effects. In platelets, a synergistic enhancement of platelet aggregation has been described during interactions involving ADP and thromboxane (Grant and Scrutton, 1980). Moreover, subthreshold amounts of NO potentiate the platelet-inhibitory effects of prostacyclin (Radomski, Palmer and Moncada, 1987d). Inhbition of cGMP-inhibited cAMP phosphodiesterase by NO has been proposed to account for the synergistic interactions between these two mediators (Maurice and Haslam, 1990).

REGULATION OF LEUKOCYTE FUNCTION BY ENDOTHELIUM

Leukocytes and Leukocyte Recruitment

Although less numerous than platelets, leukocytes are nucleate blood cells that are several times larger in size than platelets. The major role of blood leukocyte system is to participate in various inflammatory and immune reactions including the response to the endothelial injury. This response of leukocytes is known as leukocyte recruitment.

Leukocyte recruitment

This is the multistep cascade of events that leads to leukocyte presence at the site of injury. Whereas platelet recruitment occurs both at the arterial and venous vasculature, leukocyte recruitment is thought to take place mainly in the postcapillary venules (Perry and Granger, 1991).

The initiating signal for the leukocyte recruitment may be delivered by various factors including histamine and thrombin, however the subsequent steps of this process appear to be similar (Kubes, 1995). Blood leukocytes travelling in flowing blood at a relatively high speed make an initial contact with the endothelial cell lining (tethering), and then move along the endothelium at greatly reduced velocity by a process described as rolling.

Both leukocyte tethering and rolling are believed to be mediated by up-regulation of the selectin family of adhesion receptors involving the reactions dependent on P-, L- and E-selectin (Kubes, 1995). Once leukocytes begin to roll, they can then firmly adhere and finally emigrate out of the vasculature. The firm adhesion is mediated by leukocyte integrins, particularly by the CD11/CD18 receptor (Kishimoto and Anderson, 1992). The

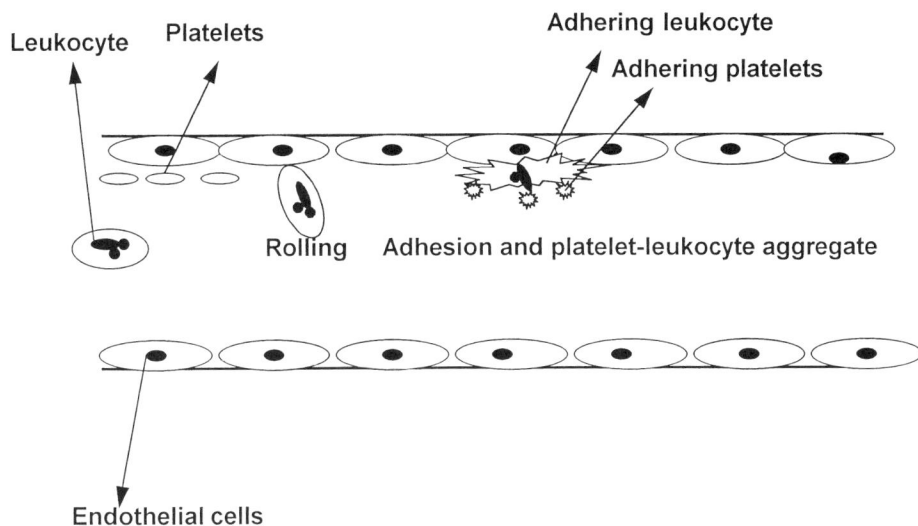

Figure 5.6   The leukocyte recruitment in the postcapillary venules.

factors that trigger leukocyte recruitment also cause increased expression of endothelial selectins (Kubes, 1995), and blockade of these receptors by monoclonal antibodies down-regulates the cascade of leukocyte adhesion to the endothelial lining. Interestingly, some of leukocyte recruitment appears to be preceded by mast cell activation, histamine release and activation of P-selectin (Gaboury *et al.*, 1995).

Leukocyte recruitment is often associated with formation of leukocyte-platelet aggregates (Figure 5.6). These interactions appear to be P-selectin-dependent (McEver, 1991).

Regulation of leukocyte recruitment

The studies using inhibitors of NOS and NO donors have shown that NO may play a crucial role in regulation of leukocyte recruitment. Indeed, inhibition of NOS with L-NAME increased leukocyte rolling and adhesion, both effects attenuated by cGMP analogues and NO donors (Kubes, Suzuki and Granger, 1991). L-NAME-induced increase in leukocyte adhesion may involve the release of chemoattractants such as platelet activating factor and leukotriene B_4 (Kubes, 1995).

THE INTERACTIONS BETWEEN RED CELLS AND THE ENDOTHELIUM

Oxygen-carrying function is the main physiological task of red cells that has direct impact on the endothelial function. Recently, it has been suggested that in addition to oxygen, red cell haemoglobin may serve as a carrier molecule for NO in the nitrosohaemoglobin complex (Stamler *et al.*, 1997)

Red Cells and Shear Stress

The importance of red cells for shear stress including changes in platelet function is best exemplified by the facts that anaemia (characterised by low red cell number and/or erythrocyte dysfunction) and polycythaemia (characterised by increased number of erythrocytes) affect haemostasis. Indeed, bleeding is an important characteristic of anaemia and thrombosis often complicates polycythaemia (Pearson, 1987).

Adhesion of Red Cells to the Endothelial Lining

As discussed before, under physiological conditions red cells occupy central position in the blood stream and have limited opportunity to interact directly with the endothelial lining. However, erythrocytes altered by the diseased processes such as sickle cell anaemia show increased adhesiveness to the endothelial cells (Wick and Eckman, 1996). This increase is probably mediated via activation of interactions with adhesion proteins including high molecular weight von Willebrand factor and thrombospondin (Wick and Eckman, 1996).

CONCLUSIONS

Research over the past 25 years has revolutionised the views on the role of endothelial cells in regulation of vessel wall reactivity and haemostasis. It is now clear that the endothelial cells are the "maestro" of the vasculature and play a key role in regulation of vascular haemostasis.

ACKNOWLEDGEMENTS

This work was supported by a grant from Medical Research Council of Canada. MWR is a Scholar of the Heritage Foundation for Medical Research.

REFERENCES

Aharony, D., Smith, J.B. and Silver, M.J. (1981) Regulation of arachidonate-induced platelet aggregation by the lipoxygenase product, 12-hydroperoxyeicosatetraenoic acid. *Biochem. Biophys. Acta*, **718**, 193–200.

Beckman, J. and Tsai, J.H. (1994) Reactions and diffusion of nitric oxide and peroxynitrite. *The Biochemist*, **16**, 8–10.

Bode-Boger, S.M., Boger, R.H., Galland, A. and Frolich, J.C. (1998) Differential inhibition of human platelet aggregation and thromboxane A_2 formation by L-arginine in vivo and in vitro *Naunyn-Schmiedeberg,s Arch Pharmacol*, **357**, 143–150.

Born, G.V. (1985). Adenosine diphosphate as a mediator of platelet aggregation in vivo: an editorial view. *Circulation*, **72**, 741–746.

Buchanan, M.R. and Brister, S.J. (1991) Anithrombotics and the lipoxygenase pathway. In *Antithrombotics* edited by A.G. Herman, pp. 159–180. Dordrecht, Kluwer Academic Publishers

Bunting, S., Moncada, S. and Vane, J.R. (1983). The prostacyclin-thromboxane A_2 balance: Pathophysiological and therapeutic implications *Br. Med. Bull.*, **39**, 271–276.

Cooke, J.P., Rossitch Jr, E., Andon, N.A., Loscalzo, J. and Dzau, V.J. (1991) Flow activates an endothelial potassium channel to release an endogenous nitrovasodilator. *J. Clin. Invest.*, **88**, 1663–1671.

De Clerck, F., Van Gorp, L., Beetens, J., Verheyen, A. and Janssen, P.A. (1990). Arachidonic acid metabolites, ADP and thrombin modulate occlusive thrombus formation over extensive arterial injury in the rat. *Blood Coagulation & Fibrinolysis*, **1**, 247–258.

Ferreira, S.H., Moncada, S. and Vane J.R. (1973) The blockade of local generation of prostaglandins explains the analgesic action of aspirin. *Agents & Actions*, **3**, 385.

Freedman, J.E., Loscalzo, J., Barnard, M.R., Alpert, C., Keaney, J.F., Jr and Michelson, A.D. (1997) Nitric oxide released from activated platelets inhibits platelet recruitment. *J. Clin. Invest.*, **100**, 350–356.

Furchgott, R.F. and Zawadzki, J.V. (1980) The obligatory role of endothelial cells in the relaxation of arterial smooth muscle by acetylcholine. *Nature*, **288**, 373–376.

Gaboury, J.P., Johnston, B., Niu, X-F. and Kubes, P. (1995) Mechanisms underlying acute mast cell-induced leukocyte rolling and adhesion in vivo. *J. Immunol.*, **154**, 804–813.

Ginsberg, M.H., Loftus, J.C. and Plow, E.F. (1988) Cytoadhesins, integrins and platelets. *Thromb. Haemost.*, **59**, 1–6.

Grant, J.A. and Scrutton, M.C. (1980) Positive interaction between agonists in the aggregation response of human blood platelets: interaction between ADP, adrenaline and vasopressin. *Br. J. Haematol.*, **44**, 109–125.

Hamberg, M., Svensson J. and Samuelsson, B. (1975) Thromboxanes: a new group of biologically active compounds derived from prostaglandin endoperoxides. *Proc. Natl. Acad. Sci. USA*, **72**, 2994–2998.

Kishimoto, T.K. and Anderson, D.C. (1992) The role of integrins in inflammation. In *Inflammation: basic principles and clinical correlates* edited by J.I. Gallin, Goldstein, I.M., Snyderman, R. Pp. 353–406, New York, Raven Press.

Kubes, P. (1995) Nitric oxide: a homeostatic regulator of leukocyte-endothelial cell interactions. In *Nitric oxide: a modulator cell-cell interactions in the microcirculation*, edited by P. Kubes, pp. 19–42. Austin, Texas, R.G. Landes.

Kubes, P., Suzuki, M. and Granger, D.N. (1991) Nitric oxide: an endogenous modulator of leukocyte adhesion. *Proc. Natl. Acad. Sci. USA*, **88**, 4651–4655.

Marcus, A.J. and Safier, L.B. (1993) Thromboregulation: multicellular modulation of platelet reactivity in hemostasis and thrombosis. *FASEB J.*, **7**, 516–520.

Maurice, D.H. and Haslam, R.J. (1990) Molecular basis of the synergistic inhibition of platelet function by nitrovasodilators and activators of adenylate cyclase: inhibition of cyclic AMP breakdown by cyclic GMP. *Mol. Pharmacol.*, **37**, 671–681.

McEver, R.P. (1991) GMP-140: a receptor for neutrophils and monocytes on activated platelets and endothelium. *J. Cell Biochem.*, **45**, 156–161.

Moncada, S. (1982) Biological importance of prostacyclin. *Br. J. Pharmacol.*, **76**, 3–31.

Moncada, S., Gryglewski, R.J., Bunting, S. and Vane, J.R. (1976) An enzyme isolated from arteries transforms prostaglandin endoperoxides to an unstable substance that inhibits platelet aggregation *Nature*, **263**, 663–665.

Moro, M.A., Darley-Usmar, V., Goodwin, D.A., Read, N.G., Zamora-Pino, R., Feelisch, M. *et al.* (1994). Paradoxical fate and biological action of peroxynitrite on human platelets. *Proc. Natl. Acad. Sci. USA*, **91**, 6702–6706.

Nowak, J., Radomski, M., Kaijser, L. and Gryglewski, R.J. (1981) Conversion of exogenous arachidonic acid to prostaglandins in the pulmonary circulation in vivo. A human and animal study. *Acta Physiol. Scand.*, **112**, 405–411.

Patrono C. and Renda, G. (1997) Platelet activation and inhibition in unstable coronary syndroms *Am. J. Cardiol.*, **80**, 17E-20E.

Pearson, T. (1987) Rheology of the absolute polycythaemias *Baillieres Clinical hematology*, **1**, 637–664.

Perry, M.A. and Granger, D.N. (1991) Role of CD11/CD18 in shear rate-dependent leukocyte-endothelial cell interactions in cat mesenteric venules. *J. Clin. Invest.*, **87**, 1798–1804.

Puri, R.N. and Colman, R.W. (1997) Immunoaffinity method to identify aggregin, a putative ADP receptor in human blood platelets. *Arch. Biochem. Biophys.*, **347**, 263–270.

Radomski, M.W. (1995) Nitric oxide-biological mediator, modulator and effector molecule. *Ann. Med.*, **27**, 321–330.

Radomski, M.W., Palmer, R.M.J. and Moncada, S. (1987a) The role of nitric oxide and cGMP in platelet adhesion to vascular endothelium. *Biochem. Biophys. Res. Commun.*, **148**, 1482–1489.

Radomski, M.W., Palmer, R.M.J. and Moncada, S. (1987b) Endogenous nitric oxide inhibits human platelet adhesion to vascular endothelium. *Lancet*, **2**, 1057–1058.

Radomski, M.W., Palmer, R.M.J. and Moncada, S. (1987c) Comparative pharmacology of endothelium-derived relaxing factor, nitric oxide and prostacyclin in platelets. *Br. J. Pharmacol.*, **92**, 181–187.

Radomski, M.W., Palmer, R.M.J. and Moncada, S. (1987d) The anti-aggregating properties of vascular endothe-lium: interactions between prostacyclin and nitric oxide. *Br. J. Pharmacol.*, **92**, 639–646.

Radomski, M.W., Palmer, R.M.J. and Moncada, S. (1990). An L-arginine/nitric oxide pathway present in human platelets regulates aggregation. *Proc. Natl. Acad. Sci. USA*, **87**, 5193–5197.

Radomski, M.W. and Salas, E. (1995). Biological significance of nitric oxide in platelet function. In *Nitric oxide: a modulator of cell-cell interactions in the microcirculation* edited by P. Kubes, pp. 43–74. Heidelberg, Springer Verlag.

Salvemini, D., de Nucci, G., Sneddon, J.M. and Vane, J.R. (1989) Superoxide anions enhance platelet adhesion and aggregation. *Br. J. Pharmacol.*, **97**, 1145–1150.

Sawicki, G., Salas, E., Murat, J., Miszta-Lane, H. and Radomski, M.W. (1997) Release of gelatinase A from human platelets mediates aggregation. *Nature*, **386**, 616–619.

Sawicki, G., Sanders, E.J., Salas, E., Wozniake, M., Rodrigo, J. and Radominski, M.W. (1988) Localization and translocation of MMP-2 during aggregation of human platelets. *Thromb. Haemost.*, **80**, 836–839.

Schafer, A.I. (1996) Aniplatelet therapy. *Am. J. Med.*, **101**, 199–209.

Slack, M.S., Cui, Y. and Tiritto, V.T. (1993) The effects of flow on blood coagulation and thrombosis. *Thromb. Haemost.*, **70**, 129–134.

Stamler, J.S., Jia, L., Eu, J.P., McMahon, T.J., Demchenko, I.T., Bonaventura, J. *et al.* (1997) Blood flow regulation by S-nitrosohemoglobin in the physiological oxygen gradient *Science*, **276**, 2034–2037.

Villa, L.M., Salas, E., Darley-Usmar, V.M., Radomski, M.W. and Moncada, S. (1994) Peroxynitrite induces both vasodilatation and impaired vascular relaxation in the rat isolated perfused heart. *Proc. Natl. Acad. Sci. USA*, **91**, 12383–12387.

White, J.G. (1988). Platelet membrane ultrastructure and its changes during platelet activation. In *Platelet Membrane Receptors: Molecular Biology, Immunology, Biochemistry and Pathology* edited by G.A. Jamieson, pp. 1–32. New York, Alan R. Liss, Inc.

Wick, T.M. and Eckman, J.R. (1996) Molecular basis of sickle cell-endothelial cell interactions *Curr. Opinion Hematol.*, **3**, 118–124.

Yao, S.K., Ober, J.C., Krishnaswami, A., Ferguson J.J., Anderson, H.V., Golino, P., Buja, L.M. and Willerson, J.T. (1992) Endogenous nitric oxide protects against platelet aggregation and cyclic flow variations in stenosed and endothelium-injured arteries. *Circulation*, **86**, 1302–1309.

Yang, Z., Arnet, U., Bauer, E., von Segesser, L., Siebenmann, R., Turina, M. and Luscher, T.F. (1994) Thrombin-induced endothelium-dependent inhibition and direct activation of platelet-vessel wall interaction. Role of prostacyclin, nitric oxide, and thromboxane A_2. *Circulation*, **89**, 2266–2272.

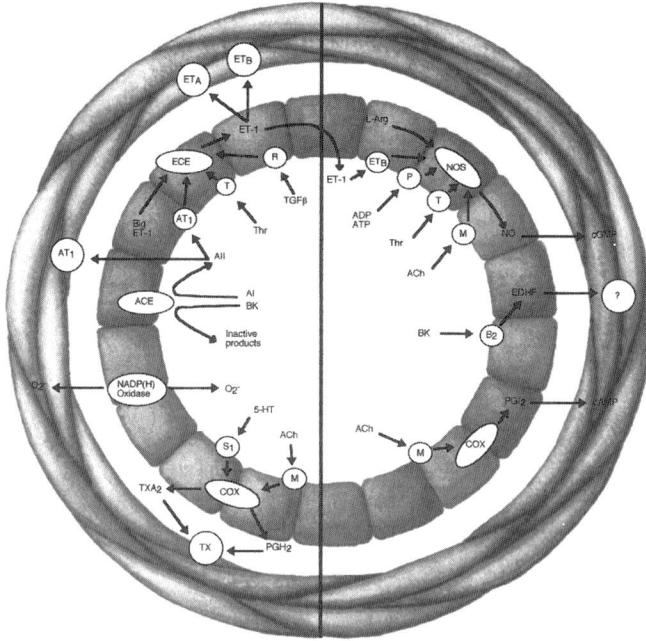

C
O
N
T
R
A
C
T
I
O
N

R
E
L
A
X
A
T
I
O
N

Cardiovascular disease still accounts for the majority of morbidity and mortality in Western countries. Most forms of cardiovascular disease involve atherosclerotic vascular changes in the coronary, cerebral, renal and peripheral circulation leading to angina pectoris and myocardial infarction, stroke, renal failure and claudicatio.

The endothelium is a monolayer of cells lying on the vascular wall, which for years was considered to be only a protective barrier. In the past two decades, however, it has been shown that the endothelium plays indeed an active role in the regulation of vascular smooth muscle cell function and tone. The endothelium is in a strategic anatomical position within the blood vessel wall, located between the circulating blood and vascular smooth muscles. It can respond to mechanical and hormonal signal from the blood. Of particular importance is the fact, that the endothelium is a source of mediators which can, in a predominantly paracrine fashion, modulate the contractile state and proliferative responses o vascular smooth muscle cells, platelet function, coagulation as well as monocyte adhesion. The important role taken by the endothelium in the control of vascular tone is do to its capacity to release both vasodilating and vasoconstricting substances (Lüscher, 1990; Yanagisawa et al., 1988; Furchtgott, 1980).

The endothelium plays a protective role as it prevents adhesion of circulating blood cells, keeps the vasculature in a vasodilated state and inhibits vascular smooth muscle proliferation and migration. In disease states, on the other hand, endothelial dysfunction contributes to enhanced vasoconstrictor responses, adhesion of platelets and monocytes as well as proliferation and migration of vascular smooth muscle cells, all as events known to occur in atherosclerosis.

Endothelial function is impaired in certain cardiovascular conditions including atherosclerosis and in the presence of risk factors such as including atherosclerosis (Zeiher et al., 1993), diabetes (Johnstone, 1993), smoking (Zeiher et al., 1995), hypercholesterolemia (Chowienczyk et al., 1992, Drexler et al., 1991, Creager et al., 1992), aging (Taddei et al., 1995), menopause (Taddei et al., 1996a) and hypertension (Taddei et al., 1993, Taddei et al., 1997c, Vallance et al., 1992a).

Hypertension is an important risk factor for the development of cardiovascular disease (MacMahon et al., 1990) (Figure 6.1). It is associated with an increase in pressure on the arterial side of the circulation, most due to elevated peripheral resistance, determined by the contractile state of the resistance arteries with a diameter of 200 μm or less. The resistance arteries are influenced by neuronal stimulation (in particular from the sympathetic nervous system), by circulating hormones and paracrine and autocrine mechanisms with the blood vessel wall.

Normally the vessel wall is in a constant state of vasodilation due to the basal formation of nitric oxide (NO) by endothelial cells (Rees *et al.*, 1989). This dominant vasorelaxing propriety of the endothelium may act as a compensatory mechanism in an attempt to limit vascular resistance. On the other hand, the vascular endothelium might be involved directly to increase peripheral resistance, via an enhanced release of constricting factors and/ or a decreased release of relaxing factors. Furthermore, the endothelium may importantly contribute to the vascular complications of hypertension, as it becomes dysfunctional (Blot *et al.*, 1994; Palmer *et al.*, 1992).

6 Endothelial Dysfunction and Hypertension

Roberto Corti, Isabella Sudano, Christian Binggeli, Eduardo Nava, Georg Noll and Thomas F. Lüscher

Division of Cardiology, University Hospital Zürich, CH-8091 Zurich, Switzerland

THE PHYSIOLOGIC FUNCTION OF THE ENDOTHELIUM

The endothelium is an organ with a very active secreting activity (Figure 6.1). Endothelial cells release numerous vasoactive substances (i.e. nitic oxide; NO, endothelin; ET, prostacyclin and endothelium-derived hyperpolarising factor; EDHF) which regulate vascular smooth muscle and trafficking blood cells. In response to many stimuli, such as increased shear forces exerted by increased blood flow (Lüscher, 1990; Yanagisawa *et al.*, 1988; Furchtgott, 1980) the endothelium releases NO, which is a potent vasodilator which also inhibits cellular growth and migration. In addition, NO possesses antiatherogenic and thromboresistant proprieties by preventing platelet aggregation and adhesion.

NO is formed from L-arginine by oxidation of its guanidine-nitrogen terminal (Palmer *et al.*, 1988; Moncada, 1992). The catalyzing enzyme NO-synthase (NOS) is constitutively expressed and exists in several isoforms in endothelial cells, platelets, macrophages, vascular smooth muscle cells, and the brain. NO is rapidly inactivated by free radicals, so that its half-life is only of few seconds.

Endothelial cells also release the vasoconstrictor ET. ET exists in three isoforms: ET-1, ET-2 and ET-3. Endothelial cells produce ET-1 exclusively (Yanagisawa *et al.*, 1988).

Translation of messenger RNA (mRNA) generates pre-pro-endothelin, which is covered to big ET; its conversion to the nature peptide ET-1 by the ET-converting enzyme (ECE) is necessary for the development of full vascular activity. The circulating levels of ET-1 are very low; this suggests that little of the peptide is formed physiologically, which may be due to the absence of stimuli for production of ET, the presence of potent inhibitory mechanisms, or its preferential release abluminally toward smooth muscle cells (Wagner *et al.*, 1992).

Due to degradation by endopeptidases in the plasma, lung, and kidney circulating ET-1 has a short half-time, of about 4 to 7 minutes (de Nucci *et al.*, 1988).

Endothelin can stimulate the release and action of NO via a distinct endothelial receptor (ET$_B$-receptor). This explains why ET causes a transient vasodilation at lower concentrations, which precedes its pressor effect (Yanagisawa *et al.*, 1988; Wright and Fozard, 1988; Kiowski *et al.*, 1991).

Moreover in pathological conditions, the endothelium can also produce other endothelium-derived contracting factors (EDCFs), which are mainly cyclooxygenase-dependent

Correspondence: Thomas F. Lüscher, M.D., F.E.S.C., Professor and Head of Cardiology, University Hospital, CH-8091 Zurich, Switzerland. Tel: 0041 1 255 22 16; Fax: 0041 1 255 44 01.

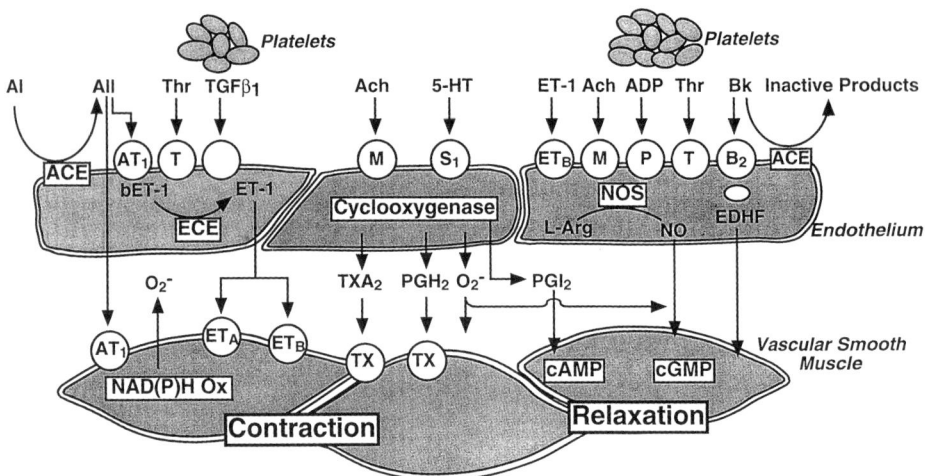

Figure 6.1 Endothelium-derived vasoactive substances: Various blood- and platelet-derived substances can activate specific receptors (open circles) on the endothelial membrane to release relaxing factors such as nitric oxide (NO), prostacyclin (PGI2) and a hyperpolarizing factor (EDHF). Furthermore contracting factors are released such as ET (ET1), angiotensin (A), and thromboxane AII (TXAII) as well as prostaglandin H2 (PGH2). O_2^- = superoxide (from Lüscher, Noll, Braunwald's Heart Disease 1997).

prostanoids (i.e. thromboxane A_2: TXA_2, and prostaglandine H_2: PGH_2) or superoxide anions.

EXPERIMENTAL MODELS OF HYPERTENSION

Inhibitors of NO production cause endothelium-dependent contractions of isolated arteries (Tschudi *et al.*, 1994b), decrease blood flow (Vallance *et al.*, 1989; Joannides *et al.*, 1995a) and induce pronounced and sustained hypertension when infused intravenously or given orally in vivo (Rees *et al.*, 1989; Moreau *et al.*, 1995).

Alteration in endothelium-dependent relaxation in hypertension is not uniform and depends on the model of hypertension as well as the vascular bed studied (Figure 6.2). In some vascular beds of hypertensive rats such as the aorta, mesenteric, carotid and cerebral vessels, endothelium-dependent relaxation is impaired (Luscher, 1994a; Luscher and Vanhoutte, 1986a; Dohi *et al.*, 1990; Diederich *et al.*, 1990; Dohi *et al.*, 1991; Luscher, 1992; Luscher, 1994b; Kung and Luscher, 1995; Lüscher, 1986). In contrast, in coronary and renal arteries of spontaneously hypertensive rats, endothelial function does not seem to be affected by high blood pressure (Tschudi *et al.*, 1991; Luscher, 1991). On the other hand, depending on the animal model and the vascular bed endothelium-dependent contractions have been documented. Since cyclooxygenase inhibitors and thromboxane receptor antagonists can inhibit this response, the most likely contractile factors are thomboxane A_2 and/or prostaglandin H_2 (Kung and Luscher, 1995; Noll *et al.*, 1997a).

Acute, pharmacologically-induced elevations in blood pressure cause an increased release of NO and, on the other hand, drops in pressure are followed by a decreased

Genetic Hypertension   Salt-related Hypertension

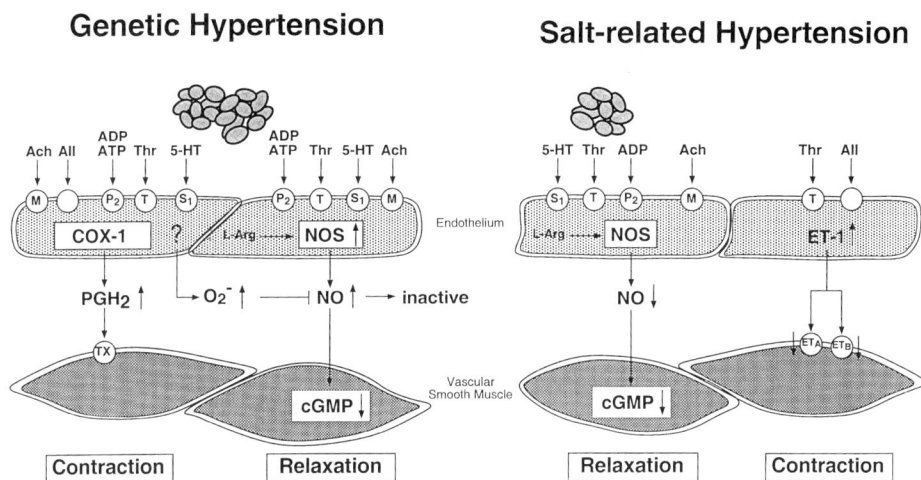

Figure 6.2 Endothelial dysfunction in experimental hypertension: While in the SHR (left panel) NOS is up-regulated and NO is inactivated by O$_2$-, in salt-related hypertension (Dahl rats) NO is produced in lesser amounts, while the ET system is up-regulated. Abbreviations as Figure 2 (from Lüscher, Noll, Braunwald's Heart Disease 1997).

production, suggesting that high blood pressure up-regulates NO production and vice versa (Nava, 1994). The mechanism by which high blood pressure leads to an increased production of NO is not clear yet. It is known that the release of NO by endothelial cells can be altered by changes in blood flow (Joannides *et al.*, 1995a, Buga *et al.*, 1991) and that mRNA and protein for cNOS can be induced by mechanical forces (Sessa *et al.*, 1994). It is likely that not only shear stress, but also other mechanical factors such as blood pressure itself and pulsatile stretch (Hishikawa and Luscher, 1997) contribute to this phenomenon.

The activity of cNOS is higher in mesenteric resistance arteries obtained from spontaneous hypertensive rats (SHRs) compared to age-matched normotensive rats (Noll *et al.*, 1997b). Moreover, the concentration of the oxidative product of NO, nitrate, measured by HPLC and capillary electrophoresis, is higher in the hypertensive rats as compared to their normotensive controls (Noll *et al.*, 1997b). In contrast, prehypertensive young SHRs exhibit similar nitrate levels as age-matched normotensive controls. These observations demonstrate that the basal release of NO is increased in rats with spontaneous hypertension and that this increased production is directly related to the increased blood pressure of the animals.

Further studies demonstrated that the level of cyclic GMP in mesenteric resistance arteries is similar in SHR and in Wystar-Kyoto rats (WKY) (Noll *et al.*, 1997b). Moreover, the NO-dependent vasodilator tone, assessed by the blood pressure effects of L-nitroarginine methylester (L-NAME), is not higher in hypertensive rats as would be expected in a situation in which the production of NO is increased. The capacity of vascular smooth muscle cells of hypertensive rats to respond to NO, on the other hand, must be fully maintained as organic nitrates lower blood pressure in a similar fashion in both strains of rats and relaxations to sodium nitroprusside are enhanced in this condition (Diederich *et al.*, 1990). These studies indicate that the endogenously-produced NO is increased in spontaneous hypertension, but is not able to properly raise cyclic GMP levels in the vascular smooth muscle cells of these animals.

Hence, it appears that in spontaneous hypertension, an additional unknown event takes place that blunts the hemodynamic actions of NO. The hypertrophied and fibrotic intimal layer of hypertensive vessels may represent a physical barrier for NO. The chemical environment that NO has when released can also determine its fate. In this line, oxidative stress has been proposed to play a role in the pathogenesis of some cardiovascular diseases including hypertension (Ohara *et al.*, 1993, Nabel *et al.*, 1988, Nakazono *et al.*, 1991). Recent experimental evidence using a porphyrinic microsensor for direct measurement of NO has demonstrated that in the presence of superoxide dismutase, NO release from isolated resistance vessels is improved in the stroke-prone SHR (Tschudi *et al.*, 1996). Thus, a higher production of oxidative radicals like superoxide anion by a dysfunctional NO-synthase (Cosentino *et al.*, 1998) or a diminished activity of superoxide dismutase may account for an increased degradation of NO.

NO production might be heterogeneously affected in different forms of hypertension (Lüscher, 1988). Indeed, in Dahl salt-sensitive rats endothelium-dependent relaxations are impaired, but not those to sodium nitroprusside (Lüscher, 1988; Luscher *et al.*, 1987a). In contrast to spontaneous hypertension, in salt-sensitive hypertension no release of vasoconstrictor prostanoids can be demonstrated (Luscher *et al.*, 1987a). This suggests that a decreased NO production could contribute to the pathogenesis of this form of hypertension.

Similar experimental findings suggest that ET may be differently involved in different forms of hypertension. In fact, in some animal models of hypertension, such as the DOCA salt hypertensive rat, ET receptor blockade causes marked reductions in blood pressure, which is also associated with regression of vascular hypertrophy (Li *et al.*, 1994, Schiffrin *et al.*, 1995). In keeping with that, ET-1 secretion is augmented in cultured endothelial cells from DOCA-salt hypertensive rats (Takada *et al.*, 1996). Accordingly, ET antagonists lower blood pressure in salt-depleted monkeys (Clozel *et al.*, 1993).

However, the effects of ET-antagonism in other experimental models of hypertension, notably SHRs, are less clear. In the SHR, both circulating and vascular ET as well as ET tissue content of the renal medulla are reduced (Li and Schiffrin, 1995a; Kitamura *et al.*, 1989). In contrast, in the stroke-prone SHR the ET axis is activated and ET antagonism significantly reduces blood pressure and prevents cardiac and vascular hypertrophy (Stasch *et al.*, 1995; Chillon *et al.*, 1996). As in Dahl salt-sensitive rats, ET levels are increased and ET-antagonists lower blood pressure, this indicates that the ET system is particularly activated in severe, salt sensitive (low-renin) hypertension (Barton *et al.*, 1998).

In one-kidney, one-clip (low-renin) hypertension (Li *et al.*, 1996) and two-kidney, two-clip acute renal failure (Ruschitzka *et al.*, 1998) the circulating and tissue ET-1 system is activated. At variance, the ET expression is augmented only in the late phase of two-kidney, one-clip Goldblatt (high-renin) hypertension resembling true renovascular hypertension and activation of the renin-angiotensin-system in man (Sventek, 1996). In contrast, two weeks administration of angiotensin II increases the production of ET in the blood vessel wall of the rat (Moreau *et al.*, 1997a; d'Uscio *et al.*, 1997). Most interestingly, selective ET_A receptor antagonism reduces blood pressure and in particular vascular hypertrophy (Moreau *et al.*, 1997a) and endothelial dysfunction (d'Uscio *et al.*, 1997) under these experimental conditions. These data strongly suggest that ET antagonists may be of particular value in conditions of increased activity of the renin-angiotensin system. This is consistent with additional hypotensive effects of ET antagonists in hypertensive dogs already treated with an ACE inhibitor (Donckier *et al.*, 1997). However, the discrepancy between the beneficial effects of selective ET_A receptor blockade in angiotensin II induced hypertension and the lack of effects in two-kidney, one-clip model is difficult to reconcile, but may be due to a lesser extent of activation of the tissue (rather than circulating) renin-angiotensin and ET system.

In NO-deficient hypertension induced by L-NMMA or L-NAME, ET production is enhanced, but the peptide is only involved in the early, but not chronic phase of hypertension (Sventek *et al.*, 1997; Moreau *et al.*, 1997b). Recently, a role of ET was also suggested in fructose-fed hypertensive rats exhibiting hyperinsulinemia and insulin resistance, as chronic combined ET blockade reduces blood pressure in this experimental model of hypertension (Verma *et al.*, 1995). Interestingly, hepatic overexpression of preproET-1 in rats also resulted in elevation of blood pressure that was reduced by an ET_A antagonist (Niranjan, 1996).

To further elucidate the role of ET, transgenic and gene knock-out rats have been developed. Endothelin-2 transgenic rats exhibit elevated ET plasma levels, but do not develop hypertension (possibly because of the activation of compensatory vasodilator mechanisms) (Hocher *et al.*, 1996). Surprisingly, ET-1 gene knock-out mice are actually hypertensive (Kurihara, 1993b). It is likely that the small increase in blood pressure in ET knock-out mice is related to hypoxia and in turn activation of the sympathetic nervous system in these animals. The finding that ET knock-out rats have profound malformations of the throat indicates that the peptide may be importantly involved in the development of these organs (Kurihara, 1993a).

Other vasoactive mediators are also candidates to contribute to endothelial dysfunction in hypertension. Indeed, the responses to angiotensin I and II are increased in spontaneously hypertensive rats (Tschudi and Luscher, 1995), and in addition platelets and platelet-derived substances (ADP, ATP, serotonin), known to stimulate the formation of EDCFs (Luscher and Vanhoutte, 1986b), may lead to increased peripheral vascular resistance and also to complications of hypertension.

HYPERTENSION IN MAN

Role of Nitric Oxide

Experiments in humans have demonstrated a diminished basal and stimulated NO production (Taddei *et al.*, 1993; Kiowski *et al.*, 1991; Calver *et al.*, 1992; Panza *et al.*, 1990). The decrease in forearm blood flow induced by L-NMMA (reflecting basal NO formation) is smaller in hypertensive than in normotensive patients (Calver *et al.*, 1992). L-NMMA infusion however provides only indirect evidence and no analytical determinations of basal NO production. Benjamin and co-workers demonstrated that indeed plasma levels of NO are reduced in patients with essential hypertension (Forte *et al.*, 1997) (Figure 6.3). The stimulated release of NO as assessed by the vasodilator effects of acetylcholine in the forearm circulation of patients with essential, renovascular or endocrine hypertension are reduced in all but one study (Forte *et al.*, 1997; Panza *et al.*, 1990; Panza *et al.*, 1993).

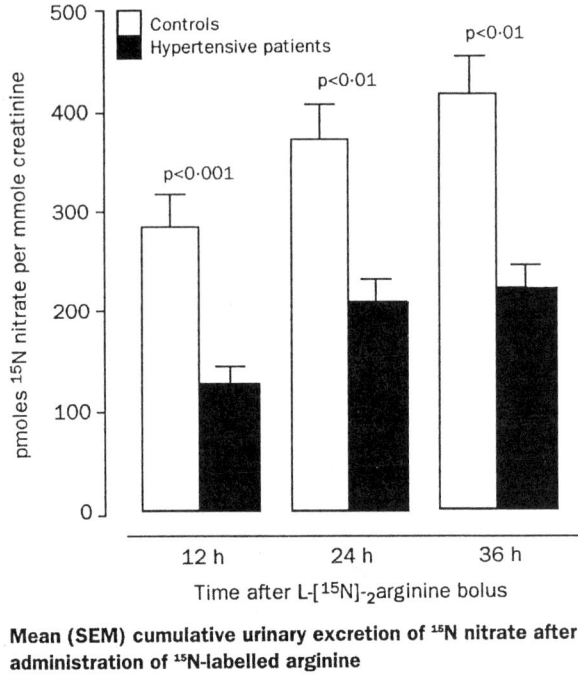

Mean (SEM) cumulative urinary excretion of ^{15}N nitrate after administration of ^{15}N-labelled arginine

Figure 6.3 Plasma nitric oxide levels in normotensive controls and patients with essential hypertension (from Forte et al., Lancet, 1997).

The reasons for the negative results in the study by Cockcroft *et al.* is very likely due to the relatively low dosages of acetylcholine infused and/or heterogeneity in endothelial dysfunction in different patients. Similar findings have been obtained in the coronary circulation, particularly in the presence of left ventricular hypertrophy (Zeiher *et al.*, 1993; Cockcroft *et al.*, 1994).

In patients with essential hypertension, the impaired response to acetylcholine in the forearm circulation can be improved by indomethacin, suggesting that cyclooxygenase-dependent vasoconstrictor prostanoids also contribute to impaired endothelium-dependent relaxation in hypertensive patients (Taddei *et al.*, 1993). Moreover, besides endothelium-derived contracting factors (EDCFs) such as TXA_2 and PGH_2, oxygen-free radicals can play an important role in endothelial dysfunction in hypertension (Tschudi *et al.*, 1996; Taddei *et al.*, 1998a).

Interestingly, there appears to be a difference in endothelial dysfunction between primary and secondary forms of hypertension. In essential hypertension a normalization of blood pressure values do not improve the endothelial function, in contrast to what append with secondary forms of hypertension (Taddei *et al.*, 1993). Moreover, in essential hypertension no correlation between blood pressure values and degree of endothelial dysfunction was found (Taddei *et al.*, 1993).

NO plays an important role in renal function. Indeed, the kidney is extremely sensitive to NO inhibition as very low doses of L-arginine analogues, which do not affect blood pressure, diminish diuresis, natriuresis and renal plasma flow (Lahera *et al.*, 1991; Salazar *et al.*, 1992). It is possible that in some forms of hypertension, minimal alterations in the renal production of NO, which do not alter endothelium-dependent relaxations, lead to systemic hypertension due to a change in the management of body fluids by the kidney. Moreover, it has been recently shown that renal failure is also associated with an accumulation of an endogenous inhibitor of NO synthesis, asymmetrical dimethylarginine (Vallance *et al.*, 1992b), which could also explain the increase in peripheral resistance and hypertension observed in these patients.

Role of Endothelin

Whether or not ET is involved in hypertension is still controversial (Luscher *et al.*, 1993). Because of its vasoconstrictor action and its effects on vascular hypertrophy, ET-1 has also been implicated in the pathogenesis and/or the maintenance of hypertension. However, whether ET production is altered in human hypertension remains still elusive (Luscher *et al.*, 1993; Vanhoutte, 1993). Although few studies found increased plasma levels of ET in hypertensives, many others found no differences as compared to controls. Interestingly, patients with ET-secreting hemangioendotheliomas have huge increases in plasma ET and are hypertensive (Yokokawa *et al.*, 1991). Plasma ET concentrations are also elevated in women with preeclampsia (Kamoi *et al.*, 1990; Schiff *et al.*, 1992; Nova *et al.*, 1991). Increased ET levels in African-Americans who often present with severe and salt-sensitive (low-renin) hypertension, point to the fact that severity of the blood pressure increase as well as salt-sensitivity (as suggested by the experimental models; see above) are important denominators for the activation of the ET system in hypertension (Ergul *et al.*, 1996). However, circulating ET may not reflect local levels of the peptide, as in the blood vessel wall ET is primarily released abluminally (Wagner *et al.*, 1992). Recent studies, using inhibitors of the ET converting enzyme or ET receptor antagonists, suggest that ET does contribute to blood pressure elevation in certain forms of hypertension in laboratory animals and in humans (McMahon *et al.*, 1991; Nishikibe *et al.*, 1993; Haynes *et al.*, 1996).

Infusion of exogenous ET does increase blood pressure in experimental animals and in man (Pernow *et al.*, 1996). In healthy subjects, ET-1 receptor antagonism as well as infusion of the inhibitor of ET converting enzyme phosphoramidone increases forearm blood flow and lowers blood pressure indicating a role of ET-1 in the regulation of vascular tone (Kiowski *et al.*, 1991; Haynes and Webb, 1994; Kaasjager *et al.*, 1997). In essential hypertensives, ET-1-induces an increase in blood pressure and systemic vascular resistance, while cardiac index and natriuresis is reduced (Kaasjager *et al.*, 1997). Moreover, the vasoconstrictor response to ET is increased in the human hand vein circulation of patients with essential hypertension (Haynes *et al.*, 1995). Interestingly, normotensive offsprings of hypertensive parents exhibit enhanced plasma ET responses to mental stress indicating that genetically determined activation of the ET system is already present at this early stage of disease (Noll *et al.*, 1996). Furthermore, ET-1 gene expression is enhanced in small arteries of patients with moderate to severe hypertensive, whereas expression is similar in control subjects and untreated mild hypertensives (Schiffrin *et al.*, 1997). In line with these studies chronic therapy with the ET_A-/ET_B-receptor antagonist bosentan does indeed lower blood pressure in patients with essential hypertension to quite a similar degree as the ACE-inhibitor enalapril (Krum *et al.*, 1998) (Figure 6.4) (see below).

Systolic Blood Pressure   Diastolic Blood Pressure

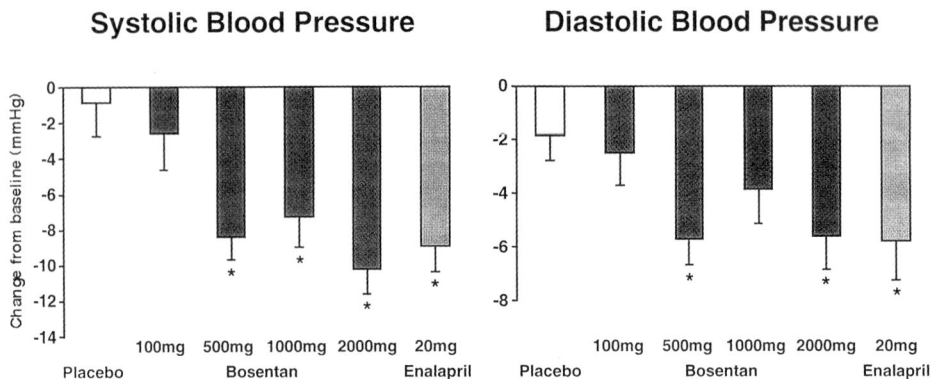

Figure 6.4 Effects of the ET_A-/ET_B-receptor antagonist bosentan in patients with essential hypertension as compared to the ACE-inhibitor enalapril (from Krum H et al. N. Engl. J. Med. 1998).

ANTIHYPERTENSIVE THERAPY AND ENDOTHELIAL FUNCTION

Endothelial dysfunction is a common feature of several pathological processes including hypertension, hyperlipidemia and atherosclerosis. Drugs that could improve the endothelial function or enhance alternative pathways to substitute for the alterations of the release of endothelial mediators may have a potential advantage in the treatment of these pathological conditions. Several antihypertensive agents can prevent and reverse impaired endothelium-dependent relaxations in large conduit arteries (Luscher *et al.*, 1987b; Tschudi *et al.*, 1994a) as well as in resistance arteries of hypertensive rats (Dohi *et al.*, 1994). Since diuretics, calcium antagonists, ACE-inhibitors and angiotensin (AII) receptor antagonists improve or normalize the endothelial dysfunction in hypertensive rats, the blood pressure lowering properties of these agents appear to be involved in this effect. However, additional pressure-independent effects of the various drugs cannot be ruled out and are described for each class of pharmacological agents.

Nitrates

Nitrovasodilators such as nitroglycerin, sodium nitroprusside and linsidomine exert their vasodilator effects by releasing NO from their molecule (Feelisch, 1987), therefore acting through a mechanism identical with that of endogenously produced NO. Hence, these drugs may be particularly useful at sites of reduced vascular formation of NO such as diseased coronary arteries. Of particular interest is the fact that the endogenous production of NO reduces the sensitivity of the blood vessel wall to nitrates and nitrovasodilators (Pohl, 1987). Conversely, in human arteries devoid of endothelium, the concentration-relaxation curve to linsidomine is shifted to the left (Luscher *et al.*, 1989; Joannides *et al.*, 1995b). This could ensure a more selective action of these drugs to dysfunctional vascular beds.

In addition to be more effective in sites of reduced NO formation, the nitrovasodilators seem to be more effective in larger epicardial coronary arteries rather than in smaller coronary arteries (Sellke *et al.*, 1990). This may be due to the fact that small coronary arteries are unable to transform nitroglycerin into active molecules. Since adenosine, which is primarily relaxing small vessels, may precipitate ischemia, this selective efficacy of nitrovasodilators may have important clinical implications. A clinically important drawback, however, is the fact that most nitrovasodilators are prone to tolerance.

Calcium Antagonists

Under acute conditions, calcium antagonists do not affect the release of endothelium-derived vasoactive substances, although the production of these factors in endothelial cell is associated with an increase in intracellular calcium (Vanhoutte, 1988). Indeed, endothelial cells do not appear to possess voltage-operated calcium channels. During their chronic administration, however, these antihypertensive agents improve endothelial function in experimental animals as well as in patients with essential hypertentsion (Tschudi *et al.*, 1994a; Takase *et al.*, 1996; Kung *et al.*, 1995a; Taddei *et al.*, 1997b; Taddei *et al.*, 1997a). Since this beneficial effect is also observed in L-NAME-induced hypertension, in which the synthesis of NO is inhibited, alternative pathways of endothelium-dependent relaxation (i.e. enhanced release of hyperpolarizing factor; EDHF) may be involved (Takase *et al.*, 1996; Kung *et al.*, 1995a). Calcium antagonists may also facilitate the effects of endothelium-derived relaxing factors at the level of vascular smooth muscle, as suggested by an enhanced sodium nitroprusside-induced relaxation under certain conditions (Kung *et al.*, 1995b). Moreover, some calcium-antagonists (verapamil, nifedipine, amlodipine, isradipine and lacidipine) show antioxidant activity in vitro which may be relevant to improve endothelial function in the presence of increased oxidative stress. In contrast, for diltiazem no antioxidant effect could be found (Lupo *et al.*, 1994).

In addition, calcium antagonists seem to interfere with the vasoconstrictor effects of ET and cyclooxygenase-derived contracting factors (Noll *et al.*, 1991). Indeed, in the porcine coronary artery ET receptors are linked to voltage-operated calcium channels via a G-protein and calcium-antagonists attenuate ET-induced vasoconstriction in this blood vessel (Goto *et al.*, 1989). Furthermore, in the human forearm circulation, ET-1 induces potent contraction, which is prevented by nifedipine and verapamil (Kiowski *et al.*, 1991). However, there may be regional differences, as calcium antagonists are ineffective in inhibiting ET-induced contraction in some vessels, like the internal mammary artery (Yang *et al.*, 1990). It is not clear, however, if the usual clinical doses of calcium antagonists are sufficient to antagonize endogenous ET-induced contraction.

Angiotensin-converting Enzyme Inhibitors

The angiotensin-converting enzyme (ACE) is mainly located on the endothelial cell membrane where it transforms angiotensin I into angiotensin II and breaks down bradykinin, a potent stimulator of the L-arginine and cyclooxygenase pathways (Palmer *et al.*, 1987). Therefore, ACE inhibitors not only prevent the formation of a potent vasoconstrictor with proliferative properties, but also increases the local concentration of bradykinin and, in turn the production of NO and prostacyclin (Wiemer *et al.*, 1991). This latter effect may participate in the protective effects of the ACE inhibitors by improving local blood flow and preventing platelet activation. Accordingly, pretreatment of human saphenous vein and coronary artery with an ACE inhibitor enhances endothelium-dependent relaxation to bradykinin (Yang *et al.*, 1993; Auch Schwelk *et al.*, 1992). The decreased degradation of bradykinin could therefore explain the improved endothelial function observed with ACE inhibitors in normotensive and particularly in hypertensive rats (Dohi *et al.*, 1994; Kahonen *et al.*, 1995; Bossaller *et al.*, 1992). However, the improvement of the endothelial function by ACE-inhibitors in L-NAME-induced hypertension suggest that they may also enhance other endothelium-dependent mechanisms (i.e. enhanced release of EDHF), since the activity of the enzyme NOS is inhibited in this experimental model (Kung *et al.*, 1995a; Takase *et al.*, 1996). The mechanism used by ACE-inhibitors also seem to require some time to develop and cannot be reproduce in acute conditions, again suggesting that another mechanism than ACE inhibition, which is rapid, is involved (Takase *et al.*, 1996).

In contrast to the striking improvements obtained in experimental models of hypertension, data of studies in hypertensive patients are still controversial. ACE-inhibitors seem to improve endothelial function in subcutaneous arteries (Schiffrin and Deng, 1995), epicardial arteries [Mancini, 1996 #134] and renal circulation (Mimran *et al.*, 1995). In the forearm circulation, on the other hand, treatment with captopril and enalapril (Creager and Roddy, 1994) or cilazapril (Kiowski *et al.*, 1996) failed to improve vasodilation to a muscarinic agonist, while lisinopril selectively improves the vasodilating response to bradykinin without restoring NO bioavailability (Taddei *et al.*, 1998b). The reasons for this discrepancy between results obtained in experimental models of hypertension and studies in hypertensive patients are not clear at the present time. This discrepancy may originate from the fact that endothelial dysfunction may be treated at a much later stage in patients than in the rat.

Alternatively, prolonged therapy may be required to restore the endothelial function in hypertensive patients. Interestingly, in patients with coronary artery disease 6 months treatment with the ACE-inhibitor quinapril improved endothelial function of epicardial coronary arteries (Mancini *et al.*, 1996).

Angiotensin II Receptor Antagonists

The recently developed AII-receptor antagonists (AT_1-receptor blockers) may have advantages, as they are more potent inhibitors of the angiotensin II vasoconstrictor axis than ACE-inhibitors (Timmermans *et al.*, 1993). These new drugs are also not associated with cough, a side effect of ACE-inhibitors generally attributed to the diminished breakdown of bradykinin. However, if indeed the concomitant stimulation of the L-arginine NO pathway by bradykinin also proves to be an important property of ACE-inhibitors, AII-receptor antagonists would lack this beneficial effect.

In the experimental model of angiotensin II-induced hypertension losartan enhanced endothelial-dependent relaxation to acetylcholine and prevents the increase in tissue ET-1 content, suggesting that AT_1 receptor blocker can modulate tissue ET-1 in vivo (d'Uscio *et al.*, 1998). In addition, losartan also blocks the angiotensin-induced production of oxygen-derived free radicals in this model (Rajagopalan *et al.*, 1996).

In aortic rings obtained from SHR, prolonged antihypertensive treatment with losartan reverses endothelial dysfunction not only by enhancing NO-dependent relaxation but also by reducing formation of cycloaxygenase-dependent EDCFs (Rodrigo *et al.*, 1997). Similar results were also found with captopril as an antihypertensive agent (Rodrigo *et al.*, 1997).

The effects of AII antagonists on endothelial dysfunction in human hypertension are still unknown. As expected in normotensive and hypertensive men the intrabrachial

Table 6.1 Endothelin Receptor Antagonists

Drug	Receptor	Proposed	Development Indication Status
ABT 627	ETA	Prostate cancer	phase I/II
BMS 193884	ETA	CHF	phase I
BMS 20794	ETA	CHF	preclinical
Bosentan	ETA /ETB	CHF	phase II/III
EMD 94246	ETA	CHF, hypertension	preclinical
L 743929	ETA		preclinical
LU-135252	ETA	CHF, hypertension	phase I/II
PD 142893	ETA /ETB	CV diseases	preclinical
PD 145065	ETA /ETB		preclinical
PD 56707	ETA		preclinical
PD 159433	ETA		preclinical
Ro 61-1790	ETA	SAH	preclinical (only iv)
S 0139	ETA	Hypertension	preclinical
SB 209670	ETA	CV diseases	phase I
SB 217242	ETA /ETB	COPD	phase I
T0115	ETA	Hypertension	preclinical
TAK-044	ETA(/B)	CAD, SAH, RTR	phase II (only iv)
TBC 11251	ETA	CHF, PPH	phase I/II
ZD 1611		COPD, PPH	preclinical

infusion of an AT_1-receptor blocker (losartan) inhibits the angiotensin II-induced vasocon-striction (Baan *et al.*, 1998).

Endothelin Antagonists

In recent years a large number of ET-receptor antagonists have been developed (Table 6.1). In many experimental models of hypertension, these molecules are not effective to lower blood pressure (Li and Schiffrin, 1995b). As discussed above, they have a modest anti-hypertensive efficacy in DOCA-salt hypertensive rats, although they have a more profound effect on to prevent vascular hypertrophy (Li *et al.*, 1994). These antagonists also are able to prevent L-NAME-induced hypertension acutely and early on in the chronic phase of blood pressure elevation but eventually a similar increase in blood pressure is noted in spite of ET_A-/ET_B-blockade [70]. This is surprising if one considers the negative feedback exerted by NO on ET-1 release that is blocked in this model (Richard *et al.*, 1995; Donckier *et al.*, 1995).

As studies in humans demonstrated that ET_A- as well as combined ET_A-/ET_B-receptor blockade increases blood flow in the forearm (Haynes and Webb, 1994) as well as in the human skin microcirculation ET must contribute to the regulation of the cardiovascular system also in man (Wenzel *et al.*, 1994). Indeed, intravenous infusion of the ET_A-/ET_B-receptor antagonist TAK-044 in humans with preserved left ventricular function lowers peripheral vascular resistance, blood pressure and increases cardiac output and heart rate (Sutsch, 1997). Similarly, intravenous infusion of the $ET_{A/B}$-receptor antagonist bosentan in patients with coronary artery disease lowers blood pressure under acute conditions

(Wenzel, 1998). Bosentan does exhibit a pronounced antihypertensive activity in patients with essential hypertension similar to that exerted by the ACE-inhibitor enalapril (Figure 6.5) (Krum *et al.*, 1998). This strongly suggests that indeed ET is involved in human hypertension and may provide a new therapeutic approach to treat this condition. Of particular interest obviously in this context is the fact that ET antagonist are particularly efficacious in reversing vascular functional and structural changes in experimental hypertension.

The type of antagonist (ET_A-receptor selective or combined $ET_{A/B}$-receptor antagonists) that would be more effective is still a matter of debate. Most of the first generation molecules were specific ET_A-receptor antagonists (Ihara *et al.*, 1991; Itoh *et al.*, 1993; Sogabe *et al.*, 1993). Combined $ET_{A/B}$-receptor antagonists are now also available (Clozel *et al.*, 1993; Clozel *et al.*, 1994; Luscher, 1993; Ikeda *et al.*, 1994; Ohlstein *et al.*, 1994) and they may prove to be better therapeutic agents, since in several blood vessels, including humans vessels, both ET_A- and ET_B-receptors mediate vasoconstriction (Sumner, 1992; Seo *et al.*, 1994; Ihara *et al.*, 1991). However, these non-selective antagonists also block the endothelium-dependent vasodilatation of ET, by blocking ET_B-receptors (Clozel *et al.*, 1994; Takase *et al.*, 1995). Indeed, in L-NAME hypertension selective ET_A-blockers (i.e. LU135252) improve endothelial dysfunction (in spite of no antihypertensive efficacy) (D'uscio, 1998), while the $ET_{A/B}$-receptor antagonist bosentan is ineffective (Moreau *et al.*, 1997b). Endothelin converting enzyme inhibitors may also prove to be valuable therapeutic agents, although their development lags behind the receptor antagonists.

CONCLUSIONS

Endothelial dysfunction does occur in experimental and human hypertension and may be relevant both for development of high blood pressure as well as for its cardiovascular complications (Figure 6.5). Endothelial dysfunction involves a decreased basal and stimulated release of nitric oxide, enhanced release of endothelium-derived contracting factors such as thromboxane A_2/prostaglandin H_2 and most likely also ET-1. Whether or not it is a primary or secondary phenomenon as suggested by experimental models and studies in patients with secondary hypertension remains uncertain. Indeed, in essential hypertensive patients endothelial dysfunction may represent at least in part a primary phenomenon as it already occurs in normotensive offspring of hypertensive (Luscher, 1994a; Dohi *et al.*, 1990; Lockette *et al.*, 1986; Taddei *et al.*, 1996b).

Endothelial dysfunction in hypertensive patients may be particularly important it is further aggravated by other risk factors such as hypercholesteremia, smoking and diabetes, very much along with the cardiovascular complications.

Blood pressure lowering does normalize endothelial dysfunction in experimental models of hypertension, while in patients with essential hypertension it is may be more difficult to achieve. Studies with a larger number of patients and prolonged treatment periods are required to determine to what extent there are differences between different antihypertensive drugs and their capacity to reverse endothelial dysfunction. Furthermore, studies are required to link endothelial dysfunction with the clinical events occurring later in the disease process.

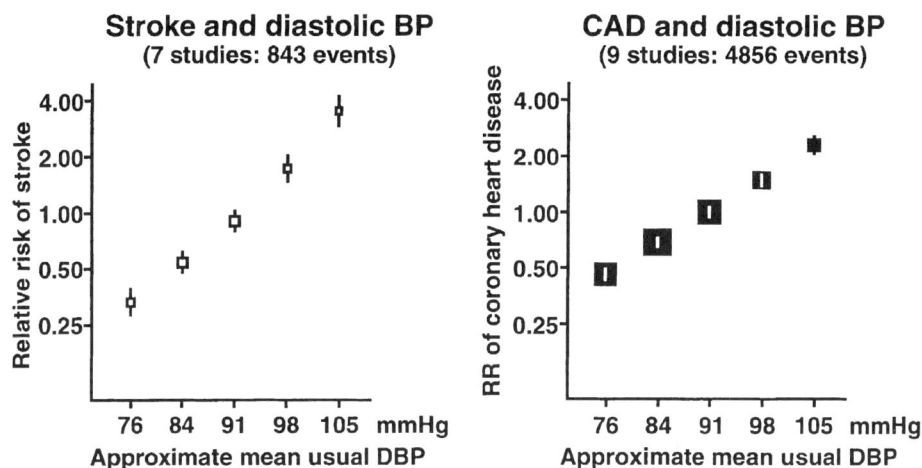

Figure 6.5 Relation of high blood pressure with cardiovascular complications.

ACKNOWLEDGEMENTS

Swiss National Foundation (TF Lüscher Nr. 32.52069.97), G Noll (Nr. 32.52690.97) and the Swiss Heart Foundation.

REFERENCES

Auch Schwelk, W., Bossaller, C., Claus, M., Graf, K., Grafe, M. and Fleck, E. (1992) Local potentiation of bradykinin-induced vasodilation by converting-enzyme inhibition in isolated coronary arteries. *J. Cardiovasc. Pharmacol.*, **20**, S62–67.

Baan, J., Chang, P.C., Vermeij, P., Pfaffendorf, M. and vanZwieten, P.A. (1998) Effects of angiotensin II and losartan in the forearm of patients with essential hypertension. *Journal of Hypertension.*, **16**, 1299–1305.

Barton, M., d'Uscio, L.V., Shaw, S., Meyer, P., Moreau, P. and Luscher, T.F. (1998) ET(A) receptor blockade prevents increased tissue endothelin-1, vascular hypertrophy, and endothelial dysfunction in salt-sensitive hypertension. *Hypertension*, **31**, 499–504.

Blot, S., Arnal, J.F., Xu, Y., Gray, F. and Michel, J.B. (1994) Spinal cord infarcts during long-term inhibition of nitric oxide synthase in rats. *Stroke*, **25**, 1666–1673.

Bossaller, C., Auch Schwelk, W., Weber, F., Gotze, S., Grafe, M., Graf, K. and Fleck, E. (1992) Endothelium-dependent relaxations are augmented in rats chronically treated with the angiotensin-converting enzyme inhibitor enalapril. *J. Cardiovasc. Pharmacol.*, **20**, S91–95.

Buga, G.M., Gold, M.E., Fukuto, J.M. and Ignarro, L.J. (1991) Shear stress-induced release of nitric oxide from endothelial cells grown on beads. *Hypertension*, **17**, 187–193.

Calver, A., Collier, J., Moncada, S. and Vallance, P. (1992) Effect of local intra-arterial NG-monomethyl-L-arginine in patients with hypertension: the nitric oxide dilator mechanism appears abnormal. *J. Hypertens.*, **10**, 1025–1031.

Chillon, J.M., Heistad, D.D. and Baumbach, G.L. (1996) Effects of endothelin receptor inhibition on cerebral arterioles in hypertensive rats. *Hypertension*, **27**, 794–798.

Chowienczyk, P.J., Watts, G.F., Cockcroft, J.R. and Ritter, J.M. (1992) Impaired endothelium-dependent vasodilation of forearm resistance vessels in hypercholesterolaemia [see comments]. *Lancet*, **340**, 1430–1432.

R. Corti et al.

Clozel, M., Breu, V., Burri, K., Cassal, J.M., Fischli, W., Gray, G.A., Hirth, G., Loffler, B.M., Muller, M., Neidhart, W. *et al.* (1993) Pathophysiological role of endothelin revealed by the first orally active endothelin receptor antagonist. *Nature*, **365**, 759–761.

Clozel, M., Breu, V., Gray, G. A., Kalina, B., Loffler, B.M., Burri, K., Cassal, J.M., Hirth, G., Muller, M., Neidhart, W. *et al.* (1994) Pharmacological characterization of bosentan, a new potent orally active nonpeptide endothelin receptor antagonist. *J. Pharmacol. Exp. Ther.*, **270**, 228–235.

Cockcroft, J.R., Chowienczyk, P.J., Benjamin, N. and Ritter, J.M. (1994) Preserved endothelium-dependent vasodilatation in patients with essential hypertension [see comments] [published erratum appears in *N. Engl. J. Med.* 1995 May 25;332(21):1455]. *N. Engl. J. Med.*, **330**, 1036–1040.

Cosentino, F., Patton, S., d'Uscio, L.V., Werner, E.R., Werner Felmayer, G., Moreau, P., Malinski, T. and Luscher, T.F. (1998) Tetrahydrobiopterin alters superoxide and nitric oxide release in prehypertensive rats. *J. Clin. Invest.*, **101**, 1530–1537.

Creager, M.A., Gallagher, S.J., Girerd, X.J., Coleman, S.M., Dzau, V.J. and Cooke, J.P. (1992) L-arginine improves endothelium-dependent vasodilation in hypercholesterolemic humans. *J. Clin. Invest.*, **90**, 1248–1253.

Creager, M.A. and Roddy, M.A. (1994) Effect of captopril and enalapril on endothelial function in hypertensive patients. *Hypertension*, **24**, 499–505.

d'Uscio, L.V., Moreau, P., Shaw, S., Takase, H., Barton, M. and Luscher, T.F. (1997) Effects of chronic ETA-receptor blockade in angiotensin II-induced hypertension. *Hypertension*, **29**, 435–441.

d'Uscio, L.V., Shaw, S., Barton, M. and Luscher, T.F. (1998) Losartan but not verapamil inhibits angiotensin II-induced tissue endothelin-1 increase: role of blood pressure and endothelial function. *Hypertension*, **31**, 1305–1310.

d'Uscio, L.V., M.P., Shaw, S., Barton, M. and Lüscher, T.F. (1998) Effects of selective, non-peptide ETa-antagonist on endothelial function in chronic inhibition of nitric oxide synthesis. *Br. J. Pharmacol.*, submitted.

de Nucci, G., Thomas, R., D'Orleans Juste, P., Antunes, E., Walder, C., Warner, T.D. and Vane, J.R. (1988) Pressor effects of circulating endothelin are limited by its removal in the pulmonary circulation and by the release of prostacyclin and endothelium-derived relaxing factor. *Proc. Natl. Acad. Sci. USA*, **85**, 9797–9800.

Diederich, D., Yang, Z.H., Buhler, F.R. and Luscher, T.F. (1990) Impaired endothelium-dependent relaxations in hypertensive resistance arteries involve cyclooxygenase pathway. *Am. J. Physiol.*, **258**, H445–451.

Dohi, Y., Criscione, L. and Luscher, T.F. (1991) Renovascular hypertension impairs formation of endothelium-derived relaxing factors and sensitivity to endothelin-1 in resistance arteries. *Br. J. Pharmacol.*, **104**, 349–354.

Dohi, Y., Criscione, L., Pfeiffer, K. and Luscher, T.F. (1994) Angiotensin blockade or calcium antagonists improve endothelial dysfunction in hypertension: studies in perfused mesenteric resistance arteries. *J. Cardiovasc. Pharmacol.*, **24**, 372–379.

Dohi, Y., Thiel, M.A., Buhler, F.R. and Luscher, T.F. (1990) Activation of endothelial L-arginine pathway in resistance arteries. Effect of age and hypertension. *Hypertension*, **16**, 170–179.

Donckier, J., Stoleru, L., Hayashida, W., Van Mechelen, H., Selvais, P., Galanti, L., Clozel, J. P., Ketelslegers, J.M. and Pouleur, H. (1995) Role of endogenous endothelin-1 in experimental renal hypertension in dogs. *Circulation*, **92**, 106–113.

Donckier, J.E., Massart, P.E., Hodeige, D., Van Mechelen, H., Clozel, J.P., Laloux, O., Ketelslegers, J.M., Charlier, A.A. and Heyndrickx, G.R. (1997) Additional hypotensive effect of endothelin-1 receptor antagonism in hypertensive dogs under angiotensin-converting enzyme inhibition. *Circulation*, **96**, 1250–1256.

Drexler, H., Zeiher, A.M., Meinzer, K. and Just, H. (1991) Correction of endothelial dysfunction in coronary microcirculation of hypercholesterolaemic patients by L-arginine. *Lancet*, **338**, 1546–1550.

Ergul, S., Parish, D.C., Puett, D. and Ergul, A. (1996) Racial differences in plasma endothelin-1 concentrations in individuals with essential hypertension [published erratum appears in Hypertension 1997 Mar;29(3):912]. *Hypertension*, **28**, 652–655.

Feelisch, M. and Noack, E.A. (1987) Correlation between nitric oxide formation during degradation of organic nitrates and activation of guanylate cyclase. *Eur. J. Pharmacol.*, **139**, 19–30.

Forte, P., Copland, M., Smith, L.M., Milne, E., Sutherland, J. and Benjamin, N. (1997) Basal nitric oxide synthesis in essential hypertension [see comments]. *Lancet*, **349**, 837–842.

Furchtgott, R.F., Z.J. (1980) The obligatory role of endothelial cells in the relaxation of arterial smooth muscle by acetylcholine. *Nature*, **288**, 373–376.

Goto, K., Kasuya, Y., Matsuki, N., Takuwa, Y., Kurihara, H., Ishikawa, T., Kimura, S., Yanagisawa, M. and Masaki, T. (1989) Endothelin activates the dihydropyridine-sensitive, voltage-dependent Ca2+ channel in vascular smooth muscle. *Proc. Natl. Acad. Sci. USA*, **86**, 3915–3918.

Haynes, W.G., Ferro, C.J., O'Kane, K.P., Somerville, D., Lomax, C.C. and Webb, D.J. (1996) Systemic endothelin receptor blockade decreases peripheral vascular resistance and blood pressure in humans. *Circulation*, **93**, 1860–1870.

Haynes, W.G., Moffat, S. and Webb, D.J. (1995) An investigation into the direct and indirect venoconstrictor effects of endothelin-1 and big endothelin-1 in man. *Br. J. Clin. Pharmacol.*, **40**, 307–311.

Haynes, W.G. and Webb, D.J. (1994) Contribution of endogenous generation of endothelin-1 to basal vascular tone [see comments]. *Lancet*, **344**, 852–854.

Hishikawa, K. and Luscher, T.F. (1997) Pulsatile stretch stimulates superoxide production in human aortic endothelial cells. *Circulation*, **96**, 3610–3616.

Hocher, B., Liefeldt, L., Thone Reineke, C., Orzechowski, H.D., Distler, A., Bauer, C. and Paul, M. (1996) Characterization of the renal phenotype of transgenic rats expressing the human endothelin-2 gene. *Hypertension*, **28**, 196–201.

Ihara, M., Saeki, T., Funabashi, K., Nakamichi, K., Yano, M., Fukuroda, T., Miyaji, M., Nishikibe, M. and Ikemoto, F. (1991) Two endothelin receptor subtypes in porcine arteries. *J. Cardiovasc. Pharmacol.*, **17**, S119–121.

Ikeda, S., Awane, Y., Kusumoto, K., Wakimasu, M., Watanabe, T. and Fujino, M. (1994) A new endothelin receptor antagonist, TAK-044, shows long-lasting inhibition of both ETA- and ETB-mediated blood pressure responses in rats. *J. Pharmacol. Exp. Ther.*, **270**, 728–733.

Itoh, S., Sasaki, T., Ide, K., Ishikawa, K., Nishikibe, M. and Yano, M. (1993) A novel endothelin ETA receptor antagonist, BQ-485, and its preventive effect on experimental cerebral vasospasm in dogs. *Biochem. Biophys. Res. Commun.*, **195**, 969–975.

Joannides, R., Haefeli, W.E., Linder, L., Richard, V., Bakkali, E.H., Thuillez, C. and Luscher, T.F. (1995a) Nitric oxide is responsible for flow-dependent dilatation of human peripheral conduit arteries in vivo. *Circulation*, **91**, 1314–1319.

Joannides, R., Richard, V., Haefeli, W.E., Linder, L., Luscher, T.F. and Thuillez, C. (1995b) Role of basal and stimulated release of nitric oxide in the regulation of radial artery caliber in humans. *Hypertension*, **26**, 327–331.

Johnstone, M.T., C.S., Scales, K.M., Cusco, J.A., Lee, B.K. and Creager, M.A. (1993) Impaired endothelial-dependent vasodilation in patients with insulin-dependent diabetes mellitus. *Circulation*, **88**, 2510–2516.

Kaasjager, K.A., Koomans, H.A. and Rabelink, T.J. (1997) Endothelin-1-induced vasopressor responses in essential hypertension. *Hypertension*, **30**, 15–21.

Kahonen, M., Makynen, H., Wu, X., Arvola, P. and Porsti, I. (1995) Endothelial function in spontaneously hypertensive rats: influence of quinapril treatment. *Br. J. Pharmacol.*, **115**, 859–867.

Kamoi, K., Sudo, N., Ishibashi, M. and Yamaji, T. (1990) Plasma endothelin-1 levels in patients with pregnancy-induced hypertension [letter]. *N. Engl. J. Med.*, **323**, 1486–1487.

Kiowski, W., Linder, L., Nuesch, R. and Martina, B. (1996) Effects of cilazapril on vascular structure and function in essential hypertension. *Hypertension*, **27**, 371–376.

Kiowski, W., Luscher, T.F., Linder, L. and Buhler, F.R. (1991) Endothelin-1-induced vasoconstriction in humans. Reversal by calcium channel blockade but not by nitrovasodilators or endothelium-derived relaxing factor. *Circulation*, **83**, 469–475.

Kitamura, K., Tanaka, T., Kato, J., Ogawa, T., Eto, T. and Tanaka, K. (1989) Immunoreactive endothelin in rat kidney inner medulla: marked decrease in spontaneously hypertensive rats. *Biochem. Biophys. Res. Commun.*, **162**, 38–44.

Krum, H., Viskoper, R.J., Lacourciere, Y., Budde, M. and Charlon, V. (1998) The effect of an endothelin-receptor antagonist, bosentan, on blood pressure in patients with essential hypertension. Bosentan Hypertension Investigators. *N. Engl. J. Med.*, **338**, 784–790.

Kung, C.F. and Luscher, T.F. (1995) Different mechanisms of endothelial dysfunction with aging and hypertension in rat aorta. *Hypertension*, **25**, 194–200.

Kung, C.F., Moreau, P., Takase, H. and Luscher, T.F. (1995a) L-NAME hypertension alters endothelial and smooth muscle function in rat aorta. Prevention by trandolapril and verapamil. *Hypertension*, **26**, 744–751.

Kung, C.F., Tschudi, M.R., Noll, G., Clozel, J.P. and Luscher, T.F. (1995b) Differential effects of the calcium antagonist mibefradil in epicardial and intramyocardial coronary arteries. *J. Cardiovasc. Pharmacol.*, **26**, 312–318.

Kurihara, Y., K.H., Suzuki, H,, Kodama, T., Maemura, K., Oda, H., Yishikawa, T. and Yazaki, Y. (1993a) Targeted disruption of mouse endothelin-1 gene (I): lethality and craniofacial anomaly in homozygotes. *Circulation*, **88**, I182.

Kurihara, Y., K.H., Kuwaki, T., Maemura, K., Cao, W.H., Suzuki, H., Kodama, T., Yasu-yoshi, O., Kumada, M. and Yazaki, Y. (1993b) Targeted disruption of mouse endothelin-1 gene (II): elevated blood pressure in heterozygotes. *Circulation*, **88**, I332.

Lahera, V., Salom, M.G., Miranda Guardiola, F., Moncada, S. and Romero, J.C. (1991) Effects of NG-nitro-L-arginine methyl ester on renal function and blood pressure. *Am. J. Physiol.*, **261**, F1033–1037.

Li, J.S., Knafo, L., Turgeon, A., Garcia, R. and Schiffrin, E.L. (1996) Effect of endothelin antagonism on blood pressure and vascular structure in renovascular hypertensive rats. *Am. J. Physiol.*, **271**, H88–93.

Li, J.S., Lariviere, R. and Schiffrin, E.L. (1994) Effect of a nonselective endothelin antagonist on vascular remodeling in deoxycorticosterone acetate-salt hypertensive rats. Evidence for a role of endothelin in vascular hypertrophy. *Hypertension*, **24**, 183–188.

Li, J.S. and Schiffrin, E.L. (1995a) Chronic endothelin receptor antagonist treatment of young spontaneously hypertensive rats. *J. Hypertens.*, **13**, 647–652.

Li, J.S. and Schiffrin, E.L. (1995b) Effect of chronic treatment of adult spontaneously hypertensive rats with an endothelin receptor antagonist. *Hypertension*, **25**, 495–500.

Lockette, W., Otsuka, Y. and Carretero, O. (1986) The loss of endothelium-dependent vascular relaxation in hypertension. *Hypertension*, **8**, Ii61–66.

Lupo, E., Locher, R., Weisser, B. and Vetter, W. (1994) In vitro antioxidant activity of calcium antagonists against LDL oxidation compared with alpha-tocopherol. *Biochem. Biophys. Res. Commun.*, **203**, 1803–1808.

Luscher, T.F. (1991) Endothelium-derived nitric oxide: the endogenous nitrovasodilator in the human cardiovascular system. *Eur. Heart. J.*, **12**, 2–11.

Luscher, T.F. (1992) Heterogeneity of endothelial dysfunction in hypertension. *Eur. Heart. J.*, **13**, 50–55.

Luscher, T.F. (1993) Do we need endothelin antagonists? *Cardiovasc. Res.*, **27**, 2089–2093.

Luscher, T.F. (1994a) The endothelium and cardiovascular disease–a complex relation [editorial; comment]. *N. Engl. J. Med.*, **330**, 1081–1083.

Luscher, T.F. (1994b) The endothelium in hypertension: bystander, target or mediator? *J. Hypertens. Suppl*, **12**, S105–116.

Luscher, T.F., Raij, L. and Vanhoutte, P.M. (1987a) Endothelium-dependent vascular responses in normotensive and hypertensive Dahl rats. *Hypertension*, **9**, 157–163.

Luscher, T.F., Richard, V. and Yang, Z.H. (1989) Interaction between endothelium-derived nitric oxide and SIN-1 in human and porcine blood vessels. *J. Cardiovasc. Pharmacol.*, **14**, S76–80.

Luscher, T.F., Seo, B.G. and Buhler, F.R. (1993) Potential role of endothelin in hypertension. Controversy on endothelin in hypertension. *Hypertension*, **21**, 752–757.

Luscher, T.F. and Vanhoutte, P.M. (1986a) Endothelium-dependent contractions to acetylcholine in the aorta of the spontaneously hypertensive rat. *Hypertension*, **8**, 344–348.

Luscher, T.F. and Vanhoutte, P.M. (1986b) Endothelium-dependent responses to platelets and serotonin in spontaneously hypertensive rats [published erratum appears in Hypertension 1987 Apr;9(4):421]. *Hypertension*, **8**, Ii55–60.

Luscher, T.F., Vanhoutte, P.M. and Raij, L. (1987b) Antihypertensive treatment normalizes decreased endothelium-dependent relaxations in rats with salt-induced hypertension. *Hypertension*, **9**, Iii193–197.

Lüscher, T.F., R.J., Vanhoutte, P.M. (1986) Bioassay of endothelium-derived substances in the aorta of normotensive spontaneously hypertensive rats. *J. Hypertens.ion*, **4 (Suppl 6)**, 81–85.

Lüscher, T.F., V.P. (1988) In *Relaxing and contracting factors* (Ed, Vanhoutte, P.) Humana Press, Clifton, pp. 495–509.

Lüscher, T.F., V.P. (Ed.) (1990) *The endothelium: modulator of cardiovascular function,* CRC Press, Boca Raton.

MacMahon, S., Peto, R., Cutler, J., Collins, R., Sorlie, P., Neaton, J., Abbott, R., Godwin, J., Dyer, A. and Stamler, J. (1990) Blood pressure, stroke, and coronary heart disease. Part 1, Prolonged differences in blood pressure: prospective observational studies corrected for the regression dilution bias [see comments]. *Lancet*, **335**, 765–774.

Mancini, G.B., Henry, G.C., Macaya, C., O'Neill, B.J., Pucillo, A.L., Carere, R.G., Wargovich, T.J., Mudra, H., Luscher, T.F., Klibaner, M.I., Haber, H.E., Uprichard, A.C., Pepine, C.J. and Pitt, B. (1996) Angiotensin-converting enzyme inhibition with quinapril improves endothelial vasomotor dysfunction in patients with coronary artery disease. The TREND (Trial on Reversing ENdothelial Dysfunction) Study [see comments] [published erratum appears in Circulation 1996 Sep 15;94(6):1490]. *Circulation*, **94**, 258–265.

McMahon, E.G., Palomo, M.A. and Moore, W.M. (1991) Phosphoramidon blocks the pressor activity of big endothelin[1–39] and lowers blood pressure in spontaneously hypertensive rats. *J. Cardiovasc. Pharmacol.*, **17**, S29–33.

Mimran, A., Ribstein, J. and DuCailar, G. (1995) Contrasting effect of antihypertensive treatment on the renal response to L-arginine. *Hypertension*, **26**, 937–941.

Moncada, S. (1992) The 1991 Ulf von Euler Lecture. The L-arginine: nitric oxide pathway. *Acta Physiol. Scand.*, **145**, 201–227.

Moreau, P., d'Uscio, L.V., Shaw, S., Takase, H., Barton, M. and Luscher, T.F. (1997a) Angiotensin II increases tissue endothelin and induces vascular hypertrophy: reversal by ET(A)-receptor antagonist. *Circulation*, **96**, 1593–1597.

Moreau, P., Takase, H., Kung, C.F., Shaw, S. and Luscher, T.F. (1997b) Blood pressure and vascular effects of endothelin blockade in chronic nitric oxide-deficient hypertension. *Hypertension*, **29**, 763–769.

Moreau, P., Takase, H., Kung, C.F., van Rooijen, M.M., Schaffner, T. and Luscher, T.F. (1995) Structure and function of the rat basilar artery during chronic nitric oxide synthase inhibition. *Stroke*, **26**, 1922–1928.

Nabel, E.G., Ganz, P., Gordon, J.B., Alexander, R.W. and Selwyn, A.P. (1988) Dilation of normal and constriction of atherosclerotic coronary arteries caused by the cold pressor test. *Circulation*, **77**, 43–52.

Nakazono, K., Watanabe, N., Matsuno, K., Sasaki, J., Sato, T. and Inoue, M. (1991) Does superoxide underlie the pathogenesis of hypertension? *Proc. Natl. Acad. Sci. USA*, **88**, 10045–10048.

Nava, E., L.A., Wilkund, N.P. and Moncada, S. (1994) In *The biology of nitric oxide*(Ed, Feelisch M, B.R. a. M.S.) Portland press, London, pp. 179–181.

Niranjan, V, T.S., Dewit, D., Gerard, R.D. and Yanagisawa, M. (1996) Systemic hypertension induced by epatic overexpresion of human preproendothelin-1 in rats. *J. Clin. Invest.*, **98**, 2364–2372.

Nishikibe, M., Tsuchida, S., Okada, M., Fukuroda, T., Shimamoto, K., Yano, M., Ishikawa, K. and Ikemoto, F. (1993) Antihypertensive effect of a newly synthesized endothelin antagonist, BQ-123, in a genetic hypertensive model. *Life Sci*, **52**, 717–724.

Noll, G., Buhler, F.R., Yang, Z. and Luscher, T.F. (1991) Different potency of endothelium-derived relaxing factors against thromboxane, endothelin, and potassium chloride in intramyocardial porcine coronary arteries. *J. Cardiovasc. Pharmacol.*, **18**, 120–126.

Noll, G., Lang, M.G., Tschudi, M.R., Ganten, D. and Luscher, T.F. (1997a) Endothelial vasoconstrictor prostanoids modulate contractions to acetylcholine and ANG II in Ren-2 rats. *Am. J. Physiol.*, **272**, H493–500.

Noll, G., Tschudi, M., Nava, E. and Luscher, T.F. (1997b) Endothelium and high blood pressure. *Int J Microcirc Clin Exp*, **17**, 273–279.

Noll, G., Wenzel, R.R., Schneider, M., Oesch, V., Binggeli, C., Shaw, S., Weidmann, P. and Luscher, T.F. (1996) Increased activation of sympathetic nervous system and endothelin by mental stress in normotensive offspring of hypertensive parents [see comments]. *Circulation*, **93**, 866–869.

Nova, A., Sibai, B.M., Barton, J.R., Mercer, B.M. and Mitchell, M.D. (1991) Maternal plasma level of endothelin is increased in preeclampsia. *Am. J. Obstet. Gynecol.*, **165**, 724–727.

Ohara, Y., Peterson, T.E. and Harrison, D.G. (1993) Hypercholesterolemia increases endothelial superoxide anion production. *J. Clin. Invest.*, **91**, 2546–2551.

Ohlstein, E.H., Nambi, P., Douglas, S.A., Edwards, R.M., Gellai, M., Lago, A., Leber, J.D., Cousins, R.D., Gao, A., Frazee, J.S. *et al.* (1994) SB 209670, a rationally designed potent nonpeptide endothelin receptor antagonist. *Proc. Natl. Acad. Sci. USA*, **91**, 8052–8056.

Palmer, R.M., Ashton, D.S. and Moncada, S. (1988) Vascular endothelial cells synthesize nitric oxide from L-arginine. *Nature*, **333**, 664–666.

Palmer, R.M., Bridge, L., Foxwell, N.A. and Moncada, S. (1992) The role of nitric oxide in endothelial cell damage and its inhibition by glucocorticoids. *Br. J. Pharmacol.*, **105**, 11–12.

Palmer, R.M., Ferrige, A.G. and Moncada, S. (1987) Nitric oxide release accounts for the biological activity of endothelium-derived relaxing factor. *Nature*, **327**, 524–526.

Panza, J.A., Quyyumi, A.A., Brush, J.E., Jr. and Epstein, S.E. (1990) Abnormal endothelium-dependent vascular relaxation in patients with essential hypertension [see comments]. *N. Engl. J. Med.*, **323**, 22–27.

Panza, J.A., Quyyumi, A.A., Callahan, T.S. and Epstein, S.E. (1993) Effect of antihypertensive treatment on endothelium-dependent vascular relaxation in patients with essential hypertension. *J. Am. Coll. Cardiol.*, **21**, 1145–1151.

Pernow, J., Kaijser, L., Lundberg, J. M. and Ahlborg, G. (1996) Comparable potent coronary constrictor effects of endothelin-1 and big endothelin-1 in humans. *Circulation*, **94**, 2077–2082.

Pohl, U. and Busse, R. (1987) Endothelium-derived relaxant factor inhibits effects of nitrocompounds in isolated arteries. *Am. J. Physiol.*, **252**, H307–313.

Rajagopalan, S., Kurz, S., Munzel, T., Tarpey, M., Freeman, B.A., Griendling, K.K. and Harrison, D.G. (1996) Angiotensin II-mediated hypertension in the rat increases vascular superoxide production via membrane NADH/NADPH oxidase activation. Contribution to alterations of vasomotor tone. *J. Clin. Invest.*, **97**, 1916–1923.

Rees, D.D., Palmer, R.M. and Moncada, S. (1989) Role of endothelium-derived nitric oxide in the regulation of blood pressure. *Proc. Natl. Acad. Sci. USA*, **86**, 3375–3378.

Richard, V., Hogie, M., Clozel, M., Loffler, B.M. and Thuillez, C. (1995) In vivo evidence of an endothelin-induced vasopressor tone after inhibition of nitric oxide synthesis in rats. *Circulation*, **91**, 771–775.

Rodrigo, E., Maeso, R., Munoz Garcia, R., Navarro Cid, J., Ruilope, L.M., Cachofeiro, V. and Lahera, V. (1997) Endothelial dysfunction in spontaneously hypertensive rats: consequences of chronic treatment with losartan or captopril. *J. Hypertens.*, **15**, 613–618.

Ruschitzka, F., Shaw, S., Noll, G., Barton, M., Schulz, E., Muller, G.A. and Luscher, T.F. (1998) Endothelial vasoconstrictor prostanoids, vascular reactivity, and acute renal failure. *Kidney International.*, **54**, S199–S201.

Salazar, F.J., Pinilla, J.M., Lopez, F., Romero, J.C. and Quesada, T. (1992) Renal effects of prolonged synthesis inhibition of endothelium-derived nitric oxide. *Hypertension*, **20**, 113–117.

Schiff, E., Ben Baruch, G., Peleg, E., Rosenthal, T., Alcalay, M., Devir, M. and Mashiach, S. (1992) Immuno-reactive circulating endothelin-1 in normal and hypertensive pregnancies. *Am. J. Obstet. Gynecol.*, **166**, 624–628.

Schiffrin, E.L. and Deng, L.Y. (1995) Comparison of effects of angiotensin I-converting enzyme inhibition and beta-blockade for 2 years on function of small arteries from hypertensive patients. *Hypertension*, **25**, 699–703.

Schiffrin, E.L., Deng, L.Y., Sventek, P. and Day, R. (1997) Enhanced expression of endothelin-1 gene in resistance arteries in severe human essential hypertension. *J. Hypertens.*, **15**, 57–63.

Schiffrin, E.L., Sventek, P., Li, J.S., Turgeon, A. and Reudelhuber, T. (1995) Antihypertensive effect of an endothelin receptor antagonist in DOCA-salt spontaneously hypertensive rats. *Br. J. Pharmacol.*, **115**, 1377–1381.

Sellke, F.W., Myers, P.R., Bates, J.N. and Harrison, D.G. (1990) Influence of vessel size on the sensitivity of porcine coronary microvessels to nitroglycerin. *Am. J. Physiol.*, **258**, H515–520.

Seo, B., Oemar, B.S., Siebenmann, R., von Segesser, L. and Luscher, T.F. (1994) Both ETA and ETB receptors mediate contraction to endothelin-1 in human blood vessels [see comments]. *Circulation*, **89**, 1203–1208.

Sessa, W.C., Pritchard, K., Seyedi, N., Wang, J. and Hintze, T.H. (1994) Chronic exercise in dogs increases coronary vascular nitric oxide production and endothelial cell nitric oxide synthase gene expression. *Circ Res*, **74**, 349–353.

Sogabe, K., Nirei, H., Shoubo, M., Nomoto, A., Ao, S., Notsu, Y. and Ono, T. (1993) Pharmacological profile of FR139317, a novel, potent endothelin ETA receptor antagonist. *J. Pharmacol. Exp. Ther.*, **264**, 1040–1046.

Stasch, J.P., Hirth Dietrich, C., Frobel, K. and Wegner, M. (1995) Prolonged endothelin blockade reduces hypertension and cardiac hypertrophy in SHR-SP. *J. Cardiovasc. Pharmacol.*, **26**, S436–438.

Sumner, M.J., C.T., Mundin, J.M., White, D.G. and Watts, I.S. (1992) Endothelin ETA and ETB receptors mediated vascular smooth muscle contraction. *Br. J. Pharmacol.*, **107**, 858–860.

Sutsch, G., F.M., Yan, X-W., Wenzel, R., Binggeli, C., Bianchetti, M.C., Kiowski, W. and Lüscher, T.F. (1997) Acute hemodynamic and renal effect of the endothelin-1 receptor antagonist TAK-044 in patients without heart failure. *JACC*, **29 (Suppl A)**, 444A.

Sventek, P., Turgeon, A. and Schiffrin, E.L. (1997) Vascular endothelin-1 gene expression and effect on blood pressure of chronic ETA endothelin receptor antagonism after nitric oxide synthase inhibition with L-NAME in normal rats. *Circulation*, **95**, 240–244.

Sventek P, T.A. and Schiffrin, E.L. (1996) Vascular and cardiac overexpression of endothelin-1 gene in 1-kidney, one clip Goldblatt hypertensive rats but only in the late phase 2-kidney, one clip Goldblatt hypertension. *J. Hypertens.*, **14**, 57–64.

Taddei, S., Virdis, A., Ghiadoni, L., Magagna, A. and Salvetti, A. (1998a) Vitamin C improves endothelium-dependent vasodilation by restoring nitric oxide activity in essential hypertension. *Circulation*, **97**, 2222–2229.

Taddei, S., Virdis, A., Ghiadoni, L., Mattei, P. and Salvetti, A. (1998b) Effects of angiotensin converting enzyme inhibition on endothelium-dependent vasodilatation in essential hypertensive patients. *Journal of Hypertension*, **16**, 447–456.

Taddei, S., Virdis, A., Ghiadoni, L., Mattei, P., Sudano, I., Bernini, G., Pinto, S. and Salvetti, A. (1996a) Menopause is associated with endothelial dysfunction in women. *Hypertension*, **28**, 576–582.

Taddei, S., Virdis, A., Ghiadoni, L., Sudano, I., Noll, G., Lüscher, T.F. and Salvetti, A. (1997a) Nifedipine enhances endothelium-dependent relaxation and inhibits contraction to endothelin-1 and phenilephrine in human hypertension. *Circulation*, **96 (Suppl I)**, I762–I763.

Taddei, S., Virdis, A., Ghiadoni, L., Uleri, S., Magagna, A. and Salvetti, A. (1997b) Lacidipine restores endothelium-dependent vasodilation in essential hypertensive patients. *Hypertension*, **30**, 1606–1612.

Taddei, S., Virdis, A., Mattei, P., Ghiadoni, L., Fasolo, C.B., Sudano, I. and Salvetti, A. (1997c) Hypertension causes premature aging of endothelial function in humans. *Hypertension*, **29**, 736–743.

Taddei, S., Virdis, A., Mattei, P., Ghiadoni, L., Gennari, A., Fasolo, C.B., Sudano, I. and Salvetti, A. (1995) Aging and endothelial function in normotensive subjects and patients with essential hypertension. *Circulation*, **91**, 1981–1987.

Taddei, S., Virdis, A., Mattei, P., Ghiadoni, L., Sudano, I. and Salvetti, A. (1996b) Defective L-arginine-nitric oxide pathway in offspring of essential hypertensive patients. *Circulation*, **94**, 1298–1303.

Taddei, S., Virdis, A., Mattei, P. and Salvetti, A. (1993) Vasodilation to acetylcholine in primary and secondary forms of human hypertension. *Hypertension*, **21**, 929–933.

Takada, K., Matsumura, Y., Dohmen, S., Mitsutomi, N., Takaoka, M. and Morimoto, S. (1996) Endothelin-1 secretion from cultured vascular endothelial cells of DOCA-salt hypertensive rats. *Life Sci*, **59**, L111–116.

Takase, H., Moreau, P., Kung, C.F., Nava, E. and Luscher, T.F. (1996) Antihypertensive therapy prevents endothelial dysfunction in chronic nitric oxide deficiency. Effect of verapamil and trandolapril. *Hypertension*, **27**, 25–31.

Takase, H., Moreau, P. and Luscher, T.F. (1995) Endothelin receptor subtypes in small arteries. Studies with FR139317 and bosentan. *Hypertension*, **25**, 739–743.

Timmermans, P.B., Wong, P.C., Chiu, A.T., Herblin, W.F., Benfield, P., Carini, D.J., Lee, R.J., Wexler, R.R., Saye, J.A. and Smith, R.D. (1993) Angiotensin II receptors and angiotensin II receptor antagonists. *Pharmacol. Rev.*, **45**, 205–251.

Tschudi, M.R., Criscione, L. and Luscher, T.F. (1991) Effect of aging and hypertension on endothelial function of rat coronary arteries. *J. Hypertens. Suppl*, **9**, S164–165.

Tschudi, M.R., Criscione, L., Novosel, D., Pfeiffer, K. and Luscher, T.F. (1994a) Antihypertensive therapy augments endothelium-dependent relaxations in coronary arteries of spontaneously hypertensive rats. *Circulation*, **89**, 2212–2218.

Tschudi, M.R. and Luscher, T.F. (1995) Age and hypertension differently affect coronary contractions to endothelin-1, serotonin, and angiotensins. *Circulation*, **91**, 2415–2422.

Tschudi, M.R., Mesaros, S., Luscher, T.F. and Malinski, T. (1996) Direct in situ measurement of nitric oxide in mesenteric resistance arteries. Increased decomposition by superoxide in hypertension. *Hypertension*, **27**, 32–35.

Tschudi, M.R., Noll, G., Arnet, U., Novosel, D., Ganten, D. and Luscher, T.F. (1994b) Alterations in coronary artery vascular reactivity of hypertensive Ren-2 transgenic rats. *Circulation*, **89**, 2780–2786.

Vallance, P., Calver, A. and Collier, J. (1992a) The vascular endothelium in diabetes and hypertension. *J. Hypertens. Suppl*, **10**, S25–29.

Vallance, P., Collier, J. and Moncada, S. (1989) Effects of endothelium-derived nitric oxide on peripheral arteriolar tone in man [see comments]. *Lancet*, **2**, 997–1000.

Vallance, P., Leone, A., Calver, A., Collier, J. and Moncada, S. (1992b) Accumulation of an endogenous inhibitor of nitric oxide synthesis in chronic renal failure. *Lancet*, **339**, 572–575.

Vanhoutte, P. (1993) Other endothelium-derived vasoactive factors. *Circulation*, **87 (Suppl VII)**, VII9–VII17.

Vanhoutte, P.M. (1988) Vascular endothelium and Ca2+ antagonists. *J. Cardiovasc. Pharmacol.*, **12**, S21–28.

Verma, S., Bhanot, S. and McNeill, J.H. (1995) Effect of chronic endothelin blockade in hyperinsulinemic hypertensive rats. *Am. J. Physiol.*, **269**, H2017–2021.

Wagner, O.F., Christ, G., Wojta, J., Vierhapper, H., Parzer, S., Nowotny, P.J., Schneider, B., Waldhausl, W. and Binder, B.R. (1992) Polar secretion of endothelin-1 by cultured endothelial cells. *J. Biol. Chem.*, **267**, 16066–16068.

Wenzel, R.R., Noll, G. and Luscher, T.F. (1994) Endothelin receptor antagonists inhibit endothelin in human skin microcirculation. *Hypertension*, **23**, 581–586.

Wenzel, R.R., F.M., Shaw, S., Noll, G., Kaufmann, U., Schmitt, R., Jones, C.R., Clozel, M., Meier, B. and Lüscher, T.F. (1998) Hemodynamic and coronary effects of the endothelin antagonist bosentan in patients with coronary artery disease. *Circulation*, **98**, in press.

Wiemer, G., Scholkens, B. A., Becker, R. H. and Busse, R. (1991) Ramiprilat enhances endothelial autacoid formation by inhibiting breakdown of endothelium-derived bradykinin. *Hypertension*, **18**, 558–563.

Wright, C.E. and Fozard, J.R. (1988) Regional vasodilation is a prominent feature of the haemodynamic response to endothelin in anaesthetized, spontaneously hypertensive rats. *Eur. J. Pharmacol.*, **155**, 201–203.

Yanagisawa, M., Kurihara, H., Kimura, S., Tomobe, Y., Kobayashi, M., Mitsui, Y., Yazaki, Y., Goto, K. and Masaki, T. (1988) A novel potent vasoconstrictor peptide produced by vascular endothelial cells [see comments]. *Nature*, **332**, 411–415.

Yang, Z., Arnet, U., von Segesser, L., Siebenmann, R., Turina, M. and Luscher, T.F. (1993) Different effects of angiotensin-converting enzyme inhibition in human arteries and veins. *J. Cardiovasc. Pharmacol.*, **22**, S17–22.

Yang, Z., Bauer, E., von Segesser, L., Stulz, P., Turina, M. and Luscher, T.F. (1990) Different mobilization of calcium in endothelin-1–induced contractions in human arteries and veins: effects of calcium antagonists. *J. Cardiovasc. Pharmacol.*, **16**, 654–660.

Yokokawa, K., Tahara, H., Kohno, M., Murakawa, K., Yasunari, K., Nakagawa, K., Hamada, T., Otani, S., Yanagisawa, M. and Takeda, T. (1991) Hypertension associated with endothelin-secreting malignant hemangioendothelioma. *Ann Intern Med*, **114**, 213–215.

Zeiher, A.M., Drexler, H., Saurbier, B. and Just, H. (1993) Endothelium-mediated coronary blood flow modulation in humans. Effects of age, atherosclerosis, hypercholesterolemia, and hypertension. *J. Clin. Invest.*, **92**, 652–662.

Zeiher, A.M., Schachinger, V. and Minners, J. (1995) Long-term cigarette smoking impairs endothelium-dependent coronary arterial vasodilator function. *Circulation*, **92**, 1094–1100.

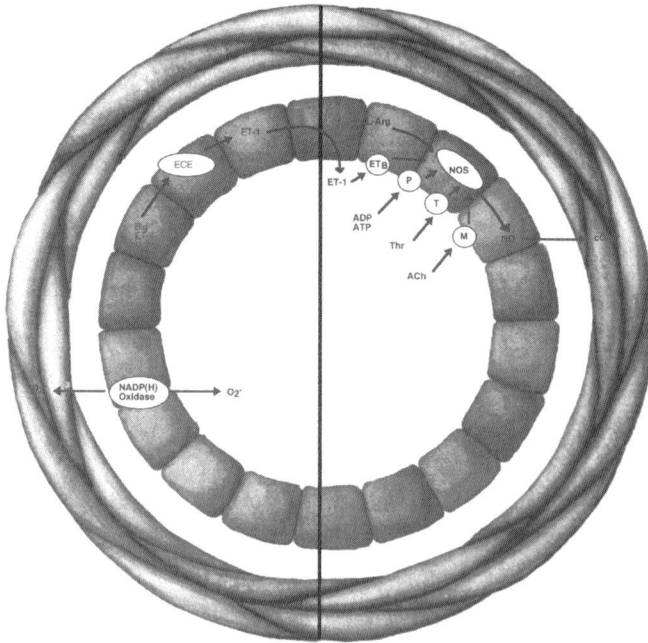

Patients with chronic heart failure are hemodynamically characterized by increased vasoconstriction and a reduced vasodilator response to exercise. In addition to various compensatory neurohumoral mechanisms there is evidence that the endothelium plays an important role in the abnormal vasodilator response. This evidence comes from studies investigating the microvascular response to regional, intrarterial administration of the endothelium-dependent vasodilator acetylcholine which found that the vasodilator response and, therefore, bioavailability of nitric oxide was impaired in the microcirculation of the leg, forearm, and myocardium of patients with chronic heart failure. The mechanisms underlying this abnormal response are not entirely clear but may reflect a muscarinic receptor abnormality. Since conduit artery vasodilation during hyperemic blood flow is also impaired and since this response is not dependent on muscarinic receptor activation this possibility appears to be unlikely. However, impaired smooth muscle responsiveness to nitric oxide stimulation, impaired L-arginine availability or utilization, endothelial release of vasoconstricting prostanoids, increased nitric oxide degradation and reduced nitric oxide synthase activity have all been implicated in this impaired response. In addition, the vasoconstrictor activity of endothelin-1 seems to play an important role in the regulation of tone in chronic heart failure although the importance of different endothelin-receptors is not clear yet.

Key words: Heart failure, endothelium, acetylcholine, nitric oxide, L-arginine, substance P, endothelin-1

7 Endothelial Control of Vascular Tone in Chronic Heart Failure

Wolfgang Kiowski[1] and Helmut Drexler[2]

[1]*Divisions of Cardiology, University Hospital Zurich, CH-8091 Zurich, Switzerland*
[2]*Medizinische Hochschule Hannover, Germany*

INTRODUCTION

It is increasingly recognized that the endothelium plays an important role in the control of vascular tone by releasing both vasodilating and vasoconstricting substances (Furchtgott and Zawadsky, 1980; Yanagisawa *et al.*, 1988; Vanhoutte, 1989; Vane *et al.*, 1990). These mechanisms are important both for the regulation of microvascular (Linder *et al.*, 1990; Kiowski *et al.*, 1991; Panza *et al.*, 1993; Haynes and Webb, 1994) and larger conduit arteries (Joannides *et al.*, 1995; Hornig *et al.*, 1996). Patients with chronic heart failure are hemodynamically characterized by increased vasoconstriction and a reduced vasodilator response to exercise (Zelis *et al.*, 1968). These abnormalities appear to be due to a number of compensatory mechanisms and some neurohumoral factors involved in this impaired vasodilator response have been studied extensively in the past. However, there is growing evidence now to suggest that the endothelium also plays an important role in the abnormal vasodilator response. In particular, the role of endothelium-derived nitric oxide has received considerable attention. The present review, therefore, summarizes the evidence for a reduced vasodilator capacity of the endothelium in chronic heart failure and focuses on the potential mechanisms underlying this abnormality. In addition, the importance of endothelial vasoconstrictor mechanisms is also discussed.

IMPAIRED ENDOTHELIUM DEPENDENT VASODILATION

The endothelium, in response to a variety of stimuli including increased shear stress during increased blood flow (Joannides *et al.*, 1995; Hornig *et al.*, 1996) or muscarinic receptor stimulation (Furchgott and Zawadzky, 1980; Linder *et al.*, 1990) releases nitric oxide (NO) which is formed intracellularly by the action of the enzyme NO-synthase from L-arginine (Palmer *et al.*, 1987). Following its release and uptake into vascular smooth muscle cells NO stimulates guanylyl cyclase to form cyclic GMP which then results in smooth muscle relaxation and vascular dilatation. Because of the ease with which endothelium-dependent, nitric oxide mediated vasodilator function can be assessed by acetylcholine or other muscarinic receptor agonists, a number of studies has used regional, e.g. intraarterial infusions of acetylcholine and measured the resultant vasodilation. Under the assumption

---

Correspondence: W. Kiowski, MD, Division of Cardiology, University Hospital Zurich, CH-8091 Zurich, Tel: ++41/1/255 23 58; Fax: ++41/1/255 44 01; E-mail: karkiw@unizh.usz.ch

that a diminished vasodilator response to acetylcholine reflects endothelial dysfunction (Linder *et al.*, 1990; Panza *et al.*, 1993) evidence for an impaired endothelium-dependent vasodilator response was found in the microcirculation of the myocardium (Treasure *et al.*, 1990), leg (Katz *et al.*, 1992), and forearm (Kubo *et al.*, 1991; Drexler *et al.*, 1992; Hirooka *et al.*, 1992; Katz *et al.*, 1993; Chin *et al.*, 1996). These data are compatible with the view that the impairment of endothelium-dependent, nitric oxide-mediated vasodilator function is a generalized phenomenon in patients with chronic heart failure. While the diminished peripheral vasodilator capacity may be important for the reduced tissue perfusion during physical exercise and, therefore, is likely to contribute to impaired exercise capacity, the disturbance of microvascular dilatation in the coronary circulation might result in ischemia and further myocardial damage and dysfunction in chronic heart failure.

The studies cited so far tested endothelium dependent vasodilator capacity but did not evaluate the importance of endothelium-derived NO in the maintenance of basal vascular tone. The regional infusion of N-monomethyl-L-arginine (L-NMMA) a selective inhibitor of NO-synthase (Moncada *et al.*, 1989) can be used to assess the importance of endothelium-derived NO for the maintenance of basal vascular tone (Vallance *et al.*, 1989). Using brachial artery infusions of L-NMMA, it was shown that resistance vessels from patients with chronic heart failure showed greater vasoconstriction in response to removal of NO-mediated vasodilation by L-NMMA as compared to control subjects (Drexler *et al.*, 1992). Therefore, in contrast to stimulated NO-dependent vasodilation, the contribution of NO to basal tone is preserved or may even be enhanced in forearm resistance vessels in patients with chronic heart failure (Drexler *et al.*, 1992). However, these results were obtained in a small group of patients and further studies are needed to better define the importance of endothelium-derived NO for the regulation of basal vascular tone in heart failure patients. Interestingly though, peripheral venous nitrate, the product of NO-metabolism in blood, has also been found to be increased in heart failure (Habib *et al.*, 1994; Winlaw *et al.*, 1994), a finding compatible with an enhanced basal nitric oxide production in heart failure.

The study described so far evaluated endothelial function of resistance vessels. Much less is known about conduit arteries. The measurement by ultrasound techniques (Celermajer *et al.*, 1993; Joannides *et al.*, 1995; Hornig *et al.*, 1996) of diameter increases of larger conduit arteries in response to increases in flow as occur e.g. during reactive hyperemia provides the possibility to study endothelium-dependent vasodilation of larger conduit arteries in humans. The validity of this approach is shown in Figure 7.1. In seven normal volunteers, flow-mediated vasodilation was assessed as diameter changes of the radial artery during hyperemia following upper arm occlusion before and after brachial artery infusion of the NO-synthase inhibitor L-NMMA (7 μmol/min) (Hornig *et al.*, 1996). As depicted in the figure, flow-dependent vasodilation (expressed as percent change from baseline vessel diameter) was signicantly reduced following NO-synthase inhibition. This finding confirms previous results (Joannides *et al.*, 1995) and demonstrates that flow-mediated vasodilation indeed is largely NO driven. Using this approach in patients with chronic heart failure it was shown that flow-dependent vasodilation was significantly attenuated in patients with chronic heart failure in the radial artery (Hornig *et al.*, 1996). Interestingly, this impairment was similar in patients with dilated and ischemic cardiomyopathy and was significantly higher in the non-dominant arm. Importantly, after blockade of NO-synthase by L-NMMA flow-dependent forearm vasodilation was similar in control subjects and patients suggesting that the impairment in patients was mostly due to reduced NO-bioavailability.

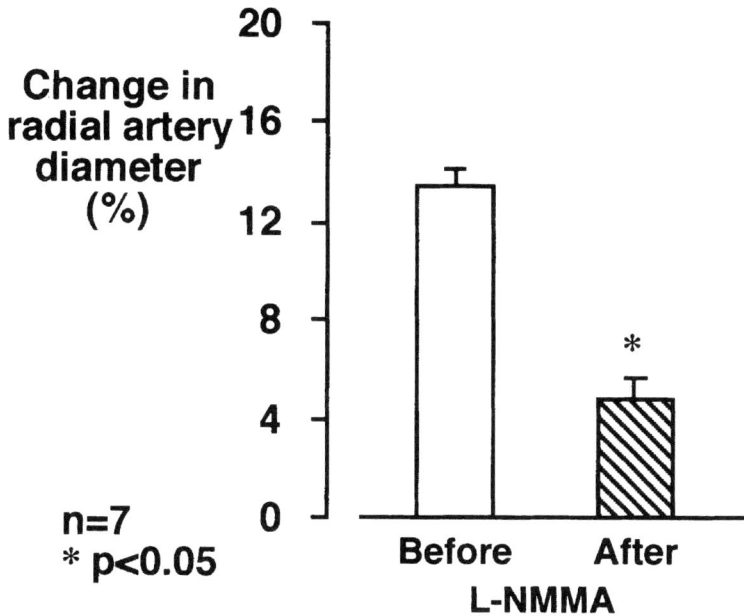

Figure 7.1 Nitric oxide dependcy of flow dependent vasodilation in normal volunteers. Flow dependent vasodilation was assessed as percent changes of radial artery diameter following release of an upper arm cuff which was inflated to suprasytolic pressure for 7 minutes in 7 normal volunteers before adn after brachial artery infusion of L-NMMA (7μmol/min) to inhibit NO synthesis. As shown, radial artery vasodilation was markedly reduced during hyperemia following L-NMMA demonstrating that flow dependent vasodilation is to alarge extent NO driven (Hornig *et al.*, 1996).

Taken together, endothelial dysfunction is not only present in the microcirculation of patients with heart failure but also in large conduit vessels. Since conduit vessels are more than just passive conduits (Ramsey and Jones, 1994) it is tempting to speculate that such an impairment of flow-mediated vasodilation during exercise might lead to increased impedance to left ventricular ejection and, thereby, contribute to the hemodynamic derangements characteristic of chronic heart failure. So far, it is not known what constitutes the primary stimulus for endothelial vasodilator dysfunction in chronic heart failure. Animal experiments suggest that endothelial dysfunction is a progressive, time-dependent process that probably plays a minor role early in heart failure. Thus, rats in whom heart failure was induced by coronary ligation and subsequent myocardial infarction demonstrated no evidence of endothelial dysfunction at week 1 but a reduced acetylcholine response of thoracic aortic rings was evident after 4 and 16 weeks as compared to sham-operated control rats (Teerlink *et al.*, 1993).

MECHANISMS OF ENDOTHELIAL VASODILATOR DYSFUNCTION

A number of factors might be responsible for the impaired NO-dependent vasodilation in chronic heart failure (Table 7.1) and many of these aspects have been studied in humans.

Table 7.1 Potential mechanisms of reduced acetylcholine responses
in chronic heart failure.

Reduced NO-synthase activity
Impaired muscarinic receptor function/signal transduction pathway
Reduced L-arginine availability/utilization
Increased nitric oxide degradation
Acetylcholine induced vasoconstrictor prostanoid release
Impaired smooth muscle responses to NO

Obviously, the reduced endothelium-dependent, NO-mediated vasodilator response to acetylcholine may be due to a reduced activity of the NO-forming enzyme NO synthase. Since pulsatile flow (Rubanyi *et al.*, 1986; Awobsi *et al.*, 1994) and the associated shear stress (Buga *et al.*, 1991; Uematsu *et al.*, 1995) are important regulators of NO-production it is tempting to speculate that the reduced cardiac output and stroke volume may be the link to impaired endothelial vasodilator function in chronic heart failure. Recent experimental evidence supports this possibility. Thus, in dogs with heart failure after 1 month of rapid ventricular pacing NO-synthase gene expression was reduced by 56% in endothelial cells from the thoracic aorta as compared to control animals. In addition, there was a marked, e.g. 70 % reduction in endothelial cell NO-synthase protein in heart failure dogs and a marked reduction in nitrate production as a measure of enzyme activity in response to stimulation by either acetylcholine or bradykinin. Interestingly, this down-regulation of NO-synthase was accompanied by a similar down-regulation of endothelial cyclooxygenase-1 in heart failure dogs (Smith *et al.*, 1996). Moreover, nitric oxide production from microvessels isolated from patients with endstage heart failure undergoing transplantation appeared to be reduced as compared to nitrite production from the microvasculature of normal hearts (Kichuk *et al.*, 1996). However, the number of normal hearts was too small to allow a statistically valid comparison. Taking together, these experimental studies indicate that stimulated endothelial NO-production may be reduced in patient with chronic heart failure, possibly through reduced gene expression of vascular endothelial NO-synthase.

Abnormalities found in studies using acetylcholine might also be explained by a defect at the muscarinic receptor level or its signal transduction pathway. Also, the majority of patients with heart failure has underlying ischemic heart disease and coronary risk factors are known to adversely affect the vascular response to acetylcholine (Creager *et al.*, 1990; Linder *et al.*, 1990; Panza *et al.*, 1993; Seiler *et al.*, 1993); however, a reduced response to acetylcholine was also found in heart failure patients with non-ischemic cardiomyopathy and without coronary risk factors (Nakamura *et al.*, 1996). Moreover, acetylcholine has direct smooth muscle contracting properties (Furchtgott and Zawadsky, 1980). Therefore, the response to acetylcholine always represents the net effects of the release of vasodilating substances from the endothelium and direct smooth muscle vasoconstriction. Accordingly, the use of a pure endothelium-dependent vasodilator without effects on vascular smooth muscle might provide a more clear picture of endothelial vasodilator capacity. Substance P is a peptide which stimulates NO-synthase through a different endothelial receptor, e.g. the tachykinin receptor (Saito *et al.*, 1991) and has no direct effects on smooth muscle cells. Substance P has been shown to result in dilation of epicardial coronary arteries and coronary microvessels in humans (Crossman *et al.*, 1989) and of forearm resistance and capacity vessels (McEwan *et al.*, 1988). This vasodilator effect can be significantly attenuated by the NO-synthase inhibitor L-NMMA in normal subjects (Panza *et al.*, 1994).

Interestingly, acetylcholine-induced dilation of forearm resistance vessels was significantly reduced in patients with heart failure whereas the increase in forearm blood flow in response to substance P was not impaired (Hirooka *et al.*, 1992). Similarly, intracoronary infusions of acetylcholine caused significantly less increases in coronary blood flow in heart failure patients as compared to control subjects whereas substance P resulted and similar increases in coronary blood flow in both groups (Holdright *et al.*, 1994). Furthermore, the epicardial vasodilator response to substance P was also similar in both groups (Holdright *et al.*, 1994). These studies, therefore, are compatible with the contention that chronic heart failure may be associated with a specific receptor abnormality of the muscarinic receptor and/or post receptor coupling mechanisms which could contribute to the observed reduction of the response to acetylcholine in patients. These data suggest also that substance P may be a better pharmacological to investigate endothelial NO-dependent vasodilation than acetylcholine. Nevertheless, results obtained with stimulation of NO-release through another, non-muscarinic receptor suggest that the abnormality in heart failure patients is not likely to be explained solely by a defect at the muscarinic receptor level. Thus, vasodilation in response to stimulation of vasopressin type-2 (V2) receptors (Hirsch *et al.*, 1989) is dependent on endothelial NO release and can be blocked by L-NMMA but not indomethacin (Liard, 1994; Tagawa *et al.*, 1995). When the V2-receptor agonist desmopressin was infused into the brachial artery of heart failure patients and control subjects the ensuing vasodilation was significantly attenuated in patients (Rector *et al.*, 1996). Moreover, inhibition of NO-synthesis by L-NMMA reduced desmopressin responses to a significantly greater extent in control subjects as compared to patients (Rector *et al.*, 1996). Accordingly, these data are compatible with the view that impaired endothelium-dependent vasodilation in patients with chronic heart failure is not limited to a defect of the muscarinic receptor or its signal transduction pathway.

EFFECTS OF L-ARGININE

In some studies, brachial artery infusions of L-arginine have been shown to augment the forearm vasodilator response to acetylcholine in normal subjects (Panza *et al.*, 1993; Hirooka *et al.*, 1994) but this finding is not universal. The effects of intraarterial L-arginine on the response to acetylcholine have also been studied in patients with heart failure (Hirooka *et al.*, 1994). L-arginine augmented the vasodilator response to acetylcholine in normal subjects except for the highest dose. In contrast, in patients with heart failure, L-arginine also augmented the vasodilator response to the highest dose of acetylcholine. Moreover, L-arginine did not affect the postischemic increase in forearm blood flow after upper arm occlusion in normal subjects; however, it significantly increased postischemic blood flow in patients with heart failure (Hirooka *et al.*, 1994). While the meaning of an enhanced vasodilator response to only the highest dose of acetylcholine in patients is somewhat uncertain the finding of an increased maximal vasodilator response after pretreatment with L-arginine during reactive hyperemia is interesting. These findings may suggest that impaired endothelium-dependent vasodilation during maximal pharmacological stimulation depends upon substrate availability/utilization and would be compatible with the view that impaired post-ischemic vasodilation results largely from a defect in release of nitric oxide from the endothelium in heart failure patients.

In contrast to results obtained with intraarterial administration of L-arginine, oral supplementation with L-arginine in doses between 5.6 and 20g/day for 4–6 weeks failed to augment the response to muscarinic receptor stimulation (Chin *et al.*, 1996; Rector *et al.*, 1996). Also, it did not enhance reactive hyperemia (Rector *et al.*, 1996). Thus, oral administration of L-arginine does not suffice to improve endothelial, NO-dependent vasodilator dysfunction in patients with heart failure. Interestingly though, oral L-arginine improved functional status as indicated by increased distances during a six minute walk test and lower scores on the Living With Heart Failure questionnaire in one (Rector *et al.*, 1996) but not the other study (Chin *et al.*, 1996). Further studies are needed to identify whether and how oral L-arginine supplementation might affect NO-dependent vasodilation in patients with heart failure. Appropriate control studies with D-arginine are also required.

ROLE OF CYTOKINES

Recent data suggest that increased levels of cytokines might be involved in the development of peripheral endothelial dysfunction in patients with heart failure. Circulating cytokines, and particularly tumor necrosis factor alpha (TNFα), are increased in severe heart failure (Wiedermann *et al.*, 1993). Experimental evidence suggests that TNFa decreases constitutive nitric oxide synthase messenger RNA in vascular endothelial cells by shortening its half-life (Yoshizumi *et al.*, 1993) and increases expression of the inducible form of nitric oxide synthase in vascular endothelial (Gross *et al.*, 1991) and smooth muscle cells (Busse and Mulsch, 1990). Moreover, it increases vascular smooth muscle production of superoxide anion (O_2^-) which decreases the half-life of nitric oxide (Matsubara and Ziff, 1986). Thus, TNFα may either stimulate or inhibit endothelium-dependent, nitric oxide-mediated vasodilation dependent on the predominant effects of either decreasing NO-bioavailability through enhanced destruction or increasing NO-bioavailability through NO synthase induction. So far, there is little data available in humans. Interestingly, TNFa concentrations were closely correlated with forearm blood flow responses to brachial artery infusions of acetylcholine (Katz *et al.*, 1994). Thus, even moderate increases in serum TNFα concentrations as found in that study may be sufficient to activate the inducible form of nitric oxide synthase and potentiate the vascular effects of constitutive nitric oxide synthase stimulation by acetylcholine (Katz *et al.*, 1994). A study in human hand veins in vivo showed that TNFα potently inhibits endothelium-dependent relaxation in these vessels (Bhagat *et al.*, 1997).

SUPEROXIDE

A further possibility which might explain endothelial dysfunction due to reduced NO-bioavailability is increased degradation of NOs. Principle among the substances involved in the breakdown process of NO is O_2^-, an avid scavenger of endothelium-derived NO

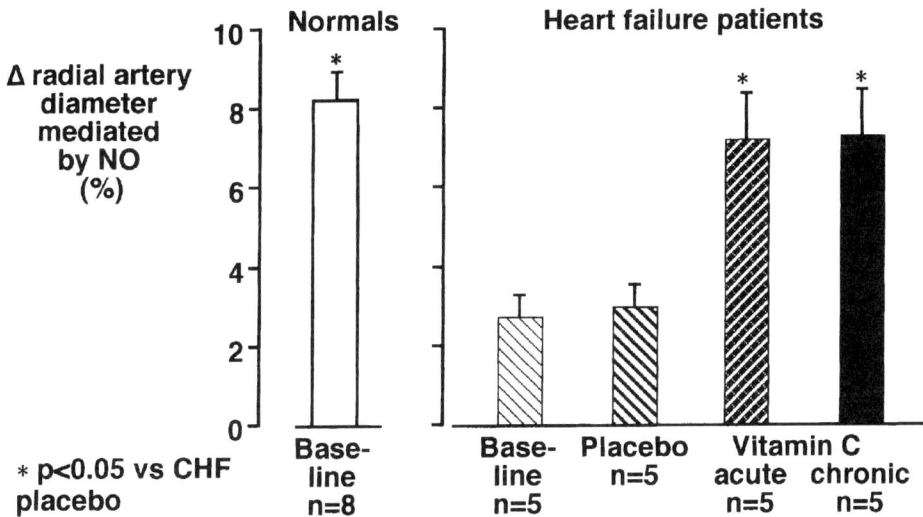

Figure 7.2   Effect of acute and chronic vitamin C administration on NO dependent component of flow mediated radial artery vasodilation in patients with heart failure. NO dependency was assessed as difference between control measurements and measurements during NO synthesis inhibition by brachial artery L-NMMA. Placebo administration did not change the markedly diminished response in heart failure patients but both acute brachial artery infusion and chronic oral supplementation with vitamin C markely improved the NO dependent component of flow dependent vasodilation so that it was no longer different from control subjects.

(Gryglewski *et al.*, 1986). Accordingly, prevention of O_2^- formation or administration of antioxidant substances which scavenge O_2^- like vitamin C (Frei *et al.*, 1989) could provide information for the importance of this potential mechanism. So far, only limited data is available in patients with heart failure. However, a recent study suggests that increased oxidative stress may well be an important contributing factor to the impaired endothelial vasodilator function of patients with chronic heart failure. Thus, as shown in Figure 7.2, the fraction of flow dependent radial artery vasodilation which is mediated by NO (as assessed by NO-synthase inhibition by brachial artery infusion of L-NMMA) was markedly impaired in patients with heart failure as compared to control subjects (Hornig *et al.*, 1998). While placebo did not change this impaired vasodilator response in patients, acute intraarterial administration of vitamin C significantly enhanced NO-mediated, flow dependent vasodilation in patients with heart failure. Moreover, oral administration of vitamin C resulted in maintenance of this improved vasodilator function (Hornig *et al.*, 1998). Therefore, these data suggest that increased NO-degradation due to oxidative stress may well be an important factor contributing to impaired NO-dependent vasodilatation in patients with heart failure. The primary stimulus for increased oxidative stress remains to be established. However, increased levels of cytokines and among them TNFα increase O_2^- release from human endothelial cells (Matsubara and Ziff, 1986). Also, angiotensin II which is high in advanced heart failure stimutes vascular O_2^- production via membrane NADH/NADPH oxidase activation (Rajagopalan *et al.*, 1996).

GUANYLYL CYCLASE RESPONSIVENESS

NO stimulates guanylyl cyclase in vascular smooth muscle cells and the resultant increase in cyclic GMP leads to vasodilatation. It is conceivable that a reduced responsiveness of this system to NO-stimulation could also contribute to endothelial vasodilator dysfunction. This possibility has been tested in most studies by assessing the vasodilator response to arterial infusion of a direct NO-donor (e.g. nitroglycerin or sodium nitroprusside) and a reduced response to nitroglycerin was observed by Zelis *et al.* in their landmark study of the peripheral circulation in, 1968 already (Zelis *et al.*, 1968). While many more recent studies did not find significant differences in the vascular responses to direct-acting NO-donors the opposite has also been reported. Thus, nitroglycerin resulted in a significantly smaller increase in mean blood flow velocity of the superficial femoral artery after intraarterial infusion in patients with heart failure as compared to control subjects (Katz *et al.*, 1992) and brachial artery infusions of nitroglycerin caused significantly less forearm resistance vessel dilatation in heart failure patients as compared to control subjects (Katz *et al.*, 1993). Therefore, a reduced vascular smooth muscle responsiveness to NO-dependent cyclic GMP-mediated vasodilation may also contribute to this apparent endothelial vasodilator dysfunction, at least in some patients.

Finally, acetylcholine also stimulates production of endothelium-derived vasoactive substances originating from the cyclooxygenase metabolic pathway (Katusic *et al.*, 1988) and an acetylcholine-mediated cyclooxygenase-dependent vasoconstrictor effect has been reported in a canine model of heart failure (Kaiser *et al.*, 1989). This possibility was tested in a study of the forearm circulation in which patients with heart failure showed a blunted response to acetylcholine as compared to control subjects (Katz *et al.*, 1993). When these experiments were repeated after cyclooxygenase inhibition with indomethacin the vasodilator response to acetylcholine was unchanged in normal subjects but significantly increased in patients (39%). Despite this improvement, the response was still significantly attenuated as compared to normal subjects (Katz *et al.*, 1993). In addition, intraarterial infusion of sodium nitroprusside in heart failure patients treated with aspirin resulted in significantly greater vasodilatation as compared to patients not pretreated with aspirin (Jeserich *et al.*, 1995). Both findings are compatible with the view that an abnormal production of cyclooxygenase dependent vasoconstricting factor(s) seems to be present in the peripheral circulation of patients with heart failure. Such an effect may blunt the vasodilatory effects of both endogenous NO liberated by e.g. acetylcholine as well as of exogenous NO derived from direct NO donors.

ENDOTHELIN DEPENDENT VASOCONSTRICTION

Endothelin-1 is the principal endothelin isoform of the family of endothelins and it is the most potent vasoconstrictor substance generated in the human vascular wall (Yanagisawa *et al.*, 1988). Plasma levels of the peptide are increased 2 to 3 fold in patients with heart failure, particularly in more advanced heart failure (Cody *et al.*, 1992; Wei *et al.*, 1994; Kiowski *et al.*, 1995) and the increase in the plasma levels is correlated with the extent of hemodynamic impairment (Kiowski *et al.*, 1991; Cody *et al.*, 1992). This is shown in Figure 7.3 which demonstrates significant correlations between plasma levels of this peptide and cardiac filling pressures, pulmonary pressure, and cardiac output in patients

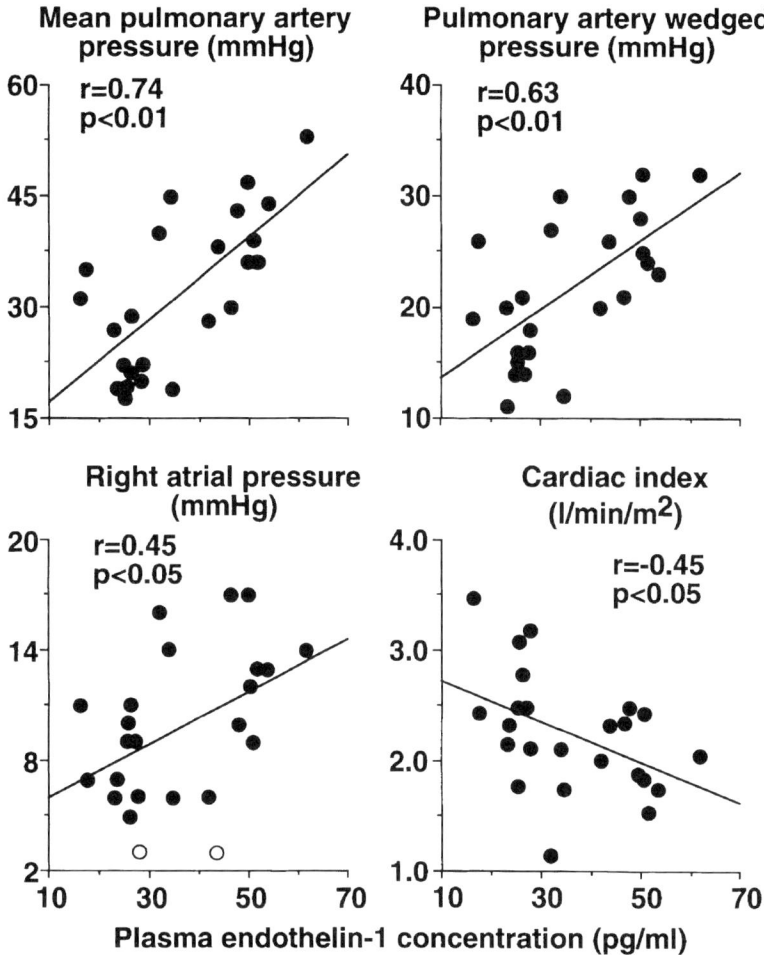

Figure 7.3 Relationships between baseline plasma endothelin-1 concentrations and the extent of pulmonary hypertension, cardiac filling pressures and cardiac output in patients with chronic heart failure. The correlations indicate that higher plasma endothelin-1 concentrations are associated with a greater degree of pulmonary hypertension, more severely elevated right and left sided cardiac filling pressures and lower cardiac output.

with symptoms according to NYHA class III. Plasma big endothelin-1 which may better reflect endothelial synthesis than the mature peptide also predicts prognosis in heart failure (Pacher *et al.*, 1996). Because endothelin-1 has also anti-natriuretic and mitogenic properties (Ito *et al.*, 1991; Sorensen *et al.*, 1994) it has been suggested that it may play an important role in the pathophysiology of chronic heart failure. Endothelin exerts its vascular effects through two principal endothelin receptor subtypes, ET_A and ET_B. Vascular smooth muscle ET_A receptors mediate endothelin-1-induced vasoconstriction (Riezebos *et al.*, 1994) and endothelial ET_B receptors mediate vasodilation through release of either prostacyclin and/or nitric oxide (Takayanagi *et al.*, 1991; Hirata *et al.*, 1993). However, there is also evidence, that vascular smooth muscle ETB receptors mediate vasoconstriction (Gray *et al.*, 1994; Seo *et al.*, 1994; Haynes *et al.*, 1995). Accordingly, stimulation of vascular ET_B presumably results in a mixture of endothelium-mediated vasodilation and direct smooth muscle vasoconstriction.

The advent of specific endothelin-receptor antagonists has provided the opportunity to investigate the role of endothelin-1 mediated vasoconstriction under normal circumstances and in patients with heart failure. As shown in normal volunteers, infusion of the ET_A receptor antagonist BQ 123 resulted in significant forearm vasodilation indicating that endothelin is involved in the regulation of basal vascular tone in normal subjects (Haynes and Webb, 1994).

In patients with heart failure, intravenous infusion of the mixed ET_A/ET_B receptor antagonist bosentan also resulted in significant hemodynamic effects (Kiowski *et al.*, 1995). As shown in Figure 7.4, endothelin receptor antagonism resulted in significant decreases of left and right heart filling pressures, arterial pressure and pulmonary artery pressure together with an increase in cardiac output as compared to placebo administration. Since heart rate did not change, the increase in cardiac output was due to an increase in stroke volume. Accordingly, calculated systemic vascular resistance was significantly decreased and, importantly, pulmonary vascular resistance was also significantly reduced. The results indicate, that blockade of endogenous endothelin-1-mediated vasoconstriction resulted in significant hemodynamic improvement in patients with advanced heart failure pointing towards the importance of this endothelial vasoconstrictor system in patients with heart failure.

So far, the role of the ET_B receptor in patients with heart failure is not clear. Thus, the vasoconstrictor response to a specific ET_B receptor agonist, sarafotoxin S6c was increased in patients with heart failure compatible with an increased sensitivity of this receptor to endothelin-1 (Love *et al.*, 1996). However, a more recent study using the ET_B receptor antagonist BQ 788 suggested that the overall effect of ET_B receptor stimulation in the human forearm may be dilatation (Verhaar *et al.*, 1998). However, more studies are needed to ascertain the specific role of ETB receptors in the control of the circulation in patients with heart failure.

CONCLUSIONS

There is abundant evidence showing that chronic heart failure is associated with impaired endothelium-dependent microvascular and larger conduit vessel vasodilator dysfunction. Although much of this impairment can be attributed to reduced NO bioavalability the precise mechanism(s) underlying this defect remain(s) to be established. The role of endothelin-1-mediated vasoconstriction and the importance of different endothelin receptors in chronic heart failure also need further study.

REFERENCES

Awobsi, M.A., M.D. Widmath, W.C. Sessa and B.E. Sumpio (1994) Cyclic strain increases endothelial nitric oxide synthase activity. *Surgery*, **116**, 439–45.

Bhagat, K, and Vallancem P. (1997) Inflammatory cytokines impair endothelium-dependent dilatation in human veins *in vivo*. *Circulation*, **96**, 3042–3047.

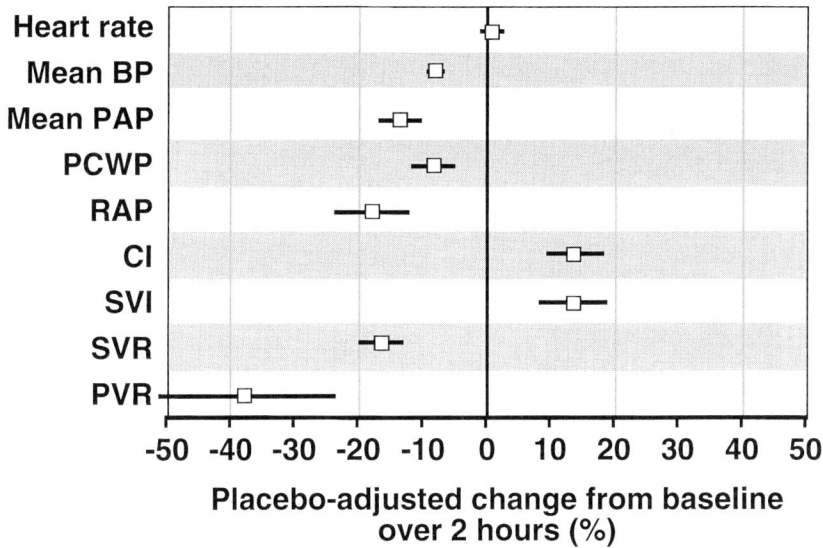

Figure 7.4 Placebo-adjusted hemodynamic changes observed over two hours in 12 patients in whom the mixed ETA/ETB receptor antagonist bosentan was given intravenously (62). Values are means and the bars indicate 95% confidence intervals of the differences against placebo-treated patients (n=12). Abbr.: BP: blood pressure; PAP: pulmonary artery pressure; PCWP: pulmonary capillary wedge pressure; RAP: right atrial pressure; CI: cardac index; SVI: stroke volume index; SVR: systemic vascular resistance; PVR: pulmonary vascular resistance

Buga, G.M., M.E. Gold, J.M. Fukuto and L.J. Ignarro (1991) Shear-stress induced nitric oxide release from endothelial cells grown on beads. *Hypertension* **17**, 187–92.

Busse, R. and A. Mulsch (1990) Induction of nitric oxide synthase by cytokines in vascular smooth muscle. *FEBS Lett.*, **275**, 87–90.

Celermajer, D.S., K.E. Sorensen, D. Georgakopoulos, C. Bull, O. Thomas, J. Robinson, *et al.* (1993) Cigarette smoking is associated with dose-related and potentially reversible impairment of endothelium-dependent dilation in healthy young adults. *Circulation*, **88**, 2149–55.

Chin, D.J., D.M. Kaye, J. Lefkovits, J. Wong, P. Bergin and G.L. Jennings (1996) Dietary supplementation with L-arginine fails to restore endothelial function in forearm resistance arteries of patients with severe heart failure. *J. Am. Coll. Cardiol.*, **27**, 1207–13.

Cody, R.J., G.J. Haas, P.F. Binkley, Q. Capers and R. Kelley (1992) Plasma endothelin correlates with the extent of pulmonary hypertension in patients with chronic congestive heart failure. *Circulation*, **85**, 504–9.

Creager, M.A., J.P. Cooke, M.E. Mendelsohn, S.J. Gallagher, S.M. Coleman, J. Loscalzo, *et al.* (1990) Impaired vasodilation of forearm resistance vessels in hypercholesterolemic humans. *J. Clin. Invest.*, **86**, 228–34.

Crossman, D.C., S.W. Larkin, R.W. Fuller, G.J. Davies and A. Maseri (1989) Substance P dilates epicardial coronary arteries and increases coronary blood flow in humans. *Circulation*, **80**, 475–84.

Drexler, H., D. Hayoz, T. Munzel, B. Hornig, H. Just, H.R. Brunner, *et al.* (1992) Endothelial function in chronic congestive heart failure. *Am. J. Cardiol.*, **69**, 1596–601.

Frei, B., L. England and B.N. Ames (1989) Ascorbate is an outstanding antioxidant in human blood plasma. *Proc. Natl. Acad. Sci. USA*, **86**, 6377–81.

Furchgott, R.F. and J.V. Zawadzki (1980) The obligatory role of endothelial cells in the relaxation of arterial smooth muscle by acetylcholine. *Nature*, **288**, 373–376.

Gray, G.A., B.M. Löffler and M. Clozel (1994) Characterization of endothelin receptors mediating contraction of rabbit saphenous vein. *Am. J. Physiol.*, **266**, H959–66.

Gross, S.S., E.A. Jaffe, R. Levi and R.G. Kilbourn (1991) Cytokine-activated endothelial cells express an isotype of nitric oxide synthase which is tetrahydrobiopterin-dependent, calmodulin-independent and inhibited by arginine analogs with a rank order of potency characteristic of activated macrophages. *Biochem. Biophys. Res. Comm.*, **178**, 823–9.

Gryglewski, R.J., R.M. Palmer and S. Moncada (1986) Superoxide anion is involved in the breakdown of endothelium-derived vascular relaxing factor. *Nature*, **320**, 454–6.

Habib, F., D. Dutka, D. Crossman, C.M. Oakley and J.G. Cleland (1994) Enhanced basal nitric oxide production in heart failure:another failed counter-regulatory vasodilator mechanism? [see comments]. *Lancet*, **344**, 371–3.

Haynes, W.G., F.E. Strachan and D.J. Webb (1995) Endothelin ETA and ETB receptors cause vasoconstriction of human resistance and capacitance vessels in vivo. *Circulation*, **92**, 357–63.

Haynes, W.G. and D.J. Webb (1994) Contribution of endogenous generation of endothelin-1 to basal vascular tone. *Lancet*, **344**, 852–4.

Hirata, Y., T. Emori, S. Eguchi, K. Kanno, T. Imai, K. Ohta, *et al.* (1993) Endothelin receptor subtype B mediates synthesis of nitric oxide by cultured bovine endothelial cells. *J. Clin. Invest.*, **91**, 1367–73.

Hirooka, Y., T. Imaizumi, S. Harada, H. Masaki, M. Momohara, T. Tagawa, *et al.* (1992) Endothelium-dependent forearm vasodilation to acetylcholine but not to substance P is impaired in patients with heart failure. *J. Cardiovasc. Pharmacol.* (suppl 12) **20**, 221–5.

Hirooka, Y., T. Imaizumi, T. Tagawa, M. Shiramoto, T. Endo, S. Ando, *et al.* (1994) Effects of L-arginine on impaired acetylcholine-induced and ischemic vasodilation of the forearm in patients with heart failure. *Circulation*, **90**, 658–68.

Hirsch, A.T., V.J. Dzau, J.A. Majzoub and M.A. Creager (1989) Vasopressin-mediated forearm vasodilation in normal humans. Evidence for a vascular vasopressin V2 receptor. *J. Clin. Invest.*, **84**, 418–26.

Holdright, D.R., D. Clarke, K. Fox, W.P. Poole and P. Collins (1994) The effects of intracoronary substance P and acetylcholine on coronary blood flow in patients with idiopathic dilated cardiomyopathy. *Eur. Heart J.*, **15**, 1537–44.

Hornig, B., N. Arakawa, C. Kohler and H. Drexler (1998) Vitamin C improves endothelial function of conduit arteries in patients with chronic heart failure. *Circulation*, **97**, 363–8.

Hornig, B., V. Maier and H. Drexler (1996) Physical training improves endothelial function in patients with chronic heart failure. *Circulation*, **93**, 210–4.

Ito, H., Y. Hirata, M. Hiroe, M. Tsujino, S. Adachi, T. Takamoto, *et al.* (1991) Endothelin-1 induces hypertrophy with enhanced expression of muscle-specific genes in cultured neonatal rat cardiomyocytes. *Circ. Res.*, **69**, 209–15.

Jeserich, M., L. Pape, H. Just, B. Hornig, M. Kupfer, T. Munzel, *et al.* (1995) Effect of long-term angiotensin-converting enzyme inhibition on vascular function in patients with chronic congestive heart failure. *Am. J. Cardiol.*, **76**, 1079–82.

Joannides, R., W.E. Haefeli, L. Linder, V. Richard, E.H. Bakkali, C. Thuillez, *et al.* (1995) Nitric oxide is responsible for flow-dependent dilatation of human peripheral conduit arteries in vivo. *Circulation*, **91**, 1314–9.

Kaiser, L., R.C. Spickard and N.B. Olivier (1989) Heart failure depresses endothelium-dependent responses in canine femoral artery. *Am. J. Physiol.*, **256**, H962–7.

Katusic, Z.S., J.T. Shepherd and P.M. Vanhoutte (1988) Endothelium-dependent contractions to calcium ionophore A23187, arachidonic acid, and acetylcholine in canine basilar arteries. *Stroke*, **19**, 476–9.

Katz, S.D., L. Biasucci, C. Sabba, J.A. Strom, G. Jondeau, M. Galvao, *et al.* (1992) Impaired endothelium-mediated vasodilation in the peripheral vasculature of patients with congestive heart failure. *J. Am. Coll. Cardiol.*, **19**, 918–25.

Katz, S.D., R. Rao, J.W. Berman, M. Schwarz, L. Demopoulos, R. Bijou, *et al.* (1994) Pathophysiological correlates of increased serum tumor necrosis factor in patients with congestive heart failure. Relation to nitric oxide-dependent vasodilation in the forearm circulation. *Circulation*, **90**, 12–6.

Katz, S.D., M. Schwarz, J. Yuen and T.H. LeJemtel (1993) Impaired acetylcholine-mediated vasodilation in patients with congestive heart failure. Role of endothelium-derived vasodilating and vasoconstricting factors [see comments]. *Circulation* **88**, 55–61.

Kichuk, M.R., N. Seyedi, X. Zhang, C.C. Marboe, R.E. Michler, L.J. Addonizio, *et al.* (1996) Regulation of nitric oxide production in human coronary microvessels and the contribution of local kinin formation. *Circulation*, **94**, 44–51.

Kiowski, W., T.F. Lüscher, L. Linder and F.R. Bühler (1991) Endothelin-1-induced vasoconstriction in humans. Reversal by calcium channel blockade but not by nitrovasodilators or endothelium-derived relaxing factor. *Circulation*, **83**, 469–75.

Kiowski, W., G. Sütsch, P. Hunziker, P. Müller, J. Kim, E. Oechslin, *et al.* (1995) Evidence for endothelin-1-mediated vasoconstriction in severe chronic heart failure. *Lancet*, **346**, 732–6.

Kubo, S.H., T.S. Rector, A.J. Bank, R.E. Williams and S.M. Heifetz (1991) Endothelium-dependent vasodilation is attenuated in patients with heart failure. *Circulation*, **84**, 1589–96.

Liard, J.F. (1994) L-NAME antagonizes vasopressin V2-induced vasodilatation in dogs. *Am. J. Physiol.*, **266**, H99–106.

Linder, L., W. Kiowski, F.R. Buhler and T.F. Luscher (1990) Indirect evidence for release of endothelium-derived relaxing factor in human forearm circulation in vivo. Blunted response in essential hypertension. *Circulation* **81**, 1762–7.

Love, M.P., W.G. Haynes, G.A. Gray, D.J. Webb and J.J.V. McMurray (1996) Vasodilator effects of endothelin-converting enzyme inhibition and endothelin ET_A receptor blockade in chronic heart failure patients treated with ACE inhibitors. *Circulation*, **94**, 2131–7.

Matsubara, T. and M. Ziff (1986) Increased superoxide anion release from human endothelial cells in response to cytokines. *J. Immunol.*, **137**, 3295–8.

McEwan, J.R., N. Benjamin, S. Larkin, R.W. Fuller, C.T. Dollery and I. MacIntyre (1988) Vasodilatation by calcitonin gene-related peptide and by substance P:a comparison of their effects on resistance and capacitance vessels of human forearms. *Circulation*, **77**, 1072–80.

Moncada, S., R.M.J. Palmer and E.A. Higgs (1989) Biosynthesis of nitric oxide from L-arginine. A pathway for the regulation of cell function and communication. *Biochem. Pharmacol.*, **38**, 1709–1715.

Nakamura, M., H. Yoshida, N. Arakawa, Y. Mizunuma, S. Makita and K. Hiramori (1996) Endothelium-dependent vasodilatation is not selectively impaired in patients with chronic heart failure secondary to valvular heart disease and congenital heart disease [see comments]. *Eur. Heart J.*, **17**, 1875–81.

Pacher, R., B. Stanek, M. Hulsmann, S.J. Koller, R. Berger, M. Schuller, *et al.* (1996) Prognostic impact of big endothelin-1 plasma concentrations compared with invasive hemodynamic evaluation in severe heart failure. *J. Am. Coll. Cardiol.*, **27**, 633–41.

Palmer, R.M.J., A.G. Ferrige and S. Moncada (1987) Nitric oxide release accounts for the biological activity of endothelium-derived relaxing factor. *Nature*, **327**, 524–526.

Panza, J.A., P.R. Casino, C.M. Kilcoyne and A.A. Quyyumi (1993) Role of endothelium-derived nitric oxide in the abnormal endothelium-dependent vascular relaxation of patients with essential hypertension. *Circulation*, **87**, 1468–74.

Panza, J.A., P.R. Casino, C.M. Kilcoyne and A.A. Quyyumi (1994) Impaired endothelium-dependent vasodilation in patients with essential hypertension:evidence that the abnormality is not at the muscarinic receptor level. *J. Am. Coll. Cardiol.* **23**, 1610–6.

Rajagopalan, S., S. Kurz, T. Munzel, M. Tarpey, B.A. Freeman, K.K. Griendling, *et al.* (1996) Angiotensin II-mediated hypertension in the rat increases vascular superoxide production via membrane NADH/NADPH oxidase activation. Contribution to alterations of vasomotor tone. *J. Clin. Invest.*, **97**, 1916–23.

Ramsey, M.W. and C.J.H. Jones (1994) Large arteries are more than passive conduits. *Br. Heart J.*, **72**, 3–4.

Rector, T.S., A.J. Bank, K.A. Mullen, L.K. Tschumperlin, R. Sih, K. Pillai, *et al.* (1996) Randomized, double-blind, placebo-controlled study of supplemental oral L-arginine in patients with heart failure [see comments]. *Circulation*, **93**, 2135–41.

Rector, T.S., A.J. Bank, L.K. Tschumperlin, K.A. Mullen, K.A. Lin and S.H. Kubo (1996) Abnormal desmopressin-induced forearm vasodilatation in patients with heart failure:dependence on nitric oxide synthase activity. *Clin. Pharmacol. Ther.*, **60**, 667–74.

Riezebos, J., I.S. Watts and P.J. Vallance (1994) Endothelin receptors mediating functional responses in human small arteries and veins. *Br. J. Pharmacol.*, **111**, 609–15.

Rubanyi, G., C.J. Romero and P.M. Vanhoutte (1986) Flow-induced release of endothelium-derived relaxing factor. *Am. J. Physiol.*, **250**, H1145–H1149.

Saito, R., S. Nonaka, H. Konishi, Y. Takano, Y. Shimohigashi, H. Matsumoto, *et al.* (1991) Pharmacological properties of the tachykinin receptor subtype in the endothelial cell and vasodilation. *Ann. NY Acad. Sci.*, **632**, 457–9.

Seiler, C., O.M. Hess, M. Buechi, T.M. Suter and H.P. Krayenbuehl (1993) Influence of serum cholesterol and other coronary risk factors on vasomotion of angiographically normal coronary arteries. *Circulation*, **88**, 2139–48.

Seo, B., B.S. Oemar, R. Siebenmann, L. von Segesser and T.F. Lüscher (1994) Both ETA and ETB receptors mediate contraction to endothelin-1 in human blood vessels. *Circulation*, **89**, 1203–8.

Smith, C.J., D. Sun, C. Hoegler, B.S. Roth, X. Zhang, G. Zhao, *et al.* (1996) Reduced gene expression of vascular endothelial NO synthase and cyclooxygenase-1 in heart failure. *Circ. Res.*, **78**, 58–64.

Sorensen, S.S., J.K. Madsen and E.B. Pedersen (1994) Systemic and renal effect of intravenous infusion of endothelin-1 in healthy human volunteers. *Am. J. Physiol.* 266:F411–8.

Tagawa, T., T. Imaizumi, M. Shiramoto, T. Endo, K. Hironaga and A. Takeshita (1995) V2 receptor-mediated vasodilation in healthy humans. *J. Cardiovasc. Pharmacol.*, **25**, 387–92.

Takayanagi, R., K. Kitazumi, C. Takasaki, K. Ohnaka, S. Aimoto, K. Tasaka, *et al.* (1991) Presence of non-selective type of endothelin receptor on vascular endothelium and its linkage to vasodilation. *FEBS Lett.*, **282**, 103–6.

Teerlink, J.R., M. Clozel, W. Fischli and J.P. Clozel (1993) Temporal evolution of endothelial dysfunction in a rat model of chronic heart failure. *J. Am. Coll. Cardiol.*, **22**, 615–20.

Treasure, C.B., J.A. Vita, D.A. Cox, R.D. Fish, J.B. Gordon, G.H. Mudge, *et al.* (1990) Endothelium-dependent dilation of the coronary microvasculature is impaired in dilated cardiomyopathy. *Circulation*, **81**, 772–9.

Uematsu, M., Y. Ohara, J.P. Navas, K. Nishida, T.J. Murphy, R.W. Alexander, *et al.* (1995) Regulation of endothelial cell nitric oxide synthase mRNA expression by shear stress. *Am. J. Physiol.* **269**, C1371–8.

Vallance, P., J. Collier and S. Moncada (1989) Effects of endothelium-derived nitric oxide on peripheral arteriolar tone in man. *Lancet*, **2**, 997–1000.

Vane, J.R., E.E. Änggard and R.M. Botting (1990) Regulatory functions of the endothelium. *N. Engl. J. Med.*, **323**, 27–36.

Vanhoutte, P.M. (1989) Endothelium and control of vascular function. State of the Art lecture. *Hypertension*, **13**, 658–67.

Verhaar, M.C., F.E. Strachan, D.E. Newby, N.L. Cruden, H.A. Koomans, T.J. Rabelink, *et al.* (1998) Endothelin-A receptor antagonist-mediated vasodilation is attenuated by inhibition of nitric oxide synthesis and by endothelin-B receptor blockade. *Circulation*, **97**, 752–6.

Wei, C.M., A. Lerman, R.J. Rodeheffer, C.G. McGregor, R.R. Brandt, S. Wright, *et al.* (1994) Endothelin in human congestive heart failure. *Circulation*, **89**, 1580–6.

Wiedermann, C.J., H. Beimpold, M. Herold, E. Knapp and H. Braunsteiner (1993) Increased levels of serum neopterin and decreased production of neutrophil superoxide anions in chronic heart failure with elevated levels of tumor necrosis factor-alpha. *J. Am. Coll. Cardiol.*, **22**, 1897–901.

Winlaw, D.S., G.A. Smythe, A.M. Keogh, C.G. Schyvens, P.M. Spratt and P.S. Macdonald (1994) Increased nitric oxide production in heart failure. *Lancet*, **344**, 373–4.

Yanagisawa, M., H. Kurihara, S. Kimura, K. Goto and T. Masaki (1988) A novel peptide vasoconstrictor, endothelin, is produced by vascular endothelium and modulates smooth muscle Ca2+ channels. *J Hypertens* Suppl **6**, 188–91.

Yoshizumi, M., M.A. Perrella, J.J. Burnett and M.E. Lee (1993) Tumor necrosis factor downregulates an endothelial nitric oxide synthase mRNA by shortening its half-life. *Circ. Res.*, **73**, 205–9.

Zelis, R., D.T. Mason and E. Braunwald (1968) A comparison of the effects of vasodilator stimuli on peripheral resistance vessels in normal subjects and in patients with congestive heart failure. *J. Clin. Invest.*, **47**, 960–70.

C
O
N
T
R
A
C
T
I
O
N

R
E
L
A
X
A
T
I
O
N

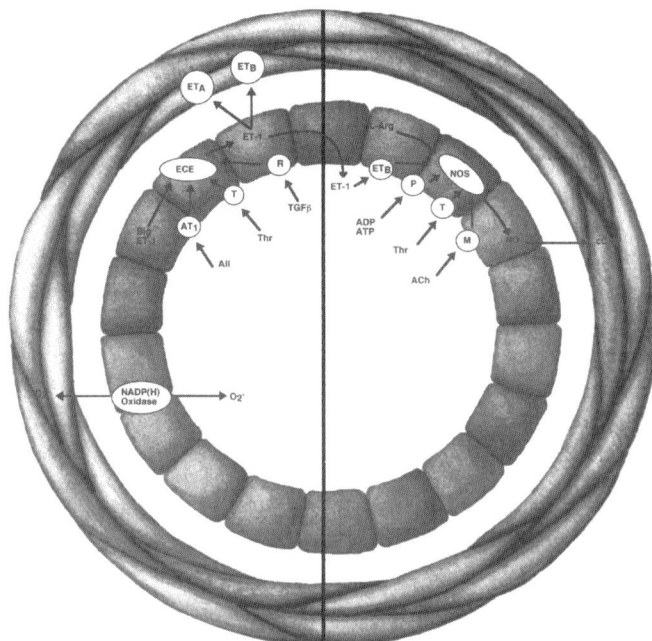

Hypercholesterolaemia is a major risk factor for the development of atherosclerosis. Increasing experimental evidence suggests that, in common with other risk factors, hypercholesterolaemia may adversely affect endothelial function. One molecule of particular interest in the prevention of atheroma is the endothelial mediator nitric oxide. Nitric oxide prevents adhesion of leucocytes to the endothelium, inhibits platelet activation and prevents smooth muscle cell growth. In the presence of hypercholesterolaemia the bioactivity of endothelium-derived nitric oxide is reduced and superoxide is generated. Work in our laboratory indicates that increasing nitric oxide by providing excess L-arginine retards the development of early atheroma in animal models. Furthermore, the arginine might enhance apoptosis in intimal lesions. Studies in patients have indicated that endothelium-dependent dilatation is lost in the presence of hypercholesterolaemia and this can be restored by lipid lowering treatment. Again L-arginine seems to exert a beneficial effect. This might be because it reverses tonic inhibition of nitric oxide synthase caused by accumulation of the endogenous inhibitor asymmetric dimethylarginine. This chapter describes the evidence that the nitric oxide pathway is central to the link between hypercholesterolaemia and atherogenesis in humans.

Key words: Nitric oxide, endothelium, cholesterol, LDL cholesterol, ADMA, superoxide anion, monocytes, plaque rupture.

3 Hypercholesterolemia, Atherosclerosis, and the NO Synthase Pathway

John P. Cooke[1] and Mark A. Creager[2]

[1]*Division of Cardiovascular Medicine, Stanford University, Stanford, CA 94305 5406, USA*
[2]*Vascular Medicine and Atherosclerosis Unit, Brigham & Women's Hospital and Harvard Medical School, Boston, MA 02115, USA*

CHOLESTEROL AND ATHEROGENESIS

In Europe and the United States, atherosclerosis affecting the coronary or carotid arteries is the greatest cause of morbidity and annual mortality. Elevated levels of serum cholesterol correlate with the clinical prevalence of atherosclerosis. Myocardial infarctions are much less common in countries where dietary intake of cholesterol and serum cholesterol levels are low, than in countries where individuals consume a Western diet and have higher cholesterol levels. (Keys, 1980; Kannel, Castelli and Gordon, 1979). Other major risk factors are hypertension, diabetes mellitus, tobacco use and family history of premature atherosclerosis. Additional determinants include sedentary state, lipopotein(a), and hyperhomocysteinemia.

Atherosclerosis has been said to be a "response to injury" of the endothelium. However, the mechanism(s) by which hypercholesterolemia and the other risk factors alter the endothelium have not been determined. Recent data from our laboratories and others indicates that there may be a common pathway by which these risk factors affect endothelial function. Experimental models of hypertension, hypercholesterolemia, diabetes mellitus or tobacco exposure, are characterized by common endothelial abnormalities; increased generation of superoxide anion and reduced bioactivity and/or synthesis of endothelium derived nitric oxide (Rajagopalan *et al.*, 1996; Ohara, Petersen and Harrison, 1993; Tesfamariam and Cohen, 1992; Tsao *et al.*, 1995). These aberrations of endothelial function represent an oxidative stress to the cells and occurs within minutes to hours of exposure to the noxious stimuli. Oxidative stress perturbs the cell membrane, and increases endothelial permeability. Moreover, the increased endothelial elaboration of oxygen-derived free radicals activates an oxidant sensitive transcriptional pathway mediated by nuclear factor κ B(NFκB). Usually NFκB is found in the cytoplasm in an inactive state, bound to its protein inhibitor IκBα (Murohara, *et al.*, 1994; Mercurio *et al.*, 1997). However, oxidative stress activates an IκB kinase, leading to phosphorylation, ubiquitination and ultimately degradation of IκBα (Mercurio, *et al.*, 1997; Woronicz *et al.*, 1997). This leaves NFKB free to translocate to the nucleus, where it induces the expression of adhesion molecules (eg. vascular cell adhesion molecule or VCAM-1) that participate in monocyte adhesion and infiltration (Cybulsky and Gimbrone, 1991; Berliner *et al.*, 1995; Tsao *et al.*, 1996; Tsao *et al.*, 1997; Ross, 1997). The expression of these adhesion molecules and chemokines

Correspondence: John P. Cooke, MD, PhD, Division of Cardiovascular Medicine, Stanford University School of Medicine, 300 Pasteur Drive, Stanford, CA 94305-5406, USA.

may explain the observation that within several days of a high cholesterol diet, monocytes adhere to the endothelium, particularly at intercellular junctions (Ross, 1997).

The monocytes migrate into the subendothelium, where they begin to accumulate lipid and become foam cells. This is the earliest event in the formation of the fatty streak. These activated monocytes (macrophages) release mitogens and chemoattractants that recruit additional macrophages as well as vascular smooth muscle cells into the lesion. As foam cells accumulate in the subendothelial space they distort the overlying endothelium, and eventually may even rupture through the endothelial surface (Ross, 1997). In these areas of endothelial ulceration, platelets adhere to the vessel wall, releasing epidermal growth factor, platelet-derived growth factor, and other mitogens and cytokines that contribute to smooth muscle migration and proliferation. These factors induce smooth muscle cells in the vessel wall to proliferate and migrate into the area of the lesion. These vascular smooth muscle cells undergo a change in phenotype from a "contractile" cell to a "secretory" cell. These secretory vascular smooth muscle cells elaborate extracellular matrix (eg. elastin), which transforms the lesion into a fibrous plaque. Extracellular matrix may contribute significantly to growth of the lesion. A genetic variant of the stromelysin promoter causes reduced degradation of extracellular matrix (Ye *et al.*, 1996). Extracellular matrix accumulates and leads to accelerated progression of atherosclerosis (Ye *et al.*, 1996). The smooth muscle cells may also become engorged with lipid to form foam cells. The lesion grows with the recruitment of more cells, the elaboration of extracellular matrix, and the accumulation of lipid until it is transformed from a fibrous plaque to a complex plaque.

The complex plaque typically is characterized by a fibrous cap which overlies a necrotic core. The necrotic core is composed of cell debris and cholesterol, and contains a high concentration of the thrombogenic tissue factor, secreted by macrophages. In later stage lesions, calcification may occur. Calcifying vascular cells in the vessel wall can transform into osteoblast-like cells and secrete bone proteins such as osteopontin (Parhami *et al.*, 1997). Microscopic examination of these areas reveals histology very similar to bone tissue. Oxidized lipoprotein stimulates the elaboration of bone protein by these vascular cells. Intriguingly, oxidized lipoprotein reduces bone formation by osteoblasts (Parhami *et al.*, 1997). This recent finding may account for the clinical observation that some patients with atherosclerosis (typically elderly women) appear (by X-ray) to have nearly as much calcium in their aorta as in their spine.

Plaque rupture is the most common cause of myocardial infarctions. Rupture of the complex plaque exposes the flowing blood to the highly thrombogenic constituents of the plaque (the foam cells, which elaborate tissue factor). Microscopic examination of the ruptured plaque generally reveals that the rupture has occurred at the shoulder of the lesion. In this area the fibrous cap can be seen to be thinned. Immunohistochemical studies reveal an intense concentration of macrophages in the area, which are elaborating copious amounts of metalloproteinases (Knox *et al.*, 1997). These macrophages appear to be undermining the fibrous cap, weakening it and predisposing to its rupture under the stress of hemodynamic forces. Ruptured plaques generally have a greater macrophage density than stable plaque (Lendon *et al.*, 1991). The determinants of plaque rupture include the size, eccentricity and composition of the necrotic core. The larger, more eccentric or more fluid the necrotic core, the greater the mechanical stress on the fibrous cap (Loree *et al.*, 1994). Fluidity of the necrotic core is determined by the proportion of cholesterol ester (greater fluidity) and crystalline cholesterol (less fluidity) contained in the core (Loree *et al.*, 1994).

An important determinant of plaque rupture is the thickness of the fibrous cap (Loree *et al.*, 1994). The fibrous cap, weakened and thinned by the degradative action of the macrophages, ruptures under the stress of hemodynamic forces. With rupture of the plaque, the luminal blood comes into contact with highly thombogenic tissue factor deposited by macrophages in the necrotic core (Marmur *et al.*, 1996). Under the influence of tissue factor, thrombus forms in the fissures of the lesion. The thrombus often extends into, and may occlude, the lumen. Plaque rupture and thrombus formation is the most common cause of heart attack and stroke. Furthermore, as the thrombus organizes it can contribute to growth of the lesion and increase the symptoms of the patient.

As can be gleaned from the above discussion, plaque growth and rupture is an inflammatory process, with the participation of vascular adhesion molecules and chemokines, and monocyte adherence and infiltration. Inflammation of the fibrous cap leads to plaque rupture. The causative factors initiating inflammation of the fibrous cap are unknown. However, there is mounting circumstantial evidence that implicates infection in acute coronary syndromes (Benditt, Barrett and McDougall, 1983; Libby, Egan and Skarlatos, 1997). There is seroepidemiological and immunohistochemical evidence that infectious agents such as cytomegalovirus, herpes virus, or chlamydia pneumoniae are associated with atherosclerotic vascular disease and vascular events (Melnick, Adam and DeBakey, 1990; Gratton *et al.*, 1989; Saikku *et al.*, 1988; Thom *et al.*, 1992; Muhlstein *et al.*, 1996; Minick *et al.*, 1979). Such infections may trigger plaque rupture by increasing hemodynamic stress (e.g., tachycardia and increased cardiac output that may accompany a febrile illness) or may directly affect the vascular biology of the plaque. Infection localizing to the plaque may activate endothelial cells to express adhesion molecules, may stimulate vascular cells to undergo proliferation, and/or induce resident inflammatory cells to elaborate cytokines that promote further local inflammation (Libby, Egan and Skarlatos, 1997). Endothelial cells infected with CMV or herpes virus express leukocyte adhesion molecules which may participate in monocyte and T-cell recruitment (Sedmark *et al.*, 1995; Etingin, Silverstein and Hajjar, 1991).

NITRIC OXIDE: AN ENDOGENOUS ANTI-ATHEROGENIC MOLECULE

As we shall see, endothelium-derived nitric oxide (NO) is a potent inhibitor of the processes that lead to development of an atherosclerotic plaque. NO is a product of the metabolism of L-arginine by the endothelial isoform of NO synthase (Moncada and Higgs, 1995). NO is a potent endogenous vasodilator exerting its actions in the same way as do exogenous nitrovasodilators such as nitroglycerin. NO released from the endothelium diffuses to the subjacent vascular smooth muscle and activates soluble guanylate cyclase within the vascular smooth muscle, leading to the production of cyclic guanosine monophosphate (cGMP). This cyclic nucleotide is the second messenger for the action of endothelium-derived NO as well as exogenous nitrovasodilators, and it activates cGMP-dependent kinases and phosphatases that mediate vascular smooth muscle relaxation. NO is not only a potent vasodilator but also has important effects on circulating blood elements and on the vascular wall that may protect the vessel from the development of atherosclerosis. NO inhibits platelet adherence and aggregation (Radomski, Palmer and Moncada, 1987; Stamler *et al.*, 1989; Pohl and Busse, 1989). Together, the endothelial products NO and prostacyclin confer a resistance to platelet-vessel wall interaction. NO exerts its effect on platelet reactivity in part by stimulating intra-platelet production of cGMP which subsequently phosphorylates proteins which regulate platelet activation and adherence (Pohl *et al.*, 1994). Platelets themselves contain small amounts of NO synthase and are capable of generating NO which may act as a brake on their activation (Radomski, Palmer and Moncada, 1990; Mehta *et al.*, 1995).

NO also prevents adherence of leukocytes to the endothelium (Kubes *et al.*, 1991). This salutary effect of NO was first discovered using models of ischemia-reperfusion. When the coronary artery of an experimental animal is ligated, this induces ischemia of the myocardium subserved by that vessel. When the ligature is released, the ensuing reperfusion is associated with further injury to the myocardium which is in part due to the adherence and infiltration of neutrophils, and the concomitant release of oxygen-derived free radicals. The adherence of leukocytes and subsequent reperfusion injury can be markedly inhibited by the simultaneous perfusion of the coronary artery by exogenous NO donors (Johnson *et al.*, 1990).

Atherosclerosis begins with an alteration in the adhesiveness of the endothelium for circulating monocytes. The production of NO can be modulated in the laboratory; a reduction in its synthesis accelerates atherosclerosis, whereas an increase in its synthesis suppresses and can even reverse atherosclerosis. One explanation for the observation is that NO inhibits monocyte adherence to the endothelium. Indeed, the ability of monocytes to bind to endothelial cells *in vitro* is inhibited by exogenous NO in a dose-dependent manner (Bath *et al.*, 1991). Moreover, within minutes of exposure to NO, endothelial cells become more resistant to monocyte adherence (Tsao *et al.*, 1995). Because of the rapid time course, this effect of NO must be due to inhibition of signaling pathways involved in adhesion-perhaps by a cGMP-dependent mechanism. More chronic exposure to NO suppresses gene expression of adhesion molecules (such as VCAM), and chemokines (such as MCP-1) involved in monocyte adhesion and infiltration. By contrast, inhibition of NO synthesis increases the expression of endothelial proteins required for monocyte adhesion (Zeiher *et al.*, 1995). NO appears to exert its effects on gene expression by blocking the activation of specific transcriptional proteins (such as nuclear factor kB) (Figure 8.1) (Tsao *et al.*, 1996; Tsao *et al.*, 1997).

Accumulating evidence supports the hypothesis that NO exerts its effect on monocyte adherence and accumulation in part by modulating the activity of the redox-responsive transcriptional pathways described earlier (Tsao *et al.*, 1996; Tsao *et al.*, 1997; Marui *et al.*, 1993; DeCaterina *et al.*, 1995; Spiecker, Peng and Liao, 1997). Vascular cells exposed to oxidized lipoprotein or cytokines begin to elaborate superoxide anion (Tsao *et al.*, 1996; Tsao *et al.*, 1997). This generation of reactive oxygen species is associated with the transcription of a number of genes participating in atherogenesis including MCP-1 mRNA. MCP-1 increases chemotactic activity of the vessel wall for monocytes (Tsao *et al.*, 1997). These effects are all suppressed by the NO-donor, DETA-NO. NO may reduce the half-life of MCP-1 mRNA, as well as reducing its transcription. NO exerts its effects on MCP-1 expression in cytokine-stimulated HUVECs in a cGMP-independent fashion.

It is well established that hypercholesterolemia reduces the bioactivity of endothelium-derived NO (Heistad *et al.*, 1984; McLenahan *et al.*, 1991). (see below). In parallel, the endothelium begins to generate superoxide anion (Ohara, Petersen and Harrison, 1993). This increased endothelial generation of superoxide anion induced by hypercholesterolemia in rabbit models can be reversed by placing the animal on a low cholesterol diet (Ohara

Hypercholesterolemia
Diabetes Mellitus
Hypertension
Smoking

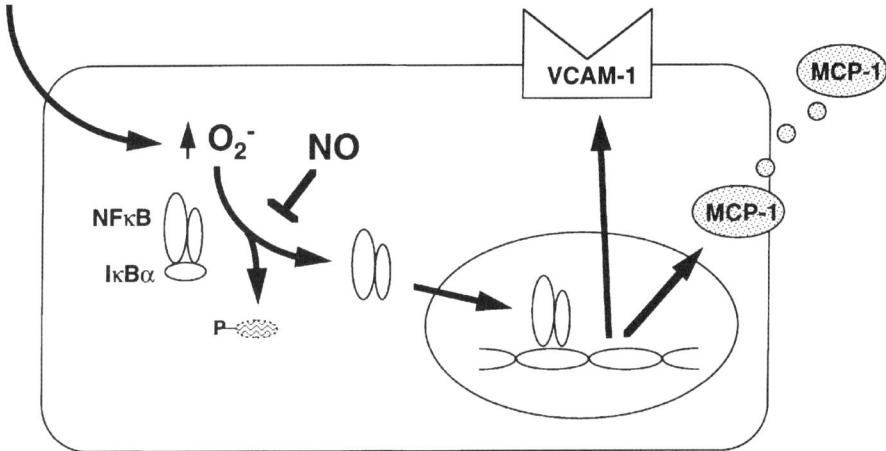

Figure 8.1 Atherosclerotic risk factors such as hypercholesterolemia, hypertension, tobacco, and diabetes mellitus lead to increased free radical production and decreased nitric oxide activity in endothelial cells. This endothelial dysfunction not only has acute effects on vascular tone, but also chronic effects on vessel structure. Increased superoxide anion leads to activation of NFkB via phosphorylation and degradation of the inhibitor protein IκBa. NFκB is then free to translocate into the nucleus to initiate transcription of proatherogenic genes such as VCAM-1 and MCP-1. Nitric oxide can inhibit these processes by inhibiting superoxide production, directly scavenging superoxide anions, as well as increasing the transcription and activity of IκBα. Moreover, since NO is a paracrine factor, it can have important inhibitory effects on circulating leukocytes and underlying smooth muscle cells.

et al., 1995). The reduction in superoxide anion generation is associated with an improvement in endothelium-dependent vasodilator function. In addition to inducing the generation of superoxide anion, hypercholesterolemia causes a decline in tissue glutathione levels, and thereby increases susceptibility to oxidative damage (Ma *et al.*, 1997). This alteration in endothelial redox state triggers the oxidant-sensitive transcriptional cascade that results in the activation of genes encoding molecules that regulate endothelial adhesiveness (Tsao *et al.*, 1996; Marui *et al.*, 1993). The cytokine-induced activation of VCAM-1 and MCP-1 in cultured endothelial cells is suppressed by antioxidants or NO donors (Tsao *et al.*, 1996; Tsao *et al.*, 1997). This effect of NO appears to be due in part to stabilization and/or increased expression of IκBα, which complexes with NFκB to inhibit its transcriptional activity (Peng, Libby and Liao, 1995; Spiecker, Peng and Liao, 1997).

NO may act by reducing intracellular oxidative stress. There are several possible mechanisms by which NO may reduce oxidative stress. NO can scavenge superoxide anion, although the product of this reaction, peroxynitrite anion, is itself a highly reactive free radical (Radi *et al.*, 1991). However, it is possible that peroxynitrite anion could subsequently nitrosylate sulfhydryl groups to form S-nitrosothiols (Radi *et al.*, 1991). This class of molecules is known to induce vasodilation, inhibit platelet aggregation, and interfere with leukocyte adherence to the vessel wall (Stamler *et al.*, 1992). Another mechanism by which NO may ameliorate oxidative stress is by terminating the autocatalytic chain of lipid peroxidation that is initiated by oxidized LDL or intracellular generation of oxygen-derived free radicals. Indeed, exogenous NO inhibits copper-induced oxidation of LDL cholesterol, causing a lag in the formation of conjugated dienes (Hogg *et al.*, 1993). Finally, NO may directly suppress the generation of oxygen-derived free radicals by nitrosylating, and thereby inactivating oxidative enzymes. This hypothesis is supported by the observation that the generation of superoxide anion by stimulated neutrophils is reduced by their exposure to exogenous NO (Clancy *et al.*, 1992). This is due to the inactivation of NADPH oxygenase, a multimeric enzyme, with cytosolic and particulate components. The particulate component is vulnerable to nitrosylation by NO (either at its heme moiety or sulfhydryl group), which prevents its association with the cytosolic component, and reconstitution of the active enzyme. A similar phenomenon may occur in endothelial cells. This would explain the observation of Niu and colleagues, that antagonism of endogenous NO production increases oxidative stress in HUVECs, as demonstrated using redox-sensitive fluorophores (Niu, Smith and Kubes, 1994). Furthermore, Pagano and colleagues (1993) have shown that exogenous NO donors inhibit the generation of superoxide anion by the endothelium of rabbit thoracic aortae treated *ex vivo* with antagonists of superoxide dismutase. Thus, NO appears to exert its effects, in part, by reducing intracellular oxidative stress, thereby defusing oxidant-triggered transcription.

NO also regulates the growth of vascular smooth muscle cells. *In vitro*, NO-donors inhibit the proliferation of vascular smooth muscle cells; this effect is mimicked by exogenous administration of 8-bromo-cGMP, a stable analog of the second messenger of NO action (Garg *et al.*, 1989). Other agents such as atrial natriuretic peptide which increase the intracellular levels of cGMP inhibit proliferation of vascular smooth muscle cells in culture. Does NO inhibit the proliferation of vascular smooth muscle cells *in vivo*? Some initial studies indicate that NO does indeed play an important role in controlling vascular growth. In experimental animal models, chronic inhibition of NO synthesis causes hyperreactivity to vasoconstrictors, and medial thickening, in the coronary microvasculature (Numaguchi *et al.*, 1995; Ita *et al.*, 1995). These effects are not mediated by the hypertension that is induced by NOS antagonists, because co-administration of hydralazine to normalize blood pressure does not reverse the effects of NOS antagonists upon microvasculature structure and function (Numaguchi *et al.*, 1995) In a number of disease states where the release of NO is reduced or abolished, such as restenosis, hypercholesterolemia, and hypertension, there is an increase in the proliferation of vascular smooth muscle cells within the media and the intima. In experimental animal models, augmentation of endogenous NO synthesis (as with administration of the NO precursor L-arginine), inhibits "restenosis" (myointimal hyperplasia) after balloon angioplasty; the effect of L-arginine is blocked by antagonists of NO synthase (McNamara *et al.*, 1993; Tarry and Makhoul, 1994). In one study, after subjecting the carotid artery to balloon angioplasty, the vessel was transfected with a plasmid construct containing the gene encoding NO synthase. This gene transfer had the effect of enhancing NO synthesis locally and inhibiting myointimal hyperplasia (von der Leyen *et al.*, 1995). Also, intramural administration of a single dose of L-arginine, at the time of balloon angioplasty, can markedly inhibit myointimal hyperplasia 2–4 weeks later (Schwarzacher *et al.*, 1997). This effect of L-arginine is associated with increased local production of NO, probably due to utilization of L-arginine by induced NO synthase in vascular smooth muscle cells in the injured area.

Proliferation of vascular smooth muscle cells, as well as monocyte adherence and infiltration, platelet adherence, and aggregation, are key processes involved in atherogenesis. Because endothelium-derived NO inhibits each of these processes, we have proposed that NO is an endogenous anti-atherogenic molecule (Cooke *et al.*, 1992; Cooke and Tsao, 1993; Cooke and Tsao, 1994; Cooke and Dzau, 1997). Therefore an endothelial injury or alteration which results in a reduction in NO activity could promote atherogenesis.

EFFECT OF NO RESTORATION ON ATHEROGENESIS

If a reduction in NO activity promotes atherogenesis, a restoration of NO activity might be expected to retard progression of the disease. Cooke and colleagues tested the hypothesis that increased synthesis of NO by the vessel wall would inhibit atherosclerosis. New Zealand White rabbits were placed on normal or high cholesterol diet; some of the animals on the high cholesterol diet also received supplemental dietary arginine or methionine to increase the intake of these amino acids six-fold (Cooke *et al.*, 1992; Wang *et al.*, 1994). After 10 weeks, the thoracic aortae and coronary arteries were harvested for studies of vascular reactivity and histomorphometry. Supplemental dietary arginine did not alter the lipid profile nor any hemodynamic parameters in the hypercholesterolemic animals; the only difference between the hypercholesterolemic groups was that plasma arginine was doubled in the hypercholesterolemic animals receiving dietary arginine supplementation. As expected, NO-dependent vasodilation to acetylcholine was inhibited in the hypercholesterolemic animals receiving vehicle. By contrast hypercholesterolemic animals receiving L-arginine had an improvement in NO-dependent vasodilation. The improvement of NO-dependent vasodilation in hypercholesterolemic animals receiving L-arginine was associated with a striking effect on vascular structure. Chronic L-arginine administration reduced the surface area of the thoracic and abdominal aorta involved by lesions in the hypercholesterolemic animals (Figure 8.2) (Cooke *et al.*, 1992). In the left main coronary artery the differences were even more striking; no lesions at all were observed in the hypercholesterolemic animals receiving L-arginine (Wang *et al.*, 1994). The mechanism by which dietary arginine inhibits atherogenesis appeared to be due, in part, to an inhibition of monocyte-endothelial cell interaction in the hypercholesterolemic animals. Boger and colleagues have shown that administration of L-arginine to hypercholesterolemic rabbits increases vascular NO production, and reduces superoxide anion elaboration (1995). As mentioned previously, the balance between nitric oxide and superoxide anion may have profound effects on the activation of genes regulating monocyte adhesion and accumulation. Indeed, after exposure to a high cholesterol diet the thoracic aorta of hypercholesterolemic animals elaborates more superoxide anion (Ohara, Petersen and Harrison, 1993; Boger *et al.*, 1995) and has increased adhesiveness for monocytes (Tsao *et al.*, 1994). Compared with the thoracic aortae from normal animals, those from hypercholesterolemic rabbits manifested a three-fold increase in the number of adherent cells. The increase in cell binding is attenuated in thoracic aortae from hypercholesterolemic animals receiving dietary arginine supplements (Figure 8.3) (Tsao *et al.*, 1994). This is associated with an increase in the release of NO from these tissues, as well as a reduced

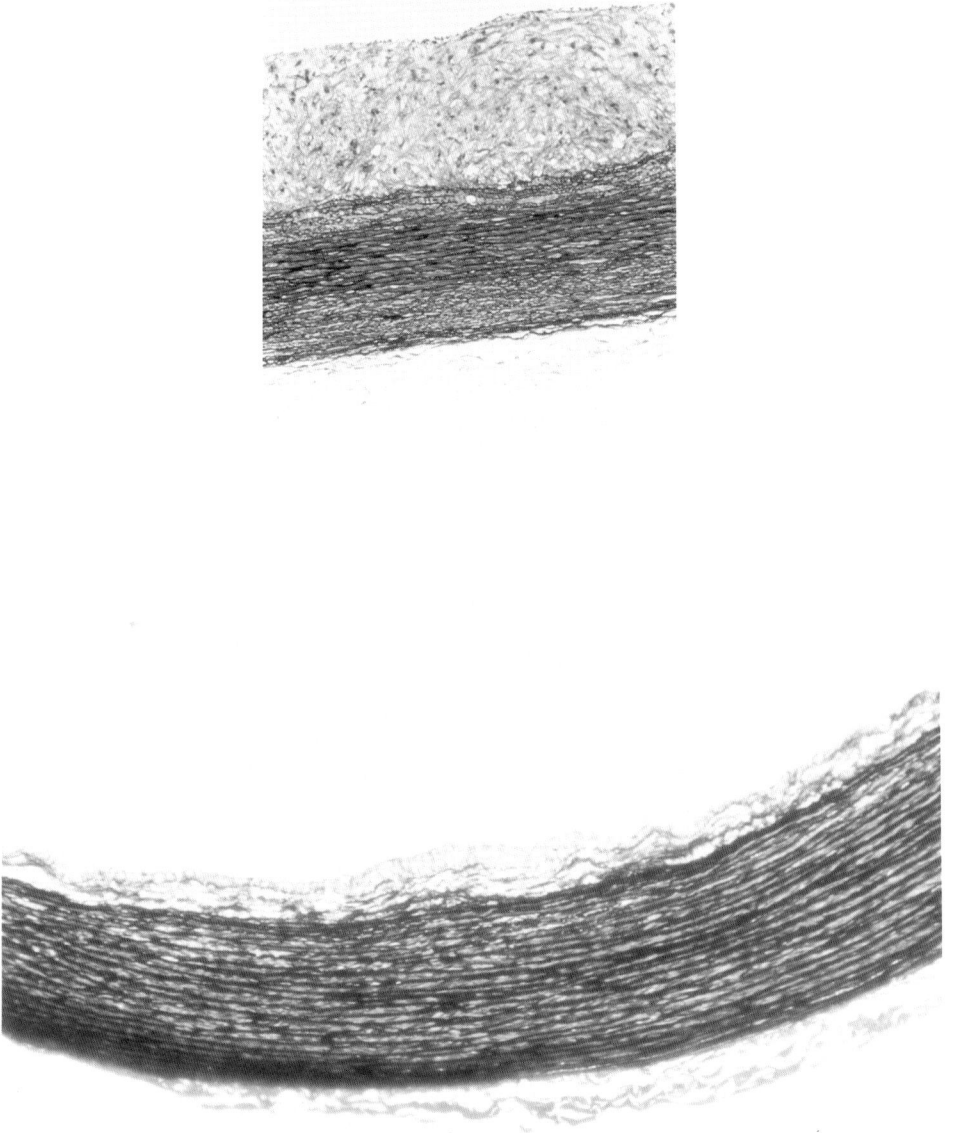

Figure 8.2 Photomicrographs of abdominal aorta from New Zealand White rabbits fed a high cholesterol diet in the absence (top panel) or the presence (bottom panel) of supplemental dietary arginine.

vascular expression of MCP-1 (Tsao *et al.*, 1996). The number of adherent cells is dramatically increased in normocholesterolemic animals receiving the NO synthase antagonist nitro-arginine (Figure 8.3) (Tsao *et al.*, 1994). The increased adhesiveness of the endothelium in the nitro-arginine treated animals is associated with increased vascular expression of MCP-1 (Tsao *et al.*, 1996). Thus, reductions in NO activity by hypercholesterolemia, or inhibition of NO synthesis, is associated with increased monocyte-endothelial cell binding, possibly due to the increased expression of specific chemokines and adhesion molecules. Indeed, a number of investigators have now shown that administration of NO synthase antagonists accelerates atherosclerosis in animal models (Cayette *et al.*, 1994; Naruse *et al.*, 1994; Aji *et al.*, 1997).

Apoptosis has been reported to occur in vascular cells of human atherosclerotic plaque (Isner *et al.*, 1995; Geng and Libby, 1995). Administration of arginine restores NO activity in hypercholesterolemic rabbits, and is associated with regression of pre-existing lesions (Candipan *et al.*, 1996). Preliminary studies indicate that enhancement of vascular NO activity causes regression by inducing apoptosis of macrophages within the lesion. Factors involved in the initiation and regulation of apoptosis in atherosclerosis have not been fully elucidated, but immunohistochemical studies provide evidence for several proteins known to participate in apoptosis, including p53 (Bennett, Evan and Schwarz, 1995). Among the myriad pathways that may be involved, there is accumulating evidence to implicate L-arginine/NO synthase (Messmer, Lapetina and Brune, 1995). Cytokine-mediated activation of iNOS induces peroxynitrate generation and apoptosis of macrophages *in vitro* (Ischiropoulos, Zhu and Beckman, 1992; Lin *et al.*, 1995). The effect of iNOS activation is augmented by additional L-arginine, and attenuated by antagonists of NO synthase. Therefore it is possible that enhancement of vascular NO protection could induce apoptosis and regression of pre-existing lesions. Accordingly, in a study by Wang et al, male New Zealand White rabbits were fed a 0.5% cholesterol diet for 10 weeks and subsequently placed on 2.5% L-arginine HCl in the drinking water. The cholesterol diet was continued for two weeks, at which time the aortae were harvested for histological studies. As observed by Hoechst staining, L-arginine treatment increased the number of apoptotic cells (largely macrophages) in the intimal lesions by three-fold (Wang *et al.*, in press). In subsequent studies, aortae were harvested for *ex vivo* examination. Aortic segments were incubated in cell culture medium for 4 to 24 hours with modulators of NO synthase pathway. The tissues were then collected for histological studies, and the conditioned medium collected for measurement of nitrogen oxides by chemiluminescence. Addition of sodium nitroprusside to the medium caused a time-dependent increase in apoptosis of vascular cells (largely macrophages) in the intimal lesion. L-arginine (10^{-3} M) had an identical effect on apoptosis, which was associated with an increase in NO released into the medium. These effects were not mimicked by D-arginine, and they were antagonized by the NO synthase inhibitor, L-nitro-arginine (10^{-4} M). The effect of L-arginine was not influenced by an antagonist of cGMP-dependent protein kinase, nor was the effect mimicked by the agonist of protein kinase G or 8 Br cyclic GMP. These results indicated that supplemental L-arginine induces apoptosis of macrophages in intimal lesions by its metabolism to nitric oxide, which acts via a cyclic-GMP independent pathway. Previous studies have suggested that apoptosis induced by iNOS activity may be mediated in part by mechanisms independent of cGMP (Bennett, Evan and Schwarz, 1995; Messmer, Lapetina and Brune, 1995). Furthermore, these studies explain a previous observation that supplementation of dietary arginine induces regression of atheroma in this animal model (Candipan *et al.*, 1996). It is likely

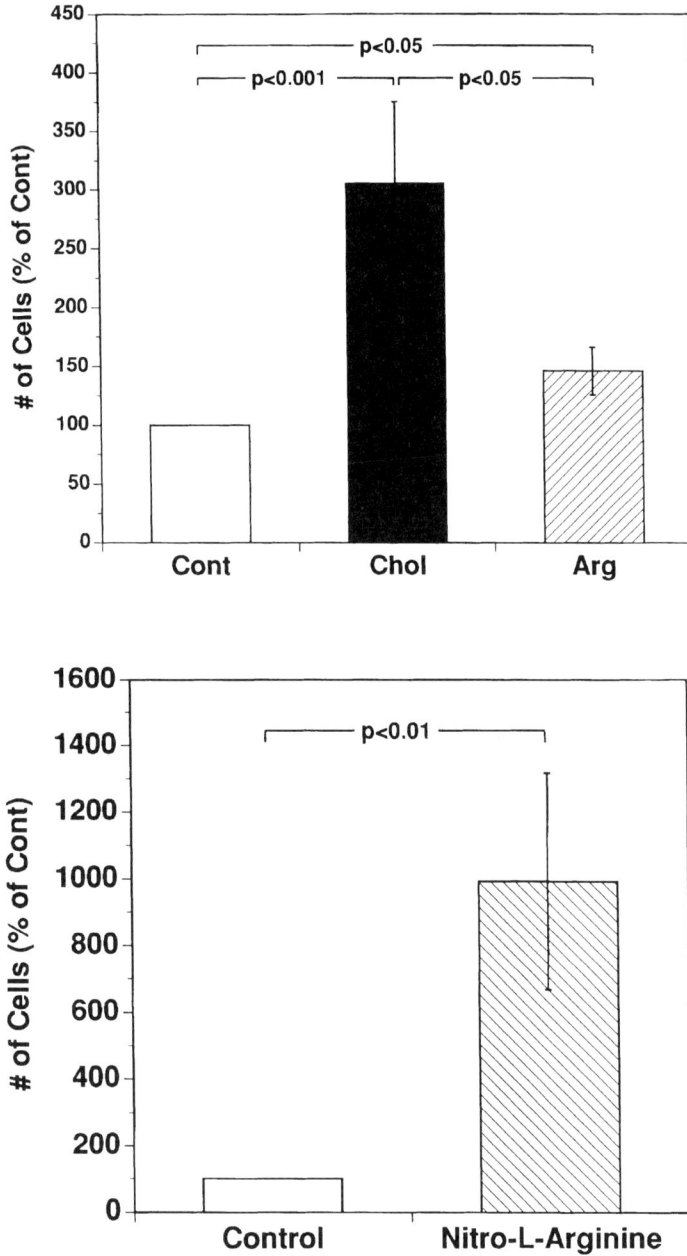

Figure 8.3 Histograms illustrating the results of an ex vivo functional binding assay. New Zealand White rabbits were fed a high cholesterol or normal chow diet. Some hypercholesterolemic animals received oral L-arginine supplementation, whereas some normocholesterolemic animal received L-nitro arginine (the NOS antagonist) in their drinking water. After two weeks the thoracic aortae were harvested. In an ex vivo functional binding assay with monocytoid cells, the thoracic aortae of hypercholesterolemic animals had greater adhesiveness for monocytoid cells, than did the aortae from normocholesterolemic animals (top panel). L-arginine supplementation reduced endothelial adhesiveness. By contrast the administration of L-nitro arginine to normocholesterolemic animals markedly increased the adhesiveness of their aortae (bottom panel).

that iNOS expressed by cells within the lesion is responsible for the effect of L-arginine. Indeed, previous immunohistochemical studies have detected iNOS in the intimal macrophages and vascular smooth muscle cells of human atherosclerotic plaque (Buttery *et al.*, 1996). These are activated cells which also produce superoxide anion. In this milieu, the product of iNOS is quickly transformed into peroxynitrite anion, a highly reactive free radical. Peroxynitrite anion is cytotoxic and induces apoptosis (Ischiropoulos, Zhu and Beckman, 1992; Lin *et al.*, 1995). Peroxynitrite anion can also affect cell function by nitrosylating tyrosine residues that are involved in the signal transduction of transmembrane receptors. Using monoclonal antibodies directed against nitrotyrosine, evidence of peroxynitrite formation has been observed in human atherosclerotic plaque (Beckman *et al.*, 1994). This is relevant to the study described above, since peroxynitrite anion is likely the NO species mediating the apoptosis observed in the aforementioned study.

The activation of iNOS may have complex effects on the evolution of atherosclerotic plaque. By inducing cell death, iNOS activation may contribute to the development of the "necrotic core" of complex lesions. One might also speculate that iNOS may be involved in the characteristic atrophy of the media beneath atheroma or the dissolution of the fibrous cap by activated macrophages, as peroxynitrite anion or other NO donors may induce apoptosis of vascular smooth muscle (Pollman *et al.*, 1996) Furthermore, peroxynitrite anion may reduce collagen formation by vascular cells, and activate metalloproteinases which degrade extracellular matrix (Rajagopalan *et al.*, 1996). These actions of peroxynitrite anion would contribute to plaque instability and have led some to explore antagonism of iNOS as a potential therapeutic avenue. However, it is likely that such a strategy would have unintended consequences. Antagonism of iNOS activity could promote platelet aggregation, leukocyte adherence, vasoconstriction, and proliferation of vascular smooth muscle cells and macrophages. We speculate that iNOS may be in fact a countervailing force in the accretion of atherosclerotic plaque. Furthermore, by reducing proliferation and by promoting apoptosis of macrophages in the lesion, iNOS activation may lead to plaque stabilization and even regression. It is worthy of emphasis that both macrophages and vascular smooth muscle cells contribute to the intimal lesion in the balloon-injured hypercholesterolemic rabbits, but it was largely the macrophages that appear to undergo apoptosis in an animal model (Wang *et al.* in press). The contrarian concept that iNOS may act as a brake on vascular inflammation and lesion development has recently received strong experimental support. Rat aortic allografts develop significant increases in intimal thickness associated with an increase in inducible NO synthase expression (Shears *et al.*, 1997). Inhibition of NO synthase activity (using a NO antagonist) or expression (using cyclosporine), increases intimal thickening. By contrast, adenoviral-mediated iNOS gene transfer abolishes allograft vasculopathy in this model (Shears *et al.*, 1997). This report is consistent with the notion that the product of iNOS activation may suppress recruitment of inflammatory cells, reduce their proliferation and/or enhance their apoptosis to act as a countervailing force in vascular inflammation.

EFFECT OF HYPERCHOLESTEROLEMIA ON NITRIC OXIDE

The mechanism whereby hypercholesterolemia causes atherosclerosis may be mediated, in part, by its effect on the availability and subsequent activity of endothelium-derived NO. The bioavailability of endothelium-derived NO can be assessed in experimental models and in humans by administering pharmacologic agents that result in activation of eNOS and by assessing the vasorelaxant response of arteries in vitro or vasorelaxation in vivo. Endothelium-dependent relaxation, mediated by nitric oxide, is abnormal in arteries isolated from hypercholesterolemic animals (Verbeuren, 1986; Cohen *et al.*, 1988; Bossaller *et al.*, 1987). Incubation of normal aortic rings with LDL cholesterol, particularly modified or oxidized LDL cholesterol, blunts endothelium-dependent relaxation (Andrews *et al.*, 1987; Jacobs, Plane and Bruckdorfer, 1990; Takahashi *et al.*, 1990; Simon, Cunningham and Cohen, 1990). A specific component of oxidized LDL, lysophosphatidylcholine, inhibits endothelium-dependent relaxation when it is placed in the incubation medium (Yokoyama *et al.*, 1990; Kugiyama *et al.*, 1990; Flavahan, 1993). Also, endothelium-dependent vasodilation to parenteral administration of acetylcholine is reduced in peripheral resistance vessels of hypercholesterolemic models *in vivo* (Osborne *et al.*, 1989; Girerd *et al.*, 1990). Thus, abnormal endothelium-dependent vasodilation occurs in vascular preparations exposed to cholesterol and in vessels of hypercholesterolemic animals, even in the absence of atherosclerosis. These observations are consistent with the premise that reduced availability of NO in hypercholesterolemic states precedes, and may contribute to, atherosclerosis.

Endothelium-dependent vasodilation is abnormal also in hypercholesterolemic humans. Creager *et al.* (1990) reported that methacholine-induced, endothelium-dependent vasodilation is impaired in forearm resistance vessels of hypercholesterolemic patients who have no clinical manifestations of atherosclerosis. (Figure 8.4). These observations have been confirmed by others (Celermajer *et al.*, 1992; Casino *et al.*, 1993; Stroes *et al.*, 1995). Hypercholesterolemia has been shown to impair endothelium-dependent vasodilation in coronary arteries, even in the absence of atherosclerosis (Zeiher *et al.*, 1991). Moreover, the cholesterol concentration correlates with the vasoconstrictive response to acetylcholine in epicardial coronary arteries of patients with coronary atherosclerosis (Vita *et al.*, 1990). There even appears to be an inverse relationship between cholesterol and endothelium-dependent vasodilation among healthy subjects whose cholesterol levels are less than 200 mg/dl (Steinberg *et al.*, 1997).

Conversely, lipid lowering therapy improves endothelial function in experimental models and in patients with hypercholesterolemia. Harrison *et al.* (1987) found that dietary restriction of cholesterol restored endothelium-dependent relaxation of atherosclerotic iliac arteries excised from cynomolgus monkeys previously fed a high cholesterol diet. In humans, lipid lowering therapy has been reported to improve endothelium-dependent vasodilation of coronary and peripheral arteries. Six months of treatment with cholestyramine has been shown to enhance endothelium-dependent vasodilation of angiographically normal epicardial coronary arteries of hypercholesterolemic patients (Leung, Lau and Wong, 1993). Similarly, HMG Co-A reductase inhibitors, administered for six months improved endothelium-dependent vasodilation of atherosclerotic coronary arteries of hypercholesterolemic patients (Egashira *et al.*, 1994; Treasure *et al.*, 1995). Also, HMG Co-A reductase inhibitors administered for one to three months, have been reported to improve endothelium-dependent vasodilation in forearm resistance vessels (Stroes *et al.*, 1995; Leung, Lau and Wong, 1993; O'Driscoll, Green and Taylor, 1997). Even acute reduction of cholesterol with LDL apheresis has been shown to improve endothelium-dependent vasodilation in the forearm of patients with hypercholesterolemia, within hours of treatment (Tamai *et al.*, 1997). Thus, reduction of cholesterol appears to restore the

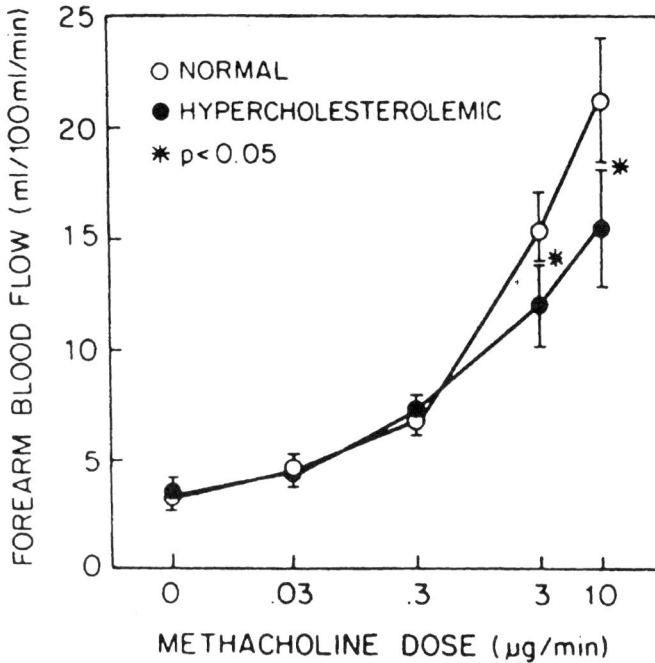

Figure 8.4 The effect of hypercholesterolemia on endothelium-dependent vasodilation. The forearm blood flow response to the endothelium-dependent vasodilator, methacholine choloride, was less in hypercholesterolemic subjects than in age-matched healthy subjects. Adapted from Creager *et al.* (ref. 99).

vasoactive regulating component of endothelial function, in vessels with and without atherosclerosis, and it might do so within a short period of time.

Multiple mechanisms may account for the abnormalities in the L-arginine/NO pathway induced by hypercholesterolemia. These include: reduced availability of L-arginine; abnormalities of endothelial receptor-G protein coupling; decreased levels of cofactors such as tetrahydrobiopterin; reduced nitric oxide synthase expression and activity; increased degradation of NO by superoxide anion or oxidized lipoproteins or, a circulating inhibitor of NO synthase (Cooke and Dzau, 1997).

L-arginine and Asymmetric Dimethylarginine

Several years ago we found that L-arginine, but not D-arginine improved endothelium-dependent vasodilation to acetylcholine in the hindlimb of cholesterol fed rabbits (Girerd *et al.*, 1990). We further observed that endothelium-dependent relaxation from thoracic aortae excised from cholesterol-fed rabbits treated with L-arginine is greater than those from rabbits treated with vehicle (Cooke *et al.*, 1991). Subsequently, we studied hypercholesterolemic humans and found that acute parenteral administration of L-arginine, but not D-arginine, improves endothelium-dependent vasodilation of forearm resistance vessels (Creager et al., 1992). L-arginine also acutely improves endothelium-dependent vasodilation of coronary resistance vessels of hypercholesterolemic humans Drexler et al., 1991). Clarkson et al. (1996) reported that L-arginine administered orally for four weeks, improved flow-mediated endothelium-dependent vasodilation of the brachial artery of hypercholesterolemic patients.

The mechanism by which L-arginine exerts its beneficial effects is likely through enhanced production of endothelium-derived NO. Until recently it was not apparent how exogenous L-arginine could increase NO elaboration in hypercholesterolemia, given the low K_m of the purified endothelial NO synthase (2.9 µM) (Pollock et al., 1991), and the relatively high physiological plasma concentrations of L-arginine (60–100 µmol/L), which are not reduced in hypercholesterolemia. This "L-arginine paradox" may be due, in part, to the presence of endogenous competitive NOS inhibitors like asymmetric dimethylarginine(ADMA) (Figure 8.3). Recently, ADMA has been characterized to be an endogenous, competitive inhibitor of NO synthase (Vallance et al., 1992). ADMA has been shown to be synthesized by human endothelial cells (Fickling et al., 1993). The plasma level of ADMA is elevated in hypercholesterolemic rabbits (Bode-Boger et al., 1996). as well as in hypercholesterolemic and atherosclerotic humans (Boger et al., 1997; Boger et al. , in press) concomitantly with impaired endothelial NO elaboration. ADMA plasma levels, which are 1.0 ± .0.1 µmol/L in healthy humans, are elevated to 2.2 ± .0.2 µmol/L in hypercholesterolemic individuals (Boger et al., 1997). In elderly patients with peripheral arterial disease and generalized atherosclerosis, ADMA levels range from 2.5 to 3.5 µmol/L, corresponding to the severity of the vascular disease (Boger et al., 1997). Cooke and colleagues have examined the effect of ADMA upon cultured endothelial cells (Boger et al., under review). Incubation of endothelial cells with ADMA (at concentrations that are observed in hypercholesterolemic humans) inhibit NO production. This effect is associated with increased endothelial superoxide radical elaboration and NFkB activation, resulting in enhanced MCP-1 expression and endothelial adhesiveness for monocytes. These effects of ADMA are reversed by L-arginine.

Intracellular ADMA levels within endothelial cells may be higher than those values observed in plasma. Support for this speculation comes from studies of regenerated endothelium. After balloon angioplasty and endothelial denudation, the endothelial mon-olayer regenerates. However, these cells are morphologically abnormal, displaying a polygonal, cuboidal appearance; poorly developed intercellular junctions, with occasional gaps between endothelial cells; and a lack of the usual alignment with the flow field (Shimokawa, Aarhus and Vanhoutte, 1987; Weidinger et al., 1990; Weidinger et al., 1991). Furthermore, endothelium mediated vasodilation is impaired in vessels with a regenerated endothelium (Shimokawa, Aarhus and Vanhoutte, 1987; Weidinger et al., 1990; Weidinger et al., 1991). A recent study has shed light on the latter observation. Isolation of endothelial cells from the luminal lining that has regenerated after balloon injury of the rabbit iliac artery, reveals intracellular levels of ADMA that are three-fold higher than those observed in endothelial cells from uninjured vessel (Azumi et al., 1995).

The observation that endothelial cells are a source of ADMA corroborates the previous finding by Fickling et al. (1993) that human endothelial cells release ADMA and SDMA. ADMA competitively inhibits endothelial NO synthesis in an autocrine manner. This observation is in line with previous findings that exogenous ADMA concentrations be-tween 1 and 10 µM affect the activity of the NO synthase in the vasculature of rat mesentery tissue and rat brain (Kurose et al., 1995; Faraci, Brian and Heistad, 1995). Faraci et al.

(1995) calculated an IC_{50} value for the inhibition of NO production in rat cerebellar homogenate by ADMA of $1.8 \pm .1$ µM, and Fickling *et al.* (1993) reported that 2 and 10 µM ADMA inhibited nitrite production in LPS-simulated J774 macrophages by 17 and 33% respectively. Taken together, these data suggest that ADMA may be a potential autocrine regulator of endothelial NO synthase.

The source of ADMA in endothelial cells is currently unclear. Dimethylarginines are likely the result of degradation of methylated proteins (McDermott, 1976; MacAllister *et al.*, 1996). The specific enzyme S-adenosylmethionine: protein arginine N-methyltransferase (protein methylase I) has been shown to methylate internal arginine residues in a variety of polypeptides, yielding N^G-monomethyl-L-arginine, N^G, N^G-dimethyl-L-arginine, and N^G, $N^{G'}$-dimethyl-L-arginine upon proteolysis. The metabolism of ADMA, but not SDMA, occurs via hydrolytic degradation to citrulline by the enzyme dimethylarginine dimethylaminohydrolase (DDAH) (MacAllister *et al.*, 1994). Inhibition of DDAH causes a gradual vasoconstriction of vascular segments, which is reversed by L-arginine (MacAllister *et al.*, 1996). This latter finding suggests that regulation of intra-cellular ADMA levels affects NO synthase activity. Intriguingly, Cooke and colleagues have developed preliminary data indicating that low-density lipoprotein increases endothe-lial cell ADMA elaboration. Although the mechanism by which LDL may affect ADMA formation or metabolism is currently unknown, this finding suggests that elevated ADMA levels may mediate some of the effects of LDL on the endothelium in an autocrine manner. Most importantly, as described above, within the concentration range found in cultured endothelial cells (5–40 µM), ADMA can induce some of the pathophysiological changes of the endothelium that occur in hypercholesterolemia.

G Protein Receptor Coupling, NO Synthase and Cofactors

Hypercholesterolemia may interfere with G protein endothelial receptor coupling, thereby inhibiting initial steps in signal transduction that ultimately activate NO synthase. Native LDL reduces bradykinin stimulated production of NO by inhibiting guanine nucleotide binding proteins which couple the receptor to activation of NO synthase (Liao and Clark, 1995). Hypercholesterolemia may impair endothelium-dependent relaxation of porcine coronary artery rings via a pertussis toxin-sensitive, G_i protein-dependent pathway (Shimokawa and Vanhoutte, 1989). Flavahan (1993) found that lysophosphatidylcholine, which is associated with oxidized LDL, inhibited endothelium-dependent relaxation in-duced by serotonin, but not bradykinin or ADP, in porcine coronary rings indicating that lysophosphatidylcholine selectively inhibits a Gi protein-dependent pathway. In contrast, Tanner *et al.* (1991) reported that oxidized LDL caused endothelial impairment which could be reversed by L-arginine.

Tetrahydrobiopterin (BH_4) is a cofactor for NO synthase and is involved in the electron transfer required for synthesizing NO from L-arginine (Gorren *et al.*, 1996; Klatt *et al.*, 1992). BH_4 increases production of NO by eNOS and decreases the generation of superoxide anion(Wever *et al.*, 1997). Rabelink and colleagues suggested that relative depletion of BH_4 may contribute to endothelial dysfunction since intra-arterial administration of BH_4 improved endothelium-dependent vasodilation to serotonin in hypercholesterolemic pa-tients (Stroes *et al.*, 1997).

Liao *et al.* (1995) reported that oxidized LDL, but not native LDL, decreases levels of NO synthase mRNA. This effect of oxidized LDL on expression of eNOS is mediated via inhibition of transcription as well as an increase in post-transcriptional degradation (Liao *et al.*, 1995).

Oxidant Stress

As noted previously, NO is inactivated by superoxide anion, generating peroxynitrite (Gryglewski, Palmer and Moncada, 1986; Mugge *et al.*, 1991; Rubanyi and Vanhoutte, 1986). Superoxide anion production is increased three to five-fold in cholesterol-fed rabbits compared to control rabbits (Ohara, Petersen and Harrison, 1993; Boger *et al.*, 1995). Chin *et al.* (1992) found that the lipid component of oxidized LDL inactivated NO released from bovine aortic endothelial cells. In addition, hypercholesterolemia impairs glutathione-mediated detoxification of peroxynitrite anion (Ma *et al.*, 1997). This reduces NO regeneration from peroxynitrite and also makes vascular tissue more susceptible to oxidative injury.

In animal models of hypercholesterolemia, antioxidant treatment with either superoxide dismutase, oxypurinol, α-tocopherol or probucol, restores endothelium-dependent relaxation implicating degradation of NO by superoxide anion as a cause of endothelial dysfunction (Ohara, Petersen and Harrison, 1993; Mugge *et al.*, 1991; White *et al.*, 1994; Keaney *et al.*, 1994; Keaney *et al.*, 1995; Levine *et al.*, 1996). Comparable observations have been made in hypercholesterolemic humans. The addition of probucol, a lipid lowering drug with antioxidant properties, to lovastatin improved endothelium-dependent vasodilation of epicardial coronary arteries in patients with hypercholesterolemia and coronary artery disease to a greater extent than the combination of lovastatin and cholestyramine despite similar reduction in LDL cholesterol (Anderson *et al.*, 1996). This difference was attributed to the antioxidant properties of probucol since improvement in endothelial function correlated with prolongation of the lag phase of copper-induced LDL oxidation cholesterol (Anderson *et al.*, 1996). Antioxidant treatment with vitamin C also improves endothelium-dependent vasodilation in hypercholesterolemic patients in peripheral vessels devoid of atherosclerosis. Vitamin C is a water soluble antioxidant that is capable of scavenging superoxide anion. Ting *et al.* (1997) found that acute intra-arterial administration of vitamin C enhanced the endothelium-dependent vasodilator response to methacholine chloride in hypercholesterolemic subjects (Figure 8.5). Similarly, Levine *et al.* (1996) reported that oral ingestion of vitamin C acutely improved flow-mediated endothelium-dependent vasodilation of the brachial artery of hypercholesterolemic patients with coronary artery disease. Taken together, there is compelling evidence that superoxide anion production is increased in hypercholesterolemia and contributes importantly to reduced bioavailability of endothelium-derived nitric oxide and subsequent reduction of endothelium-dependent vasodilation. The oxidative enzymes that are responsible for endothelial secretion and superoxide anion may include xanthine oxidase, NADPH oxygenase, or even NO synthase, under certain conditions (e.g., arginine depletion). NOS may itself generate superoxide anion (Pou *et al.*, 1992; Pritchard *et al.*, 1995; Huk *et al.*, 1997), an observation that may explain both the beneficial effects at anti oxidants and those of L-arginine.

SUMMARY

Vascular disease begins with an alteration in the endothelium which is characterized by an increase in intracellular oxidative stress, and the activation of oxidant-response genes

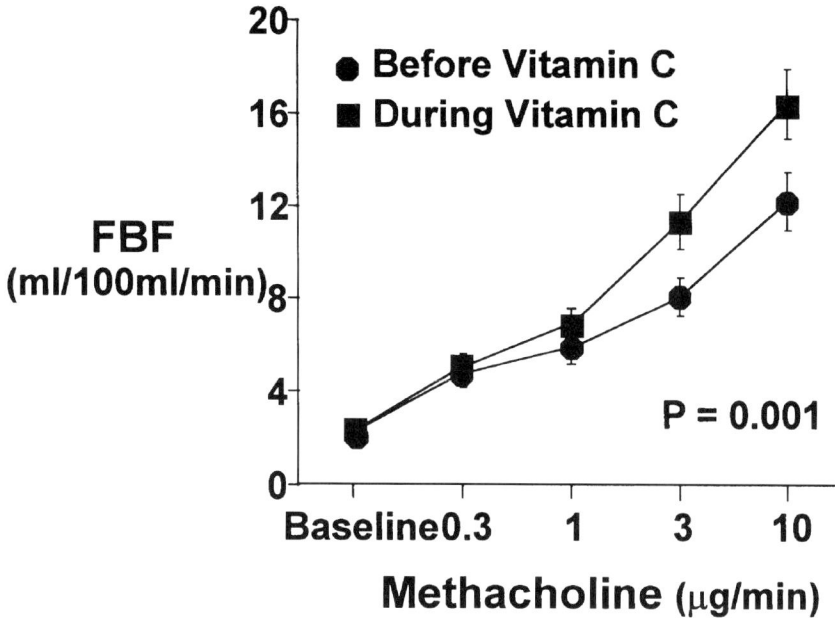

Figure 8.5   The effect of the antioxidant, vitamin C, on endothelium-dependent vasodilation in hypercholesterolemic subjects. Vitamin C enhanced the forearm blood flow response to methacholine, implicating oxidant stress, i.e., superoxide anion, as a cause of endothelial dysfunction in hypercholesterolemia. Adapted from Ting *et al.* (ref. 153).

regulating the expression of adhesion molecules and chemokines. These changes promote interaction of the endothelium with circulating blood elements. Monocyte infiltration and foam cell formation ensue, followed by further endothelial dysfunction and damage which precipitates platelet adherence and proliferation of vascular smooth muscle. These key processes in atherogenesis are opposed by nitric oxide. NO suppresses the expression and signaling of adhesion molecules involved in monocyte adhesion to the vessel wall, and inhibits platelet adherence and vascular smooth muscle cell proliferation. The NO synthase pathway is perturbed by hypercholesterolemia and other metabolic disorders that predispose to atherosclerosis. It is likely that basic insights regarding the mechanisms of endothelial dysfunction will lead to new therapeutic strategies to halt the progression, or induce regression, of atherosclerosis.

REFERENCES

Aji, W., Ravalli, S., Szabolcs, M., Jiang, X.C., Sciacca, R.R., Michler, R.E. and J.P. Cannon (1997) L-arginine prevents xanthoma development and inhibits atherosclerosis in LDL receptor knockout mice. *Circulation* **95**, 430–437.

Anderson, T.J., Meredith, I.T., Charbonneau, F., Yeung, A.C., Frei, B., Selwyn, A.P. and P. Ganz. (1996) Endothelium-dependent coronary vasomotion relates to the susceptibility of LDL to oxidation in humans. *Circulation*, **93** (9), 1647–50.

Andrews, H.E, Bruckdorfer, K.R., Dunn, R.C. and M. Jacobs (1987) Low-density lipoproteins inhibit endothelium-dependent relaxation rabbit aorta. *Nature*, **327**, 237–239.

Azumi, H., Sato, J., Hamasaki, H., Sugimoto, A., Isotani, E. and S. Obayashi Accumulation of endogenous inhibitors for nitric oxide synthesis and decreased content of L-arginine in regenerated endothelial cells. *British Journal of Pharmacology*, **115**, 1001–04.

Bath, P.M.W., Hassall, D.G., Gladwin, A.M, Palmer, R.M. and J.F. Martin (1991) Nitric oxide and prostacyclin. Divergence of inhibitory effects on monocyte chemotaxis and adhesion to endothelium in vitro. *Arteriosclerosis and Thrombosis*, **11**, 254–260.

Beckman, J.S., Ye, Y.Z., Anderson, P.G., Chen, J., Accavitti, M.A., Tarpey, M.M. and C.R. White (1994) Extensive nitration of protein tyrosines in human atherosclerosis detected by immunohistochemistry. *Biol. Chem. Hoppe Seyler*, **375**, 81–88.

Benditt, E.P., Barrett, T. and J.K. McDougall (1983) Viruses in the etiology of atherosclerosis. *Proceedings of the National Academy of Sciences USA*, **80**, 6386 6389.

Bennett, M.D., Evan, G.I. and S.M. Schwarz (1995) Apoptosis of rat vascular smooth muscle cells is regulated by p53-dependent and independent pathways. *Circulation Research*, **77**, 266–273.

Berliner, J.A., Navab, M., Fogelman, A.M., Frank, J.S., Demer, L.L., Edwards, P.A., Watson, A.D. and A.J. Lusis (1995) Atherosclerosis: Basic mechanisms. Oxidation, inflammation, and genetics. *Circulation*, **91**, 2488–2496.

Bode-Böger, S.M., Böger, R.H., Kienke, S., Junker, W. and J.C. Frölich (1996) Elevated L-arginine/dimethylarginine ratio contributes to enhanced systemic NO production by dietary L-arginine in hypercholesterolemic rabbits. *Biochemical and Biophysical Research Communications*, **219**, 598–603.

Böger, R.H., Bode-Böger, S.M., Mügge, A., Kienke, S., Brandes, R., Dwenger, A. and J.C. Frölich. (1995) Supplementation of hypercholesterolaemic rabbits with L-arginine reduces the vascular release of superoxide anions and restores NO production. *Atherosclerosis*, **117**, 273–284.

Böger, R.H., Bode-Böger, S.M., Szuba, A, Tsao, P.S., Chan, J., Tangphao, O., Blaschke, T. and J.P. Cooke (1998) ADMA: A Novel Risk Factor for Endothelial Dysfunction. Its Role in Hypercholesterolemia. *Circulation*, **96S**, A173.

Böger, R.H., Bode-Böger S.M., Thiele, W., Junker, W., Alexander, K. and J.C. Frolich (1997) Biochemical evidence for impaired nitric oxide synthesis in patients with peripheral arterial occlusive disease. *Circulation*, **95**, 2068–2074.

Böger R.H., Bode-Böger S.M., Tsao, P.S., Lin P.S., Chan, J.R. and J.P. Cooke (under review) An Endogenous Inhibitor of nitric oxide synthase regulates endothelial adhesiveness for monocytes.

Bossaller, C., Yamamoto, H., Lichtlen, P.R. and P.D. Henry (1987) Impaired cholinergic vasodilation in the cholesterol-fed rabbit in vivo. *Basic Research in Cardiology* **82**, 396–404.

Buttery, L.D., Springall, D.R., Chester D.R., Evans, T.J., Standfield, E.N., Parums, D.V., Yacoub, M.H. and J.M. Polak (1996) Inducible nitric oxide synthase is present within human adrosclerotic lesions and promotes the formation and activity of peroxynitrite. *Laboratory Investigation* **75**, 77–85.

Candipan, R. C., Wang, B., Buitrago, R., Tsao, P.S., Cooke, J.P. (1996) Regression or progression, Dependency upon vascular nitric oxide. *Arteriosclerosis, Thrombosis and Vascular Biology*, **16**, 44–50.

Casino, P.R., Kilcoyne, C.M., Quyyumi, A.A., Hoeg, J.M. and Panza, J.A. (1993) The role of nitric oxide in endothelium-dependent vasodilation of hypercholesterolemic patients. *Circulation* **88**, 2541–2547.

Cayatte, A.J., Palacino, J.J., Horten, K. and R.A. Cohen (1994) Chronic inhibition of nitric oxide production accelerates neointima formation and impairs endothelial function in hypercholesterolemic rabbits. *Arteriosclerosis and Thrombosis*, **14**, 753–759.

Celermajer, D.S., Sorensen, K.E., Gooch, V.M., Speigelhalter, D.J., Miller, O.W., Sullivan, I.D., Lloyd, J.K. and J.E. Deanfield (1992) Non-invasive detection of endothelial dysfunction in children and adults at risk of atherosclerosis. *Lancet,* **340**, 1111–1115.

Chin, J.H., Azhar, S. and B.B. Hoffman (1992) Inactivation of endothelial derived relaxing factor by oxidized lipoproteins. *Journal of Clinical Investigation*, **89**, 10–18.

Clancy, R.M., Leszczynska, P., Piziak, J. and S.B. Abramson (1992) Nitric oxide, an endothelial cell relaxation factor, inhibits neutrophil superoxide anion production via a direct action on, N. A. DPH oxidase. *Journal of Clinical Investigation*, **90**, 1116–1121.

Clarkson P, Adams, M.R., Powe, A.J., Donald, A.E., McCredie, R., Robinson, J., McCarthy, S.N., Keech, A, Celermajer, D.S. and J.E. Deanfield (1996) Oral L arginine improves endothelium-dependent dilation in hypercholesterolemic young adults. *Journal of Clinical Investigation*, **97**(8), 1989–94.

Cohen, R.A., Zitnay, K.M., Haudenschild, C.C. and L.D. Cunningham (1988) Loss of selective endothelial cell vasoactive functions in pig coronary arteries caused by hypercholesterolemia. *Circulation Research,* **63**, 903–910.

Cooke, J.P., Andon, N.A., Girerd, X.J., Hirsch, A.T. and M.A. Creager (1991) Arginine restores cholinergic relaxation of hypercholesterolemic rabbit thoracic aorta. *Circulation,* **83**, 1057–1062.

Cooke, J.P., Singer D.R., Tsao, P., Zera P., Rowan R.A. and M.E. Billingham Antiatherogenic effects of L-arginine in the hypercholesterolemic rabbit. *Journal of Clinical Investigation,* **90**, 1168–1172.

Cooke, J.P. and P.S. Tsao (1993) Cytoprotective effects of nitric oxide. *Circulation,* **88**, 2151–2154.

Cooke, J.P. and P.S. Tsao (1994) Is NO an endogenous anti-atherogenic molecule? *Arteriosclerosis and Thrombosis,* **14**, 653–655.

Cooke, J.P. and V.J. Dzau (1997) Derangements of the nitric oxide synthase pathway, L-arginine, and cardiovascular disease (editorial). *Circulation,* **96**, 379–382.

Creager, M.A., Cooke, J.P., Mendelsohn, M.E., Gallagher, S.J., Coleman, S.M., Loscalzo, J. and V.J. Dzau (1990) Impaired vasodilation of forearm resistance vessels in hypercholesterolemic humans. *Journal of Clinical Investigation,* **86**, 228–234.

Creager, M.A., Gallagher, S.J., Girerd, X.J., Coleman, S.M., Dzau, V.J. and J.P. Cooke (1992) L-arginine improves endothelium-dependent vasodilation in hypercholesterolemic humans. *Journal of Clinical Investigation,* **90**, 1248–1253.

Cybulsky, M.I. and M.A. Gimbrone, Jr. (1991) Endothelial expression of a mononuclear leukocyte adhesion molecule during atherogenesis. *Science,* **251**, 788–791.

DeCaterina, R., Libby, P., Peng, H-B., Thannickal, V.J., Rajavashisth, T.B., Gimbrone, M.A., Jr, Shin, W. S. and J.K. Liao (1995) Nitric oxide decreases cytokine-induced endothelial activation. *Journal of Clinical Investigation,* **96**, 60–68.

Drexler, H., Zeiher, A.M., Meinzer, K. and H. Just (1991) Correction of endothelial dysfunction in coronary microcirculation of hypercholesterolaemic patients by L- arginine. *Lancet,* **338** (8782–8783), 1546–50.

Egashira, K., Hirooka, Y., Kai, H., Sugimachi, M., Suzuki, S., Inou, T. and A. Takeshita (1994) Reduction in serum cholesterol with pravastatin improves endothelium dependent coronary vasomotion in patients with hypercholesterolemia. *Circulation,* **89**, 2519–2524.

Etingin, O.R., Silverstein, R.L. and D.P. Hajjar (1991) Identification of a monocyte receptor on herpes virus-infected endothelial cells. *Proceedings of the National Academy of Sciences USA,* **88**, 7200–7203.

Faraci, F.M., Brian, J.E. and D.D. Heistad (1995) Response of cerebral blood vessels to an endogenous inhibitor of nitric oxide synthase. *American Journal of Physiology,* **269**, H1522–H1527.

Fickling , S.A., Leone, A.M., Nussey, S.S., Vallance, P. and G.S.J. Whitley (1993) Synthesis of N^G, N^G dimethylarginine by human endothelial cells. *Endothelium,* **1**, 137– 140.

Flavahan, N.A. (1993) Lysophosphatidylcholine modified G protein-dependent signaling in porcine endothelial cells. *American Journal of Physiology,* **264**, H722 H727.

Garg, U.C. and A. Hassid (1989) Nitric oxide-generating vasodilators and 8–bromo cyclic guanosine monophosphate inhibit mitogenesis and proliferation of cultured rat vascular smooth muscle cells. *Journal of Clinical Investigation,* **83**, 1774–1777.

Geng, Y.J. and P. Libby (1995) Evidence for apoptosis in advanced human atheroma, colocalization with interleukin-1b-converting enzyme. *American Journal of Pathology,* **147**, 251–266.

Girerd, X.J., Hirsch, A.T., Cooke, J.P., Dzau, V.J. and M.A. Creager (1990) L- arginine augments endothelium-dependent vasodilation in cholesterol-fed rabbits. *Circulation Research,* **67**, 1301–1308.

Gorren, A.C., List, B.M., Schrammel, A., Pitters, E., Hemmens, B., Werner, E.R., Schmidt, K. and B. Mayer (1996) Tetrahydrobiopterin-free neuronal nitric oxide synthase, evidence for two identical highly anticooperative pteridine binding sites. *Biochemistry,* **35**, 16735–16745.

Gratton, M.T., Moreno-Cabral, C.E., Starnes, V.A., Oyer, P.E., Stinson, E.B. and N.E. Shumway (1989) Cytomegalovirus infection is associated with cardiac allograft rejection and atherosclerosis. *Journal of the American Medical Association,* **261**, 3561–3566.

Gryglewski, R.J., Palmer, R.M. and S. Moncada (1986) Superoxide anion is involved in the breakdown of endothelium-derived vascular relaxing factor. *Nature,* **320**, 454–6.

Harrison, D.G., Armstrong, M.L., Freiman, P.C. and D.D. Heistad (1987) Restoration of endothelium-dependent relaxation by dietary treatment of atherosclerosis. *Journal of Clinical Investigation,* **80**, 1808–1811.

Heistad, D.D., Armstrong, M.L.I., Marcus, M.L., Piegors, D.J., Mark, A.L. (1984) Augmented responses to vasoconstrictor stimuli in hypercholesterolemic and atherosclerotic monkeys. *Circulation Research,* **43**, 711–718.

Hogg, N., Kalyanaramer, B., Joseph, J., Struck, A. and S. Parthasarathy (1993) Inhibition of low-density lipoprotein oxidation by nitric oxide. Potential role in atherogenesis. *FEBS Letters*, **334**, 170–174

Huk, I., Nanobashvili, J., Neumayer, C., Punz, A., Mueller, M., Afkhampour, K., *et al.* (1997) L-arginine treatment alters the kinetics of nitric oxide and superoxide release and reduces ischemia reperfusion injury in skeletal muscle. *Circulation*, **95**, 667–675.

Ischiropoulos, H., Zhu, L. and J.S. Beckman (1992) Peroxynitrite formation from macrophage-derived nitric oxide. *Archives of Biochemistry and Biophysics*, **298**, 446– 451.

Isner, J.M., Kearney, M., Bortman, S. and J. Passeri (1995) Apoptosis in human atherosclerosis and restenosis. *Circulation*, **91**, 2703 2711.

Ito, A., Egashira, K., Kadokami, T., Fukumoto, Y., Takayanagi. T., Nakaike, R., *et al.* (1995) Chronic inhibition of endothelium-derived nitric oxide synthesis causes coronary microvascular structural changes and hyperreactivity to serotonin in pigs. *Circulation*, **92**, 2636–2644.

Jacobs, M., Plane, F. and K.R. Bruckdorfer (1990) Native and oxidized low-density lipoproteins have different inhibitory effects on endothelium-derived relaxing factor in the rabbit aorta. *British Journal of Pharmacology*, **100**, 21–26.

Johnson III, G. Tsao, P.S., Mulloy, D., Lefer, A.M. (1990) Cardioprotective effects of acidified sodium nitrite in myocardial ischemia with reperfusion. *Journal of Pharmacology and Experimental Therapeutics*, **252**, 35–41.

Keaney, Jr, J.F., Gaziano, J.M., Xu, A., Frei, B., Curran-Celentano, J., Shwaery, G.T., Loscalzo, J., Vita, J.A. (1994) Low-dose alpha-tocopherol improves and high-dose alpha-tocopherol worsens endothelial vasodilator function in cholesterol-fed rabbits. *Journal of Clinical investigation*, **93**, 844–51.

Keaney, Jr., J.F., Xu, A., Cunningham, D., Jackson, T., Frei, B. and J.A. Vita (1995) Dietary probucol preserves endothelial function in cholesterol-fed rabbits by limiting vascular oxidative stress and superoxide generation. *Journal of Clinical Investigation*, **95**, 2520–9.

Kubes, P., Suzuki, M. and D.N. Granger (1991) Nitric oxide, an endogenous modulator of leukocyte adhesion. *Proceedings of the National Academy of Sciences, USA*, **88**, 4651–4685.

Kannel, W.B., Castelli, W.P. and T. Gordon (1979) Cholesterol in the prediction of atherosclerotic disease. *Annals of Internal Medicine*, **90**, 85–91.

Keys, A. (1980) *Seven Countries*. Cambridge, Mass., Harvard University Press. Klatt, P., Heinzel, B., Mayer, B., Ambach, E., Werner-Felmayer, G., Wachter, H. *et al.* (1992) Stimulation of human nitric oxide synthase by tetrahydrobiopterin and selective binding of the cofactor. *FEBS Letters*, **305**, 160–2.

Knox, J.B., Sukhova, G.K., Whittemore, A.D. and P. Libby (1997). Evidence for altered balance between matrix metalloproteinases and their inhibitors in human aortic disease. *Circulation*, **95**, 205–212.

Kugiyama, K., Kern , S.A., Morrisett J.D., Roberts, R. and P.D. Henry (990) Impairment of endothelium-dependent arterial relaxation by lysolecithin in modified low density lipoproteins. *Nature*, **344** (6262), 160–162.

Kurose, I., Wolf, R., Grisham, M.B. and D.N. Granger (1995) Effects of an endogenous inhibitor of nitric oxide synthesis on postcapillary venules. *American Journal of Physiology*, **268**, H2224–H2231.

Lendon, C.L., Davies, M.J., Born, G.V.R. and P.D. Richardson (1991) Atherosclerotic plaque caps are locally weakened when macrophage density is increased. *Atherosclerosis*, **87**, 87–90.

Levine, G. N., Frei, B., Koulouris, S.N., Gerhard, M.D., Keaney, Jr., J.F., and Vita, J.A. (1996) Ascorbic acid reverses endothelial vasomotor dysfunction in patients with coronary artery disease. *Circulation*, **93**, 1107–13.

Leung, W.H, Lau, C.P. and Wong, C.K. (1993) Beneficial effect of cholesterol-lowering therapy on coronary endothelium-dependent relaxation in hypercholesterolaemic patients. *Lancet*, **341**, 1496–5000.

Liao, J.K. and S.L. Clark (1995) Regulation of G-protein alpha i2 subunit expression by oxidized low-density lipoprotein. *Journal of Clinical Investigation*, **95**, 1457–63.

Liao, J.K., Shin, W.S., Lee, W.Y and S.L. Clark (1995) Oxidized low-density lipoprotein decreases the expression of endothelial nitric oxide synthase. *Journal of Biological Chemistry*, **270**, 319–324.

Libby, P., Egan, D. and S. Skarlatos (1997) Roles of infectious agents in atherosclerosis and restenosis, An assessment of the evidence and need for future research, *Circulation*, **96**, 4095–4103.

Lin, K.T., Xuie, J.Y., Nomen, M., Spur, B. and P.Y. Wong (1995) Peroxynitrite induced apoptosis in HL-60 cells. *Journal of Biological Chemistry*, **270**, 16487–16490.

Loree, H.M., Tobias, B.F., Gibson, L.J., Kamm, R.D., Small, D.M. and R.T. Lee (1994) Mechanical properties of model atherosclerotic lesion lipid pools. *Arteriosclerosis and Thrombosis*, **14**, 230–234.

Ma, X.L., Lopez, B.L., Liu, G.L., Christopher, T.A., Gao, F., Guo, Y. *et al.* (1997) Hypercholesterolemia impairs a detoxification mechanism against peroxynitrite and renders the vascular tissue more susceptible to oxidative injury. *Circulation Research*, **80**, 894–901.

MacAllister, R.J., Fickling, S.A., Whitley, G.S.J. and P. Vallance (1994) Metabolism of methylarginines by human vasculature, Implications for the regulation of nitric oxide synthesis. *British Journal of Pharmacology*, **112**, 43–48.

MacAllister, R.J., Parry, H., Kimoto, M., Ogawa, T., Russell, R.J., Hodson, H., Whitley, G.S., and P. Vallance (1996) Regulation of nitric oxide synthesis by dimethylarginine dimethylaminohydrolase. *British Journal of Pharmacology*, **119**, 1533–1540.

McDermott, J.R. (1976) Studies on the catabolism of $N^{G}0$–methylarginine, N^{G}, N^{G} dimethylarginine and N^{G}, N^{G}'-dimethylarginine in the rabbit. *Biochemical Journal*, **154**, 179–184.

McLenahan, J.M., William, J.K., Fish, R.D., Ganz, P and A.P. Selwyn (1991) Loss of flow-mediated endothelium-dependent dilatation occurs early in the development of atherosclerosis. *Circulation*, **84**, 1273–1278.

McNamara, D.B., Bedi, B., Aurora, H., Tena, L., Ignarro, L.J., Kadowitz, P.J. *et al.* (1993) L-arginine inhibits balloon catheter-induced intimal hyperplasia. *Biochemical and Biophysical Research Communications*, **193**, 291–296.

Mercurio, F., Zhu, H., Murray, B.W., Shevchenko, A., Bennett, B.L., Li, J. *et al.* (1997) IKK-1 and IKK-2: cytokine-activated IkappaB kinases essential for NF-kappaB activation. *Science*, **278**, 860–866.

Marmur, J.D., Thiruvikraman, S.V., Fyfe, B.S., Guha, A., Sharma, S.K., Ambrose, J.A., *et al.* (1996) Identification of active tissue factor in human coronary atheroma. *Circulation*, **94**, 1226–1232.

Marui, N., Offermann, M.K., Swerlick, R., Kunsch, C., Rosen, C.A., Ahmad, M., *et al.* (1993) Vascular cell adhesion molecule-1 (VCAM-1) gene transcription and expression are regulated through an antioxidant-sensitive mechanisms in human vascular endothelial cells. *Journal of Clinical Investigation*, **92**, 1866–1872.

Mehta, J.L., Chen, L.Y., Kone, B.C., Mehta, P., and P. Turner. (1995) Identification of constitutive and inducible forms of nitric oxide synthase in human platelets. *Journal of Laboratory and Clinical Medicine*, **125**, 370–377.

Melnick, J.L., Adam, E., and M.E. DeBakey (1990). Possible role of cytomegalovirus in atherogenesis. *Journal of the American Medical Association*, **263**, 2204–2207.

Messmer, U.K., Lapetina, E.G. and B. Brüne (1995) Nitric oxide-induced apoptosis in, R.A. W 264.7 macrophages is antagonized by protein kinase C- and protein kinase A activating compounds. *Molecular Pharmacology*, **47**, 757–756.

Minick, C.R., Fabricant, C.G., Fabricant, J. and M.M. Litrenta (1979) Athero- arteriosclerosis induced by infection with a herpes virus. *American Journal of Pathology*, **96**, 673–706.

Moncada, S. and E.A. Higgs (1995) Molecular mechanisms and therapeutic strategies related to nitric oxide. *FASEB Journal*, **9**, 1319–1330.

Mugge, A., Elwell, J.H., Peterson, T.E., Hofmeyer, T.G., Heistad, D.D., and D.G. Harrison (1991) Chronic treatment with polyethylene-glycolated superoxide dismutase partially restores endothelium-dependent vascular relaxations in cholesterol-fed rabbits. *Circulation Research*, **69**, 1293–300.

Mugge, A., E., Elwell, J.H., Peterson, T.E., and D.G. Harrison (1991) Release of intact endothelium-derived relaxing factor depends on endothelial superoxide dismutase activity. *American Journal of Physiology*, **260**, C219–25.

Muhlestein, J.B., Hammond, E.H., Carlquist, J.F., Radicke, E., Thomson, M.J., Karagounis, L.A., *et. al.* (1996) Increased incidence of Chlamydia species within the coronary arteries of patients with symptomatic atherosclerotic versus other forms of cardiovascular disease. *Journal of the American College of Cardiology*, **27**, 1555 1561.

Murohara, T., Kugiyama, K., Ohgushi, M., Sugiyama, S. and H. Yasue (1994) Cigarette smoke extract contacts isolated porcine coronary arteries by superoxide anion- mediated degradation of EDRF, *American Journal of Physiology* **266**, H874–880.

Naruse, K., Shimizu, K., Muramatsu, M., Toki, Y., Miyazaki, Y., Okumura, K., Hashimoto, H., and T. Ito (1994) Long-term inhibition of NO synthesis promotes atherosclerosis in the hypercholesterolemic rabbit thoracic aorta. *Arteriosclerosis and Thrombosis*, **14**, 746–752.

Niu, X.F., Smith, C.W. and P. Kubes (1994) Intracellular oxidative stress induced by nitric oxide synthesis inhibition increases endothelial cell adhesion to neutrophils. *Circulation Research*, **74**, 1133–1140.

Numaguchi, K., Egashira, K., Takemoto, M., Kadokami, T., Shmokawa, H., Sueishi, K. and A. Takeshita (1995) Chronic inhibition of nitric oxide synthesis causes coronary, M.I. crovascular remodeling in rats. *Hypertension*, **26**, 957–962.

O'Driscoll, G., Green D. and R.R., Taylor (1997) Simvastatin, an HMG-coenzyme A reductase inhibitor, improves endothelial function within 1 month. *Circulation*, **95**, 1126–1131.

Ohara, Y., Petersen, T.E., Harrison, D.G. (1993) Hypercholesterolemia increases endothelial superoxide anion production. *Journal of Clinical Investigation*, 2546–2551

Ohara, Y., Peterson, T.E., Sayegh, H.S., Subramanian, R., Wilcox, J.N. and D.G. Harrison (1995) Dietary correction of hypercholesterolemia in the rabbit normalizes endothelial superoxide anion production. *Circulation*, **92**, 898–903.

Osborne, J.A., Siegman, M.J., Sedar, A.W., Mooers, S.U. and A.M. Lefer (1989) Lack of endothelium-dependent relaxation in coronary resistance arteries of cholesterol-fed rabbits. *American Journal of Physiology*, **256**, C591–C597.

Pagano, P.J., Tornheim, K., and R.A. Cohen (1993) Superoxide anion production by rabbit thoracic aorta, effect of endothelium-derived nitric oxide. *American Journal of Physiology*, **265**, H707–712.

Parhami, F., Morrow, A.D., Balucan, J., Leitinger, N., Watson, A.D., Tintut, Y., Berliner, J.A. and L.L. Demer (1997) Lipid oxidation products have opposite effects on calcifying vascular cell and bone cell differentiation. A possible explanation for the paradox of arterial calcification in osteoporotic patients. *Arteriosclerosis, Thrombosis and Vascular Biology*, **17**, 680–687.

Peng, H.B., Libby, P., and J.K. Liao (1995) Induction and stabilization of I kappa B alpha by nitric oxide mediates inhibition of NF-kappa B. *Journal of Biological Chemistry*, **270**, 14214–14219.

Pohl, U. and R. Busse (1989) EDRF increases cyclic GMP in platelets during passage through the coronary vascular bed. *Circulation Research*, **65**, 1798–1803.

Pohl, U., Nolte, C., Bunse, A., Eigenthaler, M. and U. Walter (1994) Endothelium dependent phosphorylation of vasodilator-simulated protein in platelets during coronary passage, *American Journal of Physiology*, **266**, 606–H612.

Pollman, M.J., Yamada, T., Horiuchi, M. and G.H. Gibbons (1996) Vasoactive substances regulate vascular smooth muscle cell apoptosis, countervailing influences of nitric oxide and angiotensin II. *Circulation Research*, **79**, 748–756.

Pollock, J.S., Försterman, U., Mitchell, J.A., Warner, T.D., Schmidt, H.H., Nakane, M. and F. Murad (1991) Purification and characterization of particulate endothelium derived relaxing factor synthase from cultured an native bovine aortic endothelial cells. *Proceedings of the National Academy of Sciences USA*, **88**, 10480–10484.

Pou, S., Pou, W.S., Bredt, D.S., Snyder, S.H., and G.M. Rosen (1992) Generation of superoxide by purified brain nitric oxide synthase. *Journal of Biological Chemistry*, **267**, 24173–24176.

Pritchard, Jr., K.A., Groszek, L., Smalley, D.M., Sessa, W.C., Wu, M., Villalon, P., Wolin, M.S. and M.B. Stemerman (1995) Native low-density lipoprotein increases endothelial cell-nitric oxide synthase generation of superoxide anion. *Circulation Research*, **77**, 510–518.

Radi, R., Beckman, J.S., Bush, K.M. and B.A. Freeman, B.A. (1991) Peroxynitrite oxidation of sulfhydrils. The cytotoxic potential of superoxide and nitric oxide. *Journal of Biological Chemistry*, **266**, 4244–4250.

Radomski, M.W., Palmer, R.M.J. and S. Moncada (1987) The anti-aggregating properties of vascular endothelium interactions between prostacyclin and nitric oxide, *British Journal of Pharmacology*, **92**, 629–636.

Radomski, M.W., Palmer, R.M.J. and S. Moncada (1990) An L-arginine/nitric oxide pathway present in human platelets regulates aggregation. *Proceedings of the National Academy of Sciences USA*, **87**, 5193–5197.

Rajagopalan, S., Kurz, S., Munzel, T., Tarpey, M., Freeman, B.A., Griendling, K.K., *et al.* (1996) Angiotensin II-mediated hypertension in the rat increases vascular superoxide production via membrane, N.A. DH/NADPH oxidase activations. Contribution to alterations of vasomotor tone. *Journal of Clinical Investigation*, **97**, 1916–1923.

Rajagopalan, S., Meng, X.P., Ramasamy, S., Harrison, D.G. and Z.S. Galis (1996) Reactive oxygen species produced by macrophage-derived foam cells regulate the activity of vascular matrix metalloproteinases in vitro. Implications for atherosclerotic plaque stability. *Journal of Clinical Investigation*, **98**, 2572–2579.

Rawal, N., Rajpurohit, R., Lischwe, M.A., Williams, K.R., Paik, W.K. and S. Kim (1995) Structural specificity of substrate for S-adenosylmethionine, protein arginine N-methyltransferases. *Biochimica et Biophysica Acta*, **1248**, 11–18.

Ross, R. (1997) Cellular and molecular studies of atherosclerosis. *Atherosclerosis*, **131**, S3–4.

Rubanyi, G.M. and P.M. Vanhoutte, (1986) Superoxide anions and hyperoxia inactivate endothelium-derived relaxing factor. *American Journal of Physiology*, **250**, H822–7.

Saikku, P., Leinonen, M., Mattila, K., Ekman, M.R., Nieminen, M.S., Makela, P.H., Huttunen, J.K. and V. Valtonen (1988) Serological evidence of an association of a novel Chlamydia. TWAR with chronic coronary heart disease and acute myocardial infarction. *Lancet*, **2**, 983–986.

Schwarzacher, S.P., Lim, T.T., Wang, B., Kernoff, R.S., Niebauer, J., Cooke, J.P. and A.C. Yeung (1997) Local intramural delivery of L-arginine enhances nitric oxide generation and inhibits lesion formation after balloon angioplasty. *Circulation*, **95**, 1863–1869.

Sedmak, D.D., Knight, D.A., Vook, N.C. and J.W. Waldman (1995) Divergent patterns of ELAM-1, ICAM-1 and VCAM-1 expression on cytomegalovirus-infected endothelial cells. *Transplantation*, **58**, 1379–1385.

Shimokawa, H. AND Vanhoutte, P.M. (1989) Impaired endothelium-dependent relaxation to aggregating platelets and related vasoactive substances in porcine coronary arteries in hypercholesterolemia and atherosclerosis. *Circulation Research*, **64**, 900–914.

Shears, L.L., Kawaharada, N., Tzeng, E., Billiar, T.R., Watkins, S.C., Kovesdi, I., Lizonova, A., and S.M. Pham (1997) Inducible nitric oxide synthase suppresses the development of allograft arteriosclerosis. *Journal of Clinical Investigation*, **100**, 2033– 2042.

Shimokawa, H., Aarhus, L.L. and P.M. Vanhoutte (1987) Porcine coronary arteries with regenerated endothelium have a reduced endothelium-dependent responsiveness to aggregating platelets and serotonin. *Circulation Research*, **61**, 256–270.

Simon, B.C., Cunningham, L.D. and Cohen, R.A. (1990) Oxidized low density lipoproteins cause contraction and inhibit endothelium-dependent relaxation in pig coronary artery. *Journal of Clinical Investigation*, **86**, 75–79.

Spiecker, M., Peng, H.B. and J.K. Liao (1997) Inhibition of endothelial vascular cell adhesion molecule-1 expression by nitric oxide involves the induction and nuclear translocation of IKB(. *Journal of Biological Chemistry*, **272**, 30969–30974.

Stamler, J.S., Simon, D.I., Osborne, J.A., Mullins, M.E., Jaraki, O., Michel, T., Singel, D.J. and J. Loscalzo (1992) S-nitrosylation of proteins with nitric oxide, Synthesis and characterization of biologically active compounds. *Proceedings of the National Academy of Sciences USA*, **89**, 444–448.

Stamler, J.S., Mendelsohn, M.E., Amarante, P., Smick, D., Andon, N., Davies, P.F., Cooke, J.P. and J. Loscalzo (1989) N-acetylcysteine potentiates platelet inhibition by endothelium-derived relaxing factor. *Circulation Research*, **65**, 789–95.

Steinberg, H.O., Bayazeed, B., Hook, G., Johnson, A., Cronin, J., and A.D. Baron (1997) Endothelial dysfunction is associated with cholesterol levels in the high normal range in humans. *Circulation*, **96**, 3287–3293.

Stroes, E.S., Koomans, H.A., de Bruin, T.W. and T.J. Rabelink (1995) Vascular function in the forearm of hypercholesterolaemic patients off and on lipid-lowering medication. *Lancet*, **346** (8973), 467–71.

Stroes, E., Kastelein, J., Cosentino, F., Erkelens, W., Wever, R, Koomans, H., Luscher, T. and T. Rabelink (1997) Tetrahydrobiopterin restores endothelial function in hypercholesterolemia. *Journal of Clinical Investigation*, **99**, 41–46.

Takahashi, M., Yui, Y., Yasumoto, H., Aoyama, T., Morishita, H., Hattori, R., Kawai, C. (1990) Lipoproteins are inhibitors of endothelium-dependent relaxation of rabbit aorta. *American Journal of Physiology*, **258**, H1–H8.

Tamai, O., Matsuoka, H., Itabe, H., Wada, Y., Kohno, K. and T. Imaizumi (1997) Single LDL apheresis improves endothelium-dependent vasodilatation in hypercholesterolemic humans. *Circulation*, **95**, 76–82.

Tanner, F.C., Noll, G., Boulanger, C.M. and T.F. Luscher (1991) Oxidized low density lipoproteins inhibit relaxations of porcine coronary arteries. Role of scavenger receptor and endothelium, derived nitric oxide. *Circulation*, **83**, 2012–2020.

Tarry, J.W.C. and R.G. Makhoul (1994) L-arginine improves endothelium-dependent vasorelaxation and reduces intimal hyperplasia after balloon angioplasty. *Arteriosclerosis and Thrombosis*, **6**, 938–943.

Tesfamariam, B. and R.A. Cohen (1992) Free radicals mediate endothelial cell dysfunction caused by elevated glucose. *American Journal of Physiology*, **263**, H321–326.

Thom, D.H., Grayston, J.T., Siscovick, D.S., Wang, S.P., Weiss, N.S. and J.R. Daling (1992) Association of prior infection with Chlamydia pneumoniae and angiographically demonstrated coronary artery disease. *Journal of the American Medical Association*, **268**, 68–72.

Timimi, F.K., Ting, H.H., Haley, E.A., Roddy, M.A., Ganz, P. and M.A. Creager (1997) Vitamin C improves endothelium-dependent vasodilation in forearm resistance vessels of humans with hypercholesterolemia. *Circulation*, **95**, 2617–2622.

Treasure, C.B., Klein, J.L., Weintraub, W.S., Talley J.D., Stillabower, M.E., Kosinski, A.S., Zhang, J., Boccuzzi, S.J., Cedarholm, J.C. and R.W. Alexander (1995) Beneficial effects of cholesterol-lowering therapy on the coronary endothelium in patients with coronary artery disease. *New England Journal of Medicine*, **332**, 481– 487.

Tsao, P.S., Lewis NP, Alpert S. and J.P. Cooke, J.P. (1995) Exposure to shear stress alters endothelial adhesiveness. Role of nitric oxide. *Circulation*, **92**, 3513– 3519.

Tsao, P.S., Buitrago, R., Chang, H., Chen, U.D.I. and G.M. Reaven (1995) Effects of diabetes on monocyte-endothelial interactions and endothelial superoxide production in fructose-induced insulin-resistant and hypertensive rats (Abstract). *Circulation*, **92**, A2666.

Tsao, P.S., McEvoy, L.M., Drexler, H., Butcher, E.C. and J.P. Cooke (1994) Enhanced endothelial adhesiveness in hypercholesterolemia is attenuated by L arginine. *Circulation*, **89**, 2176–2182.

Tsao, P.S., Buitrago, R., Chan, J.R. and J.P. Cooke (1996) Fluid flow inhibits endothelial adhesiveness, Nitric oxide and transcriptional regulation of VCAM-1. *Circulation*, **94**, 1682–1689.

Tsao, P.S., Wang, B.Y., Buitrago, R., Shyy, J.Y. and J.P. Cooke (1997) Nitric oxide regulates monocyte chemotactic protein-1. *Circulation*, **96**, 934–940.

Vallance, P., Leone, A., Calver, A., Collier, J. and S. Moncada (1992) Endogenous dimethylarginine as an inhibitor of nitric oxide synthesis. *Journal of Cardiovascular Pharmacology*, **20**, S60–S62.

Verbeuren, T.J., Jordaens, F. H., Zonnekeyn, L.L., Van Hove, C.E., Coene, M. C. and A.G. Herman (1986) Effect of hypercholesterolemia on vascular reactivity in the rabbit. I. Endothelium-dependent and endothelium-independent contractions and relaxations in isolated arteries of control and hypercholesterolemic rabbits. *Circulation Research*, **58**, 552–564.

Vita, J.A., Treasure, C.B., Nabel, E.G., McLenachan, J.M., Fish, R.D., Yeung A.C., *et al.* (1990) Coronary vasomotor response to acetylcholine relates to risk factors for coronary artery disease. *Circulation*, **81**, 491–497.

von der Leyen, H.E., Gibbons, G.H., Morishita, R., Lewis, N.P., Zhang, L., Nakajima, M., *et al.* (1995) Gene therapy inhibiting neointimal vascular lesion: in vivo transfer of endothelial cell nitric oxide synthase gene. *Proceedings of the National Academy of Sciences USA*. **92**, 1137–1141.

Wang, B., Singer, A., Tsao, P.S., Drexler, H., Kosek, J. and J.P. Cooke (1994) Dietary arginine prevents atherogenesis in the coronary artery of the hyper- cholesterolemic rabbit. *Journal of the American College of Cardiology*, **23**, 452–458.

Wang, B.Y., Ho, Hoai-Ky, Lin, P.S., Schwarzacher, S.P., Pollman, M.J., Gibbons, G.H., Tsao, P.S. and J.P. Cooke, Regression of atherosclerosis, role of nitric oxide and apoptosis (*Circulation*, in press).

Weidinger, F.F., McLenachan, J.M., Cybulsky, M.I., Gordon, J.B., Rennke, H. G., Hollenberg, N.K., *et al.* (1990) Persistent Dysfunction of Regenerated Endothelium after Balloon Angioplasty of Rabbit Iliac Artery. *Circulation*, **81**, 1667–1679.

Weidinger, F.F., McLenachan, J.M., Cybulsky, M.I., Fallon, J.T., Hollenberg, N.K., Cooke, J.P., *et al.* Hypercholesterolemia enhances macrophage recruitment and dysfunction of regenerated endothelium after balloon injury of the rabbit iliac artery. *Circulation*, **84**, 755–767.

Wever, R.M.F., van Dam, T., van Rijn, N.J., de Groot, F., and T.J. Rabelink, (1997) Tetrahydrobiopterin regulates superoxide and nitric oxide generation by recombinant endothelial nitric oxide synthase. *Biochemical and Biophysical Research Communications*, **237**, 340–344.

White, C.R., Brock, T.A., Chang, L.Y., Crapo, J., Briscoe, P., Ku, D., *et al.* (1994) Superoxide and peroxynitrite in atherosclerosis. *Proceedings of the National Academy of Sciences USA*, **91**, 1044–8.

Woronicz, J.D., Xiong, G., Zhaodan, C., Rothe, M. and D.V.Goeddel (1997) 1kB Kinase-B, NF-kB Activation and Complex Formation with 1kB Kinase-A and NIK, *Science*, **278** (5339), 866–869.

Ye, S., Eriksson, P., Hamsten, A., Kurkinen, M., Humphries, S. E., and A.M. Henney (1996) Progression of coronary atherosclerosis is associated with a common genetic variant of the human stromelysin-1 promoter which results in reduced gene expression. *Journal of Biological Chemistry*, **271**, 13055–13060.

Yokoyama, M., Hirata, K., Miyake, R., Akita, H., Ishikawa, Y., and H. Fukuzaki (1990) Lysophosphatidylcholine, essential role in the inhibition of endothelium-dependent vasorelaxation by oxidized low-density lipoprotein. *Biochemistry and Biophysics Research Communications*, **168**, 301–308.

Zeiher A.M., Fisslthaler, B., Schray-Utz, B. and R. Busse (1995) Nitric oxide modulates expression of monocyte chemoattractant protein-a in cultured human endothelial cells. *Circulation Research*, **76**, 980–986.

Zeiher, A.M., Drexler, H, Wolschlager and H. Just (1991) Modulation of coronary vasomotor tone in humans. *Circulation*, **83**, 391–401.

This review focuses on studies in humans of the endothelium derived mediators thought to inhibit or promote atherothrombotic in diabetes mellitus. *In vivo* studies have examined the blood flow response of the forearm vascular bed to local brachial artery infusion of endothelium-dependent (e.g. acetylcholine) and -independent vasodilators (e.g. nitrovasodilators). A drawback of this approach, partially overcome by using selective inhibitors of nitric oxide (NO) synthase (cNOS) and cyclooxygenase (COX), is that the overall blood flow response to agonists such as acetylcholine is mediated by nitric oxide (NO), vasoactive prostanoids, endothelium-derived hyperpolarising factor (EDHF) plus other factors. In type 1 diabetes, especially in the presence of microalbuminuria, NOS inhibition produces less of an effect on basal blood flow and the acetylcholine stimulated blood flow response is less readily inhibited by NOS inhibition than in control subjects. This suggests decreased basal and stimulated release or increased inactivation of NO in type 1 diabetes. The overall response to acetylcholine may be preserved suggesting that there is a compensatory increase in vasodilator COX products or EDHF. In type 2 diabetes vasodilator responsiveness to acetylcholine is impaired but the pathway responsible has not been clearly elucidated. In both type 1 and type 2 diabetes there is evidence of impaired sensitivity of vascular smooth muscle to nitrovasodilators suggesting impaired sensitivity to endothelium-derived NO. Umbilical vein endothelial cells obtained from mothers with gestational diabetes or those from non-diabetic mothers exposed to elevated concentrations of glucose exhibit an *increase* in NO and, disproportionately, in superoxide anion (O_2^-) production. The latter, through inactivation of NO, might account for the apparent decrease in NO seen in *in vivo* studies. Other mechanisms which may account for decreased availability of NO in diabetes include other sources of free radicals (including O_2^-) which may inactivate NO, and direct inactivation of NO by advanced glycosylation end products and/or by oxidised LDL. Improvements in endothelial function have been seen after lipid lowering therapy in type 1 diabetes and after acute administration of vitamin C in type 2 diabetes. Other promising strategies include novel antioxidants, angiotensin converting enzyme inhibitors, angiotensin II receptor antagonists and agents which increase insulin sensitivity.

Key words; diabetes, EDHF, cyclooxygenase, pregnancy, arginine transport, lipoproteins.

9 Endothelial Function in Diabetes Mellitus

Jim M. Ritter[1,2], Phil J. Chowienczyk[1,2] and Giovanni E. Mann[2]

[1]Department of Clinical Pharmacology, St Thomas' Hospital, UMDS, Lambeth Palace Road London, SE1 7EH, UK
[2]Centre for Cardiovascular Biology and Medicine, GKT, School of Biomedical Sciences, King's College London, Campden Hull Road, W8 7AH, UK

INTRODUCTION

Diabetes mellitus is defined by a fasting blood glucose concentration ≥ 7.0 mmol/L or an abnormal rise in blood glucose following an oral load. The two commonest forms are type 1 (insulin dependent) diabetes, which has a strongly inherited predisposition and variable annual incidence in different countries (~1 to 2 per 100,000 in Japan, > 40 per 100,000 in Finland), and type 2 (non-insulin dependent) diabetes, which also has a strong inherited component and is much commoner than type 1 (Wolff, 1993). Type 1 diabetes is caused by an autoimmune destruction of the islets of Langerhans following an environmental insult (e.g. infection with Coxsackie virus) in genetically predisposed individuals. Patients with type 1 diabetes have a near total loss of insulin secretion, which results in increased glycogen metabolism, elevated circulating plasma glucose levels and ketoacidosis. Type 2 diabetes is more complex and in these patients hyperglycaemia is associated with varying degrees of insulin resistance and relative impairment of insulin secretion (Jaap et al., 1994). Insulin secretion declines with age and thus type 2 diabetes is increasing in our ageing population. In contrast to untreated type 1 diabetes, where circulating concentrations of insulin are low or absent, the plasma concentrations of insulin in type 2 diabetics may in the early stages of the disease be higher than in non-diabetic subjects. However, blood glucose is elevated due to impaired responses to insulin either at or beyond the level of the insulin receptor.

Gestational diabetes constitutes another type of diabetes, recognized during pregnancy, and is generally associated with a transient decrease in maternal insulin sensitivity similar to that found in patients with non-insulin dependent diabetes (Dornhorst & Beard, 1993; Eriksson, 1995). Patients with gestational diabetes have elevated levels of hemoglobin A_{1c} (a marker of protein glycation in diabetes) during early pregnancy (see Sobrevia & Mann, 1997). Such women generally include a small number with previously unrecognized diabetes or impaired glucose tolerance (Dornhorst & Beard, 1993). Treatment with diet and/or insulin has been highly successful in reducing the incidence of fetal macrosomia in gestational diabetic pregnancies.

Both type 1 and type 2 diabetes are associated with vascular complications involving small and large blood vessels. Microvascular complications of retinopathy, nephropathy and neuropathy are recognised by clinicians as specific for diabetes mellitus, but it is the

Current Addresses: Professor Jim M. Ritter and Dr Phil J. Chowienczyk, Department of Clinical Pharmacology, St. Thomas' Hospital, 9KT, School of Biomedical Sciemces, KCL, Lambeth Palace Road, London SE1 7EH, UK. Tel: 0171-928 9292 ext. 2350; Fax: 0171-401-2242.

consequences of macrovascular atheromatous disease, notably myocardial infarction and thrombotic stroke, that are principally responsible for the increased incidence of disability and premature death in these patients. Microalbuminuria is the earliest evidence of glomerular endothelial injury and is a strong risk factor for myocardial infarction in diabetic patients. The increased thromboxane synthesis in type 2 patients (Davi *et al.*, 1990) implies that physiological mechanisms involved in preventing platelet activation may be defective in such patients. Although this contrasts with type 1 patients in whom abnormal thromboxane production is not a general feature (Green *et al.*, 1988), platelets from type 1 and type 2 diabetic patients have a significantly reduced activity of NO synthase and an impaired ability to mediate vasodilation (Oskarsson *et al.*, 1996; Martina *et al.*, 1998; Rabini *et al.*, 1998). Plasma levels of von Willebrand factor (vWf) are elevated in patients with diabetic microangiopathy but are apparently not influenced by changes in plasma glucose, insulin, or growth hormone levels *per se* (Porta *et al.*, 1981). The possibility that increased circulating vWf levels may precede microalbuminuria in patients with insulin-dependent diabetes has been used as indicator for ensuing diabetic complications (Stehouwer *et al.*, 1995). Moreover, advanced glycosylated end products (AGEs) are clearly implicated in endothelial cell dysfunction in diabetic patients (Makita *et al.*, 1991; Bucala *et al.*, 1991), although modulation of the secretory pathways in endothelial cells remains to be elucidated.

The recognition that atherosclerosis is initiated by endothelial injury, and that the endothelium synthesises vasoactive factors that protect against progression of atherosclerotic disease has led to investigations of endothelial function in patients with diabetes. There have been three main approaches: *in vivo* studies of vascular responses in diabetic patients, studies using biopsies of small resistance arteries mounted in organ chambers and investigated *in vitro*, and *in vitro* studies of cultured endothelial cells. Each of these approaches has its limitations, and one of our objectives is to discuss these critically. Although our current understanding is fragmentary and incomplete, there is overwhelming evidence based on these complementary techniques that endothelial function is altered in diabetes mellitus. We have focused on human studies of the L-arginine/NO signalling pathway in view of its role in regulating basal vascular tone and the progression of atherosclerosis (Ignarro, 1990; Moncada *et al.*, 1991). We refer the reader to recent comprehensive reviews of endothelial dysfunction in experimental animal models of diabetes (Cohen, 1993; Poston & Taylor, 1995; Sobrevia & Mann, 1997).

In vivo Studies of Endothelium Derived Mediators

The role of the endothelium in diabetes mellitus has been examined *in vivo* mainly by assessing local vasodilator responses to physical or pharmacological stimuli assumed to increase endothelium derived mediators. Strain gauge plethysmography (Benjamin *et al.*, 1995) has been widely used to measure limb blood flow responses to brachial or femoral artery infusion of endothelium-dependent and -independent vasodilators. This method reflects the function of resistance vessels, the calibre of which determines limb blood flow. Forearm vasculature has been extensively studied because of its accessibility, but it is also of particular interest as it is rarely affected by atheroma (Ciccone *et al.*, 1993), facilitating a distinction between functional abnormalities and structural effects of early atherosclerosis. Muscarinic agonists such as acetylcholine and methacholine have been used extensively and assumed to act largely through the L-arginine/NO pathway (Vallance *et al.*, 1989).

However, it is likely that these agonists also stimulate release of endothelium-derived hyperpolarizing factor (Chen *et al.*, 1988; Taddei *et al.*, 1995), synthesis of vasoactive prostanoids (Shimizu *et al.*, 1993; Taddei *et al.*, 1997) and may have endothelium-independent actions on vascular smooth muscle mediated either directly or through stimulation of sensory nerves (Ralevic *et al.*, 1992). The relative contributions of these various actions to the overall vasodilator effect is likely to differ for individual muscarinic agonists. Actions of acetylcholine are more readily inhibited by the NO synthase inhibitor N^G-monomethyl-L-arginine (L-NMMA) than are those of methacholine, suggesting that acetylcholine is a more sensitive probe of the NO pathway (Chowienczyk *et al.*, 1993). However, due to its extreme instability in blood, responses to acetylcholine are highly dependent on forearm geometry and basal forearm blood flow (Chowienczyk *et al.*, 1994).

Brachial artery infusion of L-NMMA has been used to assess basal NO production, with the assumption that this is directly related to the degree of forearm vasoconstriction produced by L-NMMA (Vallance *et al.*, 1989). Use of an appropriate comparator vasoconstrictor, which does not interact with the L-arginine/NO pathway, may strengthen this approach (Calver *et al.*, 1992).

A non-invasive method for measuring endothelial function in the brachial artery has been developed by Celermajer and colleagues (Celermajer *et al.*, 1992). High resolution duplex ultrasound is used to image the brachial artery and measure changes in artery diameter in response to changes in flow. Increases in blood flow generated by reactive hyperaemia in the hand after a short period of radial arterial occlusion (ie distal to the site where the artery diameter is being measured) are followed by an increase in brachial artery diameter. This flow-mediated dilation is due, in part, to endothelium-derived NO, since it is attenuated by prior infusion of L-NMMA into the brachial artery (Joannides *et al.*, 1995). The technique is capable of detecting subtle degrees of endothelial dysfunction in young subjects (Celermajer *et al.*, 1993, 1996; Sorensen *et al.*, 1994).

Doppler fluximetry has been used to measure skin blood flow responses to local heating or iontophoresis of vasoactive drugs (Jaap *et al.*, 1994; Morris *et al.*, 1995). The role of endothelin in diabetes has been studied mainly by measuring concentrations of endothelins in plasma and urine. Responses to local infusion of endothelin-1 have been studied in type 2 diabetes. Studies involving local infusion of endothelin receptor antagonists and converting enzyme inhibitors may provide more specific information.

Type 1 Diabetes

Both basal and agonist-stimulated NO release has been investigated in forearm vasculature of patients with type 1 diabetes. Calver *et al.* (1992) reported a blunted response to L-NMMA in type 1 patients, which was confirmed by Elliott *et al.* (1993), who in addition reported that this response was most marked in patients with microalbuminuria. Stimulated responses have been examined using the muscarinic agonists acetylcholine, methacholine and carbachol. In most studies vasodilator responses to these agonists were preserved (Halkin *et al.*, 1991; Calver *et al.*, 1992; Elliott *et al.*, 1993; Smits *et al.*, 1993). However, in patients with type 1 diabetes, other groups have reported a blunted response to acetylcholine and methacholine (Mäkimattila *et al.*, 1996; Johnstone *et al.*, 1993). A potentially important difference in the study by Johnstone *et al.* (1993) was that subjects were pretreated with aspirin to inhibit cyclooxygenase, thereby inhibiting the synthesis of vasoactive prostanoids. This may reconcile their findings with those of Elliott *et al* (1993), who found

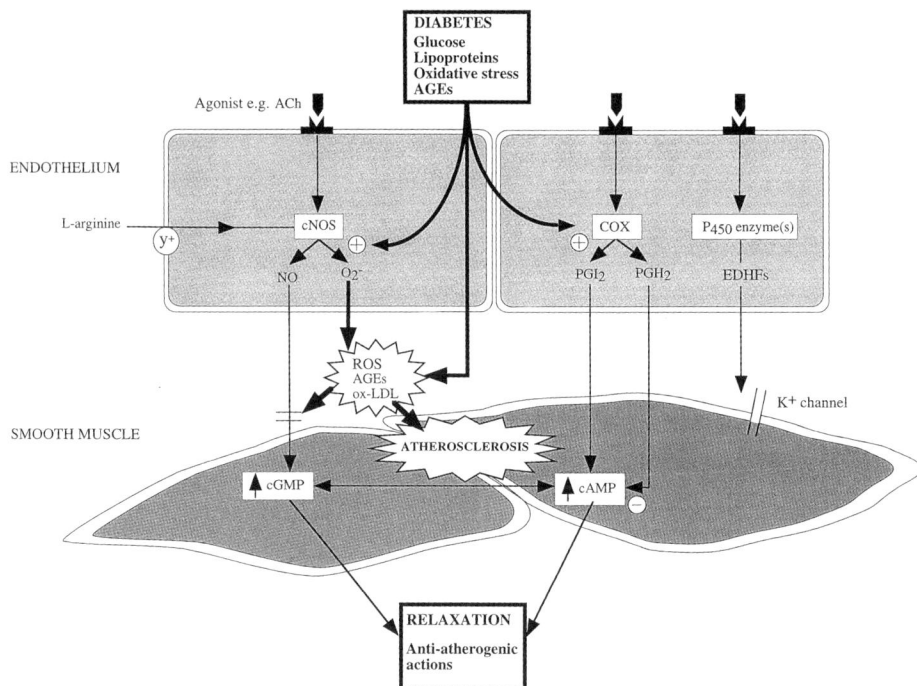

Figure 9.1 Synthesis of endothelium derived mediators. Pathways implicated in the pathogenesis of diabetic angiopathy are shown in bold. Elevated glucose and abnormal lipoproteins may increase generation of O_2^- from cNOS. COX activity may be increased but how the balance of vasodilator/vasoconstrictor products is altered in humans is unknown. The influence of diabetes on EDHF synthesis is unknown; it may be increased. ACh: acetylcholine; AGEs: advanced glycosylation end products; cAMP: cyclic adenosine monophosphate; cGMP: cyclic guanylate monophosphate; cNOS: constitutive nitric oxide synthase; COX: cyclooxygenase; EDHFs: endothelium derived hyperpolarising factors; NO: nitric oxide; O_2^-: superoxide anion; ox-LDL: oxidised low density lipoprotein; PG: prostaglandin; ROS: reactive oxygen species.

that although vasodilator responses to carbachol were preserved, the NO dependent component of the response (identified by co-infusion of L-NMMA) was reduced in patients with microalbuminuria (Figure 9.1). Taken together the results of these studies suggest the possibility that type 1 diabetes is associated with a reduced NO mediated vasodilatory response to muscarinic agonists and compensatory increase in vasodilatory prostanoids. Further studies need to compare the vasodilatory responses to muscarinic agonists in the absence and presence of cyclooxygenase inhibitors in type 1 patients and controls. Alternatively, the normal vasodilator response to carbachol but loss of inhibition by L-NMMA that were observed could be due to increased co-release of endothelium-derived hyperpolarising factor (EDHF) by muscarinic agonists in type 1 patients with endothelial dysfunction.

The apparently contradictory findings in other studies could be the consequence of differences in patient characteristics. The presence of microalbuminuria and the degree of glycaemic control are important variables (Halkin *et al.*, 1991; Elliott *et al.*, 1993; Smits *et al.*, 1993; Mäkimattila *et al.*, 1996). Methodological factors relating to choice of agonist, especially the relative magnitudes of the NO- and EDHF-mediated components, may also account for the contrasting findings in studies which have employed methacholine rather than acetylcholine (Smits *et al.*, 1993; Mäkimattila *et al.*, 1996; Johnstone *et al.*, 1993). The influence of basal blood flow may be especially important in type 1 diabetes, since basal flow increases with poor metabolic control. Flow-mediated dilation is impaired in patients with type 1 diabetes, especially in patients with microalbuminuria (Zenere *et al.*, 1995), and an inverse association between flow-mediated dilation and LDL-cholesterol has been described in patients with type 1 diabetes (Clarkson *et al.*, 1996).

Plasma concentrations of endothelin-1 have been reported to be lower in children with type 1 diabetes than in age matched controls and to be negatively associated with the duration of diabetes (Smulders *et al.*, 1994; Malamitsi-Puchner *et al.*, 1996). Elevated ET-1 immunoreactivity has been identified in the cutaneous microvasculature of patients with type 1 diabetes of recent onset, yet ET-1 immunoreactivity is reduced in patients with diabetes of longer duration and in patients with retinopathy (Properzi *et al.*, 1995).

Insulin Resistance and Type 2 Diabetes

Most studies in type 2 diabetes have focused on stimulated NO release and, unlike type 1 diabetes, there is general consensus that responses to endothelium-dependent agonists are impaired. McVeigh *et al.* (1992) first reported impaired forearm blood flow responses both to acetylcholine and GTN in patients with type 2 diabetes. Reactive hyperaemic forearm blood was similar in diabetic patients and non-diabetic control groups. Impaired vasodilator responses to acetylcholine and nitroprusside have also been described in the skin of type 2 diabetic patients (Morris *et al.*, 1995). These findings were recently extended by Williams *et al* (1996), who excluded patients with hypertension and hypercholesterolaemia, both potentially confounding factors. In this latter study of forearm blood flow, subjects were pretreated with aspirin to inhibit the endogenous production of vasoactive prostaglandins. Forearm blood flow responses to both methacholine and nitroprusside were significantly attenuated in diabetic patients compared with controls, with no significant group differences noted in responses to verapamil or hyperaemia. The use of aspirin suggests that abnormal synthesis of vasoconstrictor prostanoids cannot explain this aspect of vascular dysfunction in type 2 diabetes. Normal responses to verapamil exclude a non-specific impairment of vasodilator function. A reservation regarding the interpretation of this study is the use of methacholine, however, similar results have recently been reported using acetylcholine as an agonist (Watts *et al.*, 1996). The similarity of altered vascular responses to acetylcholine and methacholine suggests that EDHF release in response to muscarinic agonists may be impaired in subjects with type 2 diabetes, and begs a direct comparison of acetylcholine responses in type 2 diabetic patients in the absence or presence of L-NMMA. This was done by Watts *et al.*, (1996) who found that L-NMMA produced similar blunting of the response to acetylcholine in diabetics and controls, thus indirectly implicating EDHF. The study of Watts *et al.* (1996) excluded patients on oral hypoglycaemic drugs and those with autonomic neuropathy and/or proteinuria. Thus, drug therapy and autonomic dysfunction are unlikely to explain the impaired vasodilator responsiveness in type 2 diabetes, and this defect is not confined to patients with proteinuria. The impaired dilator responses to GTN or nitroprusside, as well as to acetylcholine or methacholine, in these studies again raises the question of what fractions of the observed responses to different muscarinic agonists are mediated by NO. The findings could be explained by an increased inactivation of NO and/or EDHF, resulting from an increase in endothelium-derived superoxide anion (O_2^-) production as discussed below, or an impaired sensitivity of vascular smooth muscle to NO.

Impaired vasodilator responses to NO donors such as nitroprusside or GTN in type 2 diabetic patients have not been reported by all groups. Impaired forearm vasodilator responses to methacholine, with preserved responses to nitroprusside, have recently been reported in type 2 diabetics (Ting *et al.*, 1996). Goodfellow *et al* (1996) observed impaired flow-mediated dilation of the radial artery in type 2 patients, with endothelium-independent responses to GTN unaffected. Skin vasodilator responses to iontophoresis of acetylcholine are impaired in patients with type 2 diabetes, yet responses to nitroprusside are apparently impaired only in patients with peripheral sensory neuropathy (Pitei *et al.*, 1997). We have also observed impaired endothelium-dependent vasodilation, with preserved responses to nitroprusside, in forearm vasculature of patients with type 2 diabetes of recent onset without evidence of peripheral sensory neuropathy (data not shown). One explanation is that well-controlled type 2 diabetes of recent onset is associated with an impaired release of endothelium-derived NO (or possibly endothelium-derived hyperpolarizing factor, EDHF), and that diabetes of longer duration results additionally in increased inactivation of exogenous NO and/or impaired sensitivity of vascular smooth muscle to NO. This explanation is supported by findings in patients with insulin resistance who had not developed type 2 diabetes at the time of study. Steinberg *et al* (1996) measured leg blood flow responses to femoral artery infusions of methacholine and nitroprusside in insulin sensitive controls, insulin-resistant obese subjects and patients with type 2 diabetes. The response to methacholine, but not nitroprusside, was decreased in obese and diabetic subjects. Euglycemic hyperinsulinemia increased the leg blood flow response to methacholine in control subjects, whereas responses remained blunted in obese and type 2 diabetic groups. This important study suggests that insulin-resistant obese subjects exhibit vascular dysfunction and are resistant to the potentiating effect of insulin on methacholine-induced vasodilation. It further suggests that these effects are not simply related to changes in blood glucose concentration.

Baron and colleagues had previously suggested that an impaired vasodilator action of insulin accounts for insulin resistance (Laakso *et al.*, 1990). In skeletal muscle, vasodilator actions of insulin are mediated in part by NO (Scherrer *et al.*, 1994). Thus, an impairment in endothelium-dependent vasodilation could contribute to insulin resistance by reducing delivery of insulin and glucose to skeletal muscle and hence reducing glucose uptake. Evidence against this elegant hypothesis is the observation that, in healthy volunteers, local infusion of L-NMMA attenuates the vasodilator response to insulin without reducing insulin-stimulated glucose uptake (Scherrer *et al.*, 1994). However it is possible that impaired endothelium-dependent vasodilation exacerbates insulin resistance by this mechanism in genetically predisposed subjects, in whom insulin and glucose delivery to skeletal muscle may be more critical in determining glucose uptake.

An exciting finding is that, as in hypercholesterolaemia (Ting *et al.*, 1996), impaired responsiveness to methacholine in type 1 and type 2 diabetic patients is corrected by acute administration of vitamin C (Ting *et al.*, 1996b; Timimi *et al.*, 1998). Vitamin C is likely to be acting by scavenging free radicals, and this finding therefore suggests the possibility that increased production of free radicals such as superoxide anions (O_2^-) may explain the blunted response to methacholine in type 2 diabetic patients. Dietary supplementation with fish oil also improves endothelium-dependent vasodilation in type 2 diabetes but the mechanism of action remains unresolved (McVeigh *et al.*, 1994). Ongoing trials with orally administered antioxidants suggest that vitamin E is ineffective (Gazis *et al.*, 1998). In contrast, water soluble antioxidants may prove to be more effective (Ting *et al.*, 1996ab; Timimi *et al.*, 1998).

Plasma concentrations of immunoreactive endothelin-1 have been reported to be elevated in patients with type 2 diabetes complicated by retinopathy (Morise *et al.*, 1995; Patino *et al.*, 1994), but other investigators have found no significant difference between controls and patients with and without complications (Veglio *et al.*, 1993; Bertello *et al.*, 1994). Vasoconstriction to endothelin-1 is impaired in men with type 2 diabetes (Nugent *et al.*, 1996). Thus, if impaired sensitivity to endothelin-1 does have a causal role in the pathogenesis of type 2 diabetes, it most likely results from hyperperfusion and subsequent microvascular damage (Nugent *et al.*,1996). Alternatively, reduced responsiveness to exogenous ET-1 in type II diabetes may reflect receptor down regulation secondary to increased endothelin concentrations. Studies with endothelin antagonists should help resolve these issues.

Mechanisms of Endothelial Dysfunction Suggested by Studies In Vivo

Potential causes of endothelial dysfunction that may be of relevance in either type 1 or type 2 diabetes include: lipid abnormalities, increased oxidative stress and advanced glycosylation end products (AGE) (Figure 9.2) (see Cohen, 1993; Poston and Taylor, 1995; Tribe and Poston, 1996; Sobrevia and Mann, 1997). The clinical studies described above are consistent with increased oxidative stress and inactivation of endothelium-derived NO or EDHF by free radicals such as O_2^- (Rubanyi and Vanhoutte, 1986). Oxidative stress results from an increased generation of free radicals and/or diminished cellular antioxidant defences (Halliwell, 1993). Free radical generation may be increased in diabetes as a result of glucose-induced activation of cyclooxygenase, autoxidation of glucose, increased levels of angiotensin II or an alteration in transition metal metabolism (Hawthorne *et al.*, 1989; Wolff, 1993; Ceriello *et al.*, 1993; Halliwell, 1993; Tribe and Poston, 1996; Graier *et al.*, 1996). Increased flux through the polyol pathway can deplete NADPH which is required for the glutathione-redox cycle (Cohen, 1993; Poston and Taylor, 1995; Sobrevia and Mann, 1997), and O_2^- generation is increased in hypercholesterolaemia and possibly in other forms of dyslipidaemia that are associated with type 2 diabetes, and it is possible that qualitative abnormalities of lipoproteins underlie endothelial dysfunction in diabetes (McNeill *et al.*, 1998).

Endogenous antioxidants include the enzymatic antioxidants superoxide dismutase (SOD), catalase and glutathione peroxidase and the non-enzymatic antioxidants vitamins E and C, all of which may be depleted in type 2 diabetes (Sundaram *et al.*, 1996). Reliable measures of oxidative stress *in vivo* have hitherto been lacking (Morrow and Roberts, 1996), hampering direct investigation of these hypotheses. Lipid peroxidation is a central feature of oxidant injury, and the recent discovery that this is assocated with non-enzymatic formation of isoprostanes (prostaglandin-like compounds) has led to the measurement of isoprostanes as an index of oxidative stress *in vivo* (Morrow and Roberts, 1996; Roberts and Morrow, 1997; Patrono and FitzGerald, 1997). F_2-isoprostanes have recently been shown to be elevated in patients with non-insulin dependent diabetes (Gopaul *et al.*, 1995).

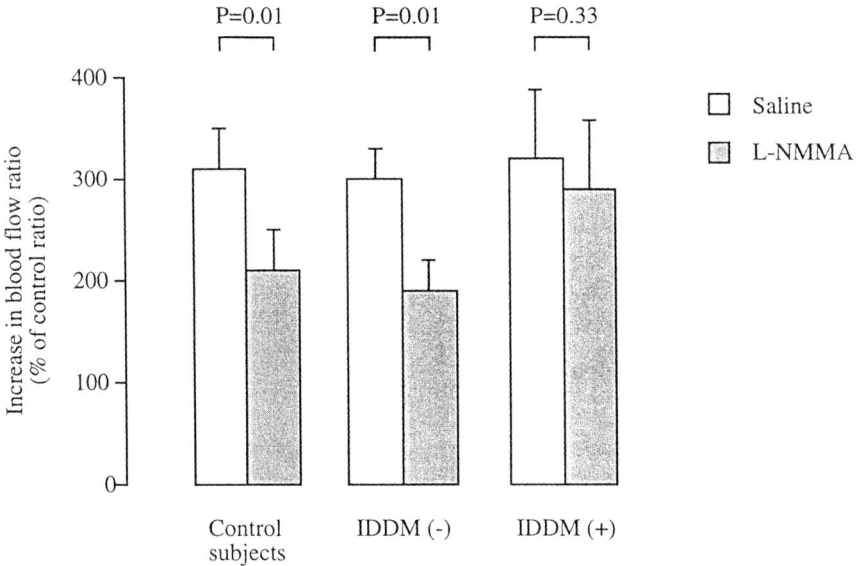

Figure 9.2   Effect of the NO synthase inhibitor L-NMMA on the forearm blood flow response to brachial artery infusion of the endothelium-dependent vasodilator carbachol in healthy control subjects and insulin-dependent diabetic patients (IDDM) with (+) and without (–) microalbuminuria. Results are the percentage increase in blood flow ratio (infused: non-infused arm) caused by carbachol 2.5 µg min^{-1} during co-infusion of saline and L-NMMA. Values are means ± SEM. Adapted from Elliott *et al.*, 1993.

Inactivation of NO and possibly other endothelium-derived mediators such as EDHF by reactive oxygen radicals provides a plausible and attractive explanation for the ability of vitamin C to improve vascular responsiveness to muscarinic agonists in acute experiments.

Effects of Diabetes on Relaxation of Isolated Human Blood Vessels In Vitro

Endothelium-dependent relaxation of resistance arteries from women with gestational diabetes is impaired compared to vessels isolated from normal pregnancies (Knock *et al.*, 1997). As relaxation to acetylcholine was similar in normal and diabetic arteries in the presence of indomethacin, these authors concluded that gestational diabetes inhibited the synthesis of dilator prostaglandins. The sensitivity of arteries to acetylcholine was reduced to a similar degree by the NO synthase inhibitor N^G-monomethyl-L-arginine in normal pregnant and gestational diabetic women, suggesting that NO synthesis was not impaired in gestational diabetes. Moreover, gestational diabetes had no effect on the sensitivity of vascular smooth muscle cells to the NO donor nitroprusside, indicating that NO-mediated vasodilator responses in maternal resistance vessels were not altered.

Similar studies using resistance arteries from human gluteal fat biopsies have shown that maximal contractile responses to noradrenaline, potassium and angiotensin II were attenuated in arterial rings from insulin-dependent diabetic patients (McNally *et al.*, 1994). Although endothelium-dependent relaxation to acetylcholine was impaired by diabetes, vasodilator responses to another endothelium-dependent agonist, bradykinin, and to nitro-prusside, were unaffected suggesting a defect in activation/coupling of the acetylcholine receptor rather than a decreased ability to synthesize and release NO. Recent studies with human saphenous vein rings have confirmed that endothelium-dependent relaxation mediated by NO is similar in rings from healthy controls or from either type 1 or type 2 diabetic patients, whereas prostanoid- and fibrinogen-mediated relaxation was abolished in rings from diabetic patients (Hicks *et al.*, 1997). Thus, impaired endothelium-dependent relaxation was attributed to a decreased prostacyclin and increased vasoconstrictor prostanoid synthesis in diabetic patients. Interestingly, this study also reported that vascular relaxation in response to the Ca^{2+} ionophore A23187 was reduced in saphenous vein rings from diabetic patients, consistent with previous findings that A23187-stimulated release of NO and stored Ca^{2+} are diminished in endothelial cells exposed to hyperglycemia (Wascher *et al.*, 1994).

The reactivity of human internal mammary artery and saphenous vein rings isolated from coronary artery disease patients with type 2 diabetes undergoing coronary artery bypass surgery is abnormal (Karasu *et al.*, 1995). In contrast to diminished constrictor responses in human gluteal resistance arteries from type 1 diabetic patients (McNally *et al.*, 1994), maximal contractile responses to noradrenaline and ET-1 were found to be increased in internal mammary artery and saphenous vein rings from coronary artery patients with type 2 diabetes (Karasu *et al.*, 1995). Endothelium-dependent relaxation of internal mammary artery rings was decreased in coronary artery patients with type 2 diabetes, whereas relaxation to nitroprusside was unaffected. Thus, NO-mediated relaxation of vascular smooth muscle cells in isolated arterial rings *in vitro* does not appear to be impaired in diabetes (McNally *et al.*, 1994; Karasu *et al.*, 1995; Knock *et al.*, 1997; Hicks *et al.*, 1997).

As described above, qualitative abnormalities in lipoproteins occur in patients with diabetes mellitus (McNeill *et al.*, 1998) Studies *in vitro* demonstrate that low density lipoproteins (LDL), and particularly oxidised LDL, powerfully inhibit endothelium-de-pendent relaxation (Jacobs *et al.*, 1990) and agonist-stimulated release of NO and prostacyclin (Jay *et al.*, 1997). In diabetes, LDL is composed of smaller, denser particles than LDL in non-diabetic subjects (Austin and Edwards, 1996). Such small dense particles are more susceptible to oxidation (Austin *et al.*, 1988; Tribble *et al.*, 1992), and indeed patients with type 2 diabetes have increased plasma antibody levels to oxidised LDL and concentrations of lipid peroxides, including 8-epi-prostaglandins (Gopaul *et al.*, 1995; Sundaram *et al.*,1996). It is possible that the altered characteristics of lipoproteins, particularly the presence of small dense LDL in diabetes, are responsible for the increased generation of O_2^-.

Diabetes-induced Alterations in NO and Prostacyclin Synthesis in Cultured Endothelial Cells

Micro- and macrovascular endothelial cells exhibit a differential responsiveness to insulin, with a much higher sensitivity detected in microvascular cells (from calf retina) than in macrovascular endothelial cells isolated from calf aorta (King *et al.*, 1983). Human en-dothelial cells in culture exhibit only a small increase in nucleic acid synthesis in response to insulin, in contrast to the marked effects of endothelial cell growth factors (King *et al.*,

1983). Human umbilical vein endothelial cells from type 1 diabetic mothers function abnormally *in vitro*, with altered cell membrane lipid composition and abnormal morphology (Cester et al., 1996), an increased rate of cell proliferation, reduced resistance to shear stress and lower rates of glucose transport compared to cells isolated from non-diabetic pregnancies (Sank et al., 1994). In contrast, umbilical vein endothelial cells, derived from diet-controlled gestational diabetic pregnancies, exhibit decreased rates of proliferation, elevated intracellular Ca^{2+} levels and a sustained membrane hyperpolarization (Sobrevia et al., 1995). Basal and histamine-stimulated synthesis of prostacyclin (and PGE_2) is markedly reduced in umbilical vein endothelial cells from gestational diabetic pregnancies (Karbowski et al., 1989; Sobrevia et al., 1995), whereas basal rates of NO production are increased (Sobrevia et al., 1995). The sustained membrane hyperpolarization and elevated levels of intracellular Ca^{2+} in endothelial cells from gestational diabetic pregnancies may explain the increased basal rates of NO production (Sobrevia et al., 1996, 1998). Although histamine evokes maximal rates of NO production in fetal endothelial cells from gestational diabetic pregnancies, sustained exposure to high glucose inhibits agonist-stimulated NO release (Sobrevia et al., 1995, 1998).

Studies in fetal endothelial cells from normal pregnancies have established that elevated glucose (EC50 11 mM) induces a time- (3–6 h) and protein synthesis dependent increase in the activity and expression of constitutive NO synthase (eNOS) (Sobrevia et al., 1996; Mann et al., 1998). Similar studies in human aortic endothelial cells have confirmed that elevated glucose (5 days) increases the release of NO and superoxide anions by 40% and 300%, respectively, perhaps explaining the imbalance in NO and O_2^- production in diabetes (Cosentino et al., 1997). Longer-term exposure (15 days) to elevated glucose apparently does not alter the expression of NO synthase in human umbilical vein endothelial cells (Mancusi et al., 1996), suggesting an adaptation in NO synthesis under conditions of sustained hyperglycemia. Although human insulin (0.1–1 nM) stimulates eNOS activity in umbilical vein endothelial cells (Sobrevia et al., 1996), activation of the L-arginine/NO pathway by insulin is markedly attenuated by elevated glucose and gestational diabetes (Sobrevia et al., 1998). Perhaps the activation of eNOS by insulin in non-diabetic endothelial cells is a consequence of stimulation of the pentose cycle by insulin, which would increase the supply of NADPH required for NO synthesis (Wu et al., 1994).

Our findings in human umbilical vein endothelial cells from gestational diabetic pregnancies are consistent with some studies reporting an increase in basal NO synthesis in diabetes (Corbett et al., 1992), but at variance with others showing an impaired synthesis of NO in diabetes (see Poston and Taylor, 1995). Direct extrapolation of findings in cultured endothelial cells to types 1 or 2 diabetic patients should be treated with caution, though many outstanding questions relating to control of NO biosynthesis in the diabetic state can now be addressed at a cellular level. Interestingly, NO synthesis and NO-mediated relaxation of isolated resistance arteries studied *in vitro* do not appear to be impaired (McNally et al., 1994; Hicks et al., 1997) in type 1 or 2 diabetic patients, whereas studies *in vivo* have pointed to impaired endothelium-dependent relaxation in diabetic patients (see section above). The finding that human insulin (and glucose) stimulate NO synthesis in cultured fetal endothelial cells is consistent with the hypothesis that insulin activates the L-arginine/NO vasodilator pathway in human skeletal muscle vasculature (see Baron, 1994; Scherrer et al., 1994; Steinberg et al., 1994).

Strategies to Improve Endothelial Function in Diabetes

Interventions that may improve endothelial function in diabetic patients include fish oil (McVeigh *et al.*, 1993, 1994), water soluble antioxidants in type 2 diabetes and lipid lowering therapy in type 1 diabetes (Mullen *et al.* 1998). Many other interventions hold therapeutic promise. Lifestyle changes to reduce insulin resistance and improve glycaemic control and dyslipidaemia through diet and exercise may be especially important. Angiotensin converting enzyme (ACE) inhibitors might improve endothelial function by preventing the generation of superoxide anions by angiotensin II and indirectly by increasing bradykinin-stimulated NO synthesis. In patients with coronary artery disease the ACE inhibitor quinapril reverses endothelial dysfunction in coronary vessels independent of alterations in plasma lipid profiles and blood pressure (Mancini *et al.*, 1996), but whether such an effect occurs in patients with diabetes is unknown. Angiotensin II (AT_1) receptor antagonists may reduce O_2^- production by blocking effects of angiotensin II on NADH/NADPH oxidase (Rajagopalan *et al.*, 1996). Moreover, statins improve endothelial function in coronary and peripheral vessels of non-diabetic patients with hypercholesterolaemia (Stroes *et al.*, 1995; Treasure *et al.*, 1995; O'Driscoll *et al.*, 1997). Promising agents for trials in diabetes include fibrates (e.g. ciprofibrate) and statins (e.g. atorvastatin), which have powerful effects on serum triglyceride and LDL-cholesterol concentrations and influence LDL particle size. Thiazolidonediones could potentially reverse endothelial dysfunction by enhancing insulin sensitivity (Nolan *et al.*, 1994) and decreasing lipoprotein oxidation (Noguchi *et al.*, 1996), providing the hepatotoxicity of troglitazone does not prove a general feature of this class of drug. Clinical trials to determine effects of many of these interventions on endothelial function are in progress. The link between endothelial dysfunction and insulin resistance suggests that interventions which improve endothelial function could also improve insulin sensitivity in type 2 diabetes.

ACKNOWLEDGEMENTS

We gratefully acknowledge the support of the Wellcome Trust, British Heart Foundation and Ministry of Agriculture, Fisheries and Food.

REFERENCES

Austin, M.A., Breslow, J.L., Hennekens, C.H., Buring, J.E., Willet, W.C. and Krauss R.M. (1988). Low-density lipoprotein subclass patterns and risk of myocardial infarction. *JAMA*, **260**, 1917–1921.

Austin, M.A. and Edwards, K.L. (1996). Small, dense low density lipoproteins, the insulin resistance syndrome and non-insulin-dependent diabetes. *Current Opinion in Lipidology*, **7**, 167–171.

Bar, R.S., Boes, M., Dake, B.L., Booth, B.A., Henley, S.A. *et al.* (1988). Insulin, insulin-like growth factors, and vascular endothelium. *American Journal Medicine* 85, 59–70.

Bar, R.S. (1992). Vascular endothelium and diabetes mellitus. In: *Endothelial Cell Dysfunction*, eds. Simionescu, N. & Simionescu, M., pp. 363–382. Plenum Press: New York.

Baron, A.D. (1994). Hemodynamic actions of insulin. *American Journal of Physiology*, **267**, F187–F202.

Benjamin, N., Calver, A., Collier, J., Robinson, B., Vallance, P. and Webb, D.J. (1995). Measuring forearm blood flow and interpreting the response to drugs and mediators. *Hypertension*, **25**, 918–923.

Bertello, P., Veglio, F., Pinna, G., *et al.* (1994). Plasma endothelin in NIDDM patients with and without complications. *Diabetes Care*, **17**, 574–577.

Bucala, R., Tracey, K.J. and Cerami, A. (1991). Advanced glycosylation products quench nitric oxide and mediate defective endothelium-dependent vasodilatation in experimental diabetes. *Journal of Clinical Investigation*, **87**, 432–438.

Calver, A., Collier, J. and Vallance, P. (1992). Inhibition and stimulation of nitric oxide in the human forearm bed of patients with insulin-dependent diabetes. *Journal of Clinical Investigation*, **90**, 2448–2554.

Celermajer, D.S., Adams, M.R., Clarkson, P., *et al.* (1996). Passive smoking and impaired endothelium-dependent arterial dilation in healthy young adults. *New England Journal of Medicine*, **334**, 150–154.

Celermajer, D.S., Sorensen, K.E., Gooch, V.M., *et al.* (1992). Non-invasive detection of endothelial dysfunction in children and adults at risk of atherosclerosis. *Lancet*, **340**, 1111–1115.

Celermajer, D.S., Sorensen, K.E., Georgakopoulos, D., *et al.* (1993). Cigarette smoking is associated with dose-related and potentially reversible impairment of endothelium-dependent dilation in healthy young adults. *Circulation*, **88**, 2149–2155.

Ceriello, A., Quatraro, A. and Giugliano, D. (1993). Diabetes mellitus and hypertension. The possible role of hyperglycaemia through oxidative stress. *Diabetologia*, **36**, 265–266.

Cester, N., Rabini, R.A., Salvolini, E., Staffolani, R., Curatola, A., Pugnaloni, A., Brunelli, M.A., Biagini, G. and Mazzanti, L. (1996). Activation of endothelial cells during insulin-dependent diabetes mellitus: A biochemical and morphological study. *European Journal of Clinical Investigation*, **26**, 569–573.

Chen, G., Suzuki, H., Weston, A.H. (1988). Acetylcholine releases endothelium-dependent hyperpolarizing factor and EDRF from rat blood vessels. *British Journal of Pharmacology*, **95**, 1165–1174.

Chowienczyk, P.J., Cockcroft, J.R. and Ritter, J.M. (1993). Differential inhibition by N^G-monomethyl-L-arginine of vasodilator effects of acetylcholine and methacholine in human forearm vasculature. *British Journal of Pharmacology*, **110**, 736–738.

Chowienczyk, P.J., Cockcroft, J.R. and Ritter, J.M. (1994).Blood flow responses to intra-arterial acetylcholine in man: effects of basal flow and conduit vessel length. *Clinical Science*, **87**, 45–51.

Ciccone, M., DiNoia, D., DiMichele, L., *et al.* (1993). The incidence of asymptomatic extracoronary athero-sclerosis in patients with coronary atherosclerosis. *International Angiology*, **12**, 25–28.

Clarkson, P., Celermajer, D.S., Donald, A.E. *et al.* (1996). Impaired vascular reactivity in insulin-dependent diabetes mellitus is related to disease duration and low density lipoprotein cholesterol levels. *Journal of American College of Cardiology*, **28**, 573–579.

Cohen, R.A. (1993). Dysfunction of vascular endothelium in diabetics mellitus. *Circulation*, **87 suppl V**, V67–V76.

Corbett, J.A., Tilton, R.G., Chang, K., Hasan, K.S., Ido, Y., Wang, J.L., Sweetland, M.A. *et al.* (1992). Aminoguanidine, a novel inhibitor of nitric oxide formation, prevents diabetic vascular dysfunction. *Diabetes*, **41**, 552–556.

Davi, G., Catalano, I., Averna, M., *et al.* (1990). Thromboxane biosynthesis and platelet function in type II diabetes mellitus. *New England Journal of Medicine*, **322**, 1769–1774.

Dornhorst, A. and Beard, R.W. (1993). Gestational diabetes: a challenge for the future. *Diabetic Medicine*, **10**, 897–905.

Elliott, T.G., Cockcroft ,J.R., Groop P-H., Viberti, G.C. and Ritter, J.M. (1993). Inhibition of nitric oxide synthesis in forearm vasculature of insulin-dependent diabetic patients: blunted vasoconstriction in patients with microalbuminuria. *Clinical Science*, **85**, 687–693.

Eriksson, U.J. (1995). The pathogenesis of congenital malformations in diabetic pregnancy. *Diabetes Metabolism Reviews*, **11**, 63–82.

Gazis, A.G., White, D.J., Page, S.R. and Cockcroft J.R. (1998). Oral vitamin E does not improve vascular endothelial function in subjects with type 2 diabetes. *Diabetic Medicine* Abstract in press

Goodfellow, J., Ramsey, M.W., Luddington, L.A., *et al.* (1996). Endothelium and inelastic arteries: an early marker of vascular dysfunction in non-insulin-dependent diabetes. *British Medical Journal*, **312**, 744–745.

Gopaul, N.K., Anggard, E.E., Mallet, A.I., Beteridge, D.J., Wolff, S.P. and Nourouz-Zadeh, J. (1995). Plasma 8–epi-PGF$_{2a}$ levels are elevated in individuals with non-insulin dependent diabetes mellitus. *FEBS Letters*, **368**, 225–229.

Graier, W.F., Simeck, S., Kukovetz, W.R. and Kostner, G.M. (1996). High D-glucose induced changes in endothelial Ca^{2+}/EDRF signalling are due to generation of superoxide anions. *Diabetes*, **45**, 1368–1395.

Green, K., Vesterqvest, O. and Grill, V. (1988). Urinary metabolites of thromboxane and prostacylin in diabetes mellitus. *Acta Endocrinologica*, **118**, 301–305.

Halkin, A., Benjamin, N., Doktor, H.S., Todd, S.D., Viberti, G.C. and Ritter JM. (1991). Vascular responsiveness and cation exchange in insulin-dependent diabetes. *Clinical Science*, **81**, 223–232.

Halliwell, B. (1993). The role of oxygen radicals in human disease, with particular reference to the vascular system. *Haemostasis*, **23**, 118–126.

Hawthorne, G.C., Bartlett, K., Hetherington, C.S. and Alberti, K.G.M.N. (1989). The effect of high glucose on polyol pathway activity and myoinositol metabolism in cultured human endothelial cells. *Diabetologia*, **32**, 163–166.

Hicks, R.C.J., Moss, J., Higman, D.J., Greenhalgh, R.M. and Powell, J.T. (1997). The influence of diabetes on the vasomotor responses of saphenous vein and the development of infra-inguinal vein graft stenosis. *Diabetes*, **47**, 113–118.

Ignarro, L.J. (1990). Biosynthesis and metabolism of endothelium-derived nitric oxide. *Annual Review of Toxicology and Pharmacology*, **30**, 535–560.

Jaap, A.J., Hammersly, M.S., Shore, A.C. and Tooke, J.E. (1994). Reduced microvascular hyperaemia in subjects at risk of developing type 2 (non-insulin dependent0 diabetes mellitus. *Diabetologia*, **37**, 214–216.

Jacobs, M., Plane, F. and Bruckdorfer, K.R. (1990). Native and oxidized low-density lipoproteins have different inhibitory effects on endothelium-derived relaxing factor in rabbit aorta. *British Journal of Pharmacology*, **100**, 21–26.

Joannides, R., Haefeli, W.E., Linder, L., *et al.* (1995). Nitric oxide is responsible for flow-dependent dilation of human peripheral conduit arteries in vivo. *Circulation*, **88**, 2511–2516.

Johnstone, M.T., Creager, S.J., Scales, K.M., Cusco, J.A., Lee, B.K. and Creager, M.A. (1993). Impaired endothelium-dependent vasodilation in patients with insulin-dependent diabetes mellitus. *Circulation*, **88**, 2510–2516.

Karbowski, B., Bauch, H.J. and Schneider, H.P.G. (1989). Funktionelle differenzierung desvaskularenendothels bei hochrisikoschwangerschaften. *Zeitschrift f r Geburtshilfe und Perinatologie*, **193**, 8–12.

Karasu, C., Soncul, H. and Altan, V.M. (1995). Effects of non-insulin dependent diabetes mellitus on the reactivity of human internal mammary artery and human saphenous vein. *Life Sciences*, **57**, 103–112.

King, G.L., Buzney, S.H.M., Kahn, C.R. and Hetu, N. (1983). Differential responsiveness to insulin of endothelial and support cels from micro- and macrovessels. *Journal of Clinical Investigation*, **71**, 974–979.

Knock, G.A., McCarthy, A.L., Lowy, C. and Poston, L. (1997). Association of gestational diabetes with abnormal maternal vascular endothelial function. *British Journal of Obstetrics and Gynaecology*, **104**, 229–234.

Laakso, M., Edelman, S.V., Brechtel, G. and Baron, A.D. (1990). Decreased effects of insulin to stimulate skeletal muscle blood flow in obese man: A novel mechanism for insulin resistance. *Journal of Clinical Investigation*, **85**, 1844–1853.

Lorenzi, M., Montisano, D.F., Toledo, S. and Barrieux, A. (1986). High glucose induces DNA damage in cultured human endothelial cells. *Journal of Clinical Investigation*, **77**, 322–325.

Makimattila, S., Virkamaki, A., Groop, P-H., *et al.* (1996). Chronic hyperglycaemia impairs endothelial function and insulin sensitivity via different mechanisms in insulin-dependent diabetes mellitus. *Circulation*, **94**, 1276–1282.

Makita, Z., Radoff, S., Rayfield, E.J., Yang, Z., Skolnik, E., Delaney, V., Friedman, E.A., Cerami, A. and Vlassara, H. (1991). Advanced glycosylation end products in patients with diabetic nephropathy. *New England Journal of Medicine*, **325**, 836–842.

Malamitsi-Puchner, A., Economou, E., Katsouyanni, K., Karachaliou, F., Delis, D. and Bartsocas CS. (1996). Endothelin 1–21 plasma concentrations in children and adolescents with insulin-dependent diabetes mellitus. *Journal of Paediatric Endocrinology and Metabolism*, **9**, 463–468.

Mancini, G.B., Henry, G.C., Macaya, C., *et al.* (1996). Angiotensin-converting enzyme inhibition with quinapril improves endothelial vasomotor dysfunction in patients with coronary artery disease. The TREND (Trial on Reversing Endothelial Dysfunction) Study. *Circulation*, **94**, 240–243.

Mancusi, G., Hutter, C., Baumgartner-Parzer, S., Schmidt, K., Schutz, W. and Sexl, V. (1996). High glucose incubation of human umbilical vein endothelial cells does not alter expression and function of either G-protein alpha subunits or of endothelial NO synthase. *Biochemical Journal*, **315**, 281–287.

Mann, G.E., Siow, R.C.M., Closs, E.I. and Sobrevia, L. (1998). Expression of human cationic amino acid transporters (hCAT) and nitric oxide synthase in human fetal endothelial cells: modulation by elevated D-glucose. In *The Biology of Nitric Oxide*, vol. 6, eds.Toda, N, Maeda, H. & Moncada, S., pp. , London: Portland Press.

Martina, V., Bruno, G.A., Trucco, E., Zumpano, M., Tagliabue, M. *et al.* (1998). Platelet cNOS activity is reduced in patients with IDDM and NIDDM. *Thrombosis and Haemostasis*, **79**, 520–522.

McNally, P.C., Watt, P.A.C., Rimmer, T., Burden, A.C., Hearnshaw, J.R. and Thurston, H. (1994). Impaired contraction and endothelium dependent relaxation in isolated resistance vessels from patients with insulin-dependent diabetes. *Clinical Science*, **87**, 31–36

McNeil, K.L, Fontana, L., Ritter, J.M., Russell-Jones, D.L. and Chowienczyk, P.J. (1998). Inhibitory effects of low-density lipoprotein on endothelium-dependent relaxation are exaggerated in men with NIDDM. *Diabetic Medicine* Abstract in press.

McVeigh, G., Brennan, G., Cohn, J., Finklestein, S., Hayes, R. and Johnston. D. (1994). Fish oil improves arterial compliance in non-insulin dependent diabetes mellitus. *Arteriosclerosis Thrombosis*, **14**, 1425–1429.

McVeigh, G.E., Brennan, G.M., Johnston, G.D., *et al.* (1992). Impaired endothelium-dependent and independent vasodilation in patients with Type 2 (non-insulin-dependent) diabetes mellitus. *Diabetologia*, **35**, 771–776.

McVeigh, G.E., Brennan, G.M., Johnston, G.D., *et al.* (1993). Dietary fish oil augments nitric oxide production or release in patients with Type 2 (non-insulin-dependent) diabetes mellitus. *Diabetologia*, **36**, 33–38.

Moncada, S., Palmer, R.M.J. and Higgs, E.A. (1991). Nitric oxide: Physiology, pathophysiology, and pharmacology. *Pharmacological Reviews*, **43**,109–142.

Morise, T., Takeuchi, Y., Kawano, M., Koni, I. and Takeda, R. (1995). Increased plasma levels of immunoreactive endothelin and von Willebrand factor in NIDDM patients. *Diabetes Care*, **18**, 87–89.

Morris, S.J., Shore, A.C. and Tooke, J.E. (1995). Responses of the skin microcirculation to acetylcholine and sodium nitroprusside in patients with NIDDM. *Diabetologia*, **38**, 1337–1344.

Morrow, J.D. and Roberts, L.J. (1996). The isoprostanes. Current knowledge and directions for future research. *Biochemical Pharmacology*, **51**, 1–9.

Mullen., M.J., Donald, A.E. Thompson, H., O'Connor, G., Thorne, S., Wright, D.J. and Deanfield, J.E. (1988). Atorvastatin but not L-arginine improves endothelial function in young subjects with insulin-dependent diabetes mellitus. *J. Am. Coll.* Cardiol., **31**, 178A.

Noguchi, N., Sakai, H., Kato, Y., Tsuchiya, J., Yamamoto, Y. and Niki, E. (1996). Inhibition of oxidation of low density lipoprotein by troglitazone. *Atherosclerosis*, **123**, 227–234.

Nolan, J.J., Ludvick, B., Beerdsen, P., Joyce, M. and Olefsky, J. (1994). Improvement in glucose tolerance and insulin resistance in obese subjects treated with troglitazone. *New England Journal of Medicine*, **331**, 1188–1193.

Nugent, A.G., McGurk, C., Hayes, J.R. and Johnston, G.D. (1996). Impaired vasoconstriction to endothelin 1 in patients with NIDDM. *Diabetes*, **45**, 105–107.

O'Driscoll, G., Green, D. and Taylor, R.R. (1997). Simvastatin, an HMG-coenzyme A reductase inhibitor, improves endothelial function within 1 month. *Circulation*, **95**, 1126–1131.

Oskarsson, H.J. and Hofmeyer, T.G. (1996). Platelets from patients with diabetes mellitus have impaired ability to mediate vasodilation. *Journal of the American College of Cardiology*, **27**, 1464–1470.

Patrono, C. and FitzGerald, G.A. (1997). Isoprostanes: potential markers of oxidative stress in atherothrombotic disease. *Arteriosclerosis, Thrombosis and Vascular Biology*, **17**, 2309–2315.

Pitei, D.L., Watkins, P.J. and Edmonds, M.E. (1997). NO-dependent smooth muscle vasodilation is reduced in NIDDM patients with peripheral sensory neuropathy. *Diabetic Medicine*, **14**, 284–290.

Porta, M., Maneschi, F., White, M.C. and Kohner, E.M. (1981). Twenty-four hour variations of von Willebrand factor and factor VIII-related antigen in diabetic retinopathy. *Metabolism*, **30**, 695–699.

Poston, L. and Taylor, P.D. (1995). Endothelium-mediated vascular function in insulin-dependent diabetes mellitus. *Clinical Science*, **88**, 245–255.

Properzi, G., Terenghi, G., Gu, X.H., *et al.* (1995). Early increase precedes a depletion of endothelin-1 but not of von Willebrand factor in cutaneous microvessels of diabetic patients. A quantitative immunohistochemical study. *Journal of Pathology*, **175**, 243–252.

Rabini, R.A., Staffolani, R., Fumelli, P., Mutus, B., Curatola, G. *et al.* (1998). Decreased nitric oxide synthase activity in platelets from IDDM and NIDDM patients. *Diabetologia*, **41**, 101–104.

Rajagopalan, S., Kurz, S., Manzel, T., *et al.* (1996). Angiotensin II-mediated hypertension in the rat increases vascular superoxide production. *Journal of Clinical Investigation*, **97**,1916–1923.

Ralevic, V., Khalil, Z., Dusting, G.J. and Helme R.D. (1992). Nitric oxide and sensory nerves are involved in the vasodilator response to acetylcholine but not calcitonin gene-related peptide in rat skin microvasculature. *British Journal of Pharmacology*, **106**, 650–655.

Roberts, L.J. and Morrow, J.D. (1997). The generation and actions of isoprostanes. *Biochimica et Biophysica Acta*, **1345**, 121–135.

Rubanyi, G.M. and Vanhoutte, P.M. (1986). Superoxide anions and hypoxia inactivate endothelium-derived relaxing factor. *American Journal of Physiology*, **250**, H822–H827.

Sank, A., Wei, D., Reid, J., Ertl, D., Nimni, M., Weaver, F., Yellin, A. & Tuan, T.L. (1994). Human endothelial cells are defective in diabetic vascular disease. *Journal of Surgical Research*, **57**, 647–653

Scherrer, U., Randin, D., Vollenweider, P., Vollenweider, L. and Nicod, P. (1994). Nitric oxide release accounts for insulin's vascular effects in humans. *Journal of Clinical Investigation*, **94**, 2511–2515.

Shimizu, K., Muramatsu, M., Kakegawa, Y., *et al.* (1993). Role of prostaglandin H_2 as an endothelium-derived contracting factor in diabetic state. *Diabetes*, **42**, 1246–1252.

Smits, P., Kapma, J-A., Jacobs M-C., Lutterman, J. and Thien, T. (1993). Endothelium-dependent vascular relaxation in patients with type I diabetes. *Diabetes*, **42**, 143–153.

Sobrevia, L., Cesare, P., Yudilevich, D.L. and Mann, G.E. (1995). Diabetes-induced activation of system y^+ and nitric oxide synthase in human endothelial cells: association with membrane hyperpolarization. *Journal of Physiology London*, **489**, 183–192.

Sobrevia, L. and Mann, G.E. (1997). Dysfunction of the endothelial nitric oxide signalling pathway in diabetes and hyperglycaemia. *Experimental Physiology*, **82**, 423–452.

Sobrevia, L., Nadal, A., Yudilevich, D.L. and Mann, G.E. (1996). Activation of L-arginine transport (system y^+) and nitric oxide synthase by elevated glucose and insulin in human endothelial cells. *Journal of Physiology*, **490**, 775–781.

Sobrevia, L., Yudilevich, D.L. and Mann, G.E. (1998). Elevated D-glucose induces insulin insensitivity in human umbilical vein endothelial cells isolated from gestational diabetic pregnancies. *Journal of Physiology London*, **506**, 219–230.

Sorensen KE, Celermajer DS, Georgakopoulos D, Hatcher G, Beteridge DJ, Deanfield JE. Impairement of endothelium-dependent dilation is an early event in children with familial hypercholesterolaemia and is related to the lipoprotein (a) level. *Journal of Clinical Investigation*, **93**, 50–55.

Stehouwer, C.D.A., Fischer, H.R.A., Van Kuijk, A.W.R., Polak, B.C.P. and Donker, A.J.M. (1995). Endothelial dysfunction precedes development of microalbuminuria in IDDM. *Diabetes*, **44**, 561–564.

Steinberg, H.O., Brechtel, G., Johnson, A., Fineberg, N. and Baron, A.D. (1994). Insulin-mediated skeletal muscle vasodilation is nitric oxide dependent. A novel action of insulin to increase nitric oxide release. *Journal of Clinical Investigation*, **94**, 1172–1179.

Steinberg, H.O., Chaker, H., Leaming, R., Johnson, A., Brechtel, G. and Baron, A.D. (1996). Obesity/insulin resistance is associated with endothelial dysfunction. Implications for the syndrome of insulin resistance. *Journal of Clinical Investigation*, **97**, 2601–2610.

Stroes, E.S.G., Koomans, H.A., de Bruin, T.W.A. and Rabelink, T.J. (1995). Vascular function in the forearm of hypercholesterolaemic patients off and on lipid-lowering medication. *Lancet*, **346**, 467–471.

Sundaram, R.K., Bhaskar, A., Vijayalingam, S., Viswanathan, M., Mohan, R. and Shanmugasundaram, K.R. (1996). Antioxidant status and lipid peroxidation in type II diabetes mellitus with and without complications. *Clinical Science*, **90**, 255–260.

Taddei, S., Verdis, A., Mattei, P., Natali, A., Ferrannini, E. and Salvetti, A. (1995). Effect of insulin on acetylcholine-induced vasodilation in normotensive subjects and patients with essential hypertension. *Circulation*, **92**, 2911–2918.

Taddei, S., Verdis, A., Mattei, P., *et al.* (1997). Hypertension causes premature aging of endothelial function in humans. *Hypertension*, **29**, 736–743.

Timimi, F.K., Ting, H.H., Haley, E.A., Roddy, M.A., Ganz, P., and Creager, M.A. (1998). Vitamin C improves endothelium-dependent vasodilation in patients with insulin-dependent diabetes mellitus. *Journal of the American College of Cardiology*, **31**, 552–557.

Ting, H.H., Timimi, F.K., Haley, E.A., Roddy, M-A., Ganz, P. and Creager, M.A. (1996a). Vitamin C restores endothelium-dependent vasodilation in patients with hypercholesterolaemia. *Circulation*, **94**, I402.

Ting, H.H., Timimi, F.K., Boles, K.S., Creager, S.J., Ganz, P. and Creager MA. (1996b). Vitamin C improves endothelium-dependent vasodilation in patients with non-insulin-dependent diabetes mellitus. *Journal of Clinical Investigation*, **97**, 22–28.

Treasure, C.B., Klein, J.L., Weintraub, W.S., *et al.* (1995). Beneficial effects of cholesterol-lowering therapy on the coronary endothelium in patients with coronary artery disease. *New England Journal of Medicine*, **332**, 481–487.

Tribble, D.L., Holl, L.G., Wood, P.D. and Krauss, R.M. (1992). Variations in oxidative susceptibility among six low density lipoprotein subfractions of differing density and particle size. *Atherosclerosis*, **93** 189–199.

Tribe, R.M. and Poston, L. (1996). Oxidative stress and lipids in diabetes: a role in endothelium vasodilator dysfunction ? *Vascular Medicine*, **1**, 195–206.

Vallance, P., Collier, J. and Moncada, S. (1989). Effects of endothelium-derived nitric oxide on peripheral arteriolar tone in man. *Lancet*, **2**, 997–1000.

Veglio, F., Bertello, P., Pinna, G., Mulatero, P., Rossi, A., Gurioli, L., Panarelli, M. and Chiandussi, L. (1993). Plasma endothelin in essential hypertension and diabetes mellitus. *Journal of Human Hypertension*, **7**, 321–325.

Wascher, T.C., Toplak, H., Krejs, G.J., Simecek, S., Kukowitz, W.R. and Graier, W.F. (1994). Intracellular mechanisms involved in D-glucose mediated amplification of agonist induced calcium response and EDRF formation in vascular endothelial cells. *Diabetes*, **43**, 984–991.

Watts, G.F., O'Brien, S.F., Silvester, W. and Millar, J.A. (1996). Impaired endothelium-dependent and independent dilation of forearm resistance arteries in men with diet treated non-insulin dependent diabetes; role of dyslipidaemia. *Clinical Science*, **91**, 567–573.

Williams, S.B., Cusco, J.A., Roddy, M.A., Johnstone, M.T. and Creager, M.A. (1996). Impaired nitric oxide-mediated vasodilation in patients with non-insulin-dependent diabetes mellitus. *Journal of American College of Cardiology*, **27**, 567–574.

Wolff, S.P. (1993). Diabetes mellitus and free radicals. Free radicals, transition metals and oxidative stress in the aetiology of diabetes mellitus and complications. *British Medical Bulletin*, **49**, 642–652.

Wu, G., Majumdar, S., Zhang, J., Lee, H. and Meininger, C.J. (1994). Insulin stimulates glycolysis and pentose cycle activity in bovine microvascular endothelial cells. *Comparative Biochemistry and Physiology _ C. Pharmacology*, **108**, 179–185.

Zenere, B.M., Arcaro, G., Saggiani, F., Rossi, L., Muggeo, M. and Lechi, A. (1995). Non-invasive detection of functional alterations of the arterial wall in IDDM patients with and without microalbuminuria. *Diabetes Care*, **18**, 975–982.

C
O
N
T
R
A
C
T
I
O
N

R
E
L
A
X
A
T
I
O
N

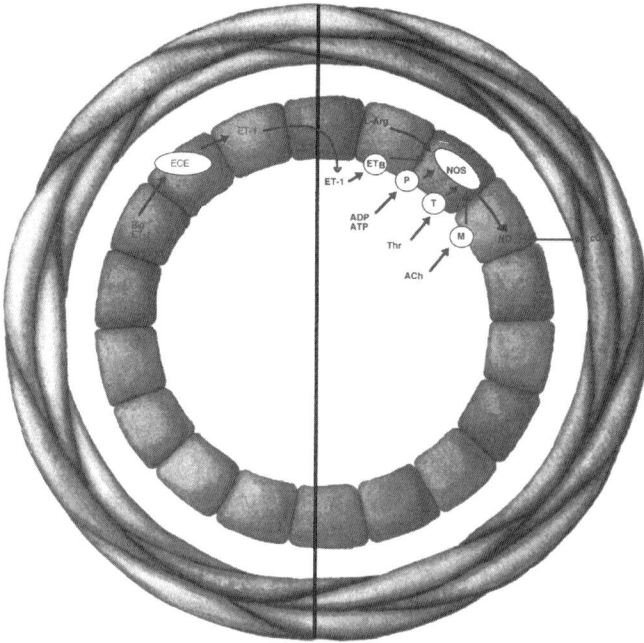

Flow-mediated dilatation of conduit arteries is dependent on intact endothelial function, and is largely (but not exclusively) mediated by endothelium-derived nitric oxide (NO). As arterial diameter can be measured accurately in humans *in vivo* by either quantitative angiography (invasive) or high resolution external vascular ultrasound (non-invasive), human studies of flow-mediated dilatation have recently been described, and have provided insights into the relationships between cardiovascular risk factors and endothelial physiology. In the last 5 years, impaired activity of endothelium-derived NO has been demonstrated in children and young adults with hypercholesterolemia, diabetes mellitus or with a history of cigarette smoking. Non-invasive assessment of flow-mediated dilatation in particular has facilitated study of arterial physiology in younger presymptomatic subjects, relatively early in the atherogenic process. This methodology has also permitted serial study of endothelium-dependent dilatation, to permit clinical trials of agents which may reverse endothelial dysfunction in high risk subjects. Amongst other interventions, oral L-arginine (the substrate of NO), vitamin C and estrogen replacement therapy have shown promise as agents that might improve arterial endothelium-dependent dilatation. As endothelium-derived NO may play a role, not only in vasodilatation, but also in regulating monocyte adhesion, platelet aggregation and smooth muscle proliferation, such agents may have potentially important effects on the development of atherosclerosis.

Key words: Endothelial function; non-invasive detection; high resolution ultrasound; risk factors.

10   Flow-Mediated Dilatation and Cardiovascular Risk

David S. Celermajer[1] and John E. Deanfield[2]

[1]*Department of Cardiology, Royal Prince Alfred Hospital, Camperdown, Sydney NSW 2050, Australia*
[2]*Cardio-Thoracic Unit, Great Ormond Street Hospital for Children, London, WC1 3JH, UK*

FLOW-MEDIATED DILATATION

Increasing blood flow through large arteries produces an increase in vessel diameter, both in animals (Hintze and Vatner, 1984) and in man (Anderson and Mark, 1989, Sinoway *et al.*, 1989, Laurent *et al.*, 1990). This phenomenon of flow-mediated dilatation (FMD) is now known to be endothelium-dependent, as a normal arterial response is critically dependent on the presence of a functionally intact endothelium (Smiesko *et al.*, 1985, Pohl *et al.*, 1986). The fact that FMD is endothelium-dependent forms the basis for *in vivo* testing of endothelial function in human subjects.

Control of vessel diameter by changes in blood flow was first proposed in 1933 (Schretzenmayr, 1933), who observed dilatation of the canine femoral artery in response to increases in femoral blood flow. Originally the mechanism of this vasodilatation was thought to be due to a retrograde wave of dilatation starting in the arterioles and spreading to the large proximal arteries; so called "ascending dilatation" (Macho *et al.*, 1981). This was soon disproved, however, after experiments in which cutting the femoral artery distal to an arteriovenous fistula failed to abolish FMD (Ingebrigtsen and Leraand, 1970; Lie *et al.*, 1970) (Figure 10.1).

It was hypothesized therefore that a local mechanism controls arterial responses to flow. Endothelial cells are ideally situated to "transduce" the signal of increased flow/shear stress into a vessel wall response via modulating the production and release of vasoactive substances. Endothelial cells in culture are known to react to changes in mechanical forces by altering metabolism (Franke *et al.*, 1984; Van Grondelle *et al.*, 1984).

Smiesko *et al.* (1985) and Pohl *et al.* (1986) have shown that FMD occurs in normal arteries and is abolished completely by techniques used to injure the endothelium (mechanical or pharmacological). Furthermore Rubanyi *et al.* (1986) showed that FMD is related to the release of an endothelium-derived relaxing factor, and Gruetter *et al.* (1981) demonstrated abolition of FMD by the concomitant administration of the EDRF inhibitor, methylene blue. Recently, Joannides *et al.* (1995) have shown that FMD in human arteries is mediated predominantly by NO release, as it can be largely attenuated by coadministration of L-NMMA, an inhibitor of NO synthesis.

There is some temporal delay between the increase in flow in an artery and the subsequent flow-associated dilatation (Coretti *et al.*, 1995). This raises the question as to whether increased flow itself mediates the dilatation, or sets up another mechanism of sustained physiologic change which in turn mediates the vasodilation (Bhagat *et al.*, 1997).

Both absolute flow increase and pulsatility are powerful stimuli for endothelium-dependent, flow-associated dilatation. The mechanism whereby endothelial cells sense an

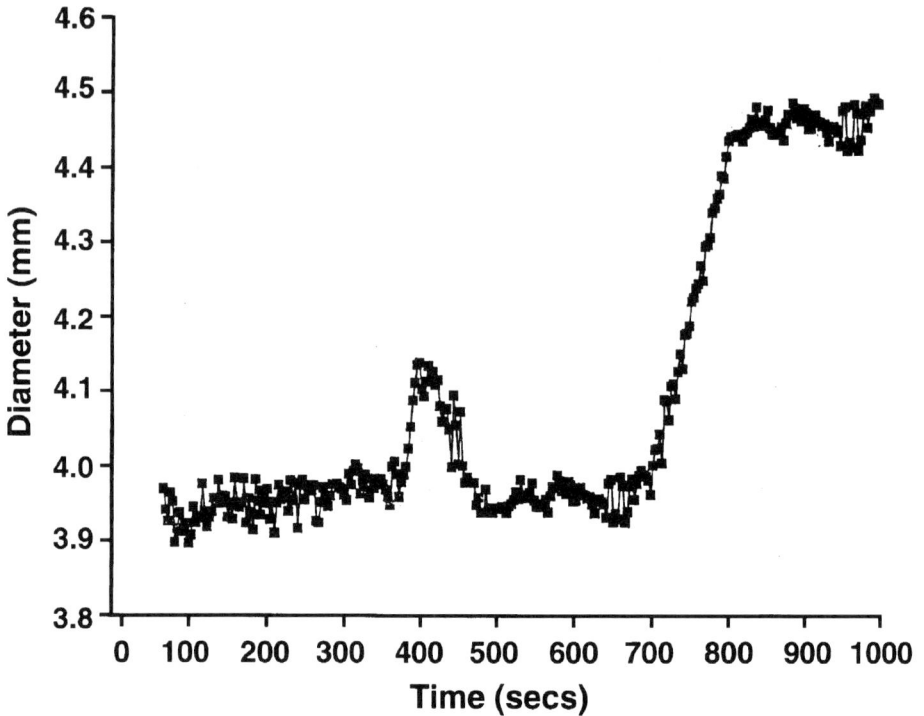

Figure 10.1 Continuous vessel diameter showing brachial artery dilatation after inflation and release of a pneumatic tourniquet on the forearm (flow mediated endothelial dependent dilatation) followed by dilatation to sublingual nitroglycerin (endothelial independent dilatation). High resolution ultrasound has been used to study conduit artery physiology (see text) non-invasively in children from 5 years of age. (Courtesy of Dr M.J. Mullen, Great Ormond Street Hospital for Children, London, UK)

increase in shear stress is not well characterized, but is probably mediated via activation of a K^+ channel in the endothelial cell membrane (Rubanyi *et al.*, 1990). Vascular relaxation may also occur by direct transmission of endothelial cell hyperpolarization to the underlying smooth muscle via gap junctions (Olesen *et al.*, 1988).

TESTING ENDOTHELIAL FUNCTION IN HUMANS

To date, most studies of endothelial function in humans have relied on measurement of endothelium-related changes in vasomotor function (even though this is only one of many aspects of endothelial physiology). In 1986, Ludmer *et al.* reported the first studies of coronary endothelium-dependent dilatation and constriction in adults with chest pain syndromes; these studies relied on invasive instrumentation of the coronary arteries, infusion of vasoactive compounds such as acetylcholine and serial measurement of arterial diameter by quantitative angiography. This and a number of subsequent studies of coronary vasomotion detected impaired endothelium-dependent dilatation in subjects with established coronary stenoses and/or with risk factors for atherosclerosis (Nabel *et al.*, 1990;

Zeiher *et al.*, 1991; Yeung *et al.*, 1991). In addition to using pharmacological agents such as acetylcholine or serotonin to stimulate endothelium-derived vasoactive mediators, some of these studies measured flow-mediated vasodilatation (induced by exercise, cardiac pacing or distal papaverine infusion — for example, see Gordon *et al.*, 1989). These early studies of coronary FMD, however, were limited to relatively small numbers of already symptomatic subjects.

Invasive studies of coronary vasomotion are clearly inappropriate for studying presymptomatic children and young adults, early in the natural history of the atherogenic process. For this reason, non-invasive testing of peripheral conduit arteries was developed. In 1992, we described a non-invasive method for measuring brachial and femoral artery FMD in humans (Celermajer *et al.*, 1992). This method had the same principles as the coronary endothelial studies described above, however the endothelial stimulus was provided by increased flow rather than by infused agents, and vessel diameter was measured by ultrasound, rather than by angiography.

THE NON-INVASIVE TECHNIQUE FOR MEASURING FMD

Serial measurements of the diameter of a target artery can be obtained using high resolution B-mode images, with commercially available 7–10 MHz linear array transducers and standard ultrasound mainframes. Only relatively superficial arteries can be well visualized using these transducers (such as the brachial, femoral or carotid arteries, but not the coronary arteries). Arterial diameter is measured at rest and during a condition of increased flow (for example, produced by distal cuff occlusion and release), resulting in flow-mediated, endothelium-dependent dilatation. The arterial diameter is usually also measured after sublingual nitroglycerin, which is an endothelium-independent dilator.

In practice, the subject lies at rest for at least 10 minutes before the first ultrasound scan (baseline). The target artery (either the superficial femoral artery just distal to the bifurcation of the common femoral, or the brachial artery 2–15 centimetres above the elbow) is scanned in longitudinal section. The center of the artery is identified when the clearest picture of the anterior and posterior intimal layers is obtained. The transmit (focus) zone is set to the depth of the near wall, in view of the greater difficulty of evaluating the near compared to far wall "m" line (the interface between media and adventitia) (Pignoli *et al.*, 1986, Nolsoe *et al.*, 1990). Our usual settings are a transmit power of –9dB, log compression 40dB and overall gain 3 dB. Individual depth and gain settings, however, are set to optimize of the lumen/arterial wall interface. Images are magnified using a resolution box function (leading to a television line width of approximately .065 mm), and machine oper ating parameters are not changed during any study. The processing curves are chosen to enhance the vessel wall/lumen interface (high contrast, crisp borders).

When a satisfactory transducer position is found, the skin is marked, and the limb remains in the same position throughout the study. Care is taken to apply the transducer without undue pressure. A resting scan is recorded and increased flow is then induced by inflation of a pneumatic tourniquet to a pressure of 250–300 mmHg for 4.5 minutes, followed by release. A second scan is taken for 30 seconds before and 90 seconds after cuff deflation, to allow measurement of FMD. Thereafter 10 minutes is allowed for vessel recovery, after which a further resting scan is taken. Sublingual nitroglycerin spray (400

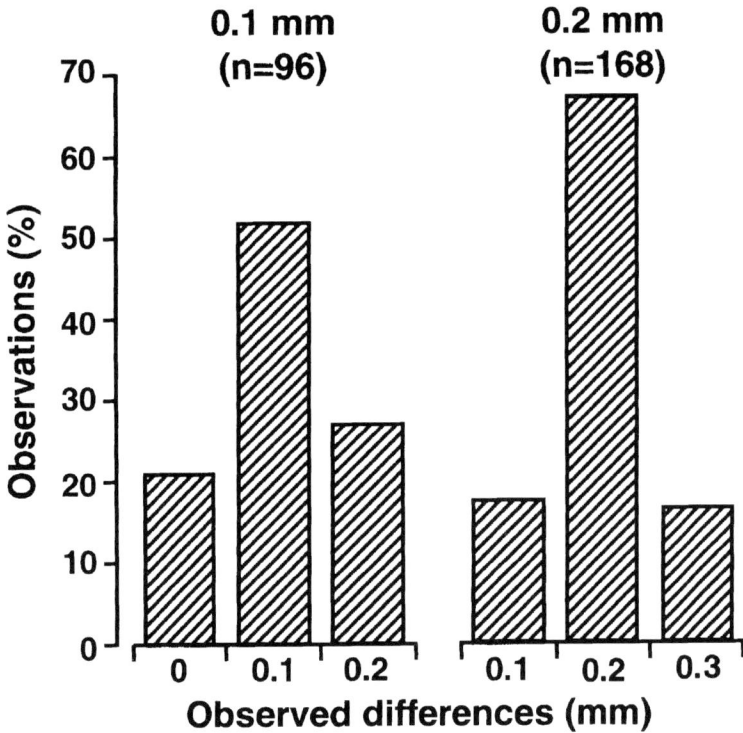

Figure 10.2 Experiments using a 'phantom' of 10 cylinders (arteries) demonstrating accuracy of ultrasound measurements of arterial diameter measurements of 0.1 mm and 0.2 mm. (From Sorensen, K.E., Celermajer, D.S., Spiegelhalter, D.J., Georgakopoulos, D., Robinson, J., Thomas, O., Deanfield, J.E. Non-invasive measurement of human endothelium dependent arterial responses: accuracy and reproducibility. *Br. Heart J.*, 1995;74:247–53. Reproduced with permission from the publishers and authors.)

mcg) is then administered, and 3–4 minutes later the last scan is performed. The electrocardiogram is monitored continuously.

Data analysis is undertaken off-line, using recorded images. Vessel diameter is measured by 2 observers, who are unaware of the condition of the subject and the stage of the experiment. The arterial diameter is measured at a fixed distance from an anatomical marker, such as a bifurcation, using ultrasonic calipers. Measurements are taken from the anterior to the posterior "m" line (the interface between the median and adventitia) at end-diastole, incident with R-wave on electrocardiogram (ECG). For the reactive hyperemia scan, diameter measurements are taken 45–60 seconds after cuff deflation. Four cardiac cycles are analysed for each scan and the measurements averaged. The vessel diameter is measured in each of the scans (that is, at rest, after reactive hyperemia, after a further 10 minutes rest and 3–4 minutes after nitroglycerin), and are then expressed as a percentage relative to the first control scan (100%).

The best arteries to study using this technique are the brachial and superficial femoral, as they are large and superficial systemic arteries in which reactive hyperemia can be induced easily. As atherosclerosis is a diffuse process, which initially often follows a parallel course in the medium-sized muscular arteries, our preliminary studies were carried out on the superficial femoral artery. When we found an inverse relationship between flow-mediated dilatation and resting vessel size in normal subjects (Celermajer *et al.*, 1992), we confined our study to arteries with diameter ≤ 6.0 mm; that is, superficial femoral arteries in children and brachial arteries in adults.

Since this initial description, many modifications have been made by different groups; all, however, rely on the accurate non-invasive measurement of flow-mediated arterial dilatation. Different devices have been used to ensure stability of the ultrasound images (e.g. stereotactic clamps) and various techniques have been reported for diameter measurement (eg B-mode, A-mode "wall tracking"). In addition, semi-automated and computerized methods for on-line or off-line diameter measurement have been described (for example, Leeson *et al.*, 1997). These modifications may have, in part, resulted in the different ranges of "normal" FMD values reported by different investigators; certain groups have reported higher (Coretti *et al.*, 1995) or lower (Joannides *et al.*, 1995) values for FMD in healthy subjects than the control values for FMD reported by our groups (Adams *et al.*, 1996). This suggests that within laboratory comparisons of FMD values for subject groups will be much easier than comparison of FMD values obtained by different groups, using various modifications of the original methodology.

ACCURACY, REPRODUCIBILITY AND LIMITATIONS OF MEASURING FMD

Accurate assessment of FMD depends on reliable detection of submillimeter differences between diameter measurements. Using arterial phantoms, 7MHz ultrasound and B-mode images, Sorensen *et al.* (1995) have demonstrated that diameter differences of 0.1–0.2 mm can be dectected reliably, in vessels measuring 2–5 mm in diameter. In this same study, FMD was relatively reproducible in 40 subjects studied on 4 occasions each (coefficient of variation < 2%). Since this time, however, Hashimoto *et al.* (1996) have demonstrated important cyclical variations of FMD in menstruating females, and Vogel *et al.* (1997) have found significant changes in FMD after a high-fat meal; for these reasons, time of cycle and post-prandial status need to be carefully controlled in any studies of serial changes in FMD.

Limitations of measuring FMD include the difficulties in maintaining stable transducer position (requiring a skilled sonographer and cooperative subject), the time consuming process of image analysis, and the lack of access to coronary arteries with external ultrasound. Furthermore, there is a wide range of "normal" values for FMD, even in healthy young subjects without identifiable vascular risk factors (Adams *et al.*, 1996). For these reasons, we do not believe that FMD is a useful screening tool for endothelial dysfunction; it does, however, remain a very useful test for studies of endothelial physiology in "at risk" populations and for examining reversibility of early arterial damage in certain groups of subjects.

Although brachial artery study *per se* is not clinically relevant, as brachial atherosclerosis is an extremely infrequent health problem, it appears that brachial FMD is well correlated with both coronary endothelial responses (Anderson *et al.*, 1995) and with the extent and severity of coronary atherosclerosis (Neunteufl *et al.*, 1997). For these reasons, study of peripheral artery FMD represents a useful surrogate for coronary abnormalities, particularly in younger subjects (Sorensen *et al.*, 1997).

THE EFFECT OF RISK FACTORS ON FMD

Impairment of conduit artery FMD is an early systemic event in children and adults with risk factors for atherosclerosis. The effect of most of the traditional risk factors on FMD has been well studied in the 1990's, illustrating impaired endothelial release of NO early in life in high risk subjects, and that most traditional vascular risk factors demonstrate a "dose-response" relationship with impaired FMD.

Active cigarette smoking is associated with reduced FMD in a dose-dependent manner, with FMD being almost completely absent in smokers with >20 pack year smoking history (Celermajer *et al.*, 1993) (Figure 10.3). Although this is potentially reversible, the degree of reversibility of smoke-related endothelial damage may be minor. Even heavy exposure to environmental tobacco smoke ("passive" smoking) has been associated with a dose-related impairment of FMD, in otherwise healthy teenagers and young adults (Celermajer *et al.*, 1996).

Familial hypercholesterolemia (FH) is associated with profound impairment of FMD, even in the first decade of life (Sorensen *et al.*, 1994) (Figure 10.4). The degree of impairment correlates with both LDL and Lp(a) levels, in young FH subjects. In non-FH subjects, however, with total cholesterol levels from 3–6.5 mmol/L, cholesterol level is only weakly (inversely) related to FMD (Celermajer *et al.*, 1994b). Less information is available about the effects of LDL, HDL, Lp(a) or triglyceride levels on FMD, in otherwise healthy, non-FH subjects. Recently, however, Lundman *et al.* (1997) have demonstrated that acute infusion of triglycerides is associated with impaired endothelium-independent vasodilatation in healthy young adults.

The relationship between blood pressure and FMD is not well studied. In a group of young adult survivors of aortic coarctation repair, exercise-related hypertension was related to an impaired nitroglycerin response, but not to FMD (Gardiner *et al.*, 1994). Studies of FMD in essential hypertensive subjects, however, have not yet been reported, possibly because of the confounding influence of anti-hypertensive agents on these tests of arterial physiology.

Data on FMD in insulin-dependent diabetics have been conflicting, with one group showing no change compared with controls (Lambert *et al.*, 1996), and another larger study showing significant impairment of both FMD and nitroglycerin responses (Clarkson *et al.*, 1996b). In this latter study, the degree of FMD impairment correlated with the duration of disease and with the LDL cholesterol level.

A positive family history of premature coronary disease is associated with a significant decrease in arterial FMD in otherwise health young subjects (Clarkson *et al.*, 1997). This is particularly marked if the index relative with premature coronary disease had no identifiable vascular risk factors themselves, consistent with a heritable component in the FMD response.

Although males have lower FMD values than females, this is accounted for completely by their larger vessel size (Celermajer *et al.*, 1992). In young subjects (< age 40 years), there appears to be no gender difference in vessel size adjusted FMD; with aging, however, there is a progressive decline in FMD, and this occurs earlier in males than in females

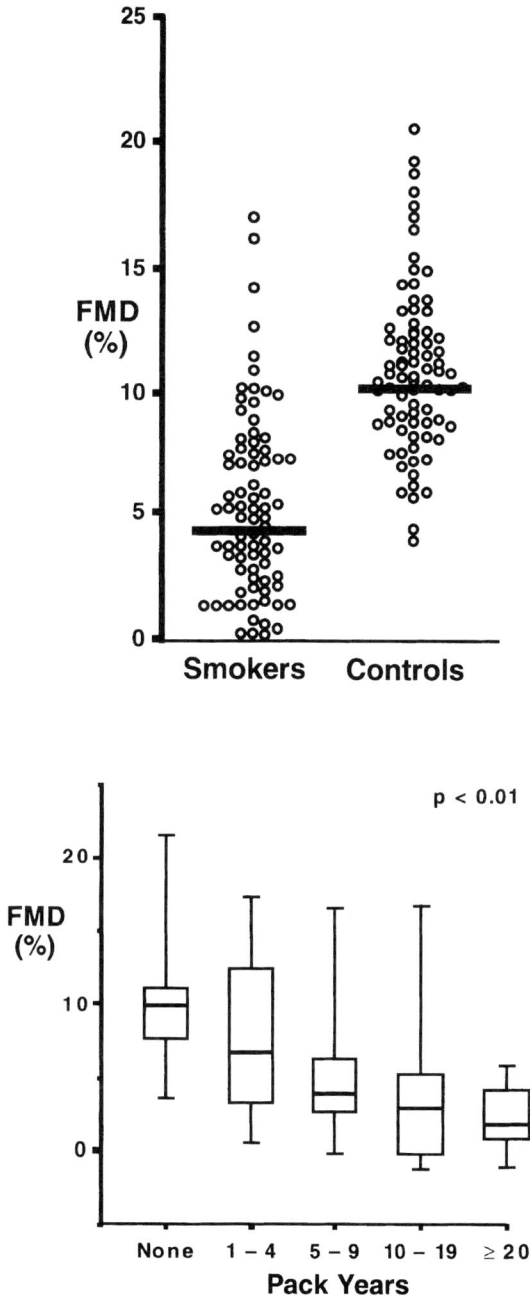

Figure 10.3 Smoking and brachial artery flow mediated dilatation. Impaired endothelial dependent dilatation is present in association with cigarette smoking in 80 young asymptomatic subjects (top) in proportion to extent and duration of smoking (bottom). (From Celermajer, D.S., Sorensen, K.E., Georgakopoulos, D., Bull, C., Thomas, O., Robinson, J., Deanfield, J.E. Cigarette smoking is associated with dose-related and potentially reversible impairment of endothelium-dependent dilation in healthy young adults. *Circulation*, 1993;88:2149–55. Reproduced with permission from the publishers and authors).

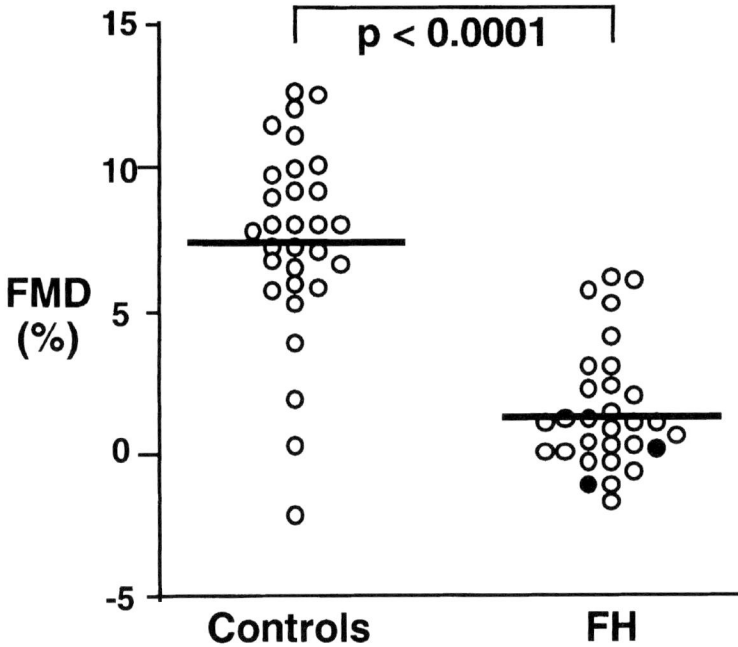

Figure 10.4 Cholesterol and femoral artery flow mediated dilatation. Impairment of endothelial dependent dilatation is A.L.ready present by the first and second decades of life (30 children age 7 to 17 years) in association with familial hypercholesterolaemia (28 heterozygotes: 2 homozygotes in black). (From Sorensen, K.E., Celermajer, D.S., Georgakopoulos, D., Hatcher, G., Betteridge, D.J., Deanfield, J.E. Impairment of endothelium-dependent dilation is an E.A.rly event in children with familial hypercholesterolemia and is related to the lipoprotein (a) level. *J. Clin. Invest.*, 1994;93:50–55. Reproduced with permission from the publishers and authors).

(Celermajer *et al.*, 1994a). This is consistent with the data of Hashimoto *et al.* (1996), who showed increased FMD during the follicular phase of normal menses in young women, suggesting an association between physiologic levels of estrogens and improved arterial FMD.

The effect of relatively newly identified risk factors for premature atherosclerosis has also been examined. The ACE "DD" genotype, which has been linked to a high risk of myocardial infarction (Cambien *et al.*, 1992), is not associated with impaired FMD (Celermajer *et al.*, 1994c). Hyperhomocysteinemia, however, is a risk factor for both premature arterial disease and for impaired FMD, in both Caucasian (Tawakol *et al.*, 1997) and Chinese subjects (Woo *et al.*, 1997a). As most endothelial function studies have been carried out in Caucasian populations, it will be important to establish whether risk factors interact with endothelial physiology in different ways in other ethnic groups. Some preliminary data, for example, suggest that aging and smoking may have less effect on FMD in Chinese compared to Caucasian subjects (Woo *et al.*, 1997b, 1997c).

Most studies have examined the effects of single risk factors on FMD, in highly selected populations. It appears, however, that in an unselected population of healthy subjects, that

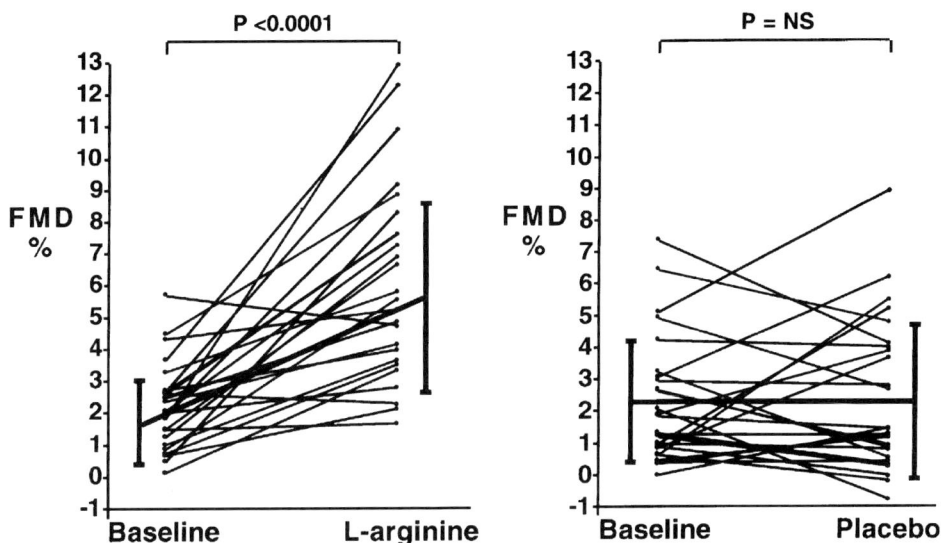

Figure 10.5 One month of dietary supplementation with L-arginine improves brachial artery endothelial function in young subjects with moderate hypercholesterolaemia studied in a double blind crossover trial. (From Clarkson, P., Adams, M.R., Powe, A.J., Donald, A.E., McCredie, R., Robinson, J., McCarthy, S.N., Keech, A., Celermajer, D.S., Deanfield, J.E. Oral L-arginine improves endothelium-dependent dilation in hypercholesterolemic young adults. *J. Clin. Invest.*, 1996;97:1989–94. Reproduced with permission from the publishers and authors).

traditional risk factors may interact with each other, to increase the risk of impaired arterial FMD (Celermajer *et al.*, 1994b). Similar observations have been made in the coronary circulation of older symptomatic subjects (Vita *et al.*, 1990).

REVERSIBILITY OF IMPAIRED FMD

Identification of subjects with impaired FMD (and therefore reduced endothelial NO release) raises the question of reversibility of this abnormality, by risk factor modification, drug therapy or other novel methods. The availability of a non-invasive technique for studying FMD, which can be performed serially, facilitates the study of such potentially FMD-enhancing strategies. Based on reproducibility data, power function analyses have been performed (Sorensen, 1995), to allow accurate calculation of sample sizes required for proposed interventional trials.

To date, 3 interventions have been shown to improve FMD with short-term oral administration. L-arginine is the substrate for endothelial nitric oxide production, and Clarkson *et al.* (1996a) have demonstrated that oral L-arginine (7 g three times daily, given for 4 weeks) improves FMD in hypercholesterolemic young adults with endothelial dysfunction (Figure 10.5). Similar beneficial effects on endothelial NO release in hypercholesterolemic rabbits have been associated with a marked decrease in aortic atherosclerosis (Cooke *et al.*, 1992).

Similarly Levine *et al.* (1996) have assessed the effect of vitamin C on arterial FMD in adults with established coronary disease, and shown acute improvement in FMD after a single oral 2g dose. The mechanism of this effect is uncertain, but may be due to ascorbate-related scavenging of oxygen-derived free radicals, with consequently increased bioavailability of endothelium-derived NO. It is not yet known, however, whether the beneficial effects of L-arginine or vitamin C will be sustained with long-term administration, or whether this will be associated with a reduction in clinically relevant endpoints.

Finally, estrogen replacement therapy (ERT) has been shown to improve FMD in postmenopausal women (Lieberman *et al.*, 1994); this may explain some of the apparent cardioprotective effects of such therapy in older females. As most older women require progesterone co-supplementation with ERT, however, to protect the uterus, prospective data on the effect of combined therapy on endothelial function are awaited; preliminary reports are reassuring (McCrohon *et al.*, 1996).

In addition to the above, trials on the effects of cholesterol lowering therapy, ACE inhibition and calcium channel blockade on arterial FMD are underway in certain high risk asymptomatic subjects; such studies may provide insights into the possible vascular protective effects of these agents, early in the natural history of atherosclerosis.

CONCLUSIONS

Non-invasive measurement of flow-mediated, endothelium-dependent arterial dilatation has provided important insights into the relationships between vascular risk factors and this important aspect of endothelial physiology, in human subjects. The availability of an ultrasound-based method for studying FMD has facilitated serial studies which have assessed strategies for improving endothelial NO release, with promising early results. Future research will focus on standardization and automation of FMD measurements, and on further trials to determine the sustainability and clinical relevance of therapies which improve FMD, in children and young adults at risk of atherosclerosis.

REFERENCES

Adams, M.R., Robinson, J., Sorensen, K.E., Deanfield, J.E. and Celermajer, D.S. (1996) Normal ranges for brachial artery flow-mediated dilation: a non-invasive ultrasound test of arterial endothelial function. *J. Vasc. Invest.*, **2**, 146–50.

Anderson, E.A. and Mark, A.L. (1989) Flow-mediated and reflex changes in large peripheral artery tone in humans. *Circulation*, **79**, 93–100.

Anderson, T.J., Uehata, A., Gerhard, M.D. *et al.* (1995) Close relationship of endothelial function in the human coronary and peripheral circulations. *J. Am. Coll. Cardiol.*, **26**, 1235–41.

Bhagat, K., Hingorani, A. and Vallance, P. (1997) Flow-associated or flow mediated dilatation? More than J.ust semantics. *Heart*, **78**, 7–8.

Cambien, F., Poirier, O., Lecerf, L. *et al.* (1992) Deletion polymorphism in the gene for angiotensin-converting enzyme is a potent risk factor for myocardial infarction. *Nature*, **359**, 641–4.

Celermajer, D.S., Sorensen, K.E., Gooch, V.M. *et al.* (1992) Non-invasive detection of endothelial dysfunction in children and adults at risk of atherosclerosis. *Lancet*, **340**, 1111–5.

Celermajer, D.S., Sorensen, K.E., Georgakopoulos, D. *et al.* (1993) Cigarette smoking is associated with dose-related and potentially reversible impairment of endothelium-dependent dilation in healthy young adults. *Circulation*, **88**, 2149–55.

Celermajer, D.S., Sorensen, K.E., Spiegelhalter, D.S., Georgakopoulos, D., Robinson, J. and Deanfield, J.E. (1994a) Aging is associated with endothelial dysfunction in healthy men years before the age-related decline in women. *J. Am. Coll. Cardiol.*, **24**, 471–6.

Celermajer, D.S., Sorensen, K.E., Bull, C., Robinson, J. and Deanfield, J.E. (1994b) Endothelium-dependent dilation in the systemic arteries of asymptomatic subjects relates to coronary risk factors and their inter-action. *J. Am. Coll. Cardiol.*, **24**, 1468–74.

Celermajer, D.S., Sorensen, K.E., Barley, J., Jeffrey, S., Carter, N. and Deanfield, J.E. (1994c) Angiotensin-converting enzyme genotype is not associated with endothelial dysfunction in subjects withour other coronary risk factors. *Atherosclerosis*, **111**, 121–6.

Celermajer, D.S., Adams, M.R., Clarkson, P. *et al.* (1996) Passive smoking and impaired endothelium-dependent dilation in healthy young adults. *N. Engl. J. Med.*, **334**, 150–4.

Clarkson, P., Adams, M.R., Powe, A.J. *et al.* (1996a) Oral L-arginine improves endothelium-dependent dilatation in hypercholesterolemic young adults. *J. Clin. Invest.* **97**,, 1989–4.

Clarkson, P., Celermajer, D.S., Sampson, M. *et al.* (1996b) Impaired vascular reactivity in insulin-dependent diabetes mellitus is related to disease duration and LDL-cholesterol levels. *J. Am. Coll. Cardiol.*, **28**, 573–9.

Clarkson, P., Celermajer, D.S., Powe, A., Donald, A.E., Henry, R.M. and Deanfield, J.E. (1997) Endothelium-dependent dilatation is impaired in young healthy subjects with a family history of premature coronary disease. *Circulation*, **96**, 3378–83.

Cooke, J.P, Singer, AH, Zera, P., Rowan, R.A. and Billingham, M.E. (1992) Antatherogenic effects of L-arginine in the hypercholesterolemic rabbit. *J. Clin. Invest.*, **90**, 168–72.

Corretti, M.C., Plotnick, G.D. and Vogel, R.A. (1995) Technical aspects of evaluating brachial artery dilation using high frequency ultrasound. *Am. J. Physiol.*, **268**, H1397–404.

Franke, R.P., Grafe, M., Schnittler, H., Seiffge, D. and Mittermayer, C. (1984) Induction of human vascular endothelial stress fibres by fluid shear stress. *Nature*, **307**, 648–9.

Gardiner, H.M., Celermajer, D.S., Sorensen, K.E. *et al.* (1994) Arterial reactivity is significantly impaired in normotensive young adults following successful repair of aortic coarctation in childhood. *Circulation*, **89**, 1745–50.

Gordon, J.B, Ganz, P., Nabel, E.G. *et al.* (1989) Atherosclerosis and endothelial function influence the coronary vasomotor response to exercise. *J. Clin. Invest.*, **83**,, 1946–52.

Gruetter, D.Y., Lyone, J.E, Kodowitz, P.J. and Ignarro, LJ. (1981) Relationship between cyclic guanosine 3'5'-monophosphate formation and relaxation of coronary arterial smooth muscle by glyceryl trinitrate, nitro-prusside, nitrite and nitric oxide: Effects of methylene blue and methemoglobin. *J. Pharmacol. Exp. Ther.*, **219**, 181–6.

Hashimoto, M., Akihita, M., Eto, M., *et al.* (1995) Modulation of endothelium-dependent flow-mediated dilatation of the brachial artery by sex and menstrual cycyle. *Circulation*, **92**, 3431–5.

Hintze, T.H. and Vatner, S.F. (1984) Reactive dilatation of large coronary arteries in conscious dogs. *Circ. Res.*, **54**, 50–7.

Ingebrigtsen, R. and Leraand, S. (1970) Dilatation of a medium-sized artery immediately after local changes of blood pressure and flow as measured by ultrasonic techniques. *Acta Physiol. Scand.*, **79**, 552–8.

Joannides, R., Haefeli, W.E., Linder, L. *et al.* (1995) Nitric oxide is responsible for flow-dependent dilatation of human peripheral conduit arteries in vivo. *Circulation*, **91**, 1314–9.

Lambert, J., Aarsen, M., Donker, A.J.M. and Stehouwer, C.D.A. (1996) Endothelium-dependent and independent vasodilation of large arteries in normoalbuminuric insulin-dependent diabetes mellitus. *Arterioscler. Thromb. Vasc. Biol.*, **16**, 705–711.

Laurent, S., Lacolley, P., Brunel, P., Lalouz, B., Pannir, B. and Safar, M. (1990) Flow-dependent vasodilation of brachial artery in essential hypertension. *Am. J. Physiol.*, **258**, H1004–11.

Leeson, P., Thorne, S., Donald, A., Mullen, M., Clarkson, P. and Deanfield, J. (1997) Non-invasive measurement of endothelial function: effect on brachial artery dilatation of graded endothelial dependent and independent stimuli. *Heart*, **78**, 22–7.

Levine, G.N., Frei, B., Kouloukis, S.N., Gerhard, M.A., Keaney, J.F. and Vita, J.A. (1996) Ascorbic acid reverses endothelial vasomotor dysfunction in patients with coronary artery disease. *Circulation*, **93**, 1107–13.

Lie, M., Sejersted, P.M. and Kiil, F. (1970) Local regulation of vascular cross section during changes in femoral arterial blood flow in dogs. *Circ. Res.*, **27**, 727–37.

Lieberman, E.H., Gerhard, M.D. Uehata, A. *et al.* (1994) Estrogen improves endothelium-dependent, flow-mediated vasodilation in postmenopausal women. *Ann. Intern. Med.*, **121**, 936–41.

Ludmer, P.L., Selwyn, A.P., Shook, T.L. *et al.* (1986) Paradoxical vasconstruction induced by acetycholine in atherosclerotic coronary arteries. *N. Engl. J. Med.*, **315**, 1046–51.

Lundman, P., Eriksson, M., Schenck-Gustafsson, K., Karpe, F. and Tornvall, P. (1997) Transient triglyceridemia decreases vascular reactivity in young, healthy men without risk factors for coronary heart disease. *Circulation*, **96**, 3266–68.

McCrohon, J.A, Adams, M.R., McCredie, R.J. *et al.* (1996) Hormone replacement therapy is associated with improved endothelium-dependent dilation in post-menopausal women. *Clin. Endocrinol.*, **45**, 435–41.

Macho, P., Hintze, T.H. and Vatner, S.F. (1981) Regulation of large coronary arteries by increases in myocardial metabolic demands in conscious dogs. *Circ. Res.*, **76**, 594–9.

Nabel, E.G., Selwyn, A.P. and Ganz, P. (1990) Large coronary arteries in humans are responsive to changing blood flow: an endothelium-dependent mechanism that fails in patients with atherosclerosis. *J. Am. Coll. Cardiol.*, **16**, 349–56.

Neunteufl, T., Katzenschlager, R., Hassan, A. *et al.* (1997) Systemic endothelial dysfunction is related to the extent and severity of coronary artery disease. *Atherosclerosis*, **129**, 111–8.

Nolsoe, C.P., Engel, U., Karstrup, S., Torp-Pederson, S., Garre, K. and Holm, H.H. (1990) The aortic wall: an in vitro study of the double-line pattern in high-resolution US. *Radiology*, **175**, 387–90.

Olesen, S.P., Clapham, D.E. and Davies, P.F. (1988) Haemodynamic shear stress activates a K+ current in vascular endothelial cells. *Nature*, **331**, 168–70.

Pignoli, P., Tremoli, E., Pili, A., Oreste, P. and Paoletti, R. (1986) Intima plus medial thickness of the arterial wall: a direct measurement with ultrasonic imaging. *Circulation*, **74**, 1399–406.

Pohl, U., Holtz, J., Busse, R. and Bassenge, E. (1986) Crucial role of endothelium in the vasodilator response to increased flow in vivo. *Hypertension*, **8**, 37–44.

Rubanyi, G.M., Romero, C. and Vanhoutte, P.M. (1986) Flow induces release of endothelium-derived relaxing factor. *Am. J. Physiol.*, **250**, 1115–9.

Rubanyi, G.M., Freay, A.D., Kauser, K., Johns, A. and Harder, D.R. (1990) Mechanoreception by endothelium: mediators and mechanisms of pressure-and flow-induced vascular responses. *Blood Vessels*, **27**, 246–57.

Schretzenmayr, A. (1933) Uber Kreislaufregulatorische Voragne an den grossen Arterien bei der Muskelarbeit. *Pflugers Arch.*, **232**, 743–8.

Sinoway, L.I., Hendreickson, C., Davidson, W.R., Prophet, S. and Zelis, R. (1989) Characteristics of flow-mediated brachial artery vasodilation in human subjects. *Circ. Res.*, **64**, 32–42.

Smiesko, V., Kozik, J. and Dolezel, S. (1985) Role of endothelium in the control of arterial diameter by blood flow. *Blood Vessels*, **22**, 247–51.

Sorensen, K.E., Celermajer, D.S., Georgakopoulos, D., Hatcher, G., Betteridge, J. and Deanfield, J.E. (1994) Impairment of endothelium-dependent dilation is an E.A.rly event in children with hypercholesterolemia, and is related to the Lp(a) level. *J. Clin. Invest.*, **93**, 50–5.

Sorensen, K.E., Celermajer, D.S., Spiegelhalter, D.J. *et al.* (1995) Non-invasive measurement of endothelium-dependent arterial responses in man: accuracy and reproducibility. *Br. Heart J.*, **74**, 247–253.

Sorensen, K.E., Kristensen, J.B. and Celermajer, D.S. (1997) Atherosclerosis in the human brachial artery. *J. Am. Coll. Cardiol.*, **29**, 318–22.

Tawakol, A., Omland, T., Wu, J.T., Gerhard, M. and Creager, M.A. (1997) Hyperhomocyst(e)inemia is associated with impaired endothelium-dependent vasodilation in humans. *Circulation*, **95**, 1119–21.

Van Grondelle, A., Worthen, G.S., Ellis, D. *et al.* (1984) A.L.tering hydrodynamic variables influences PGI$_2$ production by isolated lungs and endothelial cells. *J. Appl. Physiol.*, **57**, 388–95.

Vita, J.A., Treasure, C.B., Nabel, E.G. *et al.* (1990) Coronary vasomotor response to acetylcholine relates to risk factors for coronary artery disease. *Circulation*, **81**, 491–7.

Vogel, R.A., Corretti, M.C. and Plotnick, G.D. (1997) Effect of a single high-fat meal on endothelial function in healthy subjects. *Am. J. Cardiol.*. **79**, 350–4.

Woo, K.S., Chook, P., Lolin, Y. *et al.* (1997a) Hyperhomocysteinemia is a risk factor arterial endothelial dysfunction in humans. *Circulation*, **96**, 2542–4.

Woo, K.S., McCrohon, J.A., Chook, P. *et al.* (1997b) Chinese adults are less susceptible to age-related endothelial dysfunction than Caucasians. *J. Am. Coll. Cardiol.*, **30**, 113–8.

Woo, K.S., Robinson, J.T.C., Chook, P. *et al.* (1997c) Cigarette smoking is associated with less endothelial dysfunction in young Chinese than in young Caucasian adults. *Ann. Int. Med.*, **127**, 372–5.

Yeung, A.C., Vekshtein, V.I., Krantz, D.S. *et al.* (1991) The effects of atherosclerosis on the vasomotor response of coronary arteries to mental stress. *N. Engl. J. Med.*, **325**, 1551–6.

Zeiher, A.M., Drexler, H., Wollschlager, H. and Just, H. (1991) Modulation of coronary vasomotor tone in humans. Progressive endothelial dysfunction with different E.A.rly stages of coronary atherosclerosis. *Circulation*, **83**, 391–401.

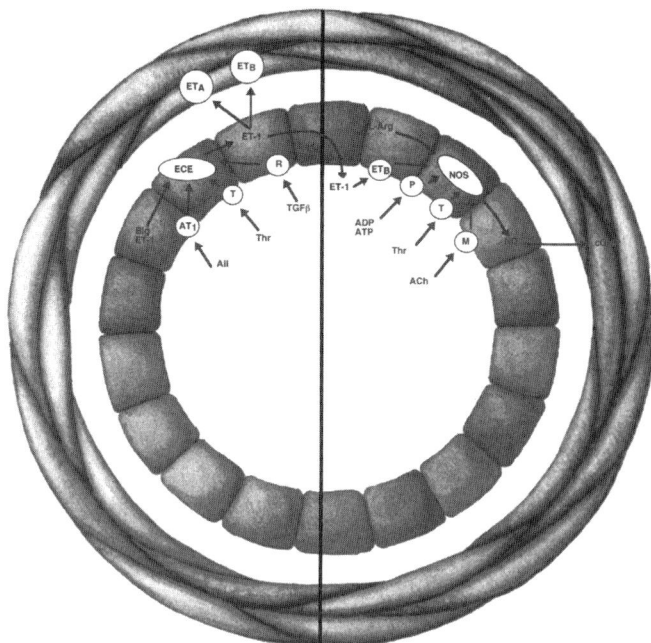

Prostaglandins, nitric oxide (NO) and endothelins are endothelium-derived mediators that have important physiological roles in the regulation of the systemic and renal vasculature. Within the kidney NO and prostaglandins have mainly vasodilator and natriuretic effects, whereas endothelin pathways predominantly increase vascular tone and reduce sodium excretion. Moreover in human renal disease there is evidence that the activity of these mediators is altered, and might contribute to systemic hypertension, reduced renal perfusion and glomerular filtration, or inflammatiry renal disease. Consequently manipulation of these mediators, either singly or in concert, might improve the treatment of renal disease and its complications.

Key words: Nitric oxide, endothelin, prostaglandins, kidney, human.

11 Endothelial Mediators and Renal Disease

Raymond J. MacAllister[1] and William B. Haynes[2]

[1]Centre for Clinical Pharmacology, The Wolfson Institute for Biomedical Research, University College London, 140 Tottenham Court Road, London W1P 9LN, UK
[2]Department of Internal Medicine, University of Iowa, Iowa City, IA 52241, USA

INTRODUCTION

As in other vascular beds, the endothelium of the renal vasculature synthesises mediators such as NO, prostaglandins and endothelin with potential to modulate blood vessel function. However, the endothelium is not the only cell type that synthesises these autacoids, and many renal parenchymal cells synthesise and respond to these mediators. This makes it difficult to be certain of the precise contribution of the endothelium to effects mediated by these agents. Moreover, the kidney is also responsible for the excretion of substances that might alter endothelial function, and as a result renal dysfunction could contribute to widespread endothelial abnormalities. In this chapter, we will review the role of endothelial mediators in the regulation of renal function with emphasis on human physiology and pathophysiology.

OVERVIEW OF THE PHYSIOLOGICAL AND PATHOPHYSIOLOGICAL ROLES OF ENDOTHELIUM-DERIVED MEDIATORS IN THE SYSTEMIC VASCULATURE

The vascular endothelium generates vasoactive mediators including nitric oxide (NO), prostanoids and endothelins (Figure 11.1). Endothelial cyclooxygenase-1 (COX-1) generates prostanoids from arachidonic acid, some of which cause smooth muscle relaxation (prostacyclin, prostaglandin E) by stimulating adenylate cyclase through G-protein coupled receptors (Coleman *et al* 1994). The vasoconstrictor prostanoid thromboxane acts on its receptors to mobilise intracellular calcium causing smooth muscle contraction. Endogenous prostanoids do not appear to modulate systemic vascular tone in healthy humans, because inhibition of COX-1 by nonsteroidal anti-inflammatory drugs (NSAIDs) has little effect on systemic vascular resistance and blood pressure (Taddei *et al.*, 1993; Pope *et al.*, 1998). However, in vascular disease, prostaglandins might be more important in the regulation of vascular tone (Taddei *et al.*, 1997; Johnson, 1997). A pathophysiological role for constrictor and pro-thrombotic prostanoids is also supported by the therapeutic effect of aspirin in patients with vascular disease (Thiemermann *et al.*, 1993).

Nitric oxide (NO) is synthesised by NO synthase in the vascular endothelium from the amino acid L-arginine. NO diffuses to the vascular smooth muscle where many of its effects are mediated by stimulation of soluble guanylate cyclase (sGC) to generate cyclic GMP (Moncada *et al.*, 1993). In contrast to COX inhibition, inhibition of NO synthesis in healthy subjects increases vascular tone and blood pressure, indicating that under

Generator cell   Target cell

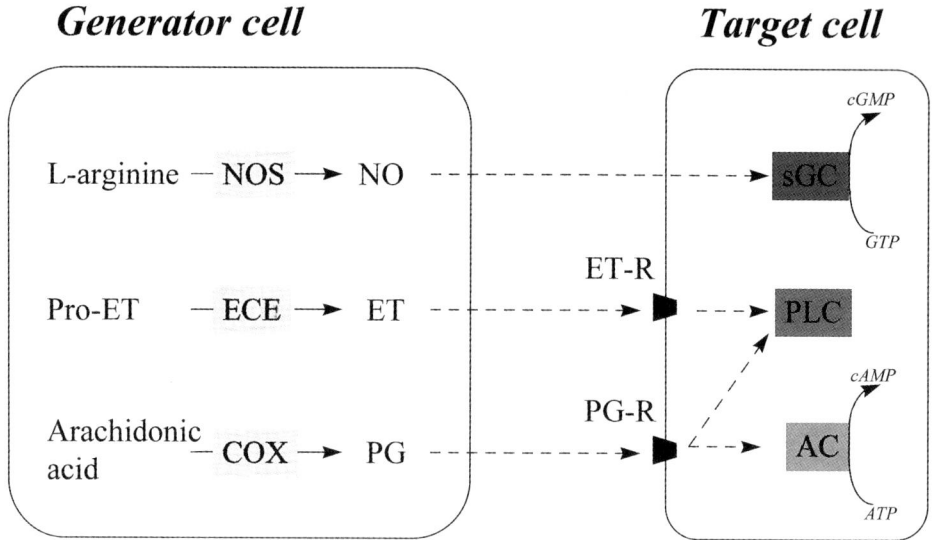

Figure 11.1 Simplified schematic of the NO, endothelin and prostaglandin pathways. In the kidney many cell types synthesise and respond to NO, endothelin and prostaglandins. NO is generated from L-arginine by the action of NO synthase enzymes (NOS) and many of its effects are a consequence of stimulation of soluble guanylate cyclase (sGC) to generate cyclic GMP (cGMP). Endothelin (ET) is produced from pro-endothelin (pro-ET) by endothelin converting enzyme (ECE). Endothelin stimulates cell surface receptors (ET-R) which mediate smooth muscle contraction by activation of phospholipase C (PLC). Cyclooxygenase enzymes (COX) metabolise arachidonic acid to a variety of prostaglandins (Pg) and these activate cell surface receptors causing constriction (mediated by PLC) or dilatation due to stimulation of adenylate cyclase (AC) to increase intracellular cyclic AMP (cAMP). In addition to these pathways, NO, endothelin and prostaglandins have been implicated in the modulation of the function of ion channels and ion pumps in the target cell membrane, and these actions could also contribute to their biological effects.

physiological conditions, there is background NO-mediated dilatation of the arterial vasculature of humans which is a determinant of blood pressure (Moncada *et al.*, 1993, see chapter 1). In addition to effects on vascular tone, NO decreases proliferation of underlying vascular smooth muscle, maintains endothelial cell impermeability, inhibits platelet aggregation and adhesion and inhibits activation and adhesion of leucocytes to endothelium (Kanwar *et al.*, 1995; Radomski *et al.*, 1993). These additional effects of NO might protect the vasculature from the development of atheroma and its complications (Cooke *et al.*, 1997). In patients at risk of developing atheroma, and in the presence of established disease, there is evidence for reduced NO-mediated effects. Whether NO-replacement therapy is anti-atherogenic remains to be determined.

The endothelins are a family of endothelium-derived peptides that possess characteristically sustained vasoconstrictor properties (Yanagisawa *et al.*, 1988). The three endothelin isopeptides are generated through cleavage of precursor big endothelin molecules to the mature peptides by endothelin converting enzymes (Inoue *et al.*, 1989). Endothelin-1 appears to be the predominant member of the family generated by vascular endothelial cells (Howard *et al.*, 1992), so our discussion focuses on this isopeptide. Endothelin exerts its actions by binding to at least two receptors, the ET_A and ET_B subtypes. In addition to its direct vascular effects, endothelin-1 has inotropic and mitogenic properties, influences salt and water homeostasis, alters central and peripheral sympathetic activity, and stimulates the renin-angiotensin-aldosterone system (Haynes *et al.*, 1998). Studies with endothelin receptor antagonists have indicated that endothelin-1 has complex opposing vascular effects mediated through vascular smooth muscle and endothelial ET_A and ET_B receptors. Endogenous generation of endothelin-1 appears to contribute to maintenance of basal vascular tone and blood pressure through activation of vascular smooth muscle ET_A receptors (Haynes *et al.*, 1994; Haynes *et al.*, 1996). At the same time, endogenous endothelin-1 acts through endothelial ET_B receptors to tonically stimulate nitric oxide formation and oppose vasoconstriction (Verhaar *et al.*, 1998).

Endothelium-derived hyperpolarising factor has also been implicated in the dilator function of the endothelium (see chapter 4). This substance has yet to be identified but appears to contribute to endothelium-dependent dilatation of the renal resistance vasculature in animals (Vargas *et al.*, 1994). Its role in humans has yet to be investigated and will not be discussed further in this chapter.

DISTRIBUTION AND BIOCHEMISTRY OF ENDOTHELIAL MEDIATORS IN THE HUMAN KIDNEY

Biochemical and molecular studies have been used to confirm that the human kidney has the capacity to synthesise and respond to NO, prostaglandins and endothelins.

Cyclo-Oxygenase Expression and Activity

Immunohistochemistry and *in situ* hybridisation has identified constitutive expression of COX-1 and COX-2 in human collecting duct, endothelial and smooth muscle cells in afferent and efferent arterioles, and in the medullary vessels (Kömhoff *et al.*, 1997). COX-1 predominates in the medulla, COX-2 in the cortex. In animals there is constitutive expression of COX-2 in the macula densa, which is increased by sodium-depletion. In contrast, there is no expression of COX-2 in the human macula densa and it remains to be determined whether human COX-2 expression can be induced. There is regional variation of prostaglandin synthesis in the human kidney; renal vessels synthesise prostacyclin and PGE_2, the glomeruli synthesise prostacyclin, the medulla mainly PGE2, and in the renal pelvis and ureters PGE_2 and thromboxane predominate (Coleman *et al.*, 1990). There is limited data on the distribution of prostaglandin receptors in the human kidney, but PGE binding sites occur at highest density in the renal medulla (Eriksson *et al.*, 1990).

NOS Distribution And Activity

Animal models suggest that all three isoforms of NOS are constitutively expressed in the kidney (Bachman *et al.*, 1995). In humans, there is evidence for expression of endothelial NOS in the renal cortex (afferent and efferent arterioles), and medullary vessels (including veins and vasa recta) (Bachman *et al.*, 1995). nNOS has been detected by immunohistochemistry in the macula densa, although staining appears to be less dense than in animals.

These areas also exhibit NADH diaphorase staining, which is consistent with NOS activity. There is NOS activity in the renal pelvis (Iversen *et al.*, 1995), and immunohistochemistry has identified nNOS positive neurones in the human uterovesical junction (Goessl *et al.*, 1995). Although human proximal tubular cells in culture synthesise NO (McLay *et al.*, 1994; McLay *et al.*, 1995) and mRNA for iNOS has been detected in these cells, (McLay *et al.*, 1994), the mRNA and protein for inducible NOS has only been detected in renal biopsy specimens from patients with glomerulonephritis, where iNOS is localised to immune cells and not the renal parenchyma (Kashem *et al.*, 1996).

Endothelin And Endothelin Receptor Distribution

Although mRNA for all three endothelin isoforms can be detected in the medulla, cortex and vasculature of human kidney, only big endothelin-1 can be detected by immunocytochemistry (Karet *et al.*, 1996). In human renal cortex, gene expression for preproendothelin-1 is particularly evident in the endothelial layer of arcuate and interlobular arteries and veins and in afferent arterioles (Pupilli *et al.*, 1994). In renal medulla, preproendothelin-1 is highly expressed in vasa recta bundles and capillaries, and collecting ducts (Pupilli *et al.*, 1994). Big endothelin-1 is cleaved to mature endothelin-1 by both neutral endopeptidase and endothelin converting enzyme in human kidney (Russell *et al.*, 1998); the latter is highly expressed in renal arteries and arterioles, the loop of Henle and collecting ducts (Pupilli *et al.*, 1997). Human renal cortex expresses three times as much mRNA for ET_A as the ET_B receptor (Karet *et al.*, 1994). In contrast, the ET_B receptor is preferentially expressed in human renal medulla (Karet *et al.*, 1994).

PHYSIOLOGICAL ROLES OF ENDOTHELIAL MEDIATORS IN THE HUMAN KIDNEY

Regulation of Vascular Tone

In humans there is functionally active background production of NO, prostaglandins and endothelin-1 in the kidney, which contributes to the regulation of resistance vessel tone and renal blood flow under basal or stimulated conditions. It is assumed that the endothelium is the source of these mediators, but this might not be the case.

Arachidonic acid (the precursor of prostaglandins) has not been given systemically to humans, but in some animal species it causes renal vasodilatation, that is blocked by NSAIDs, suggesting metabolism of arachidonic acid to dilator prostaglandins (Dibona, 1998). In humans, infusion of exogenous prostaglandins (PGE_2, prostacyclin and PGA) causes renal vasodilatation (Dibona, 1998). The role of endogenous prostaglandins has been determined in humans by inhibition of COX isoenzymes using non-steroidal anti-inflammatory drugs (NSAIDs). In healthy subjects *in vivo*, there is a small short term reduction in renal blood flow whereas long-term use of NSAIDs has no significant effect on renal blood flow. It is not clear how the kidney compensates during chronic COX inhibition (Dibona, 1998; Sedor *et al.*, 1986). The effects of NSAIDs are more marked following intra-vascular volume depletion when renal blood flow falls by 10–20% (without changes in glomerular filtration rate) (Scharschmidt *et al.*, 1986). It is possible that hypovolaemia causes neurohormonal activation, and the resulting renal effects of noradrenaline or angiotensin II are modulated by stimulation of dilator prostaglandins. In addition, there is evidence in animal models that the effects of inhibition of prostaglandin synthesis are also more marked in the presence of NO synthase inhibition (Romero *et al.*, 1992), suggesting that prostaglandin production might compensate for loss of NO-medi-ated dilatation. Whether this is also true in humans remains to be determined.

There is compelling evidence in humans that NO modulates vascular tone in the renal bed; human isolated vessels exhibit endothelium-dependent relaxation to agonists such as acetylcholine or NO donors *in vitro* (Lüscher *et al.*, 1987) and *in vivo* (Elkayam *et al.*, 1996). Moreover, inhibition of NO synthase reduces renal blood flow and glomerular filtration rate *in vivo* (Bech *et al.*, 1996; Wolzt *et al.*, 1997), suggesting that endogenous NO exerts basal vasodilatation of the resistance vasculature of the kidney.

Endothelin-1 contracts afferent and efferent arterioles equally *in vitro* (Edwards *et al.*, 1990), and thus reduces both renal plasma flow and glomerular filtration rate *in vivo* in humans (Sorensen *et al.*, 1994; Rabelink *et al.*, 1994). Endothelin-1-induced renal vaso-constriction involves ET_B receptors in the rat (Pollock *et al.*, 1993) but mainly ET_A receptors in the dog and human (Brookes *et al.*, 1994; Maguire *et al.*, 1994). Renal vasodilatation occurs following infusion of BQ-123 in dogs, suggesting that vascular generation of endothelin-1 contributes to basal renal vascular tone (Pollock *et al.*, 1993). In humans, endothelin receptor antagonists decrease effective renal vascular resistance by approximately 10%, without changing GFR (Webb *et al.*, 1998, Ferro *et al.*, 1997), suggesting that the predominant effect of endogenously generated endothelin-1 is on the efferent arteriole. These data support a physiological vasoconstrictor role for endogenous endothelin-1 through activation of ET_A receptors in human kidney.

Regulation of Sodium Excretion

In humans, the dilator prostaglandins PGE_2 and prostacyclin cause natriuresis, which might be due to their vasodilator action, or a direct effect on tubular sodium transport (Sedor *et al.*, 1986). NSAIDs reduce sodium excretion in healthy volunteers in the short term, but this does not persist with chronic usage (Sedor *et al.*, 1986). These effects of NSAIDs are more marked in the presence of sodium depletion (Sedor *et al.*, 1986). Whether sodium retention explains the pressor effects of NSAIDs in certain patient groups remains to be determined (Johnson, 1997). Other direct effects of prostaglandins on tubular function have been suggested (such as inhibition of the activity of ADH) (Sedor *et al.*, 1986), but the low incidence of electrolyte disturbance caused by NSAIDs in humans suggests that these are of minor functional importance *in vivo*.

In humans, inhibition of endogenous NO generation reduces sodium excretion (Bech *et al.*, 1996) but the mechanism of these effects have yet to be established. However, several possibilities have been suggested by animal studies. NO could have direct effects on the proximal tubule and the collecting duct to promote sodium and water loss. NO might also promote natriuresis by increasing medullary blood flow and interstitial pressure. Local inhibition of medullary synthesis of NO can lead to sodium retention and hypertension in the absence of any direct effect on the systemic vasculature (Mattson *et al.*, 1994). Micropuncture studies of animal isolated nephrons have also implicated NO synthesised by the macula densa in the modulation of the tubulo-glomerular feed-back (TGF). This negative feedback mechanism regulates sodium excretion by eliciting afferent arteriolar constriction if the sodium content of the distal tubular fluid is high, and this reduces the delivery of sodium to the tubule. NO inhibits afferent arteriolar constriction, and inhibition of NO could increase the sensitivity of this mechanism to reduce natriuresis (Wilcox *et al.*, 1992). The effects of NO on renin secretion are complex, with evidence from *in vitro* studies of stimulation and inhibition of renin secretion by NO (Reid *et al.*, 1995). However most studies of intact animals suggest that NO stimulates renin secretion (Kurtz *et al.*, 1998). In humans inhibition of NO synthesis has no effect on plasma renin in sodium-replete individuals (Dijkhorstoei *et al.*, 1998), but blocks the rise in renin that occurs following sodium depletion (Lee *et al.*, 1996). It is therefore possible that the renin-angiotensin system provides feedback control of the natriuretic effects of NO.

Several lines of animal evidence suggest an important role for renal tubular endothelin-1 in modulation of sodium and water excretion. Endothelin-1 blocks sodium and water reabsorption by inhibiting tubular Na^+/K^+-ATPase activity in the proximal tubule and collecting duct (Zeidel *et al.*, 1989), and by inhibiting the effects of ADH on tubular osmotic permeability in the collecting duct (Tomita *et al.*, 1990; Oishi *et al.*, 1991). In addition, the cAMP response of rat collecting duct cells to ADH is potentiated in the presence of specific endothelin-1 antisera, suggesting that endogenous production of endothelin-1 tonically inhibits ADH responses (Kohan & Padilla, 1993). These tubular effects also occur with ET_B receptor agonists and are not blocked by the ET_A receptor antagonist, BQ-123, suggesting that they are mediated by ET_B receptors (Kohan *et al.*, 1993; Yukimura *et al.*, 1994). Involvement of ET_B receptors is supported by the finding that ET_B knockout mice have hypertension secondary to renal sodium retention (Webb *et al.*, 1998). Taken together, these findings suggest that locally generated endothelin-1 plays a tonic physiological role in regulation of salt and water transport in the terminal nephron, with increases in local generation promoting natriuresis and diuresis via activation of ET_B receptors. When administered to humans endothelin-1 causes anti-natriuresis, presumably because ET_A receptor mediated vasoconstriction to exogenous endothelin-1 outweighs any diuretic or natriuretic effect mediated by tubular ET_B receptors (Rabelink *et al.*, 1994; Sorensen *et al.*, 1994; Bijlsma *et al.*, 1995). There are no data on the effects of selective ET_B receptor antagonism in humans.

In summary, inhibition of prostaglandin or NO synthesis, and blockade of endothelin receptors has marked functional consequences on renal blood flow and tubular function. The effects of prostaglandin synthesis inhibition are transitory, and this emphasises the requirement for chronic studies of inhibition of NO and endothelin pathways in humans. Moreover, studies of the effects of combined inhibition of endothelial mediators in humans will increase understanding of the possible interaction of these systems within the kidney.

ENDOTHELIAL MEDIATORS IN ANIMAL MODELS OF RENAL DISEASE

Methodological difficulties handicap detailed investigation of the role of endothelial mediators in human kidney disease. Most of the human studies to date have concentrated on endothelial function in the peripheral vasculature. However, the results of animal studies have implicated abnormalities of the renal NO, prostaglandin and endothelin pathways.

Increased production of constrictor and dilator prostanoids has been reported in experimental glomerulonephritis, acute renal failure, ureteric obstruction and renal transplant rejection (Coleman *et al.*, 1990), but it is unclear if they have a significant pathophysiological role; neither thromboxane antagonists nor NSAIDs have been shown to consistently improve renal function in these models (Stork *et al.*, 1986). Indeed, infusion of dilator prostaglandins has been shown to improve renal function in animal models possibly by increasing renal blood flow or by modulating the immune response (Stork *et al.*, 1986).

Reduced NO-mediated effects in the kidney have been documented in a variety of experimental renal pathologies, including salt-sensitive hypertension, obstructive nephropathy, subtotal nephrectomy, glomerular thrombosis in pregnancy and cyclosporin nephropathy (Reyes *et al.*, 1994; Wagner *et al.*, 1995). Whether the systemic NO pathway is abnormal in animal models of chronic renal failure is unclear, with evidence for normal (Wagner *et al.*, 1995), increased (Ye *et al.*, 1997; Mendez *et al.*, 1998; Aiello *et al.*, 1997) or reduced (Vaziri *et al.*, 1998) activity of the pathway. In many of these conditions, renal dysfunction is partially reversed by exogenous L-arginine, suggesting that absolute or relative substrate deficiency has a pathogenic role (Reyes *et al.*, 1994). In contrast, increased production of NO secondary to iNOS induction, have been described in glomerular inflammation (Cattell *et al.*, 1993). High concentrations of NO in these conditions might be cytotoxic, and in some studies inhibition of NO synthesis has a protective effect (Millar *et al.*, 1997).

Several factors should be considered in interpreting the results of studies of endothelin in renal disease. First, the diverse actions of endogenous endothelin-1 on renal function mean that any changes in endothelin-1 activity in models of renal disease must be interpreted cautiously. Second, the kidney is an important site for clearance of endothelin-1 (Kohno *et al.*, 1989). Thus, increases in plasma concentrations in renal disease may be due to reduced clearance of endothelin rather than increased generation. Third, urinary endothelin generation is more likely to reflect renal tubule rather than vascular generation of endothelin. Given that renal tubule generation of endothelin-1 appears to cause natriuresis, increases in urine endothelin-1 generation may be beneficial. Fourth, interpretation of changes in sensitivity to exogenous endothelin-1 may be confounded by the fact that endothelin receptor number is downregulated by increased endothelin-1 concentrations (Hirata *et al.*, 1988). Vascular responses to exogenous endothelin-1 may also be altered by vascular remodeling. Fifth, endothelin-1 appears to be primarily a locally acting paracrine substance rather than a circulating endocrine hormone. Venous plasma endothelin-1 concentrations can only be used as an approximation for endothelial synthesis of the peptide, though circulating concentrations of big endothelin may more accurately reflect endothelin-1 generation (Plumpton *et al.*, 1995; Plumpton *et al.*, 1996).

Increased circulating endothelin-1 concentrations and local endothelin-1 expression have been reported in animal models of acute renal failure (secondary to hypoxia, septic shock or radiocontrast media) (Shibouta *et al.*, 1990; Marguiles *et al.*, 1991; Firth & Ratcliffe, 1992; Krause *et al.*, 1997). Renal vasoconstriction following ischaemia is substantially ameliorated by administration of endothelin receptor antagonists (Gellai *et al.*, 1994; Chan *et al.*, 1994; Clozel *et al.*, 1993). Interestingly, in dogs, the ET_A antagonist BQ-123 does not prevent renal vasoconstriction after ischaemia (Stingo *et al.*, 1993; Brooks *et al.*, 1994), even though ET_A receptors appear to predominate in the canine renal vasculature (Karet & Davenport 1994). Ischaemia-induced renal dysfunction in the dog is prevented by $ET_{A/B}$ antagonists such as SB 209670 and L-754,142 (Brooks *et al.*, 1994; Krause *et al.*, 1997). The vasoconstrictor effects of endothelin-1 are a plausible mechanism to underlie its contribution to the pathophysiology of acute renal failure. However, the greater beneficial effects of the combined $ET_{A/B}$ antagonist in a canine model suggests that endothelin-1 may also act on non-vascular ET_B receptors, including those situated on tubular epithelial cells. There is also evidence that endothelin-1 may promote the marked neutrophil accumulation observed in renal ischaemia-reperfusion injury through induction of the CD18 integrin antigen (Espinosa et al., 1996).

Animal models of chronic renal failure are also associated with increased endothelin-1 expression (Brooks et al., 1991; Benigni et al., 1991; Orisio et al., 1993; Nakamura et al., 1995). Chronic administration of the ET_A receptor antagonist FR139317 or of the $ET_{A/B}$ antagonist bosentan prevents the development of hypertension, glomerular damage and renal insufficiency in a reduced renal mass model in rats (Benigni et al., 1993; Benigni et al., 1996). It is not known whether $ET_{A/B}$ blockade is superior to ET_A receptor blockade in prevention of progressive renal damage. Endothelin-1 might also contribute to renal damage in experimental hypertension, because treatment with endothelin antagonists reduces the rate of deterioration of renal function (Okada et al., 1995; Karam et al., 1996; Kohno et al., 1997).

ENDOTHELIAL MEDIATORS IN HUMAN DISEASE

In humans with renal disease there is evidence to suggest both increased and decreased effects of endothelial mediators locally and systemically (Figure 11.2).

Prostaglandins

There is increased urinary excretion of dilator and constrictor prostaglandins in the urine of patients with glomerulonephritis, suggesting increased renal production (Coleman et al., 1990; Patrono et al., 1986). Functionally, the dilator effect of renal prostaglandins predominates in disease states, because inhibition of COX with NSAIDs causes greater reductions in renal blood flow and glomerular filtration rate in patients with nephrotic syndrome (Vriesendorp et al., 1986)or chronic renal impairment (Patrono et al., 1986) than in healthy controls. Although NSA IDs reduce proteinuria in nephrotic syndrome, this is probably due to reduced renal blood flow and glomerular filtration, rather than an immunomodulatory effect (Vriesendorp et al., 1986).

Abnormalities of the NO Pathway

Functional studies of the peripheral vasculature suggest that there is reduced endothelium-dependent dilatation of veins (Hand et al., 1998) and conduit arteries (Kari et al., 1997) in the systemic vasculature of patients with renal failure, consistent with reduced NO-mediated effects. Whether the NO pathway in the kidney is abnormal in human renal disease remains to be determined. The direct effects of L-arginine on the renal vasculature has been used to assess endothelial function, based on the assumption that any effect of L-arginine is due to stimulation of NO production (Gaston et al., 1995; Andres et al., 1997; Campo et al., 1996). In healthy subjects, L-arginine increases renal blood flow, and glomerular filtration, effects that are diminished in hypertensive subjects (Higashi et al., 1997). However, the large doses used in most of these studies are likely to have non-specific effects on tissues. When given to achieve similar concentrations, D-arginine (which is not a substrate for NO production) also causes vasodilatation (Calver et al., 1991; MacAllister

RENAL PHYSIOLOGY		RENAL PATHOLOGY	
NO Prostaglandins ➡ vasodilatation		NO-mediated dilatation ⬇	
		Prostaglandin-mediated dilatation ⬆	
Endothelin-1 ➡ vasoconstriction		Endothelin concentration ⬆	
NO Prostaglandins ➡ natriuresis Endothelin		NO/prostaglandin-mediated inflammation ⬆	

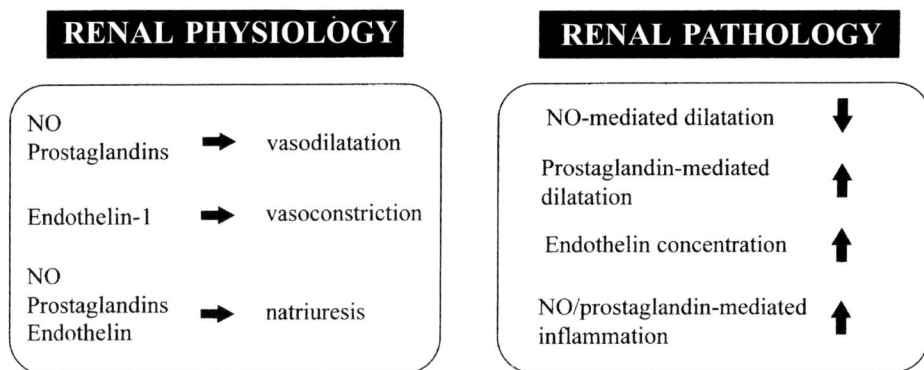

Figure 11.2 Physiological and pathophysiological roles for endothelial mediators. Alteration in the activity of the NO, prostaglandin or endothelin pathways has been documented in the kidney and the systemic vasculature of patients with renal disease.

et al., 1995; Amodeo *et al.*, 1996). Furthermore L-arginine has other NO-independent effects that could cause vasodilatation (MacAllister *et al.*, 1995; Higashi *et al.*, 1995). Therefore, it is difficult to draw clear conclusions about the NO pathway from clinical studies in which high doses of L-arginine have been used.

Impaired renal function *per se* might result in widespread effects on systemic NO pathways due to the accumulation of guanidino compounds that inhibit all three NOS isoforms. These include analogues of L-arginine, such as L-NMMA and N^GN^Gdimethyl-L-arginine (ADMA) (Vallance *et al.*, 1992). Plasma concentrations of ADMA are increased in patients with chronic renal failure (Arese *et al.*, 1995; MacAllister *et al.*, 1996), and the ratio of ADMA to L-arginine is increased approximately four-fold (MacAllister *et al.*, 1996). Therefore, it is possible that the increase in ADMA concentration could be responsible for inhibition of NO synthesis in the uraemic state. ADMA might exert direct effects on systemic vascular resistance, or because it is concentrated and excreted by the kidney, it might have local effects on renal function that could reduce sodium excretion and cause volume-dependent hypertension (Ito, 1995). The concentrations of L-NMMA circulating in healthy subjects are much lower than those of ADMA, but increased concentrations have been reported recently in patients on dialysis (Mendes-Ribeiro *et al.*, 1996). Of the other putative uraemic toxins, methylguanidine (MacAllister *et al.*, 1994) and guanidine (Sorrentino *et al.*, 1997) have been shown to be NOS inhibitors, and it is possible that uraemic plasma contains a cocktail of NOS inhibitors which act in concert to produce chronic low level inhibition of NO synthesis (Arese *et al.*, 1995).

There is also evidence for increased production of NO in humans with uraemia. Increased platelet derived NO has been implicated in the bleeding tendency of uraemia (Noris *et al.*, 1993). Induction of iNOS in immune cells has been detected in glomerulonephritis, (Kashem *et al.*, 1996), urinary tract infection, and (Wheeler *et al.*, 1997; Smith *et al.*, 1996) and renal transplant rejection (Smith *et al.*, 1996). Moreover, nitrotyrosine (a putative marker for elevated NO concentrations) has been observed in tissue from transplanted kidneys following chronic rejection (MacMillan-Crow *et al.*, 1996).

Endothelin

Circulating plasma concentrations of endothelin are increased in patients with post-ischaemic renal failure (Tomita *et al.*, 1989; Moore *et al.*, 1992, septic shock (Weitzberg *et al.*, 1991; Sanai *et al.*, 1995), and renal failure associated with radio contrast media (Marguiles *et al.*, 1991; Clarke *et al.*, 1997). Plasma concentrations of immunoreactive endothelin are also increased in patients with chronic renal failure, and are inversely proportional to renal function (Koyama *et al.*, 1989; Warrens *et al.*, 1990). Selective assays reveal that this increase is due to marked elevations in plasma endothelin-1 concentrations with little or no change in big endothelin-1 concentrations (Ferro *et al.*, 1998). Such elevations in mature endothelin-1, without changes in big endothelin-1, are most likely due to decreased renal clearance of endothelin-1 (Kohno *et al.*, 1989). Even so, in some cases, circulating endothelin-1 concentrations may reach levels that would allow endocrine actions for the peptide (Cockcroft & Webb 1989). For example, endothelin-1 concentrations in haemodialysis patients are positively correlated with arterial pressure (Miyauchi *et al.*, 1991; Tsunoda *et al.*, 1991), in contrast to essential hypertension (Davenport *et al.*, 1990). In addition to potential endocrine effects of high circulating endothelin-1 concentrations, local tissue endothelin-1 generation may also be increased in the kidney of patients with chronic renal failure, and this might explain the five-fold increase in immunoreactive endothelin in the urine of patients with chronic renal failure (Ohta *et al.*, 1991). It is possible that increased urinary endothelin excretion merely reflects a renal tubular homeostatic response to decreased sodium reabsorption. Plasma immunoreactive endothelin-1 concentrations are increased after intravenous but not subcutaneous erythropoietin administration (Carlini *et al.*, 1993; Hand *et al.*, 1995; Portoles *et al.*, 1997).

It is not possible to ascribe definite pathogenic roles to endothelin on the basis of biochemistry alone, and functional studies of patients with renal disease suggest that activity of the endothelin pathway is reduced or normal, but not increased. Although the sensitivity of veins to endothelin-1 is normal in normotensive patients with chronic renal failure, responses are reduced in hypertensive patients (Hand *et al.*, 1994). This might reflect endothelin receptor down-regulation secondary to increased concentration of endothelin-1. Patients with chronic renal failure also exhibit impaired forearm vasodilatation to brachial artery infusion of the ET_A antagonist BQ-123, suggesting reduced endothelin-1-dependent constriction (Hand *et al.*, 1995a). This could be due to ET_A receptor down-regulation, or endothelial dysfunction leading to decreased tonic ET_B receptor-mediated nitric oxide formation. Although regional vasodilatation following endothelin receptor blockade is attenuated in chronic renal failure, systemic administration of the $ET_{A/B}$ receptor antagonist TAK-044 to patients with chronic renal failure reduces blood pressure by 11% and renal vascular resistance by 10% (Ferro *et al.*, 1998). However, these patients do not appear to be more sensitive to endothelin-receptor antagonists than healthy controls (Schmetterer *et al.*, 1998).

THERAPEUTIC POSSIBILITIES

Prostaglandins

Both PGA and prostacyclin have been given to patients with acute renal failure, and although they caused dilatation, there was no improvement in renal function (Vincenti *et al.*, 1978; Ladefoged *et al.*, 1970). When used in inflammatory renal diseases, NSAIDs reduce glomerular filtration and worsen renal function, effects which have discouraged their use in these diseases (Patrono *et al.*, 1986). It has been suggested that reduced filtration might slow the progression of renal disease, but this has yet to be investigated. It is unclear if selective COX-2 inhibitors will offer any advantages over currently available NSAIDs. If manipulation of the renal prostaglandin pathways has met with limited enthusiasm as a therapeutic step, it is likely that there are beneficial effects of low-dose aspirin on the systemic vasculature of patients with renal failure.

NO

Given the evidence for the pathophysiological roles for reduced or increased NO production in experimental renal disease, there might be therapeutic advantages in increasing or decreasing the bioavailability of NO in renal disease. Dietary intervention with L-arginine has been shown to improve renal function in a variety of animal models of acute and chronic renal disease (Reyes *et al.*, 1994). This is probably a NO-mediated phenomenon, because the dosing schedules used in these studies avoid the high concentrations of L-arginine that have non-stereospecific effects. There are no studies of the effects of L-arginine in acute renal failure in humans. Nor are there any studies of chronic supplementation of L-arginine in chronic renal disease, although one study of low dose L-arginine given acutely has been shown to reduce proteinuria in patients with glomerulonephritis excretion (Wolf *et al.*, 1995). In the systemic vasculature, NO-mediated dilatation might be augmented by provision of L-arginine to antagonise the possible effects of L-arginine analogues. In patients with chronic renal failure, low-dose L-arginine (but not D-arginine) improves endothelium-dependent dilatation of human veins. Whether chronic therapy with L-arginine or NO donors might augment NO-mediated effects to improve blood pressure control and retard atherogenesis remains to be determined. In summary, given the available animal and human data there is a clear rationale for dose-ranging studies of acute and chronic L-arginine supplementation in human renal disease, to examine effects on renal and systemic vascular function. An alternative to L-arginine might be to use NO donors, but this approach has yet to be systematically investigated in animal and clinical renal disease. Selective inhibition of iNOS has promise as an anti-inflammatory therapy but advances in this area await the development of useful selective iNOS inhibitors that can be tested in renal inflammation.

Endothelin

The vasodilator and hypotensive effects of endothelin receptor antagonists in both healthy subjects (Haynes *et al.*, 1994; Haynes *et al.*, 1996) and patients with renal impairment (Hand *et al.*, 1995a; Ferro *et al.*, 1998) indicate that these agents may be therapeutically useful in chronic renal failure. In one small study of patients with acute renal failure, BQ-123 caused renal vasodilatation at doses that did not alter systemic blood pressure, but did not improve renal function (Soper *et al.*, 1998).

SUMMARY

Endothelial mediators have important physiological roles in the human kidney. NO and prostaglandins have broadly similar dilator and natriuretic roles, that contrast with the largely vasoconstrictor and anti-natriuretic effects of endothelins. In certain renal diseases of humans there is evidence for increased prostaglandin production that contributes to the maintenance of renal blood flow. In addition, there is evidence for reduced activity of the NO pathway in the systemic vasculature of patients with chronic renal failure, possibly due to reduced clearance of NOS inhibitors as renal function declines. However, it is unclear if the renal NO pathway is abnormal in renal diseases or if the pro-inflammatory effects of NO might be involved in renal disease. The role of endothelin-1 in renal disease is complex. In the kidney, there is increased endothelin-1 generation and receptor expression that could contribute to the progression of renal injury. Systemically, there are increased circulating concentrations of endothelin-1 caused by impaired renal clearance of the peptide that might cause hypertension. However, functional studies to date do not provide evidence for increased endothelin-mediated constriction in renal disease. Nonetheless it is possible that manipulation of endothelial mediators, either singly in concert, might form the basis for new therapies in the future. Such therapies will need to improve renal function or to treat the complications of atherosclerosis, the commonest cause of death in patients with renal disease.

REFERENCES

Aiello, S., Noris, M., Todeschini, M., Zappella, S., Foglieni, C., Benigni, A., *et al.* (1997) Renal and systemic nitric oxide synthesis in rats with renal mass reduction. *Kidney Int.*, **52**, 171–181.

Amodeo, C., Dichtchekenian, V., Santos, E.A., Heimann, J.C. and Zatz, R. (1996) Disparate NO2/NO3 excretion after acute arginine infusion in essential hypertension on salt restriction and overload. A non-stereospecific arginine hypotensive response? *J. Am. Soc. Nephrol.*, **7**, s1546.

Andres, A., Morales, J.M., Praga, M., Campo, C., Lahera, V., Garcia-Robles, R., *et al.* (1997) L-arginine reverses the antinatriuretic effect of cyclosporin in renal transplant patients. *Nephrol. Dial. Transplant.*, **12**, 1437–1440.

Arese, M., Strasly, M., Costamagna, C., Ghigo, D., MacAllister, R., Verzetti, G., *et al.* (1995) Regulation of nitric oxide synthesis in uraemia. *Nephrol. Dial. Transplant.*, **10**, 1386–1397.

Bachman, S., Bosse, H.M. and Mundel, P. (1995) Topography of nitric oxide synthesis by localising constitutive NO synthases in mammalian kidney. *Am. J. Physiol.*, **268**, F885–F898.

Bartholomeusz, B., Hardy, K.J., Nelson, A.S. and Phillips, P.A. (1996) Bosentan ameliorates cyclosporin A-induced hypertension in rats and primates. *Hypertension*, **27**, 1341–1345.

Bech, J.N., Nielsen, C.B. and Pedersen, E.B. (1996) Effects of systemic NO synthesis inhibition on RPF, GFR, U_{Na}, and vasoactive hormones in healthy humans. *Am. J. Physiol.*, **270**, F845–F851.

Benigni, A., Perico, N., Gaspari, F., Zoja, C., Bellizi, L., Gabanelli, M., *et al.* (1991) Increased renal endothelin production in rats with reduced renal mass. *Am. J. Physiol.*, **260**, F331–F339.

Benigni, A., Zoja, C., Corna, D. Orisio, S., Longaretti, L., Bertani, T. *et al.* (1993) Specific endothelin subtype A receptor antagonist protects against injury in renal disease progression. *Kidney Int.*, **44**, 440–445.

Benigni, A., Zola, C., Corna, D., Orisio, S., Facchinetti, D., Benati, L., *et al.* (1996) Blocking both type A and B endothelin receptors in the kidney attenuates renal injury and prolongs survival in rats with remnant kidney. *Am. J. Kidney Dis.*, **27**, 416–423.

Bijlsma , J.A., Rabelink, A.J., Kaasjager, K.A.H., Koomans, H.A. (1995). L-arginine does not prevent the renal effects of endothelin in humans. *J. Am. Soc. Nephrol.*, **5**, 1508–1516.

Bird, J.E., Giancarli, M.R., Megill, J.R. and Durham, S.K. (1996). Effects of endothelin in radiocontrast-induced nephropathy in rats are mediated through endothelin-A receptors. *J.Am. Soc. Nephrol.*, **7**, 1153–1157.

Bode-Boger, S.M., Boger, R.H., Kuhn, M., Radermacher, J. and Frolich, J.C. (1996) Recombinant human erythropoietin enhances vasoconstrictor tone via endothelin-1 and constrictor prostanoids. *Kidney Int.*, **50**, 1255–1261.

Brooks, D.P. and Contino, L.C. (1995) Prevention of cyclosporin A-induced renal vasoconstriction by the endothelin receptor antagonist SB 209670. *Eur. J. Pharmacol.*, **294**, 571–576.

Brooks, D.P. and DePalma, P.D. (1996) Blockade of radiocontrast-induced nephrotoxicity by the endothelin receptor antagonist, SB 209670. *Nephron*, **72**, 629–636.

Brooks, D.P., Contino, L.C., Storer, B. and Ohlstein, E.H. (1991) Increased endothelin excretion in rats with renal failure induced by partial nephrectomy. *Br. J. Pharmacol.*, **104**, 987–989.

Brooks, D.P., DePalma, P.D., Gellai, M., Nambi, P., Ohlstein, E.H., Elliott, J.D., *et al.* (1994). Nonpeptide endothelin receptor antagonists. III. Effect of SB 209670 and BQ123 on acute renal failure in anesthetized dogs. *J. Pharmacol. Exp. Ther.*, **271**, 769–775.

Brooks, D.P., DePalma, P.D., Pullen, M. and Nambi, P. (1994). Characterization of canine renal endothelin receptor subtypes and their function. *J. Pharmacol. Exp. Ther.*, **268**, 1091–1097.

Bunchman, T.E. and Brookshire, C.A. (1991) Cyclosporine-induced synthesis of endothelin by cultured human endothelial cells. *J. Clin. Invest.*, **88**, 310–314.

Calver, A., Collier, J. and Vallance, P. (1991) Dilator actions of arginine in human peripheral vasculature. *Clin. Sci.*, **81**, 695–700.

Campo, C., Lahera, V., Garcia-Robles, R., Cachofeiro, V., Alcazar, J.M., Andres, A., *et al.* (1996) Aging abolishes the renal response to L-arginine infusion in essential hypertension. *Kidney Int.*, **55**, S126–S128.

Carlini, R., Obialo, C.I. and Rothstein, M. (1993) Intravenous erythropoietin (rHuEPO) administration increases plasma endothelin and blood pressure in hemodialysis patients. *Am. J. Hypertens.*, **6**, 103–107.

Cattell, V. and Cook, H.T. (1993) Nitric oxide: role in the physiology and pathophysiology of the glomerulus. *Exp. Nephrol.*, **1**, 265–280.

Chan, L., Chittinandana, A., Shapiro, J.I., Shanley, P.F. and Schrier, R.W. (1994) Effect of an endothelin-receptor antagonist on ischemic acute renal failure. *Am. J. Physiol.*, **266**, F135–F138.

Clark, B.A., Kim, D. and Epstein, F.H. (1997) Endothelin and atrial natriuretic peptide levels following radiocontrast exposure in humans. *Am. J. Kidney Dis.*, **30**, 82–86.

Clozel M., Loffler B.M. and Gloor, H. (1991) Relative preservation of the responsiveness to endothelin-1 during reperfusion following renal ischemia in the rat. *J. Cardiovasc. Pharmacol.*, **17** (suppl 7), S313–S315.

Clozel, M., Breu, V., Burri, K., Cassal, J.M., Fischli, W., Gray, G.A., *et al.* (1993) Pathophysiological role of endothelin revealed by the first orally active endothelin receptor antagonist. *Nature*, **365**, 759–761.

Coleman, R.A., Kennedy, I., Humphrey, P.P.A., Bunce, K. and Lumley, P. (1990) Prostanoids and their receptors. In *Comprehensive Medicinal Chemistry*, edited by C. Hansch, P.G. Samnes, and J.B. Taylor,. pp. 643–714. Oxford: Pergamon Press

Coleman, R.A., Smith, W.L. and Narumiya, S. (1994) VIII Internatioal Union of Pharmacology classification of prostanoid receptors: properties distribution and structure of the receptors and their subtypes. *Pharmacol. Rev.*, **46**, 205–229.

Cooke, J.P. and Dzau, V.J. (1997) Derangements of the nitric oxide synthase pathway, L-arginine and cardio-vascular diseases. *Circulation*, **96**, 379–382.

Davenport, A.P., Ashby, M.J., Easton, P., Ella, S., Bedford, J., Dickerson, C., *et al.* (1990) A sensitive radio-immunoassay measuring endothelin-like immunoreactivity in human plasma: comparison of levels in patients with essential hypertension and normotensive control subjects. *Clin. Sci.*, **78**, 261–264.

Dibona, G.F. (1998) Prostaglandins and non-steroidal anti-inflammatory drugs. *Am J Med*, **80** (suppl 1A), 12–21.

Dijkhorstoei, L.T. and Koomans, H.A. (1998) Effects of a nitric oxide synthesis inhibitor on renal sodium handling and diluting capacity in humans. *Nephrol. Dial. Transplant.*, **13**, 587–593.

Edwards, R.M., Trizna, W., Ohlstein, E.H. (1990) Renal microvascular effects of endothelin. *Am. J. Physiol.*, **259**, F217–F221.

Elkayam, U., Cohen, G., Gogia, H., Mehra, A., Johnson, J.V. and Chandraratna, P.A. (1996) Renal vasodilatory effect of endothelial stimulation in patients with chronic congestive heart failure. *J. Amer. Coll. Cardiol.*, **28**, 176–182.

Eriksson, L.O., Larsson, B., Hedlund, H. and Andersson, K.E. (1990) Prostaglandin E2 binding sites in human renal tissue: characterisation and localisation by radioligand binding and autoradiography. *Acta. Physiol. Scand.*, **139**, 393–404.

Espinosa, G., Lopez-Farre, A., Cernadas, M.R., Manzarbeitia, F., Tan, D., Digiuni, E., *et al.* (1996) Role of endothelin in the pathophysiology of renal ischemia-reperfusion in normal rabbits. *Kidney Int.*, **50**, 776–782.

Ferro, C.J., Strachan, F.E., Cumming, A.D., Plumpton, C., Davenport, A.P., Haynes, W.G. *et al.* (1998) Actions of endothelin receptor antagonism in patients with chronic renal failure. *Proceedings of the 5th Endothelin Meeting*, London, 1997.

Firth, J.D. and Ratcliffe, P.J. (1992) Organ distribution of three rat endothelin messenger RNAs and the effects of ischemia on renal gene expression. *J. Clin. Invest.*, **90**, 1023–1031.

Fogo, A., Hellings, S. E, Inagami, T. and Kon, V. (1992) Endothelin receptor antagonism is protective in *in vivo* acute cyclosporin toxicity. *Kidney Int.*, **42**, 770–774.

Fujita, K., Matsumura, Y., Miyazaki, Y., Hashimoto, N., Takaoka, M. and Morimoto, S. (1996) ET_A receptor-mediated role of endothelin in the kidney of DOCA-salt hypertensive rats. *Life. Sci.*, **58**, PL1–7.

Fukuda, K., Yanagida, T., Okuda, S., Tamaki, K., Ando, T. and Fujishima, M. (1996) Role of endothelin as a mitogen in experimental glomerulonephritis in rats. *Kidney Int.*, **49**, 1320–1329.

Gaston, R.S., Schlessinger, S.D., Sanders, P.W., Barker, C.V., Curtis, J.J. and Warnock, D.G. (1995) Cyclosporine inhibits the renal response to L-arginine in human kidney transplant recipients. *J. Am. Soc. Nephrol.*, **5**, 1426–1433.

Gellai, M., Jugus, M., Fletcher, T., DeWolf, R. and Nambi, P. (1994) Reversal of postischemic acute renal failure with a selective endothelin$_A$ receptor antagonist in the rat. *J. Clin. Invest.*, **93**, 900–906.

Goessl, C., Grozdanovic, Z., Knispel, H.H., Wegner, H.E. and Miller, K. (1995) Nitroxergic innervation of the human uterovesical junction. *Urol. Res.*, **23**, 189–192.

Gomez-Garre, D., Largo, R., Liu, X.H., Gutierrez, S., Lopez-Armada, M.J., Palacios, I., *et al.* (1996) An orally active ET_A/ET_B receptor antagonist ameliorates proteinuria and glomerular lesions in rats with proliferative nephritis. *Kidney Int.*, **50**, 962–972.

Hand, M.F., Haynes, W.G, Anderton, J.L. and Webb, D.J. (1995a) Endothelin-dependent basal vascular tone is decreased in chronic renal failure. *Kidney Int.*, **48**, 1001.

Hand, M.F., Haynes, W.G. and Webb, D.J. (1998) Hemodialysis and L-arginine, but not D-arginine, correct renal failure-associated endothelial dysfunction. *Kidney Int.*, **53**, 1068–1077.

Hand, M.F., Haynes, W.G., Anderton, J.L., Winney, R. and Webb, D.J. (1994) Responsiveness of veins to endothelin-1 in chronic renal failure. *Nephrol. Dial. Transplant.*, **9**, 1349–1350.

Hand, M.F., Haynes, W.G., Johnstone, H.A., Anderton, J.L. and Webb, D.J. (1995) Erythropoietin enhances vascular responsiveness to norepinephrine in renal failure. *Kidney Int.*, **48**, 806–813.

Haynes, W.G. and Webb, D.J. (1993) The endothelin family of peptides: local hormones with diverse roles in health and disease? *Clin. Sci.*, **84**, 485–500.

Haynes, W.G. and Webb, D.J. (1994) Contribution of endogenous generation of endothelin-1 to basal vascular tone. *Lancet*, **344**, 852–854.

Haynes, W.G., Ferro, C.E., O'Kane, K., Somerville, D., Lomax, C.L. and Webb, D.J. (1996) Systemic endothelin receptor blockade decreases peripheral vascular resistance and blood pressure in man. *Circulation*, **93**, 1860–1870.

Haynes, W.G., Webb, D.J. (1998) Endothelin as a regulator of cardiovascular function in health and disease. *J. Hypertens.* (in press).

Higashi, Y., Oshima, T., Ono, N., Hiraga, H., Yoshimura, M., Watanabe, M., *et al.* (1995) Intravenous administration of L-arginine inhibits angiotensin-converting enzyme in humans. *J. Clin. Endocrinol. Metab.*, **80**, 2198–2202.

Higashi, Y., Oshima, T., Ozono, R., Matsuura, H. and Kajiyama, G. (1997) Aging and severity of hypertension attenuate endothelium-dependent renal vascular relaxation in humans. *Hypertension*, **30**, 252–258.

Hirata, Y., Yoshimi, H., Takachi, S., Yanagisawa, M. and Masaki, T. (1988) Binding and receptor down-regulation of novel vasoconstrictor endothelin in cultured rat vascular smooth muscle cells. *FEBS Lett.*, **239**, 13–17.

Hocher, B., Rohmeiss, P., Zart, R., Diekmann, F., Vogt, V., Metz, D., *et al.* (1996) Function and expression of endothelin receptor subtypes in the kidneys of spontaneously hypertensive rats. *Cardiovasc. Res.*, **31**, 499–510.

Hocher, B., Thonereineke, C., Rohmeiss, P., Schmager, F., Slowinski, T., Burst, V., *et al.* (1997) Endothelin-1 transgenic mice develop glomerulosclerosis, interstitial fibrosis, and renal cysts but not hypertension. *J. Clin. Invest.*, **99**, 1380–1389.

Hoffman, A., Grossman, E., Goldstein, D.S., Gill, J.R. and Keiser, H.R. (1990) Low urinary levels of endothelin-1 in patients with essential hypertension. *J. Am. Soc. Nephrol.*, **1**, 417.

Howard PG, Plumpton C, Davenport AP (1992) Anatomical localisation and pharmacological activity of mature endothelins and their precursors in human vascular tissue. *J. Hypertens.*, **10**, 1379–1386.

Hughes, A.K., Cline, R.C. and Kohan, D.E. (1992) Alterations in renal endothelin-1 production in the spontaneously hypertensive rat. *Hypertension,* **20**, 666–673.

Inoue, A., Yanagisawa, M., Kimura, S., Kasuya, Y., Miyauchi, T., Goto, K., *et al.* (1989) The human endothelin family: three structurally and pharmacologically distinct isopeptides predicted by three separate genes. *Proc. Natl. Acad. Sci. USA*, **86**, 2863–2867.

Ito, S. (1995) Nitric oxide and the kidney. *Curr. Opin. Nephrol. Hypertens.*, **4**, 23–30.

Iversen, H.H., Ehren, I., Gustafsson, L.E., Adolfsson, J. and Wiklund, N.P. (1995) Modulation of smooth muscle activity by nitric oxide in the human upper urinary tract. *Urol. Res.*, **23**, 391–394.

Johnson, A.G. (1997) NSAIDs and increased blood pressure; what is the clinical significance? *Drug Safety*, **17**, 277–289.

Kanwar, S. and Kubes, P. (1995) Nitric oxide is an antiadhesive molecule for leukocytes. *New Horizons*, **3**, 93–104.

Karam, H., Heudes, D., Bruneval, P., Gonzales, M.F., Loffler, B.M., Clozel, M., *et al.* (1996) Endothelin antagonism in end-organ damage of spontaneously hypertensive rats: comparison with angiotensin-converting enzyme inhibition and calcium antagonism. *Hypertension*, **28**, 379–385.

Karet F.E. and Davenport, A.D. (1994) Endothelin and the human kidney: a potential target for new drugs. *Nephrol. Dial. Transplant.*, **9**, 465–468.

Karet FE, Charnock-Jones DS, Harrison-Woolrych ML, O'Reilly G, Davenport AP. Smith SK. Quantification of mRNA in human tissue using fluorescent nested reverse-transcriptase polymerase chain reaction. *Anal. Biochem.* 220:384–90, 1994.

Karet FE. Endothelin peptides and receptors in human kidney. *Clin. Sci.* 91:267–273, 1996.

Kari, J.A., Donald, A.E., Vallance, D.T., Bruckdorfer, K.R., Leone, A., Mullen, M.J., *et al.* (1997) Physiology and biochemistry of endothelial function in children with chronic renal failure. *Kidney Int.*, **52**, 468–472.

Kashem, A., Endoh, M., Yano, N., Yamauchi, F., Nomoto, Y. and Sakai, H. (1996) Expression of inducible-NOS in human glomerulonephritis: the possible source is infiltrating monocytes/macrophages. *Kidney Int.*, **50**, 392–399.

Kitamura, K., Tanaka, T., Kato, J., Ogawa, T., Eto, T. and Tanaka, K. (1989) Immunoreactive endothelin in rat kidney inner medulla: marked decrease in spontaneously hypertensive rats. *Biochem. Biophys. Res. Commun.*, **162**, 38–44.

Kohan, D.E. and Fiedorek, F.T. (1991) Endothelin synthesis by the rat inner medullary collecting duct cells. *J. Am. Soc. Nephrol.*, **2**, 150–155.

Kohan, D.E. and Padilla, E. (1993) Osmolar regulation of endothelin-1 production by rat inner medullary collecting duct. *J. Clin. Invest.*, **91**, 1235–1240.

Kohan, D.E., Padilla, E. and Hughes, A.K. (1993) Endothelin B receptor mediates ET-1 effects on cAMP and PGE$_2$ accumulation in rat IMCD. *Am. J. Physiol.*, **265**, F670–F676.

Kohno, M., Murakawa, K., Yasunari, K., Yokokawa, K., Horio, T., Kurihara, N., *et al.* (1989) Prolonged blood pressure elevation after endothelin administration in bilaterally nephrectomized rats. *Metabolism*, **38**, 712–713.

Kohno, M., Yokokawa, K., Yasunari, K., Kano, H., Minami, M., Ueda, M., *et al.* (1997) Renoprotective effects of a combined endothelin type a type b receptor antagonist in experimental malignant hypertension. *Clin. Exp. Metab.*, **46**, 1032–1038.

Kömhoff, M., Gröne, H., Klein, T., Seyberth, H.W. and Nüsing, R.M. (1997) Localisation of cyclooxygenase-1 and -2 in adult and fetal human kidney: implication for renal function. *Am. J. Physiol.*, **272**, F460–F468.

Kon, V. and Badr, K.F. (1991) Biological actions and pathophysiologic significance of endothelin in the kidney. *Kidney Int.*, **40**, 1–12.

Kon, V., Sugiura, M., Inagami, T., Harvie, B.R., Ichikawa, I. and Hoover, R. L. (1990) Role of endothelin in cyclosporine-induced glomerular dysfunction. *Kidney Int.*, **37**, 1487–1491.

Kon, V., Yoshioka, T., Fogo. A. and Ichikawa, I. (1989) Glomerular actions of endothelin *in vivo. J. Clin. Invest.*, **83**, 1762–1767.

Kourembanas, S., Marsden, P.A., McQuillan, L.P. and Faller, D.V. (1991) Hypoxia induces endothelin gene expression and secretion in cultured human endothelium. *J. Clin. Invest.*, **88**, 1054–1057.

Koyama, H., Tabata, T., Nishzawa, Y., Inoue, T., Morii, H. and Yamaji, T. (1989) Plasma endothelin levels in patients with uraemia. *Lancet*, **333**, 991–992.

Krause, S.M., Walsh, T.F., Greenlee, W.J., Ranaei, R., Williams, D.L. Jr. and Kivlighn, S.D. (1997) Renal protection by a dual ETA/ETB endothelin antagonist, L-754,142, after aortic cross-clamping in the dog. *J. Am. Soc. Nephrol.*, **8**, 1061–1071

Kurtz, A., Gotz, K., Hamann, M. and Wagner, C. (1998) Stimulation of renin secretion by nitric oxide is mediated by phosphodiesterase 3. *Proc Natl. Acad Sci. USA*, **95**, 4743–4747.

Ladefoged, J. and Winkler, K. (1970) Haemodynamics in acute renal failure. *Scand. J. Clin. Lab. Invest.*, **26**, 83–87.

Lanese, D.M. and Conger, J.D. (1993) Effects of endothelin receptor antagonist on cyclosporin-induced vaso-constriction in isolated rat renal arterioles. *J. Clin. Invest.*, **91**, 2144–2149.

Lee, A.F.C., Kiely, D.G., Coutie, W. and Struthers, A.D. (1996) The renin response to frusemide in man is nitric oxide-dependent. *Br. J. Clin. Pharmacol.*, **42**, 652

Lopez-Ongil, S.L., Saura, M., Lamas, S., Rodriguez Puyol, M. and Rodriguez Puyol, D. (1996) Recombinant human erythropoietin does not regulate the expression of endothelin-1 and constitutive nitric oxide synthase in vascular endothelial cells. *Exp. Nephrol*, **4**, 37–42.

Lopez-Farre, A., Gomez-Garre, D., Bernabeu, F. and Lopez-Nova, J.M. (1991) A role for endothelin in the maintenance of post-ischaemic acute renal failure in the rat. *J. Physiol. (Lond.)*, **444**, 513–522.

Lüscher, T.F., Cooke, J.P., Houston, D.S., Neves, R.J. and Vanhoutte, P.M. (1987) Endothelium-dependent relaxations in human arteries. *Mayo. Clin. Proc.*, **62**, 601–606.

MacAllister, R.J., Calver, A.L., Collier, J., Edwards, C.M.B., Herreros, B., Nussey, S.S., *et al.* (1995) Vascular and hormonal response to arginine: provision of substrate for nitric oxide or non-specific effect? *Clin. Sci.*, **89**, 183–190.

MacAllister, R.J., Rambausek, M.H., Vallance, P., Williams, D., Hoffmann, K. and Ritz, E. (1996) Concentration of dimethyl-l-arginine in the plasma of patients with end-stage renal failure. *Nephrol. Dial. Transplant.*, **11**, 2449–2452.

MacAllister, R.J., Whitley, G.StJ. and Vallance, P. (1994) Effects of guanidino and uremic compounds on nitric oxide pathways. *Kidney Int.*, **45**, 737–742.

MacMillan-Crow, L.A., Crow, J.P., Kerby, J.D., Beckman, J.S. and Thompson, J.A. (1996) Nitration and inactivation of manganese superoxide dismutase in chronic rejection of human renal allografts. *Proc. Natl. Acad. Sci. USA*, **93**, 11853–11858.

Maguire, J.J., Kuc, R.E., O'Reilly, G. and Davenport A.P. (1994) Vasoconstrictor endothelin receptors characterised in human renal artery and vein in vitro. *Br. J. Pharmacol.*, **113**, 49–54.

Margulies, K.B., Hildebrand, F.L., Heublein, D.M. and Burnett, J.C. Jr. (1991) Radiocontrast increases plasma and urinary endothelin. *J. Am. Soc. Nephrol.*, **2**, 1041–1045.

Marsden, P.A., Dorfman, D.M., Collins, T., Brenner, B.M., Orkin, S.H. and Ballerman BJ. (1991) Regulated expression of endothelin 1 in glomerular capillary endothelial cells. *Am. J. Physiol.*, **261**, F117–F125.

Mattson, D.L., Lu, S., Nakanashi, K., Papanek, P.E. and Cowley, A.W., Jr. (1994) Effect of chronic renal medullary nitric oxide inhibition on blood pressure. *Am. J. Physiol.*, **266**, H1918–H1926.

McLay, J.S., Chatterjee, P.K., Mistry, S.K., Weerakody, R.P., Jardine, A.G., McKay, N.G., *et al.* (1995) Atrial natriuretic factor and angiotensin II stimulate nitric oxide release from human proximal tubular cells. *Clin. Sci.*, **89**, 527–531.

McLay, J.S., Chatterjee, P.K., Nicolson, A.G., Jardine, A.G., McKay, N.G., Ralston, S.H., *et al.* (1994) Nitric oxide production by human proximal tubular cells: a novel immnomodulatory mechanism? *Kidney Int.*, **46**, 1043–1049.

Mendes-Ribeiro, A.C., Roberts, N.B., Lane, C., Yaqoob, M. and Ellory, J.C. (1996) Accumulation of the endogenous L-arginine analogue N^Gmonomethyl-L-arginine in human end-stage renal failure patients on regular haemodialysis. *Exper. Physiol.*, **81**, 475–481.

Mendez, A., Fernandez, M., Barrios, Y., Lopez-Covella, I., Gonzalez-Mora, J.L., Del Rivero, M., *et al.* (1998) Constitutive NOS isoforms account for gastric mucosal NO overproduction in uremic rats. *Am. J. Physiol.*, **272**, G894–G901.

Millar, C.G.M. and Thiemermann, C. (1997) Intrarenal haemodynamics and renal dysfunction in endotoxaemia: effects of nitric oxide synthase inhibition. *Br. J. Pharmacol.*, **121**, 1824–1830.

Miyauchi, T., Suzuki, N., Kurihara, T., Yamaguchi, I., Sugishita, Y., Matsumoto, H., Goto, K. and Masaki, T. (1991) Endothelin-1 and endothelin-3 play different roles in acute and chronic alterations of blood pressure in patients with chronic hemodialysis. *Biochem. Biophys. Res. Commun.*, **178**, 276–281.

Moncada, S. and Higgs, A. (1993) The L-arginine-nitric oxide pathway. *N. Engl. J. Med.*, **329**, 2002–2012.

Moore, K., Wendon, J., Frazer, M., Karani, J., Williams, R. and Badr, K. (1992) Plasma endothelin immuno-reactivity in liver disease and the hepatorenal syndrome. *N. Engl. J. Med.,* **327**, 1774–1778.

Moutabarrik, A., Ishibashi, M., Fukunaga, M., Kameoka, H., Takano, Y., Kokado, Y., *et al.* (1991) FK 506 mechanism of nephrotoxicity: stimulatory effect on endothelin secretion by cultured kidney cells and tubular cell toxicity in vitro. *Tranplant. Proc.,* **23**, 3133–3136.

Nadler, S.P., Zimpelmann, J.A. and Hebert, R.L. (1992) Endothelin inhibits vasopressin-stimulated water per-meability in rat terminal inner medullary collecting duct. *J. Clin. Invest.,* **90**, 1458–1466.

Nagai, T., Akizawa, T., Nakashima, Y., Kohjiro, S., Nabeshima, K., Kanamori, N., *et al.* (1995) Effects of rHuEpo on cellular proliferation and endothelin-1 production in cultured endothelial cells. *Nephrol. Dial. Trans.,* **10**, 1814–1819.

Nakamura, T., Ebihara, I., Fukui, M., Tomino, Y. and Koide, H. (1996) Effect of a specific endothelin receptor A antagonist on glomerulonephritis of ddY mice with IgA nephropathy. *Nephron* **72**, 454–60.

Nakamura, T., Fukui, M., Ebihara, I., Osada, S., Takahashi, T., Tomino, Y. & Koide, H. Effects of a low protein diet on glomerular endothelin family gene expression in experimental focal glomerular sclerosis. *Clin. Sci.,* 88, 29–37, 1995

Nambi, P., Pullen, M., Contino, L.C. and Brooks, D.P. (1990). Upregulation of renal endothelin receptors in rats with cyclosporin A-induced nephrotoxicity. *Eur. J. Pharmacol.,* **187**, 113–116.

Nambi, P., Pullen, M., Jugus, M. and Gellai, M. (1993) Rat kidney endothelin receptors in ischemia-induced acute renal failure. *J. Pharmacol. Exp. Ther.,* **264**, 345–348.

Noris, M., Benigni, A., Boccardo, P., Aiello, S., Gaspari, F., Todeschini, M., *et al.* (1993) Enhanced nitric oxide synthesis in uremia: implications for platelet dysfunction and dialysis hypotension. *Kidney Int.,* **44**, 445–450.

Ohta, T., Hirata, Y., Shichiri, M., Kano, K., Emori, T., Tomita, K. and Marumo, F. (1991) Urinary excretion of endothelin-1 in normal subjects and patients with renal disease. *Kidney Int.,* **39**, 307–311.

Oishi, R., Monoguchi, H., Tomita, K. and Murumo, F. (1991) Endothelin-1 inhibits AVP stimulated osmotic water permeability in rat inner medullary collecting duct. *Am. J. Physiol.,* **261**, F951–F956.

Okada, M., Kobayashi, M., Maruyama, H., Takahashi, R., Ikemoto, F., Yano, M., *et al.* (1995) Effects of a selective endothelin A-receptor antagonist, BQ-123, in salt-loaded stroke-prone spontaneously hypertensive rats. *Clin. Exp. Pharmacol. Physiol.,* **22**, 763–768.

Orisio, S., Benigni, A., Bruzzi, I., Corna, D., Perico, N., Zoja, C., *et al.* (1993) Renal endothelin gene expression is increased in remnant kidney and correlates with disease progression. *Kidney Int.,* **43**, 354–358.

Patrono, C. and Pierucci, A. (1986) Renal effects of nonsteroidal anti-inflammatory drugs in chronic glomerular disease. *Am. J. Med.,* **81** (Suppl 2B), 71–83.

Plumpton, C., Ferro, C.J., Haynes, W.G., Webb, D.J. and Davenport, A.P. (1996) The increase in human plasma immunoreactive endothelin but not big endothelin-1 or its C-terminal fragment induced by systemic administration of the endothelin antagonist TAK-044. *Br. J. Pharmacol.,* **119**, 311–314.

Plumpton, C., Haynes, W.G., Webb, D.J. and Davenport, A.P. (1995) Phosphoramidon inhibition of the *in vivo* conversion of big endothelin-1 to endothelin-1 in human forearm. *Brit. J. Pharmacol.,* **116**, 1821–1828.

Pollock, D.M. and Opgenorth, T.J. (1993) Evidence for endothelin-induced renal vasoconstriction independent of ET$_A$ receptor activation. *Am. J. Physiol.,* **264**, R222–R226.

Pollock, D.M. and Opgenorth, T.J. (1994) ET$_A$ receptor-mediated responses to endothelin-1 and big endothelin-1 in the rat kidney. *Br. J. Pharmacol.,* **111**, 729–732.

Pollock, D.M. and Polakowski, J.S. (1997) ET$_A$ receptor blockade prevents hypertension associated with exog-enous endothelin-1 but not renal mass reduction in the rat. *J. Am. Soc. Nephrol.,* **8**, 1054–60.

Pope, J.E., Anderson, J.J. and Felson, D.T. (1998) A metaanalysis of the effects of nonsteroidal anti-inflammatory drugs on blood pressure. *Arch. Int. Med.,* **153**, 477–484.

Portoles, J., Torralbo, A., Martin, P., Rodrigo, J., Herrero, J.A. and Barrientos, A. (1997) Cardiovascular effects of recombinant human erythropoietin in predialysis patients. *Am. J. Kidney Dis.,* **29**, 541–8.

Pupilli, C., Brunori, M., Misciglia, N., Selli, C., Ianni, L., Yanagisawa, M., Mannelli, M., Serio, M. (1994) Presence and distribution of endothelin-1 gene expression in human kidney. *Am. J. Physiol.,* **267**, F679–87

Pupilli, C., Romagnani, P., Lasagni, L., Bellini, F., Misciglia, N., Emoto, N., Yanagisawa, M., Rizzo, M., Mannelli, M., Serio, M. (1997) Localization of endothelin-converting enzyme-1 in human kidney. *Am. J. Physiol.,* **273**, F749–56.

Rabelink, T.J., Kaasjager, K.A.H., Boer, P., Stroes, E.G., Braam, B., Koomans, H.A. (1994) Effects of endothelin-1 on renal function in humans: implications for physiology and pathophysiology. *Kidney Int.*, **46**, 376–381.

Radomski, M.W. and Moncada, S. (1993) Regulation of vascular homeostasis by nitric oxide. *Thromb. Haemostasis.*, **70**, 36–41.

Rakugi, H., Tabuchi, Y., Nakamaru, M., Nagano, M., Higashimori, K., Mikami, H., *et al.* (1990) Evidence for endothelin-1 release from resistance vessels of rats in response to hypoxia. *Biochem. Biophys. Res. Commun.*, **169**, 973–977.

Reid, I.A. and Chiu, Y.J. (1995) Nitric oxide and the control of renin secretion. *Fund. Clin. Pharmacol.*, **9**, 309–323.

Reyes, A.A., Karl, I.E. and Klahr, S. (1994) Role of arginine in health and renal disease. *Am. J. Physiol.*, **267**, F331–F346.

Romero, J.C., Lahera, V., Salom, M.G. and Biondi, M.L. (1992) Role of the endothelium-dependent relaxing factor nitric oxide on renal function. *J. Am. Soc. Nephrol.*, **2**, 1371–1387.

Roubert, P., Gillard-Roubert, V., Pourmarin, L., Cornet, S., Guilmard, C., Plas, P., *et al.* (1994) Endothelin receptor subtypes A and B are up-regulated in an experimental model of acute renal failure. *Mol. Pharmacol.*, **45**, 182–188.

Russell FD, Coppell AL, Davenport AP. In vitro enzymatic processing of radiolabelled big ET-1 in human kidney. *Biochem. Pharmacol.* 55: 697–701, 1998.

Sorensen, S.S., Madsen, J.K., Pedersen, E.B. (1994) Systemic and renal effect of intravenous infusion of endothelin-1 in healthy human volunteers. *Am. J. Physiol.* 266, F411–F418.

Sanai, L., Haynes, W.G., MacKenzie, A., Grant, I.S., Webb, D.J.: (1996) Endothelin production in sepsis and the adult respiratory distress syndrome. *Int. Care Med.*, **22**, 52–56.

Scharschmidt, L., Simonson, M. and Dunn, M.J. (1986) Glomerular prostaglandins, angiotensin II and nonsteroidal anti-inflammatory drugs. *Am. J. Med.*, **81** (Suppl 2B), 30–42.

Schmetterer, L., Dallinger, S., Bobr, B., Selenko, N., Eichler, H. and Woltz, M. (1998) Systemic and renal effects of an ET$_A$ receptor subtype-specific antagonist in healthy subjects. *Br. J. Pharmacol.*, **124**, 930–934.

Sedor, J.R., Davidson, E.W. and Dunn, M.J. (1986) Effects of non-steroidal anti-inflammatory drugs in healthy subjects. *Am. J. Med.*, **81** (Suppl 2B), 58–70.

Shibouta, Y., Suzuki, N., Shino, A., Matsumoto, H., Terashita, Z., Kondo, K., *et al.* (1990) Pathophysiological role of endothelin in acute renal failure. *Life Sci.*, **46**, 1611–1618.

Smith, S.D., Wheeler, M.A., Zhang, R., Weiss, E.D., Lorber, M.I., Sessa, W.C., *et al.* (1996) Nitric oxide synthase induction with renal-transplant rejection or infection. *Kidney Int.*, **50**, 2088–2093.

Soper, C.P., Latif, A.B. and Bending, M.R. (1998) Amelioration of hepatorenal syndrome with a selective endothelin-A antagonist. *Lancet*, **347**, 1842–1843.

Sorrentino, R., Sautebin, L. and Pinto, A. (1997) Effect of methylguanidine, guanidine and structurally related compounds on constitutive and inducible nitric oxide synthase activity. *Life Sci.*, **61**, 1283–1291.

Stingo, A. J., Clavell, A. L., Aarhus, L. L. and Burnett, J. C. Jr. (1993) Biological role for the endothelin-A receptor in aortic cross-clamping. *Hypertension,* **22**, 62–66.

Stockenhuber, F., Gottsauner-Wolf, M., Marosi, L., Kurz, R.W., Balcke, P. (1992) Plasma levels of endothelin in chronic renal failure after renal transplantation: impact of hypertension and cyclosporin A-associated nephrotoxicity. *Clin. Sci.*, **82**, 255–258.

Stork, J.E., Rahman, M.A. and Dunn, M.J. (1986) Eicosanoids in experimental and human renal disease. *Am. J. Med.*, **80** (Suppl 1A), 34–45.

Sturrock, N.D., Lang, C.C., MacFarlane, L.J., Dockrell, M.E., Ryan, M., Webb, D.J., *et al.*: Serial changes in blood pressure, renal function, endothelin and lipoprotein(a) during the first 9 days of cyclosporin therapy in males. *J. Hypertens.*, **13**, 667–673.

Sugiura, M., Inagami, T. and Kon, V. Endotoxin stimulates endothelin-release *in vivo* and in vitro as determined by radioimmunoassay. *Biochem. Biophys. Res. Commun.*, 161, 1220–1227, 1989.

Taddei, S., Virdis, A., Ghiadoni, L., Magagna, A. and Salvetti, A. (1997) Cyclooxygenase inhibition restores nitric oxide activity in essential hypertension. *Hypertension*, **29**, 274–279.

Taddei, S., Virdis, A., Mattei, P. and Salvetti, A. (1993) Vasodilation to acetylcholine in primary and secondary forms of human hypertension. *Hypertension*, **21**, 929–933.

Takeda, M., Breyer, M.D., Noland, T.D., Homma, T., Hoover, R.L., Inagami, T. *et al.* (1992) Endothelin-1 receptor antagonist: effects on endothelin- and cyclosporin-treated mesangial cells. *Kidney Int.*, **42**, 1713–1719.

Terada, Y., Tomita, K., Nonoguchi, H. and Marumo, F. (1992). Different localization of two types of endothelin receptor mRNA in microdissected rat nephron segments using reverse transcription and polymerase chain reaction assay. *J. Clin. Invest.*, **90**, 107–112.

Textor, S.C., Burnett, J.C., Romero, J.C., Canzanello, V.J., Taler, S.J., Wiesner, R., *et al.* (1995) Urinary endothelin and renal vasoconstriction with cysclosporine or FK506 after liver transplantation. *Kidney Int.*, **47**, 1426–1433.

Thiemermann, C., Mitchell, J.A. and Ferns, G.A.A. (1993) Eicosanoids and atherosclerosis. *Curr. Opin. Lipid.*, **4**, 401–406.

Tomita, K., Nonoguchi, H. and Marumo, F. (1990) Effects of endothelin on peptide-dependent cyclic adenosine monophosphate accumulation along the nephron segments of the rat. *J. Clin. Invest.*, **85**, 2014–2018.

Tomita, K., Ujiie, K., Nakanishi, T., Tomura, S., Matsuda, O., Ando, K., *et al.* (1989) Plasma endothelin levels in patients with acute renal failure. *N. Engl. J. Med.*, **321**, 1127.

Tsunoda, K., Abe, K. and Yoshinaga, K. (1991) Endothelin in hemodialysis-resistant hypertension. *Nephron*, **59**, 687–688.

Vallance, P., Leone, A., Calver, A., Collier, J. and Moncada, S. (1992) Accumulation of an endogenous inhibitor of nitric oxide synthesis in chronic renal failure. *Lancet*, **339**, 572–575.

Vargas, F., Sabio, J.M. and Luna, J.D. (1994) Contribution of endothelium-derived relaxing factors to acetyl-choline-induced vasodilatation of rat kidney. *Cardiovasc. Res.*, **28**, 1373–1377.

Vaziri, N.D., Ni, Z., Wang, X.Q., Oveisi, F. and Zhou, X.J. (1998) Downregulation of nitric oxide synthase in chronic renal insufficiency: role of excess PTH. *Am. J. Physiol.* F642–F649.

Verhaar, M.C., Strachan, F.E., Newby, D.E., Cruden, N.L., Koomans, H.A., Rabelink, T.J., *et al.* (1998) Endothelin-A receptor antagonist-mediated vasodilatation is attenuated by inhibition of nitric oxide synthesis and by endothelin-B receptor blockade. *Circulation*, **97**, 752–6.

Vincenti, F. and Goldberg, L.I. (1978) Combined use of dopamine and prostaglandin A_1 in patients with acute renal failure and hepatorenal syndrome. *Prostaglandins*, **15**, 463–472.

Vogel, V., Kramer, H.J., Backer, A., Meyer-Lehnert, H., Jelkmann, W., Fandrey, J. (1997) Effects of erythropoietin on endothelin-1 synthesis and the cellular calcium messenger system in vascular endothelial cells. *Am. J. Hypertens.*, **10**, 289–96.

Vriesendorp, R., Donker, A.J.M., De Zeeuw, D., De Jong, P.E., Van der Hem, G.K. and Brentjens, J.R.H. (1986) Effects of nonsteroidal anti-inflammatory drugs on proteinuria. *Am. J. Med.*, **81** (suppl 2B), 84–94.

Wagner, J., Wystrychowski, A., Stauss, H., Ganten, D. and Ritz, E. (1995) Decreased renal haemodynamic response to inhibition of nitric oxide synthase in subtotally nephrectomised rats. *Pflug. Archiv. Eur. J. Physiol.*, **430**, 181–187.

Warrens, A.N., Cassidy, M.J.D., Takahashi, K., Ghatei, M.A. & Bloom, S.R. (1990) Endothelin in renal failure. *Nephrol. Dial. Transplant.*, **5**, 418–422.

Webb, D.J and Cockcroft, J.R. (1989). Plasma immunoreactive endothelin in uraemia. *Lancet,,* **2**, 1211.

Webb, D.J., Monge, J.C., Rabelink, T.J. and Yanagisawa, M. (1998) Endothelin: new discoveries and rapid progress in the clinic. *Trends Pharmacol. Sci.*, (in press).

Weitzberg, E., Lundberg, J.M. and Rudehill, A. (1991) Elevated plasma levels of endothelin in patients with sepsis syndrome. *Circ. Shock,* **33**, 222–227.

Wheeler, M.A., Smith, S.D., Garcia-Cardena, G., Nathan, C.F., Weiss, R.M. and Sessa, W.C. (1997) Bacterial infection induces nitric oxide synthase in human neutrophils. *J. Clin. Invest.*, **99**, 110–116.

Wilcox, C.S., Welch, W.J., Murad, F., Gross, S.S., Taylor, G., Levi, R., *et al.* (1992) Nitric oxide synthase in macula densa regulates glomerular capillary pressure. *Proc.Natl.Acad.Sci.USA.*, **89**, 11993–11997.

Wilkes, B.M., Susin, M., Mento, P.F., Macica, C.M., Girardi, E.P., Boss, E., *et al.* (1991) Localization of endothelin-like immunoreactivity in rat kidneys. *Am. J. Physiol.*, **260**, F913–F920.

Wolf, S.C., Erley, C.M., Kenner, S., Berger, E.D. and Risler, T. (1995) Does L-arginine alter proteinuria and renal hemodynamics in patients with chronic glomerulonephritis and hypertension. *Clin. Nephrol.*, **43**, S42–S46.

Wolzt, M., Schmetterer, L., Ferber, W., Artner, E., Mensik, C., Eichler, H., *et al.* (1997) Effect of nitric oxide synthase inhibition on renal haemodynamics in humans: reversal by L-arginine. *Am. J. Physiol.*, **272**, F178–F182.

Yanagisawa, M., Kurihawa, H., Kimura, S., Tomobe,.Y., Kobayashi, M., Mitsui ,Y., *et al.* (1988) A novel potent vasoconstrictor peptide produced by vascular endothelial cells. *Nature*, **332**, 411–415.

Ye, S.H., Nosrati, S. and Campese, V.M. (1997) Nitric oxide (NO) modulates the neurogenic control of blood pressure in rats with chronic renal failure (CRF). *J. Clin. Invest.*, **99**, 540–548.

Yoshimoto, S., Ishizaki, Y., Sasaki, T., Murota, S.I. (1991) Effect of carbon dioxide and oxygen on endothelin production by cultured porcine cerebral endothelial cells. *Stroke*, **22**, 378–383.

Yukimura, T., Yamashita, Y., Miura, K., Kim, S., Iwao, H., Takai, M. *et al.* (1994) Renal vasodilating and diuretic actions of a selective endothelin ETB receptor agonist, IRL1620. *Eur. J. Pharmacol.*, **264**, 399–405.

Zeidel, M.L., Brady, H.R., Kone, B.C., Gullans, S.R. and Brenner, B.M. (1989) Endothelin, a peptide inhibitor of sodium Na^+-K^+-ATPase, in intact renal tubular endothelial cells. *Am. J. Physiol.*, **257**, C1101–C1107.

C O N T R A C T I O N

R E L A X A T I O N

Inflammation of the blood vessel wall is responsible for much of the dysfunctional systemic responses of the circulation during disease states such as septic shock. Most studies exploring the effects of inflammatory signals on vascular reactivity have been undertaken in animals. However, it is recognised that there is considerable species variation in the mechanisms of inflammation and that the results of studies *in vitro* differ from those undertaken *in vivo*. This chapter draws together the research that relates to recent work in the field of sepsis and attempts to unravel some the pathophysiological changes that are occurring within the vascular wall during this process. Reference is made to much of the human *in vitro* and *in vivo* data that relates to this condition and emphasis has been placed on trying to understand some of the changes occurring in the endothelial and smooth muscle function during this process of inflammation.

Key words: Inflammation, cytokines, endotoxin, nitric oxide, prostanoids, endothelin

12 Infection and Vascular Inflammation

Kiran Bhagat[1] and Timothy W. Evans[2]

[1]*Centre for Clinical Pharmacology, University College London, 5 University Street, London, WC1E 6JJ, UK*
[2]*Department of Critical Care Medicine, National Heart and Lung Institute, Dovehouse Street, London, SW3 6LY, UK*

Until recently, the endothelium was perceived to be a passive, metabolically inert permeability barrier whose function was primarily to contain blood and plasma. However, endothelial cells are now recognised as metabolically and physiologically dynamic, playing a primary and fundamental role in the vascular response to sepsis and systemic inflammation (Figure 12.1). In addition to responding rapidly (in seconds to minutes) to inflammatory agonists such as bradykinin and histamine; upon exposure to endotoxin or cytokines, endothelial cells undergo profound alterations of function that involve changes in gene expression and protein synthesis (Mantovani *et al.*, 1992; Pober and Cotran, 1990; Introna *et al.*, 1994). The endothelial cell responds to endotoxin via a direct, lipopolysaccharide binding protein (LBP) and a soluble CD-14 (sCD-14)-dependent pathway (Kielian and Blecha, 1995); inhibition of this binding by monoclonal antibodies against CD-14 has been shown to reduce the production of cytokines by endothelial cells *in vitro* as well as the hypotension and end organ dysfunction in primate models of endotoxaemia (Leturcq *et al.*, 1996). Most of the metabolic effects of endotoxin are mediated by endothelial and smooth muscle production of cytokines. The pro-inflammatory cytokines such as IL-1, TNF IL-6 and interferon are synthesised by cultured human endothelial cells *in vitro* (Pober and Cotran, 1990; Introna *et al.*, 1994).

The primary vasoactive substances released by endothelial cells identified to date are nitric oxide (NO), the vasoconstrictor peptide endothelins (ETs), and a variety of other vasoactive substances including the prostanoids released via the cyclooxygenase (COX) pathway of arachidonic acid metabolism.

The inflammatory response represents the reaction of the vasculature and its supporting elements to injury, and results in the formation of a protein-rich exudate, provided the injury has not been so severe as to cause actual tissue destruction. Many of the substances involved in modulating the acute inflammatory response exert their effects via cytokines. Moreover, a number of the clinical manifestations of acute bacteraemia or septicaemia are thought to be mediated by the outer wall of these organisms - exotoxin in the case of Gram-positive bacteria and endotoxin in the case of Gram-negative bacteria.

ENDOTOXIN AND VASODILATATION

Endotoxin is one of the integral components of the outer bacterial lipopolysaccharide cell wall of all Gram-negative bacteria (Rietschel *et al.*, 1994), the lipid A component being responsible for most of the molecule's toxicity. Lipopolysaccaride is highly conserved and

a)

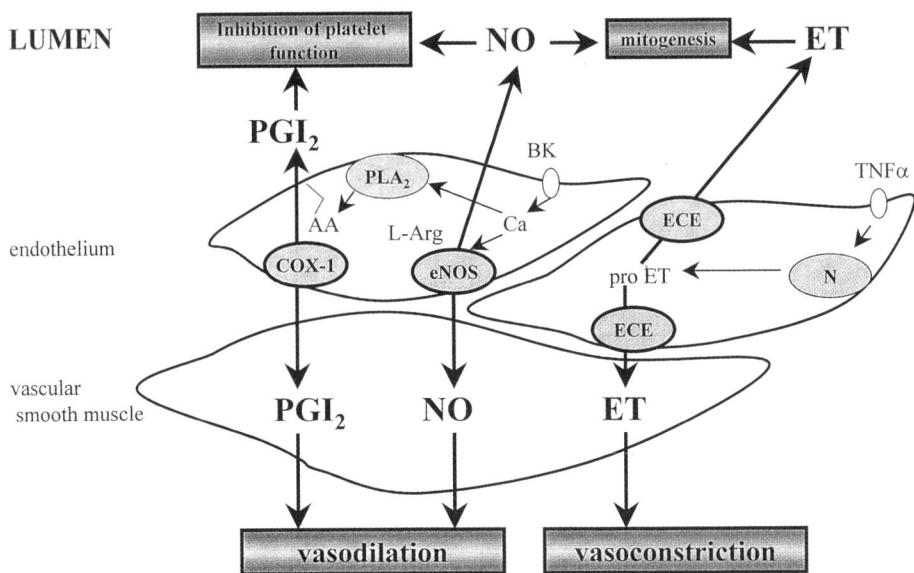

b)

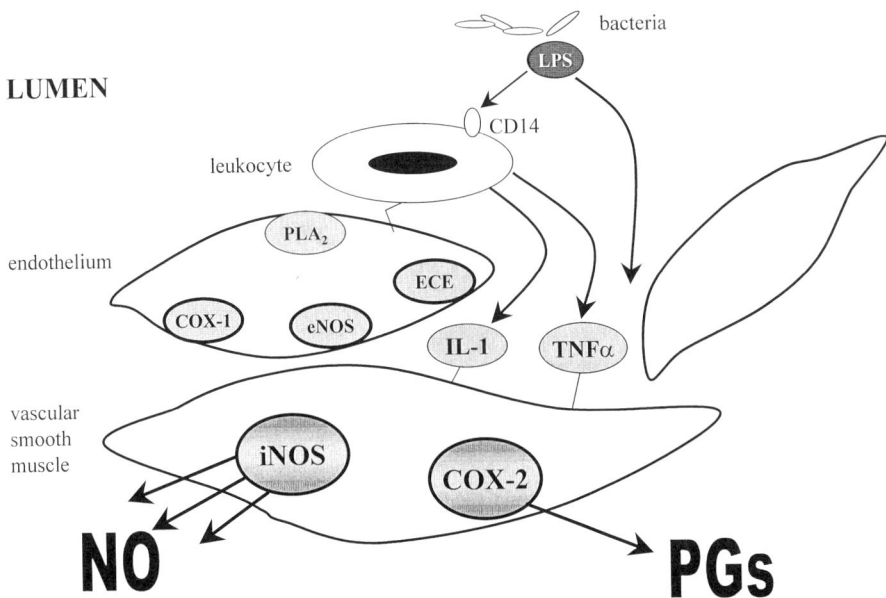

Figure 12.1 Schematic representation of endothelial/smooth muscle interaction under (top) physiological conditions and (bottom) during systemic inflammation.

appears to be essentially invariable and present in all forms of endotoxin (Luderitz *et al.*, 1966). Several lines of evidence indicate that endotoxin, and in particular lipid A, is the primary exogenous mediator in the development of vasodilatation associated with Gram-negative bacterial infections. Experimental studies have shown that injection of endotoxin results in a constellation of symptoms almost identical to those observed in Gram-negative vascular inflammation (Braude, 1980). Severe vascular wall inflammation and vasodilatation occur about 30 min. after injection of endotoxin in dogs and results in eventual vascular collapse (Snell and Parrillo, 1991).

Leukocytes (as well as other cells) respond to endotoxin by secreting a variety of cytokines that heighten the defensive responses of the host. Tumor necrosis factor (TNF) , interleukin (IL) -1 , and IL-6 are secreted acutely following stimulation of mononuclear phagocytes with endotoxin (Pugin *et al.*, 1993). These substances induce the acute phase vasodilatory inflammatory response and prime the immune system for rapid activity. The profile of cytokines secreted in response to endotoxin is now relatively well documented, but the molecules involved in the initial recognition of endotoxin are still being discovered. Recent studies have led to the recognition of at least 3 classes of molecules on leukocytes (and in some cases endothelial cells) that are receptors for endotoxin. The CD18 molecules (leukocyte integrins) bind endotoxin and participate in the phagocytic engulfment of bacteria. The scavenger (acetyl-low-density lipoprotein) receptor recognises free circulating endotoxin and mediates its uptake and degradation. A third receptor, CD14, recognises complexes of endotoxin with the serum protein lipopolysaccharide-binding protein (LBP) (Pearson, 1996; Schletter *et al.*, 1995). CD14 appears to participate in both the ingestion of, and synthetic responses to, endotoxin because blockade of CD14 with monoclonal antibodies strongly inhibits uptake of endotoxin and secretion of TNF by human mononuclear cells (Pugin *et al.*, 1993; Wright, 1991). Other endotoxin receptor molecules have also been proposed but further studies are needed to confirm their involvement (Wright, 1991).

The CD14 molecule is also present in the soluble form as it is shed from cells (Bazil and Strominger, 1991). Soluble CD14 (sCD14) can bind and increase endotoxin-induced activities, such as oxidative burst responses (Schutt *et al.*, 1991) and TNF production by whole blood cells (Haziot *et al.*, 1994). CD14 may also increase the response to endotoxin by cells that do not constitutively express this receptor: recent reports have demonstrated that activation of endothelial cells (which do not have cell-bound CD14 receptors) by endotoxin is mediated by sCD14 which can act as a receptor for endotoxin for the endothelial cell (Noel, Jr. *et al.*, 1995). The vascular smooth muscle (VSM), lacks CD14 receptors, but several studies have shown that endotoxin can act directly on VSM to induce a loss of vascular tone and consequent vasodilatation (Beasley *et al.*, 1990). Again this seems to be due to activation by sCD-14-endotoxin complex (Loppnow *et al.*, 1995).

CYTOKINES AND VASODILATATION

Cytokines are small proteins (MW 8–30000 daltons), that possess multiple biological activities. They are active in low (pico-femtomolar) concentrations and are produced primarily in response to external stimuli, for example, to endotoxin and exotoxin. Most pro-inflammatory cytokines are produced primarily in the presence of infection or disease and contribute to the immune response, inflammation and endothelial cell activation (Mantovani *et al.*, 1992). The results of many animal and human studies in which endotoxin has been administered systemically, have characterised the changes in cytokine levels that occur following injection (Cannon *et al.*, 1990; Hesse *et al.*, 1988; Klosterhalfen *et al.*, 1992; Michie *et al.*, 1988; Kuhns *et al.*, 1995). During the acute response to endotoxin administration there is a dramatic rise in 3 pro-inflammatory cytokines — TNFα, IL-1β and IL-6 (Hesse *et al.*, 1988; Klosterhalfen *et al.*, 1992; Michie *et al.*, 1988). Sixty minutes after endotoxin challenge, levels of TNFα have reached their peak, IL-6 levels increase by 90 min. and IL-1β by 120 min. TNFα has been shown to be a potent inducer of IL-1 and IL-6 release (Fong *et al.*, 1989). Moreover, IL-1β is also able to upregulate IL-6 (Shalaby *et al.*, 1989). These studies were carried out using a bolus injection of endotoxin and it is possible, however, that the alterations of cytokine release induced by a persistent infective focus may differ from the changes observed following a bolus injection of endotoxin.

Clinical studies (Hesse *et al.*, 1988; Waage *et al.*, 1989b; Waage *et al.*, 1989a; Waage *et al.*, 1987) have demonstrated that TNFα and IL-1β blood levels are significantly elevated in patients with endotoxaemia, and that an increase in IL-6 occurs during infectious episodes in humans (Waage *et al.*, 1989a). However, elevated levels of a particular cytokine in the **systemic** circulation may represent the levels required to induce changes in more distal tissues. Similarly, the lack of detection of a specific cytokine may result either from a lack of synthesis, the rapid clearance of that cytokine by binding to available receptors in target tissues, rapid renal clearance of cytokine-soluble receptor complexes, or a lack of assay sensitivity in detecting physiological levels of specific cytokines. This may explain why some studies fail to detect significant levels of IL-1β (or other pro-inflammatory cytokines) in the systemic circulation following injection of endotoxin (Kuhns *et al.*, 1995).

When injected into experimental animals, TNFα and IL-1β both induce vasodilation and shock; and act synergistically when administered together (Dinarello *et al.*, 1989). A single intravenous injection of TNFα or IL-1β administered to patients with cancer induces a sudden fall in blood pressure often requiring treatment (Chapman *et al.*, 1987; Walsh *et al.*, 1992). Healthy volunteers receiving TNFα respond in a similar fashion (van der Poll *et al.*, 1990; van der Poll *et al.*, 1992), whilst an infusion of IL-6 into patients results in fever, chills, and minor fatigue, a significant increase in C-reactive protein, fibrinogen and platelet counts but no hypotension (van Gameren *et al.*, 1994; Weber *et al.*, 1994; Weber *et al.*, 1993). The effects of three key pro-inflammatory cytokines (TNFα, IL-1β, and IL-6) will be discussed in more detail.

TNFα and Vasodilatation

TNFα is principally a macrophage/lymphocyte-derived cytokine with a broad spectrum of immunoregulatory, metabolic and pro-inflammatory activities (Beutler *et al.*, 1985; Bazzoni and Beutler, 1996). TNFα is a trimer with a total molecular weight of 52 kDa that interacts with 2 receptors — R1 and R2, which mediate cytotoxic and proliferative responses respectively. These receptors are present on nearly all cell types with the exception of erythrocytes and unstimulated T-lymphocytes. Although the presence of the receptor appears to be a prerequisite for a biological effect, there does not seem to be a correlation between the number of receptors and the magnitude of the response (Bazzoni and Beutler, 1996; Vassalli, 1992). Soluble TNFα binding proteins have also recently been characterised (Spinas *et al.*, 1992) There are 2 types, antigenically distinguishable and corresponding to the shedded extracellular domains of the 2 species of cell-bound receptor. The presence of soluble TNF-R in the serum may compete and inhibit the binding of TNFα action on cells in addition to affecting the pharmacokinetics and stability of TNFα (Suffredini *et al.*, 1995). On many cell types, even in the absence of protein production, TNFα causes the release of arachidonic acid and consequent secretion of prostanoids (Fiers, 1991). Moreover, treatment of endothelial cells with TNFα induces excessive production of PGI_2 and PGE_2, platelet activating factor (PAF) and nitric oxide (NO) (McKenna, 1990; Lamas *et al.*, 1991), all potential vasodilators both *in vitro* and *in vivo*. The addition of TNFα to many cell types induces protein synthesis following gene activation. This has been studied in more detail in endothelial and polymorphonuclear cells, monocytes and lymphocytes; which are the main targets for circulating TNFα (Fiers, 1991; Old, 1985).

Administration of TNFα depresses endothelium-dependent relaxation *in vivo* (Bhagat and Vallance, 1997a), and *in vitro* TNFα reduces the half-life of mRNA coding for nitric oxide synthase (Yoshizumi *et al.*, 1993). In addition, in patients with heart failure, significantly elevated levels of TNF have been documented (Levine *et al.*, 1990) and, in experimental heart failure, reduced gene expression of endothelial NO synthase (eNOS) and cyclo-oxygenase-1 (COX-I) activity has been reported (Smith *et al.*, 1996). It is not known whether COX-II activity contributes to these effects of TNFα, however generation of free radicals as a by-product of COX activity (Darley Usmar and Halliwell, 1996) might affect endothelial function, and in studies in animals, the endothelial dilator dysfunction that occurs during endotoxaemia is significantly restored in the presence of free radical scavengers (Siegfried *et al.*, 1992).

Phase I studies in cancer patients demonstrated that an acute bolus injection of TNFα resulted in flu like symptoms within 60–90 min (Saks and Rosenblum, 1992). Chronic infusion of this cytokine resulted in anorexia and a leucopaenia (57). In healthy volunteers acute haemodynamic changes were not detected, although a single injection of TNFα elicited rapid and sustained activation of the common pathway of coagulation, probably induced through the extrinsic route (van der Poll *et al.*, 1990; van der Poll *et al.*, 1991). Further, familial differences in endotoxin-induced TNFα release from circulating blood mononuclear cells following systemic injection of endotoxin have been shown in healthy volunteers (Derkx *et al.*, 1995). This suggests the possibility of TNFα gene polymorphism which may influence the inter-individual response to endotoxin challenge.

IL-1β and Vasodilatation

The IL-1β gene family is comprised of IL-1α, IL-1β and IL-1 receptor antagonist (IL-1Ra, 62). Each is synthesised as a precursor protein; the precursors for IL-1β (pro-IL-1α and proIL-1β) have molecular weights of 31 kDa. The proIL-1β and mature 17 kDa IL-1β are both biologically active whereas the proIL-1β requires cleavage to a 17 kDa peptide for optimal biological activity. The IL-1Ra precursor is cleaved to its mature form and secreted like most proteins. IL-1β remains cytosolic in nearly all cells, but unlike IL-1β, IL-1β is rarely found in the circulation of inflammatory fluids. There is evidence that IL-1α functions as an autocrine, intracellular messenger, particularly in cultured endothelial cells and fibroblasts (Elliott *et al.*, 1994). IL-1β also remains cytosolic in nonphagocytic cells. In mononuclear cells, however, between 40–60% is transported out of the cell. Unlike IL-1α, the IL-1β precursor requires cleavage for optimal secretion and activity. A fundamental property of IL-1β, like TNFα, is its ability to induce gene transcription of its own gene (Dinarello *et al.*, 1987) in addition to a wide variety of other genes. Cultured endothelial cells exposed to IL-1β increase the expression of adhesion molecules, which leads to the adherence of leukocytes to endothelial surfaces. These treated endothelial cells also increase production of prostaglandins, PAF, NO and synthesis of other cytokines, all of which may contribute to the acute vasodilatation seen during the acute inflammatory response. Similarly, IL-1β inhibits smooth muscle contraction and this effect appears to be largely dependent on NO production leading to increased guanylate cyclase activity (O'Neill, 1995; Bankers Fulbright *et al.*, 1996; Beasley and McGuiggin, 1994; Beasley and Eldridge, 1994). There appear to be at least two defined IL-1β cell-mediated receptors. Type I and II. In general, IL-1β binds to type I and IL-1 to type II (O'Neill, 1995; Bankers Fulbright *et al.*, 1996). Following receptor binding IL-1β has been shown to increase protein phosphorylation in cells, and much effort has been made to identify the protein kinases responsible. IL-1β causes rapid induction of a wide variety of genes that encode proinflammatory proteins as well as cytokines that initiate or augment inflammatory cell activation. The induction of these genes is regulated by IL-1β inducible transcription factors that are members of the immediate-early gene response family, including AP-1 and NFkB. These transcription factors can be activated within minutes of IL-1β receptor ligation independent of *de novo* protein synthesis. IL-1β can also induce the synthesis of components of these transcription factors later in the activation program. The mechanisms responsible for mediating synergistic functions by interaction of different IL-1β inducible transcription factors or other transcription factors is incompletely understood (O'Neill, 1995; Bankers Fulbright *et al.*, 1996).

Diminished vascular contractility of rat aorta is seen *in vitro* when the tissue is incubated with IL-1β (McKenna *et al.*, 1988; Beasley and McGuiggin, 1994; Beasley and Eldridge, 1994; Beasley *et al.*, 1991). In some studies chronic incubation with IL-1β results in a biphasic reduction in contractility (McKenna *et al.*, 1989) suggestive of the same changes that occur *in vivo* in animals and healthy volunteers injected with endotoxin. In animal studies injection of high dose IL-1β (>1μg/kg) results in acute vasodilatation, decreased systemic vascular resistance, depressed myocardial function, vascular leak and pulmonary congestion (Dinarello, 1994). IL-1β deficient mice responded normally to the systemic administration of endotoxin with no improvement in terms of mortality when compared to wild mice. However, the **local** acute phase tissue response to the inflammatory stimulus is absent when compared to IL-1β competent mice (Fantuzzi and Dinarello, 1996). This suggests that the **systemic** response to endotoxin may involve other cytokines with overlapping activities but that IL-1β appears to play a key role in the **local** inflammatory response.

Results from recent studies in humans (Bhagat and Vallance, 1999) have shown that IL-1β, can induce functionally significant hyporesponsiveness to vasoconstrictors in humans and that the effect is mediated by increased NO production. In animal models IL-1β induces NO generation through transcriptional up-regulation of iNOS. However in the experiments performed in this study no functionally active iNOS was detectable and the cause for the increased NO production was best explained by *de novo* gene transcription causing an increase in expression of GTP cyclohydrolase 1 and consequent activation of eNOS by tetrahydrobiopterin. An implication of the findings is that the endothelium is able to generate sufficient NO to cause profound vasodilatation even in the absence of expression of iNOS. The results and those of previous similar studies in cultured cells (Rosenkranz Weiss *et al.*, 1994; Werner Felmayer *et al.*, 1993; Katusic *et al.*, 1998), suggest that drugs that inhibit GTP cyclohydrolase I might be of therapeutic use to reverse local or systemic inflammatory venous dilatation

IL-6

IL-6 (MW 23 kDa) is produced by almost all cell types in response to a variety of different stimuli including endotoxin or cytokines (such as IL-1β (Content *et al.*, 1985), and TNFα (Jablons *et al.*, 1989)). The gene for IL-6 contains consensus sites for ubiquitous transcription factors such as AP-1, NF-κB, a c-fos serum-responsive element, and a cyclic AMP-responsive element (Scholz, 1996). A number of studies have shown IL-6 to be important in the regulatory production of acute phase proteins during an inflammatory response (Rusconi *et al.*, 1991; Helfgott *et al.*, 1989; Ulich *et al.*, 1991; Furukawa *et al.*, 1992). In cell culture, IL-6 does not appear to affect the prothrombotic or proinflammatory effects of IL-1β on vascular cells and it has been suggested that the key role that of IL-6 is in activating T and B lymphocytes (in addition to the systemic production of acute phase reactants). The production of IL-6 by endothelial cells supports the notion that these cells are involved in immunological pathways and in the regulation of the acute-phase response. Incubation of vascular tissue with IL-6 appears to cause an impairment on contractility in several animal studies (Ohkawa *et al.*, 1995); however, *in vitro* studies using human vessels fail to produce any changes in vessel tone (Beasley and McGuiggin, 1994). This may be due to the absence of circulating cells or intraluminal factors that may be important for IL-6 action *in vivo*.

At the time of writing, no trials looking at the effects of IL-6 receptor blockade during endotoxaemia (either in healthy volunteers or during infective states) have been published, but in both the clinical trials using anti-TNF or IL-1Ra high circulating concentrations of IL-6 predicted a worse outcome irrespective of treatment (Derkx *et al.*, 1995; Fisher, Jr. *et al.*, 1994). Recent investigations have also suggested a relationship between blood concentrations of IL-6 and poor outcome during several inflammatory conditions, including acute myocardial infarction (Neumann *et al.*, 1995), major surgical procedures (Scholz, 1996) and Kawasaki disease (Furukawa *et al.*, 1992). It remains to be determined whether IL-6 antagonism in other non-infective inflammatory diseases such as unstable angina and myocardial infarction confers any benefit in terms of outcome.

NO Release in Infection

There is considerable interest in the role of NO as a mediator of physiological and pathophysiological vasodilation. NO has been shown to be synthesized *in vitro* from the semi essential amino acid L-arginine by the membrane bound NO synthases (NOS), a process that can be inhibited by L-arginine analogues such as N^G-monomethyl-L-arginine (L-NMMA). Several distinct NOS isoforms have been identified. The enzyme is expressed constitutively in endothelium and neural tissue (e, nNOS) and is inducible in a variety of cell types (iNOS). e/n and iNOS are calcium and calmodulin dependent and independent respectively.

In healthy vessels, production of NO in the cardiovascular system occurs mainly from endothelial cells expressing eNOS. However, endotoxin, cytokines including IL-1β and TNFα, and products of Gram-positive bacteria induce the expression in endothelial and smooth muscle cells of iNOS. Induction of iNOS involves protein synthesis and is inhibited by glucocorticoids (Rees *et al.*, 1990). Interestingly the production of NO synthase by the constitutive NOS appears to be inhibited following induction of the inducible NOS and this effect appears to be mediated by a decrease in the stability of the mRNA encoding for constitutive NOS message (Yoshizumi *et al.*, 1993). Thus whereas in the healthy vessel the endothelium is the major vascular source of NO, during inflammation the whole vessel synthesises this mediator.

Patients and animals with sepsis lose peripheral vascular tone and the responsiveness of vessels to constricting agents both *in vitro* and *in vivo* is diminished (Lorente *et al.*, 1993). The incubation of bovine aortic endothelial cells with lipopolysaccharide causes a rapid release of an NO-like factor (Salvemini *et al.*, 1990; Kilbourn and Belloni, 1990), and in patients with sepsis the level of NO metabolised in plasma is significantly elevated. Infusion of NOS inhibitors in such cases can lead to a rapid and reproducible rise in systemic vascular resistance where other pressor substances are ineffective (Ochoa *et al.*, 1991; Petros *et al.*, 1991). It seems likely that the synthesis and release of NO is stimulated by this inflammatory process (Figure 12.2). Endotoxin leads to the induction of an iNOS in endothelium and underlying vascular smooth muscle as well as myocardium (Rees *et al.*, 1990; Radomski *et al.*, 1990; Fleming *et al.*, 1991a). TNFα and IL-1β can also stimulate the expression of iNOS in both endothelium and vascular smooth muscle. Patients treated with IL-2 chemotherapy excrete high levels of NO metabolites suggesting that induction of NO synthesis occurs in response to this treatment. Interestingly, TNFβ can also inhibit NO release stimulated in isolated pulmonary vessels by specific agonists (eg acetylcholine and bradykinin) although basal NO release is unaffected.

The possibility of a two stage release of NO from the blood vessel wall during sepsis has been suggested. In isolated, endotoxin-treated rat vein and pulmonary arteries, NOS inhibitors reverse the vascular hyporesponsiveness to phenylephrine. Moreover, NO-mediated hyporeactivity to noradrenaline starts within 60 minutes in a rat model of sepsis *in vivo* and is therefore too rapid to be explained by the induction of iNOS. Information regarding the latter is presently uncertain, although experiments in animal models suggest that mRNA encoding for iNOS production may be detectable as early as 20 minutes after the intra-peritoneal injection of endotoxin (Liu *et al.*, 1997). The endothelium also seems to respond immediately to the septic insult by releasing NO produced by the constitutive enzyme, eNOS. However, other studies have suggested down regulation of mRNA expression encoding for eNOS in parallel with up-regulation of iNOS mRNA (Liu *et al.*, 1997). The hypothesis that endotoxin leads to an increase in NO release from endothelium causing an early loss of vascular responsiveness *in vivo* therefore remains unproven. However, from about 3h following an endotoxic insult there is massive NO production attributable to iNOS, most probably induced in vascular smooth muscle (Fleming *et al.*, 1993). There is also evidence that endothelium is required for the NO response to be maximal. Thus, NO removal has been shown to cause significant delay in the onset of vascular hyporesponsiveness (6 hrs compared with 4 hrs) and reduced the sensitivity of rat aorta exposed to lipopolysaccharide *in vitro* (Fleming *et al.*, 1991b; Fleming *et al.*, 1993). Selective inhibitors for iNOS are now available and are the subject of intense investigation (see below).

ETs in Infection

In 1988 an endothelially-derived vasoconstrictor was cloned and sequenced following its isolation from the culture medium of porcine aortic endothelial cells and termed endothelin

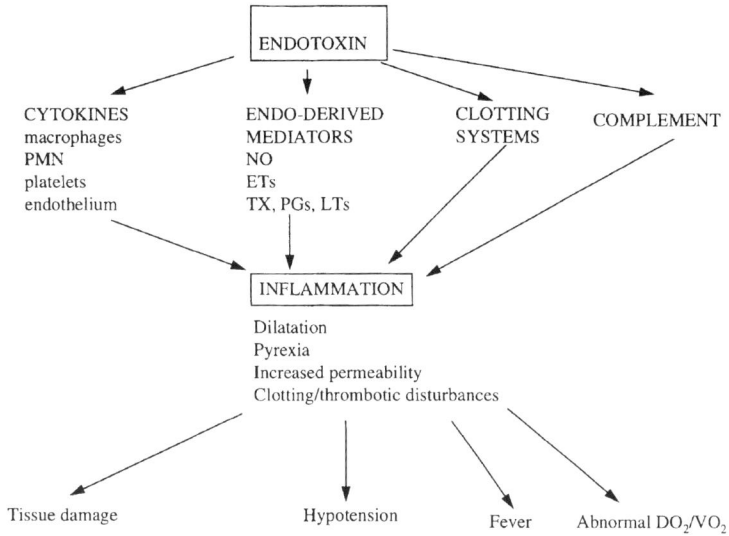

Figure 12.2 Activation of humoral and cell-mediated pathways by endotoxin resulting in inflammation and tissue damage. Abbreviations: endo-derived, endothelium derived; PMN, polymorphonuclear leucocytes; Tx, thromboxane; PGs, prostaglandins; LTs, leukotrienes.

(ET) (Yanagisawa *et al.*, 1988). The substance was found to elicit a slow onset, sustained contraction of isolated arteries from many different species. Three similar but distinct ET-related genomic loci have now been identified encoding for three similar but distinct ET molecules (ET-1, ET-2) (Inoue *et al.*, 1989). Three ET receptor subtypes probably exist although so far only two have been cloned and expressed (Arai *et al.*, 1990). ETA has the highest affinity for ET-1 compared with the other ET's and although it has wide spread expression in human in particular vascular smooth muscle, this does not include endothelial cells. ETB is non-selective, binding all three ET's which are of equipotent in displacing radiolabelled ET-1. ETB expression is also wide spread including endothelial cells. ET's probably work by increasing intra-cellular free calcium, activating phospholipase C leading to increases in inositol trisphosphate and diacylglycerol synthesis. Both are implicated in the initial rise in intra-cellular free calcium concentrations and probably underlie the initiation of ET-1 induced vascular contraction. Protein kinase C is also implicated in a second-messenger system mediating ET-1-induced contraction. ETs are not stored but are synthesised *de novo* in the endothelium. Endothelial cells exposed to agents such as adrenaline and thrombin express ET-1 mRNA. ET-1 production has also been demonstrated both a the mRNA transcription level and at that of protein release in response to angiotensin 2, thrombin and transforming growth factor beta. Mechanical shear stresses, and hypoxia also induce ET-1 production, suggesting that it has a role in modulating local blood flow in response to changes in these indices. Lastly, platelets stimulate expression of ET, mRNA and ET biosynthesis in cultured endothelial cells (Yoshizumi *et al.*, 1989; Kourembanas *et al.*, 1991; Elton *et al.*, 1992; Ohlstein *et al.*, 1991).

ET-1 is a potent vasoconstrictor in humans and many other species. ET-1 and ET-3-induced contraction of rat pulmonary arteries is enhanced by endothelial removal and NO release has been demonstrated in rat mesentery and response to ET, an effect inhibited by NO blockade. ET-3 is a more potent vasodilator that ET-1 probably because the receptor involved on the endothelium is of the ETB type which has equal affinity for all ET's. By contrast, the ETA (vasoconstrictor) receptor predominates in vascular smooth muscle and has a higher affinity for ET-1 (Hirata *et al.*, 1993). Most species demonstrate a characteristic bi-phasic response to ET infusion, with initial transient hypotension followed by sustained hypertension. The former may be partially mediated by NO (Karaki *et al.*, 1993). Specifically, ET-3 is a powerful vasodilator in isolated rat lungs under prior conditions of hypoxic vasoconstriction, although ET-2 and ET-2 are both powerful pulmonary and coronary vasoconstrictors (Karaki *et al.*, 1993). ET-1 has inotropic and chronotropic effects on cardiac muscle and can also induce proliferation in cultured vascular smooth muscle, suggesting a role in remodelling.

Under inflammatory conditions, ET's may be important mediators of vascular tonic responses. Specifically, ET release in response to endotoxin has been confirmed in *in vivo* and *in vitro*, and in endothelial cell cultures in response to a variety of cytokines and free radical species (Sugiura *et al.*, 1989). ET levels increase during endotoxaemia in many animal models and are elevated possibly in parallel with indicators of disease severity in patients with sepsis (Weitzberg *et al.*, 1991; Voerman *et al.*, 1992; Pittet *et al.*, 1991). Although the release and high circulating levels of such a potent vasoconstrictor substance would be expected to antagonise the characteristic refractory hypotension (thought to be mediator by NO) that is seen in sepsis and related syndromes, this does not seem to occur. It is possible that a complex interaction between ET's and NO explains this paradox.

COX Products in Infection

The endothelium also produces various prostanoids via the COX pathway of arachidonic acid metabolism, principle among which is the vasodilator prostacyclin which induces smooth muscle relaxation via the activation of adenylate cyclase thus increasing intracellular cyclic AMP levels. Cyclooxygenase inhibition augments certain vascular reflexes including hypoxic pulmonary vasoconstriction, implying that vasodilator prostoglandins may modulate vascular tone during hypoxaemia (Hales *et al.*, 1978; Weir *et al.*, 1976). Moreover, cytokines can stimulate prostanoid release, the vasodilator action of which is augmented by the presence of endothelium at least in the porcine coronary circulation. Thromboxane A_2 and related endoperoxides are potent vasoconstrictors which are also capable of inducing capillary permeability. ET-mediated release of certain prostanoids including prostacyclin and thromboxane has been demonstrated. Indomethacin pre-treatment of isolated vascular preparations in certain species potentiate ET-induced contraction and even increases ET-3 induced ET-1 formation in cultured endothelial cells, implying that prostacyclin is able to inhibit ET production.

Arachidonic acid is formed from cell-membrane phospholipid when the membrane is disturbed; this disturbance leads to the activation of phospholipase A_2 (Flower and Blackwell, 1976). Arachidonic acid is then available for the formation of secondary metabolites.

One class of products of the arachidonic acid cascade are the prostaglandins (PGs). The rate limiting enzyme for this prostanoid pathway is COX (Hemler and Lands, 1976). At least 2 isoforms of COX are now known to exist, one which is present as a normal constituent of healthy endothelium (COX-I) and the other which is induced in response to endotoxin and cytokines (COX-II (Hla and Neilson, 1992). The constitutive form of COX is thought to mediate many physiological functions. Its activation leads, for instance, to the production of prostacyclin which when released by the endothelium is anti-thrombogenic and vasodilatory (Moncada *et al.*, 1976). Both the smooth muscle and the endothelium have been shown to express COX-I and to generate both constrictor and dilator prostanoids. While most cells have the capacity to generate many different prostanoids there is some selectivity, with platelets and macrophages producing predominantly thromboxane A_2 (TXA$_2$) and endothelial cells PGI$_2$. PGD$_2$ is the principal prostanoid produced by mast cells and PGE$_2$ by the microvasculature (Moncada and Vane, 1978b; Moncada and Vane, 1978a).

COX-II is induced in many cells including the endothelium and smooth muscle by pro-inflammatory stimuli (O'Neill and Ford Hutchinson, 1993). Its induction is inhibited by glucocorticoids and associated with *de novo* proteins synthesis. Ferreira and others (1993) showed that using specific antisera and COX inhibitors that IL-1 is a key cytokine for the release of COX-II metabolites. The role of PGs at the site of inflammation are multiple, but their major specific action is one of vasodilatation. This is attributed particularly to PGE$_2$ and PGI$_2$. Alone these agents produce little or no vessel leakage but in combination with substances which increase vessel permeability, they markedly potentiate the formation of the resultant oedema (Williams and Morley, 1973). A role for this enzyme and its products in the pathogenesis of sepsis is supported by the finding that COX inhibitors restore blood pressure in certain animal models (Parratt and Sturgess, 1974).

In vivo human studies in which endotoxin has been injected as a bolus dose have noted acute activation of the kallikrein-kinin system and changes in the circulating prostanoid levels (DeLa Cadena *et al.*, 1993). Prior administration of a COX-inhibitor markedly reduced the acute pyrexia and malaise, but did not affect the early or late cardiovascular and haemodynamic changes associated with endotoxaemia (Martich *et al.*, 1992; Godin *et al.*, 1996).

STUDIES IN HUMANS

Investigation of the initial vascular response to sepsis in humans is difficult because of practical problems in studying patients at the onset of sepsis. Animal models based on the administration of endotoxin (or cytokines) have been used to study these early vascular events but as discussed above, these studies are limited by variation in species and end organ damage to endotoxin (Redl *et al.*, 1996). The haemodynamic alterations that occurred in normotensive and hypertensive subjects following injection of endotoxin were first documented in 1945, and demonstrated clearly the arterial and venous dilatation that occurs in endotoxaemia in humans. Similar and more extensive observations were made subsequently, comparing humans with other animals (Gilbert, 1960), and a more systematic study of the metabolic and haemodynamic changes was made by Suffredini and others in 1989. These investigators studied the temporal response of cytokine elaboration that occur after endotoxin administration as well as documenting the haemodynamic changes that ensued. Since then several studies looking at the systemic effects of cytokine administration on the cardiovascular system have also been performed (Bhagat *et al.*, 1996b; Bhagat and Vallance, 1997b; van der Poll *et al.*, 1995; Bhagat *et al.*, 1996a). Pharmacological studies have looked at the effects of prior administration of steroids, NSAIDs, cytokine antagonists and other agents on the response of the cardiovascular system to the effects of endotoxin and cytokines. However, such studies are limited by several factors associated with performing studies in healthy volunteers. Thus, administration of any systemic inflammatory agent into normal subjects is always constrained by the amount of active agent that can be safely administered. Furthermore, systemic administration of endotoxin or cytokine (or any other agent) evokes a systemic neurohumoral reflex response that confounds any attempt to study the mechanisms of changes occurring in the vasculature. To explore the cellular and mechanistic events that occur (*in vivo*) as a direct result of the local interaction of the inflammatory agent with the vessel wall, experimental models have now been developed that remove the systemic component of the acute inflammatory event (Bhagat and Vallance, 1997b).

MANIPULATION OF THE INFLAMMATORY RESPONSE

A variety of interventions have been introduced experimentally (and to a lesser extent clinically) of late designed to manipulate the inflammatory response to sepsis. Essentially, the treatment of established sepsis and associated syndromes with or without multiple organ system failure has achieved only limited success, treatment centering largely around circulatory support and attempts to maximise tissue oxygenation. Treatment strategies have now diversified and trials of non-selective NOS inhibitors are underway. The potential value of these agents is great, particularly if specific inhibition of iNOS is possible since it may allow basal NO production, which is probably important in the normal regulation of blood flow in the micro-circulation to continue unimpeded. Aminoguanidine has proved to be one such selective agent in animal models (MacAllister *et al.*, 1994; Tilton *et al.*, 1993). Inhaled nitric oxide is now in wide spread use as a means by which selective pulmonary vasodilatation may be achieved in patients with severe lung injury. The short *in vivo* half life of the gas, coupled to its delivery only to ventilated alveolar units results in a selective reduction in PVR, reduced shunt fraction and increased oxygenation (Rossaint *et al.*, 1993).

In certain animal models of sepsis, pre-treatment with steroids has been beneficial although in clinical investigations high dose steroid treatment has shown no benefit in reducing mortality in patients with established septic shock, neither have they been shown to be beneficial either as treatment or prophylaxis in patients with severe lung injury (Anonymous, 1987; Bone *et al.*, 1987). Infusion of vasodilator prostanoids has been tried both as treatment for ARDS and also to reduce mortality in patients with sepsis to little effect. Several survival rates have been improved in septic animals treated with COX inhibition, but the clinical effects of these agents have been disappointing (Holcroft *et al.*, 1986; Melot *et al.*, 1989; Bone *et al.*, 1989). Antioxidants, particularly n-acetylcysteine can neutralise oxygen free radicals and improve gas exchange haemodynamics and survival in animal models (Bernard *et al.*, 1984). Preliminary studies in patients have been encouraging. Antibodies directed against adhesion molecules are also in pre-clinical trials. Anti-ET antibodies and ET receptor antagonists would also theoretically be useful in these circumstances, in that they have been shown mainly in models of ischaemia to reduce infarct size following coronary artery ligation with reperfusion in rats.

ACKNOWLEDGEMENTS

Work supported by the British Heart Foundation.

REFERENCES

Effect of high-dose glucocorticoid therapy on mortality in patients with clinical signs of systemic sepsis. The Veterans Administration Systemic Sepsis Cooperative Study Group. (1987) *N. Engl. J. Med*, **317**, 659–665.

Arai, H., Hori, S., Aramori, I., Ohkubo, H. and Nakanishi, S. (1990) Cloning and expression of a cDNA encoding an endothelin receptor. *Nature*, **348**, 730–732.

Bankers Fulbright, J.L., Kalli, K.R. and McKean, D.J. (1996) Interleukin-1 signal transduction. *Life Sci.*, **59**, 61–83.

Bazil, V. and Strominger, J.L. (1991) Shedding as a mechanism of down-modulation of CD14 on stimulated human monocytes. *J. Immunol.*, **147**, 1567–1574.

Bazzoni, F. and Beutler, B. (1996) The tumor necrosis factor ligand and receptor families. *N. Engl. J. Med.*, **334**, 1717–1725.

Beasley, D., Cohen, R.A. and Levinsky, N.G. (1990) Endotoxin inhibits contraction of vascular smooth muscle in vitro. *Am. J. Physiol.*, **258**, H1187–92.

Beasley, D., Schwartz, J.H. and Brenner, B.M. (1991) Interleukin 1 induces prolonged L-arginine-dependent cyclic guanosine monophosphate and nitrite production in rat vascular smooth muscle cells. *J. Clin. Invest.*, **87**, 602–608.

Beasley, D. and Eldridge, M. (1994) Interleukin-1 beta and tumor necrosis factor-alpha synergistically induce NO synthase in rat vascular smooth muscle cells. *Am. J. Physiol.*, **266**, R1197–203.

Beasley, D. and McGuiggin, M. (1994) Interleukin 1 activates soluble guanylate cyclase in human vascular smooth muscle cells through a novel nitric oxide-independent pathway. *J. Exp. Med*, **179**, 71–80.

Bernard, G.R., Lucht, W.D., Niedermeyer, M.E., Snapper, J.R., Ogletree, M.L. and Brigham, K.L. (1984) Effect of N-acetylcysteine on the pulmonary response to endotoxin in the awake sheep and upon in vitro granulocyte function. *J. Clin. Invest.*, **73**, 1772–1784.

Beutler, B., Greenwald, D., Hulmes, J.D., Chang, M., Pan, Y.C., Mathison, J., Ulevitch, R. and Cerami, A. (1985) Identity of tumour necrosis factor and the macrophage-secreted factor cachectin. *Nature*, **316**, 552–554.

Bhagat, K., Collier, J. and Vallance, P. (1996a) Local venous responses to endotoxin in humans. *Circulation*, **94**, 490–497.

Bhagat, K., Moss, R., Collier, J. and Vallance, P. (1996b) Endothelial "stunning" following a brief exposure to endotoxin: a mechanism to link infection and infarction? *Cardiovasc. Res*, **32**, 822–829.

Bhagat, K. and Vallance, P. (1997a) Inflammatory cytokines impair endothelium-dependent dilatation in human veins in vivo. *Circulation*, **96**, 3042–3047.

Bhagat, K. and Vallance, P. (1997b) Inflammatory cytokines impair endothelium-dependent dilatation in humans, *in vivo. Circulation*, **96**, 3042–3047.

Bhagat, K. and Vallance, P.J. (1999) Induction of nitric oxide synthase in humans, in vivo; Endothelial nitric oxide synthase masquerading as inducible nitric oxide synthase. *Cardiovasc. Res*, **41**, 754–764.

Bone, R.C., Fisher, C.J., Jr., Clemmer, T.P., Slotman, G.J., Metz, C.A. and Balk, R.A. (1987) A controlled clinical trial of high-dose methylprednisolone in the treatment of severe sepsis and septic shock. *N. Engl. J. Med.*, **317**, 653–658.

Bone, R.C., Slotman, G., Maunder, R., Silverman, H., Hyers, T.M., Kerstein, M.D. and Ursprung, J.J. (1989) Randomized double-blind, multicenter study of prostaglandin E1 in patients with the adult respiratory distress syndrome. Prostaglandin E1 Study Group. *Chest*, **96**, 114–119.

Braude, A.I. (1980) Endotoxic immunity. *Adv. Intern. Med.*, **26**, 427–445.

Cannon, J.G., Tompkins, R.G., Gelfand, J.A., Michie, H.R., Stanford, G.G., van der Meer, J.W., Endres, S., Lonnemann, G., Corsetti, J., Chernow, B. and et al (1990) Circulating interleukin-1 and tumor necrosis factor in septic shock and experimental endotoxin fever. *J. Infect. Dis.*, **161**, 79–84.

Chapman, P.B., Lester, T.J., Casper, E.S., Gabrilove, J.L., Wong, G.Y., Kempin, S.J., Gold, P.J., Welt, S., Warren, R.S., Starnes, H.F. and et al (1987) Clinical pharmacology of recombinant human tumor necrosis factor in patients with advanced cancer. *J. Clin. Oncol.*, **5**, 1942–1951.

Content, J., De Wit, L., Poupart, P., Opdenakker, G., Van Damme, J. and Billiau, A. (1985) Induction of a 26–kDa-protein mRNA in human cells treated with an interleukin-1–related, leukocyte-derived factor. *Eur. J. Biochem.*, **152**, 253–257.

Darley Usmar, V. and Halliwell, B. (1996) Blood radicals: reactive nitrogen species, reactive oxygen species, transition metal ions, and the vascular system. *Pharm. Res.*, **13**, 649–662.

DeLa Cadena, R.A., Suffredini, A.F., Page, J.D., Pixley, R.A., Kaufman, N., Parrillo, J.E. and Colman, R.W. (1993) Activation of the kallikrein-kinin system after endotoxin administration to normal human volunteers. *Blood*, **81**, 3313–3317.

Derkx, H.H.F., Bruin, K.F., Jongeneel, C.V., De Waal, L.P., Brinkman, B.M.N., Verweij, C.L., HouwingDuistermaat, J.J., Rosendaal, F.R. and Van Deventer, S.J.H. (1995) Familial differences in endotoxin-induced TNF release in whole blood and peripheral blood mononuclear cells in vitro; relationship to TNF gene polymorphism. *Journal of Endotoxin Research*, **2**, 19–25.

Dinarello, C.A., Ikejima, T., Warner, S.J., Orencole, S.F., Lonnemann, G., Cannon, J.G. and Libby, P. (1987) Interleukin 1 induces interleukin 1. I. Induction of circulating interleukin 1 in rabbits in vivo and in human mononuclear cells in vitro. *J. Immunol.*, **139**, 1902–1910.

Dinarello, C.A., Okusawa, S. and Gelfand, J.A. (1989) Interleukin-1 induces a shock-like state in rabbits: synergism with tumor necrosis factor and the effect of cyclooxygenase inhibition. *Prog. Clin. Biol. Res.*, **286**, 243–263.

Dinarello, C.A. (1994) The biological properties of interleukin-1. *Eur. Cytokine. Netw.*, **5**, 517–531.

Elliott, M.J., Maini, R.N., Feldmann, M., Long Fox, A., Charles, P., Bijl, H. and Woody, J.N. (1994) Repeated therapy with monoclonal antibody to tumour necrosis factor alpha (cA2) in patients with rheumatoid arthritis. *Lancet*, **344**, 1125–1127.

Elton, T.S., Oparil, S., Taylor, G.R., Hicks, P.H., Yang, R.H., Jin, H. and Chen, Y.F. (1992) Normobaric hypoxia stimulates endothelin-1 gene expression in the rat. *Am. J. Physiol.*, **263**, R1260–4.

Fantuzzi, G. and Dinarello, C.A. (1996) The inflammatory response in interleukin-1 beta-deficient mice: comparison with other cytokine-related knock-out mice. *J. Leukoc. Biol.*, **59**, 489–493.

Ferreira, S.H., Lorenzetti, B.B. and Poole, S. (1993) Bradykinin initiates cytokine-mediated inflammatory hyperalgesia. *Br. J. Pharmacol.*, **110**, 1227–1231.

Fiers, W. (1991) Tumor necrosis factor. Characterization at the molecular, cellular and in vivo level. *FEBS Lett.*, **285**, 199–212.

Fisher, C.J., Jr., Dhainaut, J.F., Opal, S.M., Pribble, J.P., Balk, R.A., Slotman, G.J., Iberti, T.J., Rackow, E.C., Shapiro, M.J., Greenman, R.L. and et al (1994) Recombinant human interleukin 1 receptor antagonist in the treatment of patients with sepsis syndrome. Results from a randomized, double-blind, placebo-controlled trial. Phase III rhIL-1ra Sepsis Syndrome Study Group. *JAMA*, **271**, 1836–1843.

Fleming, I., Gray, G.A., Schott, C. and Stoclet, J.C. (1991a) Inducible but not constitutive production of nitric oxide by vascular smooth muscle cells. *Eur. J. Pharmacol.*, **200**, 375–376.

Fleming, I., Julou Schaeffer, G., Gray, G.A., Parratt, J.R. and Stoclet, J.C. (1991b) Evidence that an L-arginine/nitric oxide dependent elevation of tissue cyclic GMP content is involved in depression of vascular reactivity by endotoxin. *Br. J. Pharmacol.*, **103**, 1047–1052.

Fleming, I., Gray, G.A. and Stoclet, J.C. (1993) Influence of endothelium on induction of the L-arginine-nitric oxide pathway in rat aortas. *Am. J. Physiol.*, **264**, H1200–7.

Flower, R.J. and Blackwell, G.J. (1976) The importance of phospholipase-A2 in prostaglandin biosynthesis. *Biochem. Pharmacol.*, **25**, 285–291.

Fong, Y., Tracey, K.J., Moldawer, L.L., Hesse, D.G., Manogue, K.B., Kenney, J.S., Lee, A.T., Kuo, G.C., Allison, A.C., Lowry, S.F. and et al (1989) Antibodies to cachectin/tumor necrosis factor reduce interleukin 1 beta and interleukin 6 appearance during lethal bacteremia. *J. Exp. Med.*, **170**, 1627–1633.

Furukawa, S., Matsubara, T., Yone, K., Hirano, Y., Okumura, K. and Yabuta, K. (1992) Kawasaki disease differs from anaphylactoid purpura and measles with regard to tumour necrosis factor-alpha and interleukin 6 in serum. *Eur. J. Pediatr.*, **151**, 44–47.

Gilbert RP (1960) Mechanisms of the haemodynamic effects of endotoxin. *Physiol. Rev.*, **40** 245–279.

Godin, P.J., Fleisher, L.A., Eidsath, A., Vandivier, R.W., Preas, H.L., Banks, S.M., Buchman, T.G., Suffredini, A.F., Redl, H., Schlag, G., Bahrami, S. and Yao, Y.M. (1996) Experimental human endotoxemia increases cardiac regularity: results from a prospective, randomized, crossover trial Animal models as the basis of pharmacologic intervention in trauma and sepsis patients. *Crit. Care Med.*, **20**, 487–492.

Hales, C.A., Rouse, E.T. and Slate, J.L. (1978) Influence of aspirin and indomethacin on variability of alveolar hypoxic vasoconstriction. *J. Appl. Physiol.*, **45**, 33–39.

Haziot, A., Rong, G.W., Bazil, V., Silver, J. and Goyert, S.M. (1994) Recombinant soluble CD14 inhibits LPS-induced tumor necrosis factor-alpha production by cells in whole blood. *J. Immunol.*, **152**, 5868–5876.

Helfgott, D.C., Tatter, S.B., Santhanam, U., Clarick, R.H., Bhardwaj, N., May, L.T. and Sehgal, P.B. (1989) Multiple forms of IFN-beta 2/IL-6 in serum and body fluids during acute bacterial infection. *J. Immunol.*, **142**, 948–953.

Hemler, M. and Lands, W.E. (1976) Purification of the cyclooxygenase that forms prostaglandins. Demonstration of two forms of iron in the holoenzyme. *J. Biol. Chem.*, **251**, 5575–5579.

Hesse, D.G., Tracey, K.J., Fong, Y., Manogue, K.R., Palladino, M.A., Jr., Cerami, A., Shires, G.T. and Lowry, S.F. (1988) Cytokine appearance in human endotoxemia and primate bacteremia. *Surg. Gynecol. Obstet.*, **166**, 147–153.

Hirata, Y., Emori, T., Eguchi, S., Kanno, K., Imai, T., Ohta, K. and Marumo, F. (1993) Endothelin receptor subtype B mediates synthesis of nitric oxide by cultured bovine endothelial cells. *J. Clin. Invest.*, **91**, 1367–1373.

Hla, T. and Neilson, K. (1992) Human cyclooxygenase-2 cDNA. *Proc. Natl. Acad. Sci. USA*, **89**, 7384–7388.

Holcroft, J.W., Vassar, M.J. and Weber, C.J. (1986) Prostaglandin E1 and survival in patients with the adult respiratory distress syndrome. A prospective trial. *Ann. Surg.*, **203**, 371–378.

Inoue, A., Yanagisawa, M., Kimura, S., Kasuya, Y., Miyauchi, T., Goto, K. and Masaki, T. (1989) The human endothelin family: three structurally and pharmacologically distinct isopeptides predicted by three separate genes. *Proc. Natl. Acad. Sci. USA*, **86**, 2863–2867.

Introna, M., Colotta, F., Sozzani, S., Dejana, E. and Mantovani, A. (1994) Pro- and anti-inflammatory cytokines: interactions with vascular endothelium. *Clin. Exp. Rheumatol.*, **12 Suppl 10**, S19–23.

Jablons, D.M., Mule, J.J., McIntosh, J.K., Sehgal, P.B., May, L.T., Huang, C.M., Rosenberg, S.A. and Lotze, M.T. (1989) IL-6/IFN-beta-2 as a circulating hormone. Induction by cytokine administration in humans. *J. Immunol.*, **142**, 1542–1547.

Karaki, H., Sudjarwo, S.A., Hori, M., Sakata, K., Urade, Y., Takai, M. and Okada, T. (1993) ETB receptor antagonist, IRL 1038, selectively inhibits the endothelin-induced endothelium-dependent vascular relaxation. *Eur. J. Pharmacol.*, **231**, 371–374.

Katusic, Z.S., Stelter, A. and Milstien, S. (1998) Cytokines stimulate GTP cyclohydrolase I gene expression in cultured human umbilical vein endothelial cells. *Arterioscler. Thromb. Vasc. Biol.*, **18**, 27–32.

Kielian, T.L. and Blecha, F. (1995) CD14 and other recognition molecules for lipopolysaccharide: a review. *Immunopharmacol.*, **29**, 187–205.

Kilbourn, R.G. and Belloni, P. (1990) Endothelial cell production of nitrogen oxides in response to interferon gamma in combination with tumor necrosis factor, interleukin-1, or endotoxin. *J. Natl. Cancer Inst.*, **82**, 772–776.

Klosterhalfen, B., Horstmann Jungemann, K., Vogel, P., Flohe, S., Offner, F., Kirkpatrick, C.J. and Heinrich, P.C. (1992) Time course of various inflammatory mediators during recurrent endotoxemia. *Biochem. Pharmacol.*, **43**, 2103–2109.

Kourembanas, S., Marsden, P.A., McQuillan, L.P. and Faller, D.V. (1991) Hypoxia induces endothelin gene expression and secretion in cultured human endothelium. *J. Clin. Invest.*, **88**, 1054–1057.

Kuhns, D.B., Alvord, W.G. and Gallin, J.I. (1995) Increased circulating cytokines, cytokine antagonists, and E-selectin after intravenous administration of endotoxin in humans. *J. Infect. Dis.*, **171**, 145–152.

Lamas, S., Michel, T., Brenner, B.M. and Marsden, P.A. (1991) Nitric oxide synthesis in endothelial cells: evidence for a pathway inducible by TNF-alpha. *Am. J. Physiol.*, **261**, C634–41.

Leturcq, D.J., Moriarty, A.M., Talbott, G., Winn, R.K., Martin, T.R. and Ulevitch, R.J. (1996) Antibodies against CD14 protect primates from endotoxin-induced shock. *J. Clin. Invest.*, **98**, 1533–1538.

Levine, B., Kalman, J., Mayer, L., Fillit, H.M. and Packer, M. (1990) Elevated circulating levels of tumor necrosis factor in severe chronic heart failure. *N. Engl. J. Med*, **323**, 236–241.

Liu, S.F., Barnes, P.J. and Evans, T.W. (1997) Time course and cellular localization of lipopolysaccharide-induced inducible nitric oxide synthase messenger RNA expression in the rat in vivo. *Crit. Care Med.*, **25**, 512–518.

Loppnow, H., Stelter, F., Schonbeck, U., Schluter, C., Ernst, M., Schutt, C. and Flad, H.D. (1995) Endotoxin activates human vascular smooth muscle cells despite lack of expression of CD14 mRNA or endogenous membrane CD14. *Infect. Immun.*, **63**, 1020–1026.

Lorente, J.A., Landin, L., Renes, E., De Pablo, R., Jorge, P., Rodena, E. and Liste, D. (1993) Role of nitric oxide in the hemodynamic changes of sepsis. *Crit. Care Med.*, **21**, 759–767.

Luderitz, O., Staub, A.M. and Westphal, O. (1966) Immunochemistry of O and R antigens of Salmonella and related Enterobacteriaceae. *Bacteriol. Rev.*, **30**, 192–255.

MacAllister, R.J., Whitley, G.S. and Vallance, P. (1994) Effects of guanidino and uremic compounds on nitric oxide pathways. *Kidney Int.*, **45**, 737–742.

Mantovani, A., Bussolino, F. and Dejana, E. (1992) Cytokine regulation of endothelial cell function. *FASEB J.*, **6**, 2591–2599.

Martich, G.D., Parker, M.M., Cunnion, R.E. and Suffredini, A.F. (1992) Effects of ibuprofen and pentoxifylline on the cardiovascular response of normal humans to endotoxin. *J. Appl. Physiol.*, **73**, 925–931.

McKenna, T.M., Reusch, D.W. and Simpkins, C.O. (1988) Macrophage-conditioned medium and interleukin 1 suppress vascular contractility. *Circ. Shock*, **25**, 187–196.

McKenna, T.M., Lueders, J.E. and Titius, W.A. (1989) Monocyte-derived interleukin 1: effects on norepinephrine-stimulated aortic contraction and phosphoinositide turnover. *Circ. Shock*, **28**, 131–147.

McKenna, T.M. (1990) Prolonged exposure of rat aorta to low levels of endotoxin in vitro results in impaired contractility. Association with vascular cytokine release. *J. Clin. Invest.*, **86**, 160–168.

Melot, C., Lejeune, P., Leeman, M., Moraine, J.J. and Naeije, R. (1989) Prostaglandin E1 in the adult respiratory distress syndrome. Benefit for pulmonary hypertension and cost for pulmonary gas exchange. *Am. Rev. Respir. Dis.*, **139**, 106–110.

Michie, H.R., Manogue, K.R., Spriggs, D.R., Revhaug, A., O'Dwyer, S., Dinarello, C.A., Cerami, A., Wolff, S.M. and Wilmore, D.W. (1988) Detection of circulating tumor necrosis factor after endotoxin administration. *N. Engl. J. Med.*, **318**, 1481–1486.

Moncada, S., Gryglewski, R., Bunting, S. and Vane, J.R. (1976) An enzyme isolated from arteries transforms prostaglandin endoperoxides to an unstable substance that inhibits platelet aggregation. *Nature*, **263**, 663–665.

Moncada, S. and Vane, J.R. (1978a) Prostacyclin, platelet aggregation, and thrombosis. In: Anonymous *Platelets: a mulitdisciplinary approach*, pp. 239–258. London: Raven Press

Moncada, S. and Vane, J.R. (1978b) Prostacyclin (PGI2), the vascular wall and vasodilatation. In: Vanhoutte, P. and Leusen, I. (Eds.) *Mechanisms of vasodilatation*, pp. 107–121. Basel: Karger

Neumann, F.J., Ott, I., Gawaz, M., Richardt, G., Holzapfel, H., Jochum, M. and Schomig, A. (1995) Cardiac release of cytokines and inflammatory responses in acute myocardial infarction. *Circulation*, **92**, 748–755.

Noel, R.F., Jr., Sato, T.T., Mendez, C., Johnson, M.C. and Pohlman, T.H. (1995) Activation of human endothelial cells by viable or heat-killed gram-negative bacteria requires soluble CD14. *Infect. Immun.*, **63**, 4046–4053.

O'Neill, G.P. and Ford Hutchinson, A.W. (1993) Expression of mRNA for cyclooxygenase-1 and cyclooxygenase-2 in human tissues. *FEBS Lett.*, **330**, 156–160.

O'Neill, L.A. (1995) Interleukin-1 signal transduction. *Int. J. Clin. Lab. Res.*, **25**, 169–177.

Ochoa, J.B., Udekwu, A.O., Billiar, T.R., Curran, R.D., Cerra, F.B., Simmons, R.L. and Peitzman, A.B. (1991) Nitrogen oxide levels in patients after trauma and during sepsis. *Ann. Surg.*, **214**, 621–626.

Ohkawa, F., Ikeda, U., Kanbe, T., Kawasaki, K. and Shimada, K. (1995) Inflammatory cytokines and rat vascular tone. *Clin. Exp. Pharmacol. Physiol. Suppl.*, **1**, S169–71.

Ohlstein, E.H., Storer, B.L., Butcher, J.A., Debouck, C. and Feuerstein, G. (1991) Platelets stimulate expression of endothelin mRNA and endothelin biosynthesis in cultured endothelial cells. *Circ. Res*, **69**, 832–841.

Old, L.J. (1985) Tumor necrosis factor (TNF). *Science*, **230**, 630–632.

Parratt, J.R. and Sturgess, R.M. (1974) The effect of indomethacin on the cardiovascular and metabolic responses to E. coli endotoxin in the cat. *Br. J. Pharmacol.*, **50**, 177–183.

Pearson, A.M. (1996) Scavenger receptors in innate immunity. *Curr. Opin. Immunol.*, **8**, 20–28.

Petros, A., Bennett, D. and Vallance, P. (1991) Effect of nitric oxide synthase inhibitors on hypotension in patients with septic shock. *Lancet*, **338**, 1557–1558.

Pittet, J.F., Morel, D.R., Hemsen, A., Gunning, K., Lacroix, J.S., Suter, P.M. and Lundberg, J.M. (1991) Elevated plasma endothelin-1 concentrations are associated with the severity of illness in patients with sepsis. *Ann. Surg.*, **213**, 261–264.

Pober, J.S. and Cotran, R.S. (1990) Cytokines and endothelial cell biology. *Physiol. Rev.*, **70**, 427–451.

Pugin, J., Ulevitch, R.J. and Tobias, P.S. (1993) A critical role for monocytes and CD14 in endotoxin-induced endothelial cell activation. *J. Exp. Med.*, **178**, 2193–2200.

Radomski, M.W., Palmer, R.M. and Moncada, S. (1990) Glucocorticoids inhibit the expression of an inducible, but not the constitutive, nitric oxide synthase in vascular endothelial cells. *Proc. Natl. Acad. Sci. USA*, **87**, 10043–10047.

Redl, H., Schlag, G., Bahrami, S. and Yao, Y.M. (1996) Animal models as the basis of pharmacologic intervention in trauma and sepsis patients. *World J. Surg.*, **20**, 487–492.

Rees, D.D., Cellek, S., Palmer, R.M. and Moncada, S. (1990) Dexamethasone prevents the induction by endotoxin of a nitric oxide synthase and the associated effects on vascular tone: an insight into endotoxin shock. *Biochem. Biophys. Res. Commun.*, **173**, 541–547.

Rietschel, E.T., Kirikae, T., Schade, F.U., Mamat, U., Schmidt, G., Loppnow, H., Ulmer, A.J., Zahringer, U., Seydel, U., Di Padova, F. and et al (1994) Bacterial endotoxin: molecular relationships of structure to activity and function. *FASEB J.*, **8**, 217–225.

Rosenkranz Weiss, P., Sessa, W.C., Milstien, S., Kaufman, S., Watson, C.A. and Pober, J.S. (1994) Regulation of nitric oxide synthesis by proinflammatory cytokines in human umbilical vein endothelial cells. Elevations in tetrahydrobiopterin levels enhance endothelial nitric oxide synthase specific activity. *J. Clin. Invest.*, **93**, 2236–2243.

Rossaint, R., Falke, K.J., Lopez, F., Slama, K., Pison, U. and Zapol, W.M. (1993) Inhaled nitric oxide for the adult respiratory distress syndrome. *N. Engl. J. Med*, **328**, 399–405.

Rusconi, F., Parizzi, F., Garlaschi, L., Assael, B.M., Sironi, M., Ghezzi, P. and Mantovani, A. (1991) Interleukin 6 activity in infants and children with bacterial meningitis. The Collaborative Study on Meningitis. *Pediatr. Infect. Dis. J.*, **10**, 117–121.

Saks, S. and Rosenblum, M. (1992) Recombinant human TNF-alpha: preclinical studies and results from early clinical trials. *Immunol. Ser.*, **56**, 567–587.

Salvemini, D., Korbut, R., Anggard, E. and Vane, J. (1990) Immediate release of a nitric oxide-like factor from bovine aortic endothelial cells by Escherichia coli lipopolysaccharide. *Proc. Natl. Acad. Sci. USA*, **87**, 2593–2597.

Schletter, J., Brade, H., Brade, L., Kruger, C., Loppnow, H., Kusumoto, S., Rietschel, E.T., Flad, H.D. and Ulmer, A.J. (1995) Binding of lipopolysaccharide (LPS) to an 80–kilodalton membrane protein of human cells is mediated by soluble CD14 and LPS-binding protein. *Infect. Immun.*, **63**, 2576–2580.

Scholz, W. (1996) Interleukin 6 in diseases: cause or cure? *Immunopharmacology*, **31**, 131–150.

Schutt, C., Schilling, T. and Kruger, C. (1991) sCD14 prevents endotoxin inducible oxidative burst response of human monocytes. *Allerg. Immunol. Leipz.*, **37**, 159–164.

Shalaby, M.R., Waage, A., Aarden, L. and Espevik, T. (1989) Endotoxin, tumor necrosis factor-alpha and interleukin 1 induce interleukin 6 production in vivo. *Clin. Immunol. Immunopathol.*, **53**, 488–498.

Siegfried, M.R., Ma, X.L. and Lefer, A.M. (1992) Splanchnic vascular endothelial dysfunction in rat endotoxemia: role of superoxide radicals. *Eur. J. Pharmacol.*, **212**, 171–176.

Smith, C.J., Sun, D., Hoegler, C., Roth, B.S., Zhang, X., Zhao, G., Xu, X.B., Kobari, Y., Pritchard, K., Jr., Sessa, W.C. and Hintze, T.H. (1996) Reduced gene expression of vascular endothelial NO synthase and cyclooxygenase-1 in heart failure. *Circ. Res.*, **78**, 58–64.

Snell, R.J. and Parrillo, J.E. (1991) Cardiovascular dysfunction in septic shock. *Chest*, **99**, 1000–1009.

Spinas, G.A., Keller, U. and Brockhaus, M. (1992) Release of soluble receptors for tumor necrosis factor (TNF) in relation to circulating TNF during experimental endotoxinemia. *J. Clin. Invest.*, **90**, 533–536.

Suffredini, A.F., Reda, D., Banks, S.M., Tropea, M., Agosti, J.M. and Miller, R. (1995) Effects of recombinant dimeric TNF receptor on human inflammatory responses following intravenous endotoxin administration. *J. Immunol.*, **155**, 5038–5045.

Sugiura, M., Inagami, T. and Kon, V. (1989) Endotoxin stimulates endothelin-release in vivo and in vitro as determined by radioimmunoassay. *Biochem. Biophys. Res Commun.*, **161**, 1220–1227.

Tilton, R.G., Chang, K., Hasan, K.S., Smith, S.R., Petrash, J.M., Misko, T.P., Moore, W.M., Currie, M.G., Corbett, J.A., McDaniel, M.L. and et al (1993) Prevention of diabetic vascular dysfunction by guanidines. Inhibition of nitric oxide synthase versus advanced glycation end-product formation. *Diabetes*, **42**, 221–232.

Ulich, T.R., Watson, L.R., Yin, S.M., Guo, K.Z., Wang, P., Thang, H. and del Castillo, J. (1991) The intratracheal administration of endotoxin and cytokines. I. Characterization of LPS-induced IL-1 and TNF mRNA expression and the LPS-, IL-1-, and TNF-induced inflammatory infiltrate. *Am. J. Pathol.*, **138**, 1485–1496.

van der Poll, T., Buller, H.R., ten Cate, H., Wortel, C.H., Bauer, K.A., van Deventer, S.J., Hack, C.E., Sauerwein, H.P., Rosenberg, R.D. and ten Cate, J.W. (1990) Activation of coagulation after administration of tumor necrosis factor to normal subjects. *N. Engl. J. Med.*, **322**, 1622–1627.

van der Poll, T., Levi, M., Buller, H.R., van Deventer, S.J., de Boer, J.P., Hack, C.E. and ten Cate, J.W. (1991) Fibrinolytic response to tumor necrosis factor in healthy subjects. *J. Exp. Med.*, **174**, 729–732.

van der Poll, T., van Deventer, S.J., Hack, C.E., Wolbink, G.J., Aarden, L.A., Buller, H.R. and ten Cate, J.W. (1992) Effects on leukocytes after injection of tumor necrosis factor into healthy humans. *Blood*, **79**, 693–698.

van der Poll, T., Fischer, E., Coyle, S.M., Van Zee, K.J., Pribble, J.P., Stiles, D.M., Barie, P.S., Buurman, W.A., Moldawer, L.L. and Lowry, S.F. (1995) Interleukin-1 contributes to increased concentrations of soluble tumor necrosis factor receptor type I in sepsis. *J. Infect. Dis.*, **172**, 577–580.

van Gameren, M.M., Willemse, P.H., Mulder, N.H., Limburg, P.C., Groen, H.J., Vellenga, E. and de Vries, E.G. (1994) Effects of recombinant human interleukin-6 in cancer patients: a phase I-II study. *Blood*, **84**, 1434–1441.

Vassalli, P. (1992) The pathophysiology of tumor necrosis factors. *Annu. Rev. Immunol.*, **10**, 411–452.

Voerman, H.J., Stehouwer, C.D., van Kamp, G.J., Strack van Schijndel, R.J., Groeneveld, A.B. and Thijs, L.G. (1992) Plasma endothelin levels are increased during septic shock. *Crit. Care Med.*, **20**, 1097–1101.

Waage, A., Halstensen, A. and Espevik, T. (1987) Association between tumour necrosis factor in serum and fatal outcome in patients with meningococcal disease. *Lancet*, **1**, 355–357.

Waage, A., Brandtzaeg, P., Halstensen, A., Kierulf, P. and Espevik, T. (1989a) The complex pattern of cytokines in serum from patients with meningococcal septic shock. Association between interleukin 6, interleukin 1, and fatal outcome. *J. Exp. Med.*, **169**, 333–338.

Waage, A., Halstensen, A., Shalaby, R., Brandtzaeg, P., Kierulf, P. and Espevik, T. (1989b) Local production of tumor necrosis factor alpha, interleukin 1, and interleukin 6 in meningococcal meningitis. Relation to the inflammatory response. *J. Exp. Med.*, **170**, 1859–1867.

Walsh, C.E., Liu, J.M., Anderson, S.M., Rossio, J.L., Nienhuis, A.W. and Young, N.S. (1992) A trial of recombinant human interleukin-1 in patients with severe refractory aplastic anaemia. *Br. J. Haematol.*, **80**, 106–110.

Weber, J., Yang, J.C., Topalian, S.L., Parkinson, D.R., Schwartzentruber, D.S., Ettinghausen, S.E., Gunn, H., Mixon, A., Kim, H., Cole, D. and et al (1993) Phase I trial of subcutaneous interleukin-6 in patients with advanced malignancies. *J. Clin. Oncol.*, **11**, 499–506.

Weber, J., Gunn, H., Yang, J., Parkinson, D., Topalian, S., Schwartzentruber, D., Ettinghausen, S., Levitt, D. and Rosenberg, S.A. (1994) A phase I trial of intravenous interleukin-6 in patients with advanced cancer. *J. Immunother. Emphasis. Tumor Immunol.*, **15**, 292–302.

Weir, E.K., McMurtry, I.F., Tucker, A., Reeves, J.T. and Grover, R.F. (1976) Prostaglandin synthetase inhibitors do not decrease hypoxic pulmonary vasoconstriction. *J. Appl. Physiol.*, **41**, 714–718.

Weitzberg, E., Lundberg, J.M. and Rudehill, A. (1991) Elevated plasma levels of endothelin in patients with sepsis syndrome. *Circ. Shock*, **33**, 222–227.

Werner Felmayer, G., Werner, E.R., Fuchs, D., Hausen, A., Reibnegger, G., Schmidt, K., Weiss, G. and Wachter, H. (1993) Pteridine biosynthesis in human endothelial cells. Impact on nitric oxide-mediated formation of cyclic GMP. *J. Biol. Chem.*, **268**, 1842–1846.

Williams, T.J. and Morley, J. (1973) Prostaglandins as potentiators of increased vascular permeability in inflammation. *Nature*, **246**, 215–217.

Wright, S.D. (1991) Multiple receptors for endotoxin. *Curr. Opin. Immunol.*, **3**, 83–90.

Yanagisawa, M., Kurihara, H., Kimura, S., Tomobe, Y., Kobayashi, M., Mitsui, Y., Yazaki, Y., Goto, K. and Masaki, T. (1988) A novel potent vasoconstrictor peptide produced by vascular endothelial cells. *Nature*, **332**, 411–415.

Yoshizumi, M., Kurihara, H., Sugiyama, T., Takaku, F., Yanagisawa, M., Masaki, T. and Yazaki, Y. (1989) Hemodynamic shear stress stimulates endothelin production by cultured endothelial cells. *Biochem. Biophys. Res. Commun.*, **161**, 859–864.

Yoshizumi, M., Perrella, M.A., Burnett, J.C., Jr. and Lee, M.E. (1993) Tumor necrosis factor downregulates an endothelial nitric oxide synthase mRNA by shortening its half-life. *Circ. Res.*, **73**, 205–209.

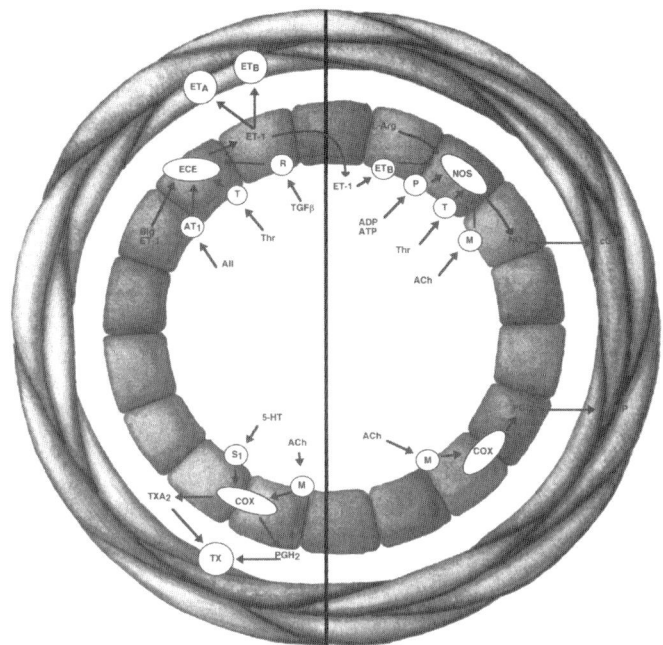

The vascular endothelial cell monolayer is likely to play a major role in the maintenance of healthy pregnancy. Through enhanced synthesis of vasodilator prostaglandins and nitric oxide the endothelium is now considered to contribute to vasodilation of many vascular beds, and so to accommodation of the greatly expanded plasma volume without elevation of blood pressure. Because of ethical considerations, investigation of endothelial function in human pregnancy necessarily has been less extensive than in non-pregnant subjects, and much emphasis has been placed on animal studies. This review concentrates upon human pregnancy and attempts to bring together the confusing and often disparate literature. The interesting possibility that synthesis of nitric oxide in pregnancy may be controlled by circulating estrogens is discussed, together with novel suggestion of a vasodilatory role for an endothelium derived hyperpolarizing factor. Preeclampsia undoubtedly originates in the feto-placental unit, but dysfunction of the maternal vascular endothelium in pre-eclampsia is now so widely reported that the maternal preeclamptic syndrome is considered to be a disease of the endothelium. The evidence for maternal endothelial cell activation and reduced endothelium dependent dilation of the maternal arteries is reviewed and various hypotheses to explain this phenomenon are addressed. More importantly, consideration is given to the possible therapeutic interventions which may offer endothelial protection and so amelioration, or prevention, of this life-threatening disorder.

Key words: Pregnancy, thrombosis, preeclampsia, ADMA, oestrogen, EDHF, LDL.

13 The Endothelium in Human Pregnancy

Lucilla Poston[1] and David J. Williams[2]

[1]Department of Obstetrics and Gynaecoloy, GKT, St. Thomas' Hospital, Lambeth Palace Road, London, SE1 7EH, UK
[2]Department of Obstetrics and Gynaecology, Imperial College School of Medicine, Chelsea and Westminster Hospital, Fulham Road, London, SW10 9NH, UK

INTRODUCTION

During healthy human pregnancy, the peripheral circulation of the mother undergoes profound vasodilatation. Cardiac output increases by 50% secondary to the fall in peripheral vascular resistance (Figure 13.1). Combined with an increase in circulating blood volume these changes are geared towards increasing blood flow to the developing fetus. Blood flow to several maternal organs also increases dramatically during pregnancy. For example, blood flow to the kidneys goes up by 80%, to the skin of the hands and feet by over 200%, and most importantly, to the uterine arteries by 1000%. Conversely, blood flow to the maternal liver does not change, whilst that to the brain actually falls. Unravelling the mechanism of these widespread, but heterogeneous vascular changes in human pregnancy has proved challenging.

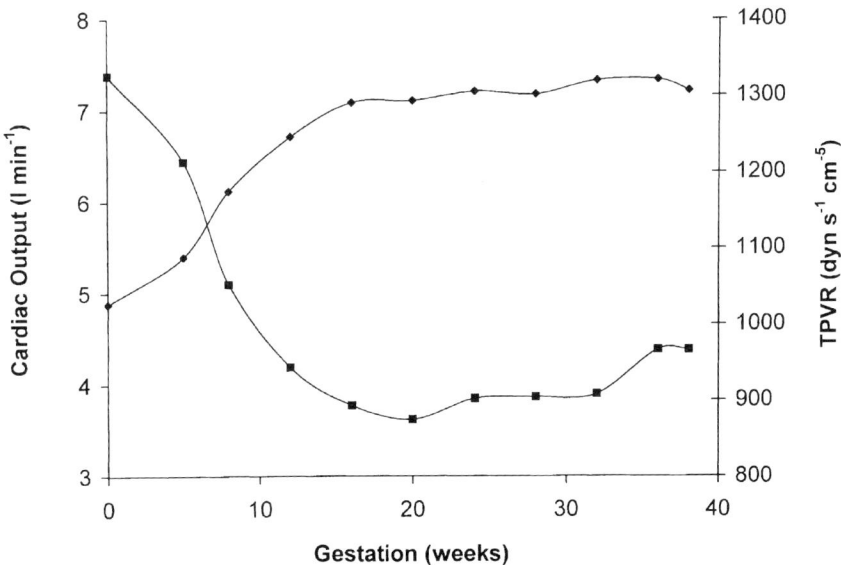

Figure 13.1 The fall in total peripheral vascular resistance (TPVR; squares) is maximal by approximately 16 weeks gestation and coincides with the rise in cardiac output (diamonds). Measured by Doppler and cross-sectional echocardiography at the aortic, pulmonary and mitral valves. Adapted with permission from Robson, S.C., Hunter, S., Boys, R.J. *et al.* (1989) *American Journal of Physiology*, **256**, H1060–H1065.

247

The study of pregnant women is difficult because of the unknown, but critical vulnerability of a developing fetus to pharmacological manipulation. Many investigators have therefore concentrated upon gestationally related changes in cardiovascular function of a variety of animals — with frequently conflicting results. There is compelling evidence for a gestational increase in vasodilator (and therefore anticoagulant) activity of the endothelium. Yet coupled to this, is the fact that on balance human pregnancy is a prothrombotic state. Furthermore, there is controversy concerning the proposal that the corollary to 'upregulation' of the endothelium in healthy pregnancy is dysfunctional endothelium in pre-eclampsia. Animal studies are of little use in the study of pre-eclampsia — a disease unique to humans in which many of the normal vascular responses to pregnancy appear to fail. This chapter attempts to summarise the diverse and often complicated literature which describes the role of the endothelium in the cardiovascular changes of healthy human pregnancy, and why pregnant women seem particularly vulnerable to diseases in which endothelial cell activation is central to their pathophysiology.

THE CARDIOVASCULAR SYSTEM IN HEALTHY PREGNANCY

Cardiovascular changes have been recognised as early as five to six weeks after the last menstrual period i.e. three to four weeks after conception (Robson *et al.*, 1989; Chapman *et al.*, 1998). At this time an increase in heart rate and ventricular function combine with a fall in total peripheral vascular resistance and increase in plasma volume to significantly increase maternal cardiac output (Robson *et al.*, 1989; Brown and Gallery, 1994). This trend persists until 24 weeks gestation by which time cardiac output has increased by approximately 45%.

Arterial dilatation in the first trimester creates a relatively underfilled state associated with a fall in blood pressure (MacGillivray, Rose and Rowe, 1969; Redman, 1995). Renal vasodilatation, which may even precede peripheral vasodilatation (Davison, 1984), leads to a rise in effective renal plasma flow (ERPF) up to 80% above pre-pregnancy levels by the end of the second trimester (Dunlop, 1981). The glomerular filtration rate (GFR) increases by 50% (Davison and Baylis, 1995). Despite the increase in renal blood flow, the vasodilated, but relatively under-filled peripheral and renal vasculature is probably partly responsible for upregulation of the renin-angiotensin-aldosterone (RAA) axis (Schrier and Briner, 1991; Brown and Gallery, 1994), although hormonal influences undoubtedly play a role. Oestrogens are well known to increase renin substrate — angiotensinogen (Schunkert, Danser, Hense et al, 1997). Indirect assays of plasma renin activity (PRA) during pregnancy, in the presence of elevated angiotensinogen, invariably report greater levels of angiotensin I and therefore higher levels of PRA (Brown and Gallery, 1994). However, postmenopausal women on oestrogen replacement therapy (HRT) have suppressed renin levels as estimated using a specific monoclonal renin antibody, (despite elevated angiotensinogen levels) compared with those not on (HRT) (Schunkert *et al.*, 1997). It is possible therefore, that in normal pregnancy elevated circulating oestrogen levels will stimulate angiotensinogen production, but circulating renin levels will be comparatively suppressed.

As a result of avid renal sodium and water retention, total extracellular fluid volume gradually increases by 6–8L with a relatively greater increase in intravascular compared with interstitial fluid volume (Brown and Gallery, 1994). At 32 weeks gestation plasma volume reaches a maximum of 40% (1.2L) above pre-pregnancy levels (Gallery, Hunyor and Gyory, 1979). Furthermore, plasma volume expansion – a characterisitic of healthy pregnancy, could enhance NO generation (Calver *et al.*, 1992). Increased plasma renin activity generates higher circulating levels of angiotensin II (ANG II), associated with reduced binding of ANG II to platelets (Baker, Broughton-Pipkin and Symonds, 1992) and a reduced pressor response to exogenous ANG II (Gant et al, 1973). However, down-regulation of ANG II receptors cannot account for the attenuated vasopressor responses also found with exogenous administration of vasopressin and noradrenaline (Chesley, 1978; Williams *et al.*, 1997). Increased activity of vasodilator substances could play this role and are discussed in detail below.

COAGULATION

Endothelium Derived Clotting Factors

In anticipation of haemorrhage at childbirth, normal pregnancy is characterised by low grade, chronic intravascular coagulation within both the maternal and utero-placental circulation (Letsky, 1995). There is evidence for increased levels of clotting factors, especially fibrinogen (Bonnar, 1987) and depression of fibrinolysis (Kruithof, Tran-Thang, Gudinchet *et al.*, 1987), although more recent studies suggest the procoagulant state of normal pregnancy is compensated by increased fibrinolysis (Sorensen, Secher and Jespersen, 1995). The endothelium is directly involved in promoting a procoagulant state in healthy pregnancy. During the third trimester, plasma levels of endothelium derived von Willebrand factor are elevated, promoting coagulation and platelet adhesion (Sorensen *et al.*, 1995). Furthermore, there is a gestational increase in endothelium production of plasminogen activator inhibitor (PAI-1) and tissue plasminogen activator (t-PA), with the effect of both inhibition and promotion of fibrinolysis, respectively. The t-PA to PAI-1 ratio remains unchanged in healthy pregnancy (Sorensen *et al.*, 1995). Thrombin generation is also increased in normal pregnancy, as are circulating levels of fibrin degradation products (FDP) (de-Boer, *et al.*, 1989; Sorensen *et al.*, 1995), although the ratio of thrombin to FDP remains unchanged. The procoagulant state of the endothelium does therefore appear to be compensated by upregulation of the fibrinolytic system (Bremme *et al.*, 1992; Sorensen *et al.*, 1995).

Nonetheless, the risk of thromboembolism increases six fold during pregnancy (Royal College of General Practioners, 1967) and is the most common cause of maternal death in the UK (Department of Health, 1998). Factors other than a procoagulant endothelium contribute to the increased incidence of thromboembolism. In late pregnancy, the gravid uterus partialy obstructs the inferior vena cava, causing venous stasis in the lower limbs. Women who have Caesarean sections are also more vulnerable to thromboembolism. Furthermore, occult thrombophilias often present clinically for the first time during pregnancy.

NITRIC OXIDE

There is much evidence from animal studies to suggest that increased activity of the L-arginine — NO pathway contributes to the generalised vasodilatation and attenuated response to exogenous vasoconstrictors characteristic of pregnancy (Chu and Beilin, 1993; Conrad *et al.*, 1993; Weiner, Knowles and Moncada, 1994). Conversely, inhibition of nitric oxide synthase in pregnant rats abolishes the refractoriness to vasopressors and produces elevation of the blood pressure and proteinuria (see later) (Yallampalli and Garfield, 1993; Allen et al, 1994). Assessment of the L-arginine — nitric oxide pathway in human pregnancy and pre-eclampsia has proved more challenging. Different methodologies and conflicting results present a confusing picture.

Normal Human Pregnancy

Cyclic guanosine monophosphate (cGMP)

The cyclic nucleotide, cGMP acts as a second messenger for NO and has been used as a surrogate marker for NOS activity. Both plasma and urinary cGMP levels, increase in pregnant rats coincidental with the gestational increase in renal blood flow (Conrad and Vernier, 1989). The same group also showed that renal vasodilatation and hyperfiltration in the pregnant rat is abolished by inhibition of nitric oxide synthase (Danielson and Conrad, 1995).

   In human pregnancy, there is agreement that urinary concentrations of cGMP increase early in pregnancy and remain elevated until term (Kopp, Paradiz and Tucci, 1977; Chapman *et al.*, 1998). Cyclic-GMP clearance increases as early as the sixth week of gestation, in parallel with a fall in plasma cGMP levels, that remain low throughout pregnancy (Chapman *et al.*, 1998). Others have found slight or significant increases in plasma cGMP during normal pregnancy, remaining high until term (Schneider *et al.*, 1996; Boccardo *et al.*, 1996). Boccardo *et al.* (1996) found cGMP levels and [^{3}H]L-citrulline production within platelets were similar between pregnant and non-pregnant women, suggesting platelets do not contribute to excessive NO production in pregnancy.

   Cyclic GMP concentrations in plasma or urine are only an indirect indication of NOS activity as responses to atrial natriuretic peptide (ANP) are also mediated through cGMP. In a meta-analysis of 53 studies it was concluded that the circulating concentration of ANP did not rise until the third trimester (Castro, Hobel and Gornbein, 1994). However, this is long after the increase in urinary cGMP, which coincides with the gestational increase in renal blood flow and glomerular filtration rate (Dunlop, 1981; Roberts, Lindheimer and Davison, 1996). It is possible that increased NOS activity has a role in renal vasodilatation during healthy human pregnancy, but it cannot be concluded from the measurement of cGMP alone that increased NOS activity plays a role in the gestational fall in systemic vascular resistance.

Oxidation Products of Nitric Oxide; Nitrite and Nitrate

Several investigators have measured the serum concentration of nitrite (NO_2^-) and nitrate (NO_3^-), or their product, NOx, during healthy pregnancy. However, most have ignored the confounding problem that concentrations of these NO metabolites are sensitive to dietary nitrogen intake. Not surprisingly, studies of normotensive pregnant women on uncontrolled nitrogen diets have shown either increased (Seligman *et al.*, 1994) or unchanged (Curtis, Gude, King *et al.*, 1995; Smarason *et al.*, 1997) plasma NO metabolites in comparison with non-pregnant women. In one study, plasma NOx levels were measured after a

12–15 hour fast, and found to be significantly elevated in normotensive pregnancy (from before 12 weeks gestation until term) compared with non-pregnant women (Nobunaga, Tokugawa and Hashimoto *et al.*, 1996). In another carefully controlled study, guanidino [N^{15}]L-arginine was infused into five healthy pregnant volunteers after being on a nitrate free diet for the preceeding week and a 12 hour fast pre-infusion. Arginine flux and nitrite/ nitrate pool turnover were higher in early compared with late pregnancy, suggesting NO production is higher in early pregnancy (Goodrum, Saade, Jahoor *et al.*, 1996).

Nitric Oxide Synthase Activity in Vivo

In vivo functional studies provide the most compelling evidence that NO synthase is upregulated in the maternal circulation during normal pregnancy. Infusion of the NO synthase inhibitor, L-NMMA, into the brachial artery caused a greater reduction in hand blood flow of pregnant compared with non-pregnant women (Williams *et al.*, 1997) (Figure 13.2). The increased efficacy of L-NMMA was observed during early pregnancy (9–15 weeks), a time when there was not yet a measurable increase in hand blood flow. This suggests that a mechanism, other than shear stress, mediates the gestational increase in NO generation. Furthermore, in late pregnancy (36–41 weeks), L-NMMA returned the elevated hand blood flow back to non-gravid levels, implicating a major role for NO in the peripheral vasodilatation of healthy pregnancy (Figure 13.2).

L-NMMA, has also been shown to induce a non-sustained venoconstriction in hand veins of healthy women in the early puerperium, but not in the same women 12–16 weeks post-partum (Ford, Robson and Mahdy, 1996). Normally, in the non-gravid state, infusion of L-NMMA, at a dose that maximally inhibits bradykinin, does not produce venoconstriction of hand veins (Vallance, Collier and Moncada, 1989). Isolated endothelial cells from hand veins of pregnant women have been shown to respond to adenine triphosphate (ATP) with a large transient increase in intracellular Ca^{2+} (Mahdy *et al.*, 1998). This response was significantly greater in endothelial cells isolated from the hand veins of healthy pregnant women compared to non-pregnant and pre-eclamptic women (Mahdy *et al.*, 1998). However, the extrapolation of evidence from the venous circulation to peripheral vascular control must be interpreted with caution.

Nitric Oxide in the Resistance Vasculature of the Maternal Circulation

Much of the evidence favouring an increase in endothelial NO synthesis in the vasculature in pregnancy is derived from studies of small arteries in animals, in which significant reduction in the EC_{50} for ACh or methacholine has repeatedly been observed. Although indicative of enhanced NO synthesis, conclusive proof is not always available as selective inhibition of NOS has often not been attempted. The reduced responsiveness to constrictor agonists observed in some small artery preparations from pregnant animals has, using NOS inhibitors, sometimes been attributed to enhanced NO synthesis. Perhaps the most convincing evidence for an increase in NOS activity comes from a study of small rat arteries which has shown an increase in expression of eNOS in pregnancy (Xu *et al.*, 1996). These confusing and often conflicting studies of NO in the vasculature in animal pregnancy have recently been covered in considerable detail by a comprehensive review (Sladek *et al.*, 1997).

Limitation in the availability of human tissue is an obvious drawback to investigation in human pregnancies. However, small arteries may be obtained from subcutaneous fat,

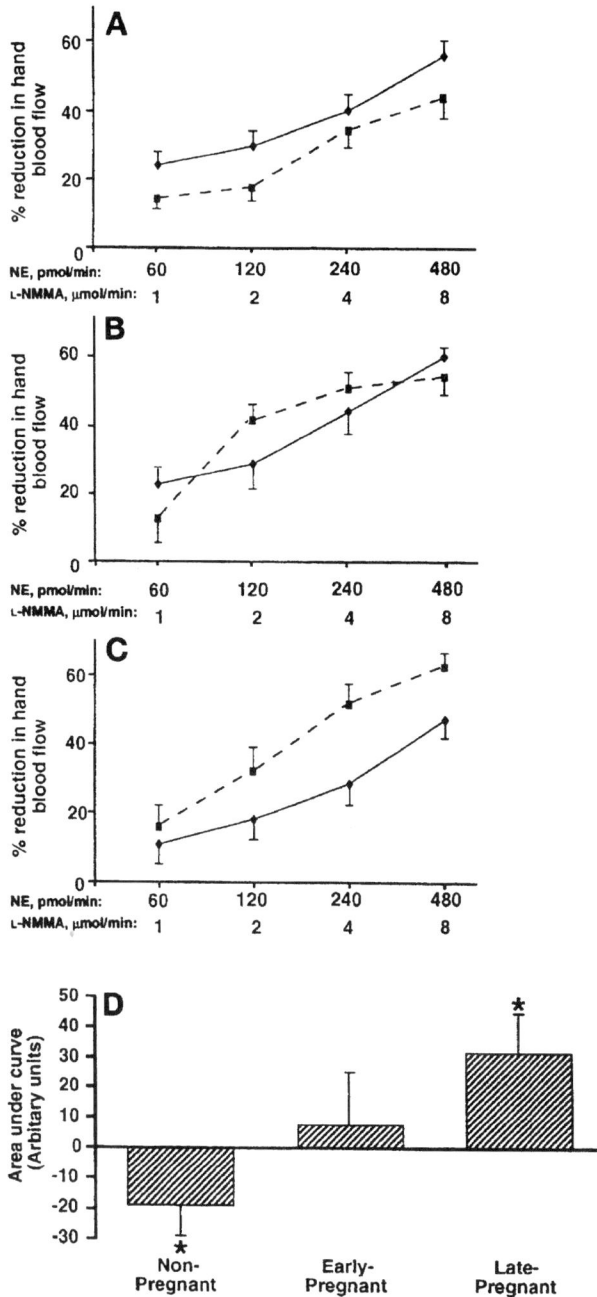

Figure 13.2 Response to brachial artery infusion of L-NMMA (dashed line) and noradrenaline (NE; solid line) on hand blood flow for nonpregnant (A), early pregnant, (9–15 weeks gestation) (B), and late pregnant (36–41 weeks gestation) subjects (C). D: area under dose response curve for noradrenaline was subtracted from area under dose response curve for L-NMMA, for each individual in each group. Response to L-NMMA increases relative to noradrenaline as pregnancy progresses. Difference between late pregnant and nonpregnant groups was significant (*P = 0.0089). With permission from Williams, D.J., Vallance, P.J.T., Neild, G.H. *et al.* (1997) *American Journal of Physiology*, **272**, H748–H752.

omentum and myometrium by biopsy at caesarean section. The development of a variety of techniques for reproducible investigation of small artery tension and diameter has greatly facilitated studies in human tissue, as small biopsies may provide adequate material for experiment. The methods most frequently used have been small vessel wire myography and small vessel perfusion myography. The wire myograph (Mulvany and Halpern, 1977) is a commercially available apparatus which enables the measurement of isometric tension in arteries as small as 150 μm internal diameter. In the perfusion myograph (Halpern *et al.*, 1984), the small artery is mounted between two fine glass cannulae and a video dimension analyser used to 'track' the movement of the vessel wall as the artery contracts or relaxes. This instrument has the advantage that pressure transducers enable the artery to be set to and maintained at a given pressure and the provision of flow pumps also enables investigation of flow mediated responses.

The blood flow to the skin is greatly enhanced in human pregnancy and investigation of this circulation may therefore provide insight into gestationally related mechanisms of dilatation. Small subcutaneous arteries can be obtained by removal of biopsies of subcutaneous fat from consenting women at caesarean section. The first study of these arteries, using the wire myograph, investigated responses to ACh in arteries of mean internal diameter 250–300 μm and showed that relaxation to ACh was no different between arteries from pregnant women and those from non-pregnant women obtained during routine abdominal surgery (McCarthy *et al.*, 1994). Interestingly, the NO synthase inhibitor, L-NMMA failed to completely inhibit relaxation to ACh, and indomethacin had little effect. The residual relaxation to ACh in the presence of the NOS inhibitor was greater in the arteries from the pregnant women and could suggest increased synthesis of an endothelium dependent dilator other than NO or PGI_2, potentially an endothelium derived hyperpolarizing factor. In a later study, the same group (Knock and Poston, 1996) found that pregnancy was associated with increased relaxation to bradykinin (BK) in small subcutaneous arteries (Figure 13.3), leading to the suggestion that elements of the signal transduction pathway for BK were preferentially affected by pregnancy.

In contrast, using arteries from the omental circulation Pascoal *et al.* (1996) concluded that neither ACh nor BK mediated relaxation was different in arteries from term pregnant women and non-pregnant women, although pregnancy was associated with increase in a novel component of BK mediated relaxation, possibly a hyperpolarising factor (Pascoal *et al.*, 1996). Another study using the wire myograph has also shown no difference in relaxation to BK in small myometrial arteries from pregnant women and from non-pregnant women obtained during hysterectomy (Ashworth *et al.*, 1997) although structurally these arteries might be expected to be very different from one another. However, Kublickiene *et al.* (1997a) have found that if mounted on a perfusion myograph, small myometrial arteries from pregnant women respond well to ACh and that this relaxation is greater than relaxation to ACh in omental arteries from term pregnant women, perhaps indicating that enhanced receptor mediated relaxation may contribute to increased myometrial blood flow in pregnancy. Taken together these studies show little consensus regarding the role of agonist stimulated NO synthesis in vasodilation of pregnancy, perhaps because of the different vascular beds studied.

There is more agreement that flow mediated NO synthesis is raised in the resistance vasculature in human pregnancy. Cockell and Poston (1997a) have shown that the subcutaneous arteries from pregnant women demonstrate a remarkably increased response to flow compared to those from non-pregnant women, which was totally inhibited by

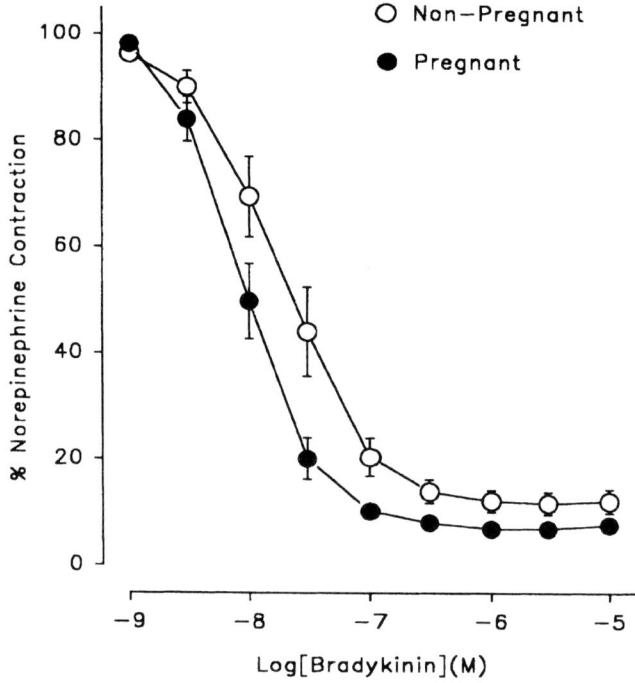

Figure 13.3 Concentration-dependent relaxation to bradykinin in isolated subcutaneous fat arteries from normotensive pregnant (solid circles) and non-pregnant women (open circles). From Knock and Poston, Am J Obstet Gynecol. 1996, 175, 1668–1674.

(a)   (b)

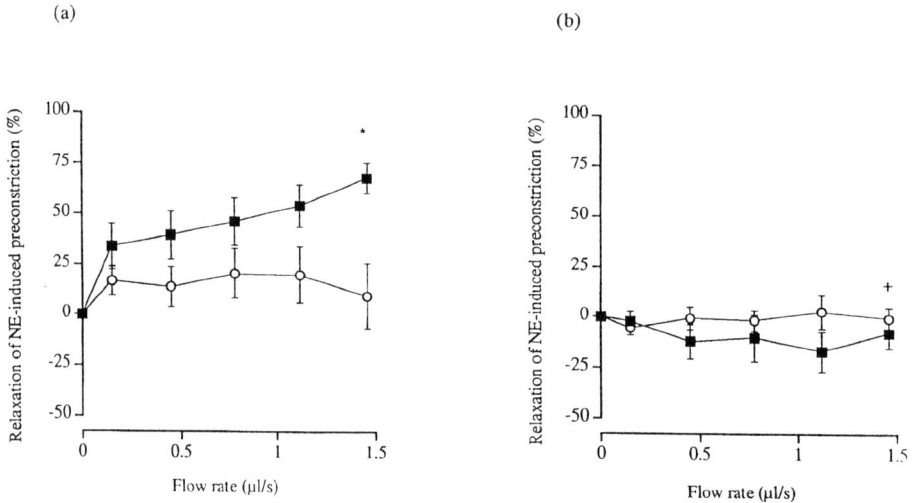

Figure 13.4a,b Percent change in NE-induced preconstriction in response to incremental increases in flow in isolated human subcutaneous arteries from normotensive pregnant women (n = 10) (a) and normotensive nonpregnant women (n = 10) (b). Black squares indicate the absence of L-NAME; open circles, the presence of L-NAME. *P<0.01 for pregnant vs pregnant with L-NAME; +P<0.01 for pregnant vs nonpregnant without L-NAME. From Cockell and Poston, Hypertension, 1997, 30 (part 1), 247–251. With permission from Cockell, and Poston (1997) *Hypertension*, **30**(part 1), 247–251.

L-NAME (Figure 13.4). This substantiated earlier work in arteries from pregnant rats (Learmont *et al.*, 1996; Cockell and Poston, 1997b) showing enhanced flow mediated dilatation, an observation recently confirmed by Ahokas *et al.*, (1997). Using the same technique, NO mediated responses to flow have been observed in small myometrial arteries from women at term (Klubickiene *et al.*, 1997b).

Role of Oestrogens in Pregnancy Associated Alteration in Nitric Oxide Synthesis

From the moment of conception and throughout healthy human pregnancy there is a steady, but massive (approximately 250 fold) rise in circulating oestradiol levels (Edouard *et al.*, 1998; Chapman, 1998). Oestrogens are now recognised to cause both endothelium-dependent and independent vasodilatation.

In vivo studies have suggested that that the acute vasodilator effects of oestrogens are endothelium dependent (Gilligan, Quyyumi and Cannon, 1994a). 17β-oestradiol infusion potentiated the reduction in coronary artery resistance evoked by acetylcholine (ACh) in women with coronary artery disease, as measured by Doppler ultrasound (Gilligan *et al.*, 1994a). Endothelium-dependence of the response was indicated as 17β-oestradiol did not affect responses to the endothelium independent vasodilators, adenosine and sodium nitroprusside. 17β-oestradiol also potentiates endothelium dependent rather than independent vasodilatation in the forearm of post-menopausal women (Gilligan *et al.*, 1994b). The immediate response was considered too short for gene transcription and protein synthesis, and it was proposed that 17β-oestradiol prolonged the half life of NO by scavenging free radicals, which would otherwise 'quench' NO. This antioxidant effect has also been demonstrated in a study in which ethinylestradiol elevated cGMP in cultured bovine aortic endothelial cells, without increasing NOS activity, NOS protein or mRNA, but simultaneously was shown to decrease synthesis of the superoxide radical (Arnal *et al.*, 1996).

Animal studies suggest involvement of the endothelium in the dilatation induced by longer term exposure to oestrogen (hours to days, i.e. appropriate to that seen during pregnancy) (Magness and Rosenfeld, 1989). The increase in uterine artery blood flow following 17β oestradiol infusion in non-pregnant sheep is blunted by simultaneous local infusion of L-nitroarginine methyl ester, suggesting oestrogen enhancement of the NO pathway (Van Buren *et al.*, 1992). The role of NO in this response has been confirmed by Veille *et al.*, (1996) who have demonstrated that after 3 days of estradiol treatment, enhanced NO mediated relaxation in the sheep uterine artery is related to greater NOS enzymatic activity. Moreover, several investigators (Hayashi *et al.*, 1995; Hishikawa *et al.*, 1995 and MacRitchie *et al.*, 1997) have shown estradiol mediated increases of NO and eNOS protein in cultured endothelial cells. 17-ß oestradiol pre-treatment of isolated arteries from the mesenteric vasculature of prepubertal female rats also induces an increase in endothelium dependent NO mediated flow induced dilatation (Cockell and Poston, 1997b).

Indirect, *in vivo* evidence from non-pregnant women, also suggests that oestrogens increase NOS activity. Women with artificially suppressed endogenous oestrogen levels, given oral oestradiol had raised plasma nitrate levels compared with those measured after placebo (Ramsay *et al.*, 1995). Consistent with this suggestion, post menopausal women given transdermal 17β-oestradiol show enhanced serum levels of nitrite and nitrate (Roselli *et al.*, 1995).

The mechanism by which oestrogen modulates NO production has not been fully elucidated, but it is known that there are 11 copies of an incomplete (half palindromic motif) estrogen response element (ERE) on the endothelial NOS gene 5' flanking 'pro-motor' region (Robinson *et al.*, 1994). Similarly, widely spaced half-palindromic motifs act synergistically in other genes to form a complete ERE and it is suggested that the occupied estrogen receptor may activate NOS-III by binding to these regions. A study of estrogen receptor knock out mice (ERKO) has confirmed a role for estrogen receptors in NO synthesis but has also produced some unexpected findings. Male homozygous estrogen receptor knockout mouse (ERKO) had lower basal release of NO than wild type males. However, female wild type mice had fewer estradiol receptors and lower basal NO release than the males and there was no difference in ACh induced relaxation between ERKO and wild type mice, amongst males or females (Rubanyi, Freay, Kauser *et al.*, 1997). It is now recognised that there are two functionally distinct estrogen receptors, alpha and beta (Gustafsson, 1997), but which of these is involved in the interrelation with NOS in the vasculature is unknown.

Acute endothelium-independent vasodilatory effects of 17β-oestradiol presumably independent of the nuclear receptor have been observed at supraphysiological levels in isolated coronary arteries from animals (Jiang *et al.*, 1991) and humans (Chester *et al.*, 1995). The mechanism of action is in part mediated by antagonism of Ca^{++} influx through voltage-gated channels in vascular smooth muscle (Jiang *et al.*, 1992) and in part through potassium channels (White *et al.*, 1995). It has been suggested that 17β-oestradiol stimulates cGMP-dependent phosphorylation of the Ca^{++} activated potassium channel, leading to potassium efflux, membrane repolarisation and relaxation of vascular smooth muscle (White *et al.*, 1995).

Prostacyclin in the Maternal Circulation

Although previously a contentious issue, it is now considered that prostacyclin (PGI_2) is more likely to play a role as a local autocoid than to contribute to the lowering of peripheral resistance in normal pregnancy. Since the half life of PGI_2 is so short, evaluation of synthesis depends on the measurement of stable metabolites. Using a sensitive and specific analytical method for the detection of the non-enzymic circulating prostacyclin metabolite, 6-oxo-$PGF_{1\alpha}$, Barrow *et al.* (1983) reported that concentrations were too low for PGI_2 to function as a circulating hormone, despite a significant upward trend during pregnancy. This conclusion was upheld by studies in pregnant animals and women in which infusion of indomethacin was shown not to affect blood pressure or peripheral resistance (Conrad and Colpoys, 1986; Sorensen *et al.*, 1992). Urinary excretion of 2,3-dinor-6-keto-PGF_{1a}, the major systemic enzymic metabolite of PGI_2, is also raised early during human pregnancy and increases with each trimester (Goodman *et al.*, 1982; Fitzgerald *et al.*, 1987). Whilst reflecting overall PGI_2 biosynthesis, measurement of this metabolite in the urine cannot discriminate between maternal and fetal sources, and may also reflect renal synthesis. In the sheep, PGI_2 biosynthesis seems to be increased preferentially in the uterine circulation during pregnancy, possibly in response to elevated AII (Magness *et al.*, 1992). Pregnancy in the ewe is also associated with a dramatic rise in the expression of COX-1 mRNA and protein in the uterine artery endothelium (Janowiak *et al.*, 1998).

Human Pregnancy is also associated with increased synthesis of the constrictor prostanoid, thromboxane (TXA_2), as assessed by measurement of its stable systemic metabolite 2,3-dinor-TXB_2. TXA_2, which in pregnancy seems to be mainly derived from platelets (Fitzgerald *et al.*, 1987b) increases 3-5 fold during gestation and remains elevated throughout (Fitzgerald *et al.*, 1987a).

Endothelium Derived Hyperpolarizing Factor (EDHF) in the Maternal Circulation

Several studies have suggested that prostaglandin and NO independent, but endothelium-dependent mechanisms of relaxation may be enhanced in human pregnancy (McCarthy *et al.*, 1994; Pascoal *et al.*, 1996), an observation supported recently in pregnant rat aorta (Bobadilla *et al.*, 1997) and from our laboratory in pregnant rat mesenteric small arteries (Gerber *et al.*, 1998). In our study the component of relaxation to acetylcholine which was insensitive to cyclooxygenase blockade or to inhibition of NO synthase, was greater in pregnant rats compared with virgin animals and totally inhibited in both groups by slight membrane depolarization (induced by modest elevation of the potassium concentration in the organ bath). This is strongly indicative of a role for enhanced synthesis of the putative endothelium derived hyperpolarizing factor (EDHF) (for review see Garland *et al.*, 1995). The exact nature of EDHF has proven to be elusive, although it is suggested by some to be a cytochrome P450 derivative.

Endothelin in the Maternal Circulation

The plasma concentration of ET-1 is not affected by normal pregnancy and is very low or undetectable in maternal plasma (Wolff *et al.*, 1997). However, ET-1 causes potent constriction of myometrial arteries from pregnant (Wolff *et al.*, 1993) and non-pregnant women (Fried and Samuelson, 1991), being more potent than all other constrictors tested (Wolff *et al.*, 1993). ET-1 was more potent than ET-3 (Wolff *et al.*, 1993) and potentially plays an important role in the regulation of uteroplacental blood flow.

ENDOTHELIAL CONTROL OF THE FETOPLACENTAL CIRCULATION

NO in the Fetoplacental Circulation

NO induced dilatation undoubtedly contributes to the maintenance of low resistance and pressure in the fetoplacental circulation. Infusion of a NOS inhibitor into a perfused human placental cotyledon leads to an increase in perfusion pressure (Gude *et al.*, 1990). Endothelial derived NOS has also been identified by immunohistochemical techniques in placental vessels (Myatt *et al.*, 1993; Buttery *et al.*, 1994) and is strongly expressed in syncytiotrophoblast, leading to the suggestion that it may play a role in dilatation of the maternal spiral artery prior to trophoblast invasion. However, the recent demonstration of lack of NOS expression in invading trophoblast would argue against that proposal (Lyall *et al.*, 1998). In our laboratory we have used the perfusion myograph to show *in vitro* flow mediated release of NO by small fetoplacental arteries (Learmont and Poston, 1996), which was inhibited by (L-NAME). Interestingly, agonists including ACh, histamine and bradykinin which are potent endothelium dependent dilators in other circulations through the release of NO, have little or no effect in small fetoplacental arteries (McCarthy *et al.*, 1994).

The syncytiotrophoblast of normal placenta synthesises corticotrophin releasing hormone (CRH), as do the fetal membranes (Perkins and Linton, 1995). Maternal serum concentrations of CRH rise in normal pregnancy and are even higher in abnormal pregnancies (Wolfe *et al.*, 1988). Despite this, it is as yet not clear whether the increased CRH levels in pregnancy have any functional significance. CRH may however, play a role in endothelium dependent dilatation in the placental circulation. In the perfused placental cotyledon preparation, CRH leads to a reduction in perfusion pressure, mediating vasodilatation through NO release (Clifton *et al.*, 1995). In contrast, we have observed that CRH has no direct vasodilatory effect on isolated small fetoplacental arteries (Dixon *et al.*, 1996) suggesting that in the intact lobule, CRH may have an indirect effect of vascular tone, potentially through release of NO from syncytiotrophoblast.

Prostacyclin in the Fetoplacental Circulation

Prostacyclin (PGI_2) applied directly to isolated stem villous arteries of the fetoplacental resistance vasculature effects considerable relaxation (Maigaard *et al.*, 1986a). More PGI_2, relative to TXB_2 is synthesised in umbilical vessels than maternal vessels (about 50:1 for umbilical veins and 20:1 for umbilical arteries, as compared with 5:1 for mammary vessels) (Benedetto *et al.*, 1987). A relatively high rate of PGI_2 production in these vessels ensures the well being of the fetus by contributing to a high blood flow and preventing thrombosis. Furthermore, more PGI_2 is produced in the juxtafetal portion of both umbilical arteries and veins compared with the juxtaplacental umbilical vessels (Benedetto *et al.*, 1987). The half life of PGI_2 in the placenta, however, is very short, as prostanoids are rapidly enzymatically inactivated in the placenta (Thaler-Dao *et al.*, 1974). Agonists which stimulate both PGI_2 and NO from the vascular endothelium e.g. histamine, consistently lead to greater NO induced relaxation than prostanoid mediated relaxation in umbilical arteries, indicating a relatively greater physiological role for NO than PGI_2 (Izumi *et al.*, 1996).

Endothelins in the Fetoplacental Circulation

The endothelins are powerful constrictors of the fetoplacental resistance vasculature (MacLean, Templeton and McGrath, 1992). Plasma levels of ET-1 in the umbilical artery are very high at birth (Nisell *et al.*, 1990) and further raised in pregnancies associated with fetal hypoxia (Ohno *et al.*, 1995). Placental arteries also develop enhanced sensitivity to endothelin in pregnancies associated with growth retardation (Liu *et al.*, 1995). Part of the constrictor activity of endothelin in the placental vasculature has been attributed to synthesis of the constrictor prostanoid TXA_2 (Howarth *et al.*, 1995). ET_A and ET_B receptors have been identified in the human placental vasculature; ET_A binding sites predominated on veins and arteries in the chorionic plate, while ET_B binding sites were present in high density in stem villi and blood vessels (Rutherford *et al.*, 1993). Endothelin is present in umbilical arteries and veins at term (Haegerstrand *et al.*, 1989), but not early in pregnancy (Sexton *et al.*, 1996). ET-1 also occurs at higher concentrations in fetal compared with maternal blood (Hakkinen *et al.*, 1992; Malamitsi-Puchner *et al.*, 1995). ET-1 and ET-3 have also been shown to contract human fetoplacental veins (Mombouli, Wasserstrum and Vanhoutte, 1993). The endothelins may therefore play a physiological role in the control of the fetoplacental circulation, particularly at the moment of childbirth when constriction of the umbilical vasculature would be beneficial. High concentrations in fetal blood could also indicate a role in fetal development as this peptide is also a growth factor.

PRE-ECLAMPSIA

Pre-eclampsia is a multi-system disorder unique to human pregnancy (Williams and de Swiet, 1997). It is classically recognised by hypertension, proteinuria and oedema, affects about three percent of primigravidae and almost invariably occurs after 20 weeks gestation. Yet these signs conceal the true identity of a disorder that may be recognised prior to the onset of hypertension and even evolve into eclampsia (convulsions), without hypertension. Pre-eclampsia remains an important cause of maternal death in Western Europe (Department of Health, 1998) and the USA (Kaunitz, Hughes, Grimes *et al.*, 1990).

Relative to the vasodilated, plasma expanded state of healthy pregnancy, pre-eclampsia is a vasoconstricted, plasma contracted condition with evidence of intravascular coagulation. The fetus is particularly vulnerable from early onset disease and the mother can succumb following an illness that can include liver haemorrhage and necrosis, acute renal failure, subendocardial necrosis, microangiopathic haemolysis, a consumptive coagulopathy and convulsions. The most common causes for maternal death are adult respiratory distress syndrome (ARDS) and intra-cerebral haemorrhage (Department of Health, 1998).

Whereas healthy maternal endothelium is crucial for the physiological adaptation to normal pregnancy, the multiple organ failure characteristic of severe pre-eclampsia is predominantly secondary to widespread endothelial cell dysfunction (Roberts and Redman, 1993). It is not yet known whether the endothelium of women destined to develop pre-eclampsia fails to adapt properly, or is damaged by unknown factors during a pre-eclamptic pregnancy. Prior to the onset of clinically identifiable disease, women destined to develop pre-eclampsia show evidence of poor placentation (Brosens, Robertson and Dixon, 1972) and high uteroplacental resistance (Bower *et al.*, 1993). How high placental vascular resistance triggers maternal endothelial dysfunction is still not completely understood.

Abnormalities of the Uterine Vasculature

In normal pregnancy, the most striking change to a maternal artery occurs within the small spiral arteries of the uterus. These terminal branches of the uterine arteries are the final pathway by which jets of oxygenated maternal blood are delivered into the intervillous space of the growing placenta. Normally, during the first trimester, the terminal branches of the spiral arterioles are invaded by placental cytotrophoblast cells (Brosens, Robertson, and Dixon, 1967; Pijnenborg *et al.*, 1980). The muscular and elastic components of the spiral arteriole wall are replaced by a fibrinoid layer of variable thickness, in which trophoblast cells are embedded. Normally, there is no disruption of the endothelium (Khong, Sawyer, and Heryet, 1992). Remodelling of the utero-placental vessels is normally complete by 18 weeks gestation (Matijevic *et al.*, 1995), but in women who develop pre-eclampsia, cytotrophoblast invasion of the spiral arteries is incomplete and high resistance vessels that retain their muscular wall, persist until term (Khong *et al.*, 1986; Zhou *et al.*, 1993). This incomplete placentation is considered to play an important role in pre-eclampsia, but as it also occurs in intra-uterine growth retardation, cannot be considered unique to pre-eclampsia. Furthermore, the endothelium of these abnormal utero-placental arteries is disrupted by intraluminal endovascular trophoblast (Khong *et al.*, 1992) (see later). Prior to any clinical symptoms or signs, doppler ultrasound of uterine arteries at 20-24 weeks gestation, can aid in the identity of women with high resistance vessels, at risk of pre-eclampsia (Bower, Bewley and Campbell, 1993).

The Sympathetic Nervous System in Pre-eclampsia

Although this chapter concentrates on the role of the endothelium in pre-eclampsia, some investigators have suggested that the pathological increase in vascular tone is secondary to changes in the autonomic nervous system. Women with pre-eclampsia have been found to have increased sympathetic nerve activity in muscle-nerve fasicles of the peroneal nerve (Schobel *et al.*, 1996) and higher plasma noradrenaline levels compared with normotensive women (Manyonda *et al.*, 1998). In this latter study, tyrosine hydroxylase activity and mRNA levels were greater in placental tissue from pre-eclamptic compared with normotensive pregnancies. It is proposed that excessive noradrenaline breaks down more triglyceride to free fatty acids, which are then oxidised to lipid peroxides (Manyonda *et al.*, 1998). The latter are cytotoxic to endothelial cells (see later). However, systemic blockade of the autonomic nervous system with tetraethylammonium chloride (TEAC) or high spinal anaesthesia causes a dramatic fall in blood pressure in normotensive pregnant women, but is much less effective in women with pre-eclampsia (Assali and Prystowsky, 1950). Despite the potential non-specificity of this method, this unique study would suggest that the hypertension of pre-eclampsia is mediated by a factor independent of the autonomic nervous system.

INVESTIGATION OF ENDOTHELIAL FUNCTION IN PRE-ECLAMPSIA

Markers of Endothelial Dysfunction

Healthy endothelial cells maintain vascular integrity, prevent platelet adhesion and influence the tone of underlying vascular smooth muscle. Damaged endothelial cells are unable to perform these three functions leading to increased capillary permeability, platelet thrombosis and increased vascular tone (Flavahan and Vanhoutte, 1995). These features are found in pre-eclampsia and suggest that the maternal syndrome is, at least in part, an endothelial disorder (Roberts *et al.*, 1989; de Groot and Taylor, 1993). Evidence of endothelial cell damage prior to clinical manifestation of pre-eclampsia can be demonstrated by the presence of markers of endothelial cell activation. Specifically, levels of fibronectin (Ballegeer *et al.*, 1989) and Factor VIII related antigen are elevated (Roberts and Redman, 1993). Furthermore, women with endothelial cell damage, secondary to pre-existing hypertension or other micro-vascular disease, have a higher incidence of pre-eclampsia than normotensive women (Ness and Roberts, 1996).

Morphological evidence of endothelial damage in pre-eclampsia can be seen in the glomerular capillaries (Figure 13.5) and in the utero-placental arteries (Figure 13.6) — a vasculopathy known as acute atherosis.

Endothelin in the Maternal Circulation in Pre-eclampsia

Women with pre-eclampsia also have higher plasma endothelin (ET-1) concentrations than women with a normal pregnancy (Wolff *et al.*, 1996a). This perhaps would be anticipated, as a marker of the generalized endothelial dysfunction, but nonetheless could contribute to vasoconstriction. Enhanced sensitivity to endothelin is unlikely to contribute to vasoconstriction of the maternal vasculature as there is no evidence to suggest an increase in

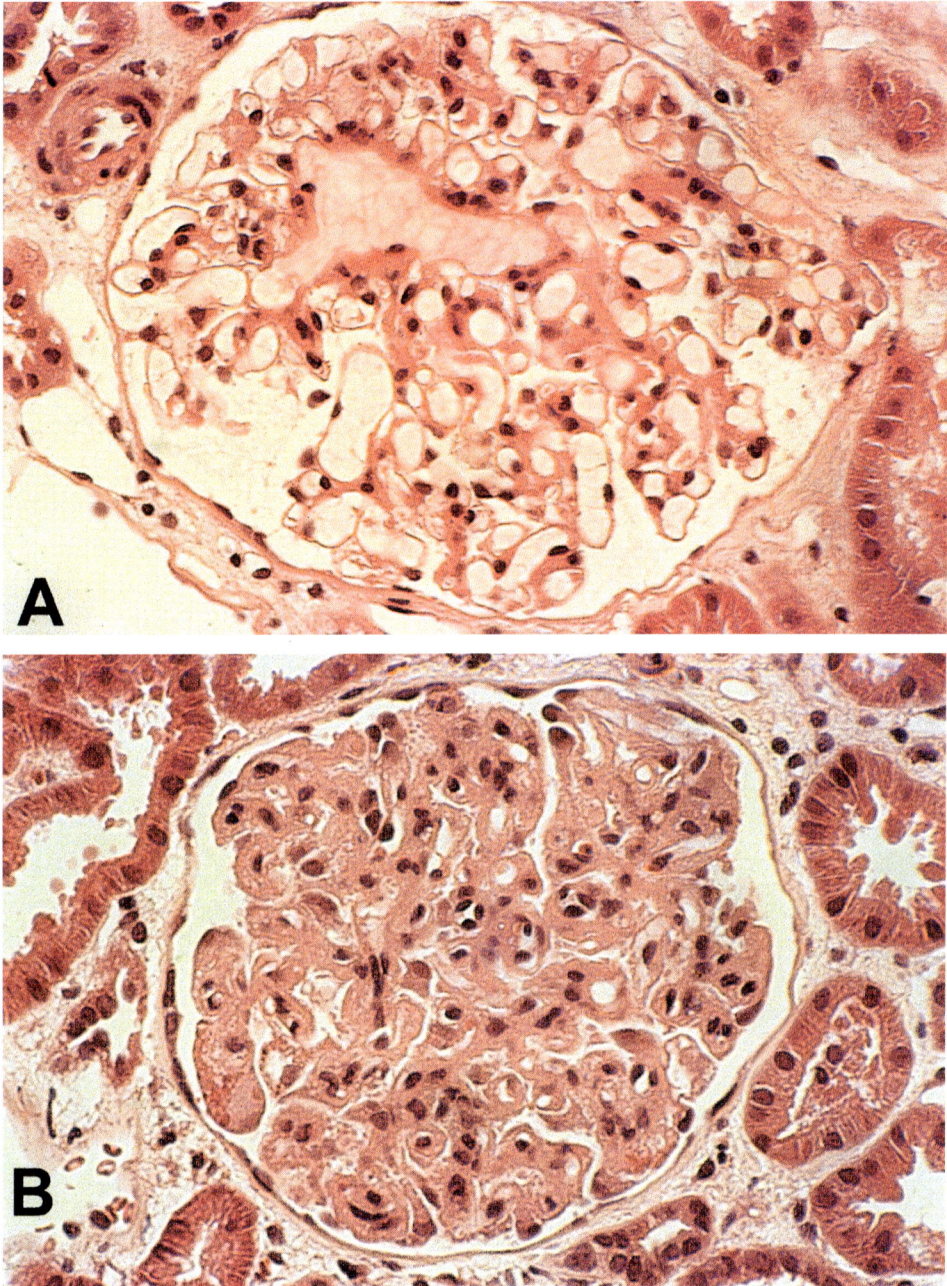

Figure 13.5 (A) Normal renal glomerulus with patent capillary lumen. (B) Glomerular endotheliosis in a 19 year old primigravida with pre-eclampsia at 25 weeks gestation. The capillary lumen have been obliterated by swollen capillary endothelial cells. Photograph kindly provided by Dr. Meryl Griffiths, Department of Pathology, University College London Medical School.

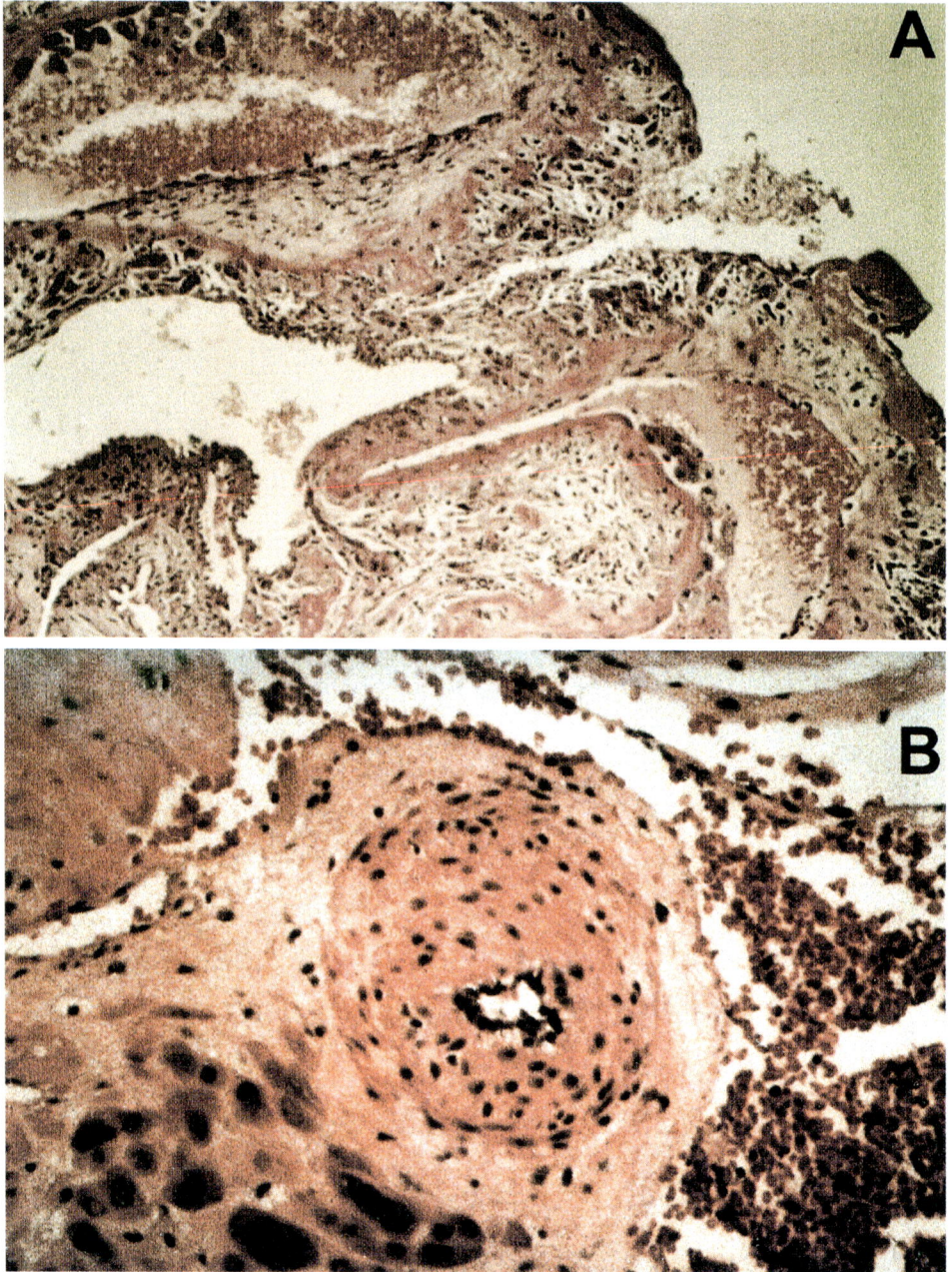

Figure 13.6 Two placental bed biopsies taken from different women at 31 weeks gestation (haematoxylin and Eosin stain, both x40). (A) Normal decidual vessel (just below middle of picture, collapsed flat due to lack of blood). There is a thin layer of media, surrounded by fibrinoid material, (B) Decidual vessel from a woman with severe pre-eclampsia and intra-uterine growth retardation. There is gross intimal hyperplasia with a much reduced lumen. In some severe cases of pre-eclampsia, (not seen in this example) the endothelium is replaced by cholesterol-laden macrophages, hence the term 'atherosis'. Photographs kindly provided by Prof. Steve Robson, Department of Fetal Medicine, Newcastle upon Tyne.

responsiveness to ET-1 in pre-eclampsia. Studies in isolated omental (Vedernikov *et al.*, 1995; Wolff *et al.*, 1996a) and myometrial arteries (Wolff *et al.*, 1996b) have shown similar responses in arteries from normal and pre-eclamptic women. One report suggests urinary excretion of ET-1 is reduced in women with pre-eclampsia (Clark *et al.*, 1997) and as endothelin is natriuretic, could be implicated as a cause of sodium retention.

Nitric Oxide in Pre-eclampsia

The L-arginine-NO pathway is an expected casualty of endothelial cell damage in pre-eclampsia. However, probably because of methodological limitations there is no consensus on whether NOS activity is altered by pre-eclampsia. Most studies have either shown no change (Cameron *et al.*, 1993; Curtis *et al.*, 1995; Silver *et al.*, 1996), or an increase (Nobunaga *et al.*, 1996; Smarason *et al.*, 1997) in circulating or urinary NO metabolites in women with pre-eclampsia, probably reflecting variable nitrate intakes. Only Seligman *et al.*, 1994 documented lower plasma NOx concentrations in women with pre-eclampsia compared with normotensive controls. Cameron *et al.* (1993) and Nobunaga *et al.* (1996) documented a correlation between systolic blood pressure and *increasing* urinary and plasma concentrations of NOx, respectively. In the latter study, volunteers were starved for 12–15 hours in an attempt to control for dietary nitrogen.

NOS is competitively inhibited by an endogenous guanidino-substituted arginine analogue, N^GN^G dimethylarginine (asymmetric dimethylarginine, ADMA). During normotensive pregnancy, the plasma concentration of ADMA is lower than non-pregnant women and those with pre-eclampsia (Fickling *et al.*, 1993). The fall in plasma ADMA concentration occurs in the first trimester and correlates with the gestational fall in blood pressure (Holden *et al.*, 1997). Furthermore, plasma ADMA levels are significantly higher in women with pre-eclampsia than gestationally matched, normotensive controls (Holden *et al.*, 1997). Consequently, endogenous inhibition of NOS by a specific inhibitor is a possible mechanism whereby NO production could be reduced in pre-eclampsia.

Nitric Oxide in the Resistance Vasculature in Pre-eclampsia

As outlined above, the lack of conformity amongst the studies investigating agonist stimulated NO in the resistance vasculature from healthy pregnant women is disappointing. However, there is much wider agreement that agonist mediated NO synthesis is reduced in small arteries from women with preeclampsia. As discussed above, pre-eclampsia is associated with many facets of endothelial cell dysfunction and the abnormal relaxation to agonists is probably yet another indication of a general endothelial cell defect. In small arteries from the subcutaneous circulation, sensitivity to both ACh (McCarthy *et al.*, 1994) and to BK (Knock *et al.*, 1996) is reduced in women with pre-eclampsia. Similarly, in the omental circulation of women with pre-eclampsia, Pascoal *et al.*, (1998) found relaxation to ACh was totally absent whilst responses to BK were unaffected when compared to normotensive controls. In small myometrial arteries from pre-eclamptic women mounted on a wire myograph, relaxation to BK was completely absent, whereas normotensive controls relaxed well (Ashworth *et al.*, 1997). In small arteries from the same circulation, but mounted on a pressurized system, maximal responses to ACh were reduced (Kublickiene *et al.*, 1998). Whilst largely in agreement as to the loss of endothelium dependent dilatation, these studies are equivocal regarding the role of NO. Knock *et al.*, (1996) and Klubickiene *et al.* (1998) have suggested that the blunting of responses to endothelium dependent dilators is due to reduced NO synthesis, but Pascoal *et al.* (1988) have suggested that the defect is NO independent. Flow mediated responses, largely NO dependent, are however, severely blunted in small subcutaneous arteries from women with preeclampsia (Cockell and Poston, 1997).

Prostanoids in the Maternal Circulation in Pre-eclampsia

In contrast to normal pregnancy, pre-eclampsia is associated with relative underproduction of PGI_2 and over abundance of TXA_2 (Fitzgerald *et al.*, 1990). The imbalance between the synthesis of these prostanoids formed the rationale for investigations of "low dose aspirin" therapy for prevention of pre-eclampsia. Whereas aspirin in excess of 80 mg/day substantially inhibits both PGI_2 and TXA_2, low or intermittent doses lead to preferential inhibition of TXA_2 biosynthesis (Ritter *et al.*, 1989), and could redress the imbalance between these prostanoids in pre-eclampsia. Selectivity for TXA_2 may lie in differential access of aspirin to platelet cyclooxygenase in the portal circulation (Pedersen and FitzGerald, 1984) with presystemic metabolism preventing access of aspirin to the systemic vasculature and placenta (for review see Dekker and Sibai, 1993) and/or in the differential affinity of aspirin for cyclooxygenase in platelets (Patrignani *et al.*, 1982). Also platelets, being anuclear, lack the capacity to regenerate cyclooxygenase, whereas endothelium can regenerate the enzyme and so maintain PGI_2 production (Ritter *et al.*, 1989). The potential benefit of aspirin had also been highlighted by the early indications of clinical improvement in patients with pre-eclampsia (Crandon and Isherwood, 1979).

The first, small placebo controlled trials which followed were supportive; the first, a double blind study, (Wallenburg *et al.*, 1986) showed that aspirin (60 mg/day) given from 28 weeks of gestation significantly lowered the incidence of pre-eclampsia in a group of 'high risk' women defined in terms of abnormal pressor responsiveness to AII. Despite several susbequent small, but supportive studies, this early promise was not fulfilled. The largest randomised trial of low dose aspirin to date, the UK based Collaborative Low Dose Aspirin Study in Pregnancy included 9364 women (CLASP Collaborative Group, 1994). Overall there was no evidence of any clinically important benefit in women who took 60 mg aspirin rather than placebo. A post hoc analysis of the CLASP data did however show that the earlier low dose aspirin was started, the greater the protection from pre-eclampsia (Broughton Pipkin *et al.*, 1996). A subsequent study of prophylactic low dose aspirin (60mg/day) given to women at high risk of pre-eclampsia from between 13–26 weeks gestation (arguably too late to be of benefit) conferred no significant benefit compared with high risk women given placebo (Caritis *et al.*, 1998).

Effect of Sera From Pre-eclamptic Women on Cultured Endothelial Cells

Several experiments have shown changes to different types of cultured endothelial cells (human umbilical vein (HUVEC), human decidual tissue, animal and fetal cells) following exposure to plasma or sera from normal pregnant or pre-eclamptic women. Initial reports tended to show that pre-eclamptic sera contained elements that were cytotoxic to endothelial cells (Rogers *et al.*, 1988; Tsukimori *et al.*, 1992). However, more recent studies have shown that pre-eclamptic sera is not cytotoxic to HUVEC (Endresen *et al.*, 1995; Zammit, Whitworth and Brown, 1996). Methods used included the release from endothelial cells of [51]Chromium, the integrity of cell membranes as judged by the number of trypan blue positive cells, incorporation of tritiated thymidine and leucine (reflecting DNA and protein synthesis) and overall cell proliferation (Endresen *et al.*, 1992).

The suggestion that pre-eclampsia was associated with reduced PGI_2 synthesis (see above) prompted investigators to determine whether a blood borne factor or factors could inhibit endothelial PGI_2 production. Surprisingly a study by Baker *et al.* (1996a) has shown that plasma from women with pre-eclampsia leads to stimulation of PGI_2 release from bovine coronary microvascular endothelial cells, followed by a subsequent fall. However, a follow up study from the same group has shown that flow across the endothelial cells (i.e. mechanical shear stress) negated the difference between the response to normal and pre-eclamptic serum (Baker *et al.*, 1996b). Another group has reported that endothelial cells isolated from maternal decidual tissue of women with pre-eclampsia synthesised an equal amount of PGI_2 compared with decidual endothelial cells from normal pregnant women (Gallery *et al.*, 1995a). In accord with Baker's first study, incubation with serum from women with preeclampsia, led to an increase in PGI_2 synthesis from maternal (decidual) and fetal (umbilical vein) endothelial cells and a further increase in PGI_2 synthesis if the decidual cells originated from women with preeclampsia (Gallery *et al.*, 1995b). In contrast, another study has shown that pre-eclamptic serum did not alter PGI_2 synthesis from HUVECs and was not cytotoxic to endothelial cells after short term incubation (Zammit *et al.*, 1996).

The conflicting results from this group of experiments are likely to represent differences in experimental method. It is very difficult to draw any other conclusion from studies that use different concentrations of serum or plasma (ranging from 2%–30%, made up in different medium), applied to different types of endothelial cells, some from the fetal circulation and some even from other animals.

Prothrombotic States

Stimulation of the coagulation cascade in response to endothelial cell damage is more likely in women who have a predisposition to thrombosis. In an uncontrolled study of women with a history of early severe pre-eclampsia, Dekker *et al.*, 1995 found a significant association with a subclinical prothrombotic state. Twenty five percent had protein S deficiency, 18% had hyperhomocysteinaemia and 16% had activated protein C resistance (Dekker *et al.*, 1995). Similarly, 14 of 158 (8.9%) women with severe pre-eclampsia were heterozygous for the factor V Leiden mutation as compared with 17 of 403 (4.2%) normotensive gravid controls (Dizon-Townson *et al.*, 1996). It is possible that during pregnancy, the normally sub-clinical prothrombotic state causes utero-placental thrombosis and placental ischaemia which then triggers the maternal syndrome of pre-eclampsia.

AII sensitivity in pre-eclampsia

The classical study of Gant *et al.*, (1973) showed that at 22–26 weeks gestation, prior to the onset of clinically identifiable pre-eclampsia, a group of predominantly black, teenage American women demonstrated greater sensitivity to an infusion of ANG II compared with women who did not develop pre-eclampsia. Compatible with underlying endothelial dysfunction, increased sensitivity to this pressor agent was acclaimed as a useful screening test for pre-eclampsia. However, the positive predictive power of this original study has not been reproducible in six other populations (Kyle *et al.*, 1995). Indeed, the most comprehensive study, involving 495 healthy nulliparous women studied at 28 weeks gestation, revealed a positive predictive value of only, 19% (Kyle *et al.*, 1995). Differences in the populations studied may explain this disparity, but overall this invasive test, originally heralded as a predictive test for pre-eclampsia is of no clinical uses in predicting pre-eclampsia in an otherwise healthy European population.

Nitric Oxide in the Fetoplacental Circulation in Pre-eclampsia

Investigations of the placentae from women with pre-eclampsia using immunohistochemical methods suggest, in contrast to the majority of studies in the *maternal* circulation, that the expression of the endothelial isoform of NOS (eNOS) is increased (Ghabour, Eis, Brockman, Pollock & Myatt, 1995). Moreover, Lyall *et al.* (1996) documented an increase in plasma NO concentrations (measured using the Greiss reaction) in fetal blood from pregnancies complicated by intrauterine growth restriction. These studies both suggest that NOS could increase as a compensatory response to poor blood flow. Theoretically, this may occur as a result of an increase in shear stress arising from elevation of vascular resistance, since shear is a stimulus to NOS activation. The function of the eNOS identified in the syncytiotrophoblast of normal placentae has yet to be explained but an intriguing, although as yet not proven, suggestion is that it may play a role in dilatation of the maternal spiral arteries during placentation. However, NO is a potent antiaggregatory agent, reduces leucocyte adhesion and has bactericidal effects. These properties may also be of benefit in the placenta.

In contrast to the studies showing enhanced NOS in the fetoplacental circulation by immunohistochemical methods, comparison of human umbilical vein endothelial cells isolated from normotensive and pre-eclamptic women has shown a similar capacity to produce NO (Orpana *et al.*, 1996). This was assessed by the accumulation of nitrate and nitrite in the culture medium and related to the number of viable endothelial cells. Interestingly the total number of viable endothelial cells and consequently the total NO production was reduced in the pre-eclamptic group (Orpana *et al.*, 1996). Extrapolation from larger fetoplacental vessels to the smaller resistance vasculature must however be made with caution.

Endothelin in the Fetoplacental Circulation in Pre-eclampsia

Elevated concentrations of ET-1 in the umbilical vein in women with pre-eclampsia have been reported to relate inversely to birth weight (Wolff *et al.*, 1996) and may suggest a pathological role for endothelin in reduction of uteroplacental blood flow. In contrast to the maternal circulation, however, a certain refractoriness seems to develop to the elevated concentrations of ET-1, as reduced sensitivity to the peptide has been observed in umbilical arteries from preeclamptic pregnancies (Bodelsson *et al.*, 1995). Again, extrapolation to the resistance vasculature and the control of placental blood flow may be inadvisable, as many differences exist in responsiveness to vasoactive agents in umbilical and small placental arteries.

Aetiology of Maternal Endothelial Dysfunction in Pre-eclampsia

How poor placentation and the resultant poor uterine blood flow leads to the maternal syndrome of pre-eclampsia, characterised by wide spread endothelial cell damage, remains

The Endothelium in Pre-eclampsia

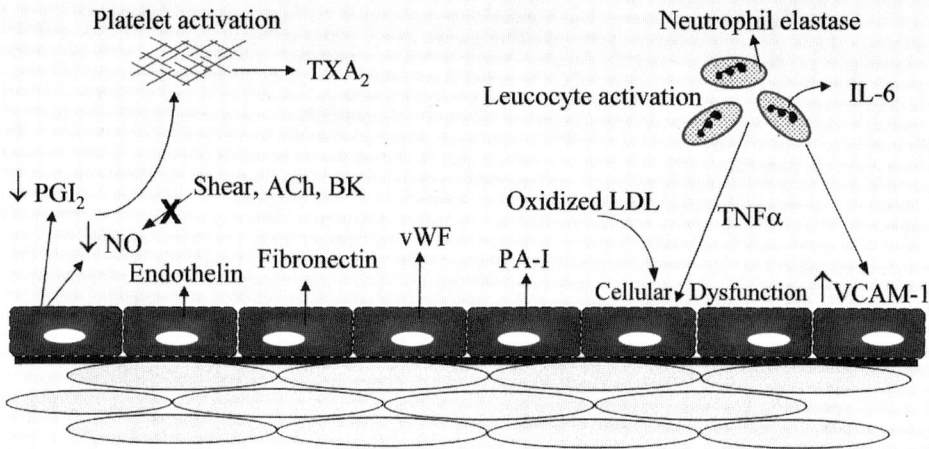

Figure 13.7 Possible mechanisms for damaged endothelium in pre-eclampsia (see text for details). Evidence for the following endothelium toxic factors exists; oxidative stress causing lipid peroxidation; cytokine activation of neutrophils with increased adhesion and toxicity to endothelium; increased markers of endothelium dysfunction (vWF and fibronectin); decreased vasodilatory response to shear stress (PEI$_2$); decreased production of prostacyclin and probably nitric oxide (NO); increased production of plasminogen activator inhibitor (PA-I) to inhibit fibrinolysis.

uncertain. Over the years there have been many theories, but currently three or four predominate (Figure 13.7). Substantial evidence now supports a role for oxidative stress. The imbalance between free radical synthesis and antioxidant capacity may arise from reduced placental perfusion coupled with dyslipidaemia. Alternative hypotheses include a role for pro-inflammatory cytokines and pro-thrombotic states. Others suggest that deported trophoblast fragments may be responsible for maternal endothelial cell damage. Before the pathogenesis of pre-eclampsia is discussed in more detail, it is worth remembering that there are other gestational syndromes, that overlap with pre-eclampsia and which have in common a dysfunctional endothelium.

These include the HELLP syndrome, an acronym for Haemolysis, Elevated Liver enzymes and Low Platelets, which overlaps with pre-eclampsia in about 20% of cases (Sibai *et al.*, 1993) and acute fatty liver of pregnancy (AFLP), which is much more rare, but also associated with pre-eclampsia in about 50% of cases (Mabie, 1992). Although the liver pathology is different between HELLP (periportal necrosis) and AFLP (central zone necrosis), the extra-hepatic manifestations are similar and include a microangiopathic haemolytic anaemia (MAHA). Indeed, pregnancy predisposes to MAHA, including both haemolytic uraemic syndrome (HUS) and thrombotic thrombocytopaenic purpura (TTP) (Sibai, Kustermann and Velasco, 1994). Although HELLP and AFLP are both improved by delivery, HUS and TTP are not. Indeed, most cases of HUS occur immediately postpartum. Patients with HUS/TTP have platelet rich thrombi occluding arterioles and capillaries, which are widespread in TTP and predominate in the kidney with HUS. It is possible that the gestational activation of the endothelium during pregnancy combined with a normally sub-clinical prothrombotic tendency predisposes to MAHA. In support of this possibility, there have been several reports of HUS/TTP in association with the anticardiolipin antibody (Kniaz, Eisenberg, Elrad *et al.*, 1992). Anticardiolipin antibodies have been associated with pre-eclampsia in some studies (Yasuda, Takakuwa and Tokunaga *et al.*, 1995), but not in others (D'Anna Scilipati, Leonardi *et al.*, 1997). It is possible that anticardiolipin antibodies are transiently present during clinically active pre-eclampsia (D.J. Williams, unpublished observation).

Postpartum acute renal failure is another well recognised condition, associated with MAHA. The thrombotic occlusion of glomerular capillaries often leads to bilateral renal cortical necrosis. This used to be a common problem following septic abortion or delivery, but is now much more rare. There is great similarity between the pathology of postpartum acute renal failure and the generalised Shwartzman reaction – an experimentally induced condition in rodents which normally follows two separate injections of endotoxin. Pregnant animals appear to be primed and require only one injection of endotoxin before developing glomerular thrombosis (Conger, Falk and Guggenheim, 1981).

Oxidative Stress and Dyslipidaemia

Free radicals may lead directly to endothelial damage through direct cytotoxicity, or indirectly through the synthesis of lipid peroxides. The evidence for oxidative stress in pre-eclampsia is substantial. Reduced plasma concentrations of vitamin C (Mikhail *et al.*, 1994), glutathione peroxidase and of superoxide dismutase (Wang and Walsh, 1996; Poranen *et al.*, 1996) have been reported. Interestingly, women with pre-eclampsia are often reported to have higher plasma levels of the anti-oxidant vitamin E, which may be associated with the response to increased oxidative stress (Uotila, Tuimala and Aarnio, 1993; Schiff *et al.*, 1996). Raised concentrations of lipid peroxides are also reported, having been evaluated by estimation of antibodies to oxidized low density lipoprotein (LDL) (Branch, 1994), plasma concentrations of malondialdehyde (Hubel *et al.*, 1996) and of the isoprostane, 8-epi-PGF_{2a} (Barden *et al.*, 1996). Isoprostanes are formed by free radical induced oxidation of arachidonic acid and are increasingly recognized as stable lipid peroxidation products which may accurately reflect oxidative damage *in vivo*.

The origin of the reactive oxygen species may lie in the placenta. Placental ischaemia accelerates trophoblast cell turnover and so increases the concentration of purines which act as substrate for xanthine dehydrogenase/oxidase (Many, Hubel and Roberts, 1996). Under hypoxic conditions, xanthine oxidase predominates over xanthine dehydrogenase to produce urate and a reactive oxygen species. This process could explain why hyperuricaemia often precedes clinically recognisable pre-eclampsia and occurs prior to any fall in GFR (Gallery, Hunyor and Gyory, 1979). Leucocyte activation, whether a primary or a secondary factor in endothelial cell activation will also contribute to oxidative stress through the generation of superoxide.

The plasma lipid profile is also abnormal in pre-eclampsia, and will also contribute to the increased generation of lipid oxidation products. Towards the end of normal pregnancy, maternal plasma levels of cholesterol and triglyceride increase by 50% and 300%, respectively (Potter and Nestel, 1979). Women with pre-eclampsia have even higher circulating levels of triglyceride, free fatty acid and total cholesterol (van den Elzen *et al.*, 1996; Hubel *et al.*, 1996), with a relative increase in LDL cholesterol. Elevated serum levels of free fatty acids have been shown to increase triglyceride synthesis in cultured endothelial cells (Endresen, 1992). Under conditions of oxidant stress and hypertriglyceridaemia, increased amounts of unsaturated fatty acids will be oxidised to lipid peroxides (Chirico *et al.*, 1993). There is also a qualitative change in LDLs in established pre-eclampsia, with a shift towards small dense particles (Hubel *et al.*, 1997), a characteristic which predisposes the LDL particle to oxidation (Chait *et al.*, 1993).

Neutrophil Activation

Neutrophils and platelets are activated in normal pregnancy and further activated in pre-eclampsia (Greer *et al.*, 1989; Zemel *et al.*, 1990). Activated neutrophils, adhere to endothelium and mediate vascular damage by the release of proteases and reactive oxygen radicals. Neutrophil elastase, a specific marker of neutrophil activation, circulates in higher concentrations in women with pre-eclampsia compared with normotensive pregnant women (Greer *et al.*, 1989). Neutrophil adhesion to the endothelium is mediated through the expression of cell adhesion molecules on the endothelial cell surface. During endothelial cell activation, expression of certain cell adhesion molecules is increased on both neutrophils and the endothelium. Neutrophil adhesion is much more marked in women with pre-eclampsia, as they have increased expression of certain cell adhesion molecules compared with healthy normotensive pregnant women (Barden *et al.*, 1997). Specifically, vascular endothelial cell adhesion molecule (VCAM-1) and E-selectin circulate in higher concentrations in women with pre-eclampsia than in normotensive pregnant women (Lyall, Greer, Boswell *et al.*, 1994).

The stimulus to neutrophil activation remains unknown, but pro-inflammatory cytokines can activate neutrophils and simultaneously increase expression of cell adhesion molecules on endothelial cells (Lyall and Greer, 1996). Leucocyte TNF-α gene expression and circulating levels of TNF-α are enhanced in pre-eclamptic patients compared with normotensive and non-pregnant women (Chen, Wilson, Wang *et al.*, 1996; Kupfermine, Peaceman, Wigton *et al.*, 1994). Furthermore, in one study the frequency of the TNF1 allele was markedly increased in pre-eclamptic patients (Chen *et al.*, 1996). TNF-α can generate reactive oxygen species, inhibit NOS, favour synthesising thromboxane A2 over prostacyclin, change endothelial cells from an anti-haemostatic to a pro-coagulant state and activate transcription of VCAM-1 (Chen *et al.*, 1996). On the basis of its biological properties therefore, TNF-α is a strong candidate for mediating endothelial damage in pre-eclampsia. A unifying hypothesis is that an ischaemic placenta over-produces inflammatory cytokines, which activate neutrophils and mediate maternal endothelial cell damage and pre-eclampsia (Conrad and Benyo, 1997).

Trophoblast Deportation

Another theory for the aetiology of pre-eclampsia suggests that cellular material from the ischaemic placenta may be shed into the maternal circulation and lead to vascular damage. Trophoblasts are deported from the placenta to the maternal circulation in normal pregnancy, and in increased numbers in pre-eclampsia (Chua *et al.*, 1991). However, due to their size, few are likely to reach the arterial circulation. Histological

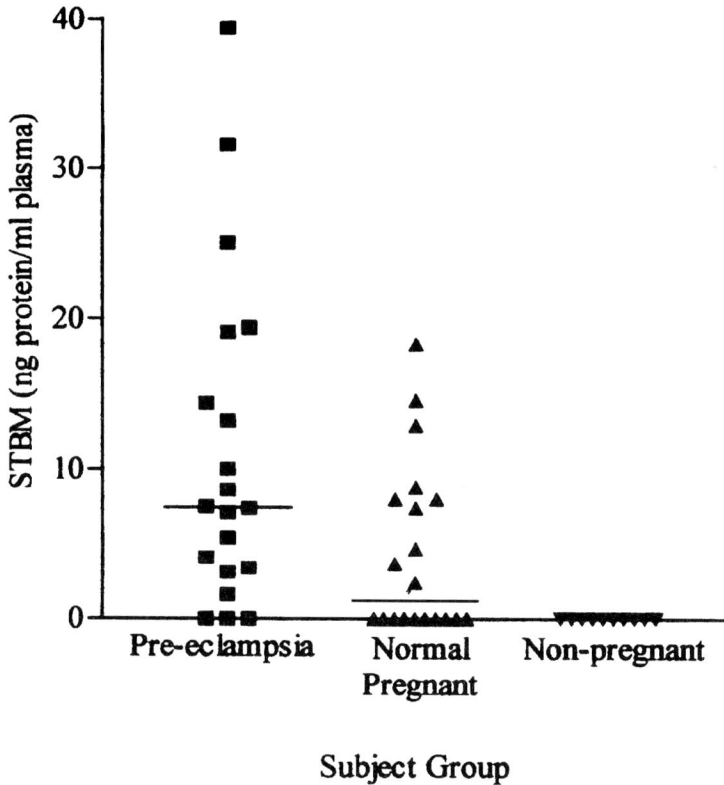

Figure 13.8 Syncytiotrophoblast microvilli detected in peripheral plasma from normal and pre-eclamptic women. Bars indicate median values. With permission from Knight *et al.* (1998), *British Journal of Obstetrics and Gynaecology*, **105**, 632–640.

evidence has suggested that the trophoblast microvilli on the surface of the placenta from women with pre-eclampsia are structurally different and fewer in number than those in normal placentae (Jones and Fox, 1980). This has led to the suggestion that trophoblast microvilli may be deported as well as the whole cells, a theory that has some strong support. Knight *et al.*, (1998) detected syncytiotrophoblast microvilli in the plasma of pregnant women by flow cytometry and fluoroimmunoassay using anti-placental alkaline phosphatase antibodies for detection. Significantly higher levels were found in women with pre-eclampsia (Figure 13.8). Concentrations were higher in uterine venous plasma than in concurrently sampled peripheral venous plasma, thus suggesting placental origin. The question remains as to whether these microvilli can inflict endothelial cell damage. Studies from the same group have shown that normal placental microvilli severely alter proliferation of cultured human umbilical venous endothelial cells (Smarason *et al.*, 1993) and in an investigation from one of our laboratories we have shown that perfusion of small sub-cutaneous arteries from pregnant women with microvilli from normal placenta can lead to a reduction in endothelium dependent relaxation (Cockell *et al.*, 1997) (Figure 13.9).

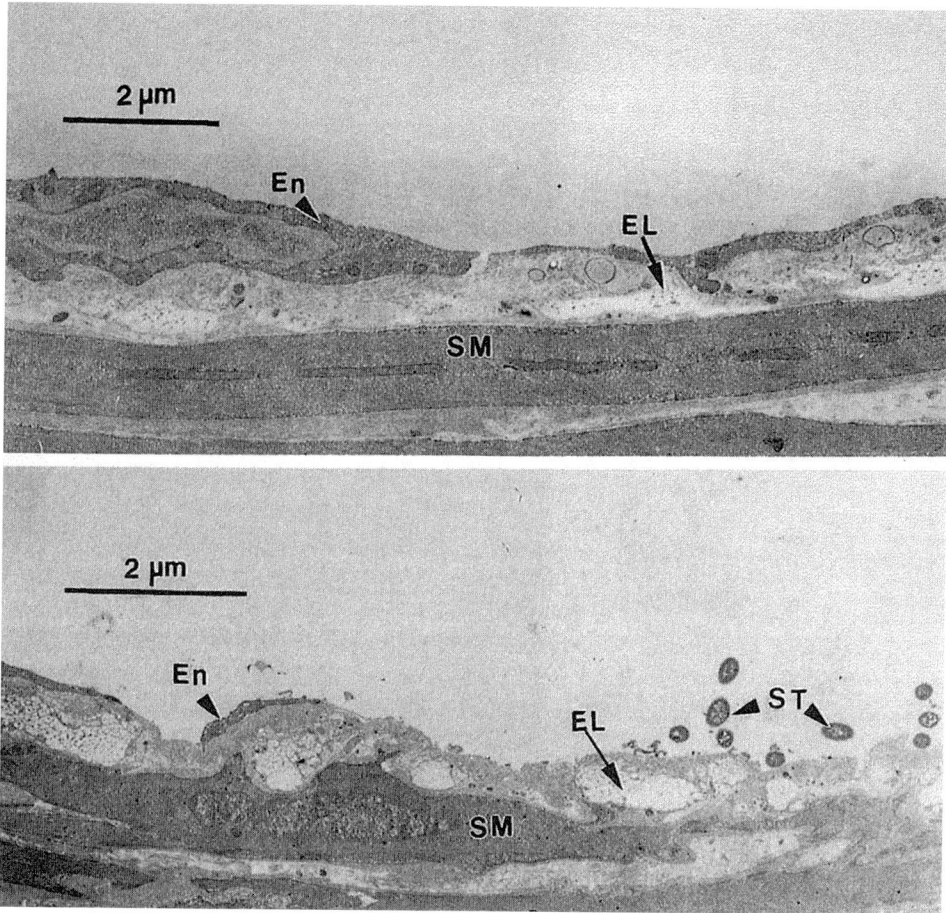

Figure 13.9 Transmission electron microscopy (magnification x 1250) of cross-section through luminal surface of maternal small subcutaneous fat artery (a) after perfusion with erythrocyte for 2 hours; (b) after perfusion with syncytiotrophoblast microvillous membrane vesicles for 2 hours. En = endothelium; SM = smooth muscle; EL = elastic; ST = syncytiotrophoblast microvillous membrane vesicles. With permission from Cockell, Learmont, Smarason *et al.* (1997) *British Journal of Obstetrics and Gynaecology*, **104**, 235–240.

CONCLUSIONS AND THE FUTURE

Despite the work of many investigators working in different fields who endeavour to understand the mechanism of pathophysiological change during pre-eclampsia, it is still unclear which are the most important pathogenic factors and which are simply innocent para-phenomenon. Conversely, attempts to understand the mechanisms of physiological change during healthy pregnancy have been relatively neglected. Interpretation of studies comparing pre-eclamptic with normotensive pregnancies would be put in context by simultaneous analysis of samples from healthy non-gravid controls. Without this information it is impossible to appreciate how much the physiological baseline has moved during healthy pregnancy.

Prevention of endothelial disruption during pre-eclampsia is a better strategy than treatment of its consequences. This may be one reason why attempts to redress the imbalance between prostacyclin and thromboxane with low dose aspirin have so far proved unhelpful in reducing the incidence of pre-eclampsis. However, it is still not clear whether a higher dose of aspirin 150mg daily, given from early in pregnancy (peri-conception) to women at high risk of pre-eclampsia would be beneficial. Furthermore, until we understand more about the activity of the L-arginine-NO pathway in pre-eclampsia, trials that supplement the NOS substrate L-arginine to women at risk of pre-eclampsia, would be premature. Donors of NO, must also be used with caution as they not only relax vascular smooth muscle, but have other, non-vascular actions such as ripening of the cervix. Antioxidants acting as free radical scavengers might be succesful as prophylaxis against endothelial cell dysfunction. Trials to compare the outcome of pregnancies in women at high risk of pre-eclampsia supplemented with a diet of antioxidants are already under way.

In conclusion, the endothelium plays a central role in the maternal adaptation to healthy human pregnancy. The peripheral circulation of the healthy mother is vasodilated and prothrombotic, but prone to endothelial cell disorders, such as pre-eclampsia — whatever it's aetiology. Delivery of the baby is the only definitive way to cure pre-eclampsia, but other endothelial disorders, such as HUS/TTP, take advantage of the prothrombotic endothelium and occur more frequently just after delivery.

REFERENCES

Ahokas, R.A., Friedman, S.A. and Sibai, B.M. (1997) Effect of indomethacin and N omega-nitro-L-arginine methyl ester on the pressure/flow relation in isolated perfused hindlimbs from pregnant and non-pregnant rats. *Journal of the Society of Gynecological Investigation*, **4**, 229–235.

Allen, R., Castro, L., Arora, C., Krakov, D., Huang, S. and Plat, L. (1994) Endothelium derived relaxing factor inhibition and the pressor response to norepinephrine in the pregnant rat. *Obstetrics and Gynecology*, **83**, 92–96.

Arnal, J.F., Clamens, S., Pechet, C., NegreSalvayre, A., Allera, C., Girolami, J.P. *et al.* (1996) Ethinylestradiol does not enhance the expression of NO synthase in bovine endothelial cells but increases the release of bioactive NO by inhibiting superoxide anion production. *Prceedings of the National Academy of Science*, **93**, 4108–4113.

Ashworth, J.R., Warren, A.Y., Baker, P.N. and Johnson IR. (1997). Loss of endothelium-dependent relaxation in myometrial resistance arteries in pre-eclampsia. *British Journal of Obstetrics and Gynaecology*, **104**, 1152–1158.

Assali, N.S. and Prystowsky, H. (1950) Studies on autonomic blockade. I. Comparison between the effects of tetraethylammonium chloride (TEAC) and high selective spinal anesthesia on blood pressure of normal and toxemic pregnancy. *Journal of Clinical Investigation*, **29**, 1354–1366.

Baker, P.N., Broughton Pipkin, F. and Symonds, E.M. (1992) Longitudinal study of platelet angiotensin II binding in human pregnancy. *Clinical Science*, **82**, 377–381.

Baker, P.N., Davidge, S.T., Barankiewicz, J. and Roberts, J.M. (1996a) Plasma of preeclamptic women stimulates then inhibits endothelial prostacyclin. *Hypertension*, **37**, 56–61.

Baker, P.N., Stranko, C.P., Davidge, S.T., Davies, P.S. and Roberts, J.M. (1996b) Mechanical stress eliminates the effects of plasma from patients with preeclampsia on endothelial cells. *American Journal of Obstetrics and Gynecology*, **174**, 730–736.

Ballegeer V., Spitz, B., Kieckens, L., Moreau, H., Van Assche, A. and Collen, D. (1989) Predictive value of increased levels of fibronectin in gestational hypertension. *American Journal of Obstetrics and Gynecology*, **161**, 432–436.

Barden, A., Beilin, L.J., Ritchie, J., Croft, K.D., Walters, B.N. and Michael, C.A. (1996). Plasma and urinary 8-iso prostane as an indicator of lipid peroxidation in preeclampsia and normal pregnancy. *Clinical Science*, **91**, 711–718.

Barden, A., Graham, D., Beilin, L.J., Ritchie, J., Baker, R., Walters, B.N. and Michael, C.A. (1997) Neutrophil CD11B expression and neutrophil activation in pre-eclampsia. *Clinical Science*, **92**, 37–44.

Barrow, S.E., Blair, I.A., Waddell, K.A., Shepherd, G.L., Lewis, P.J. and Dollery, C.T. (1983) Prostacyclin in late pregnancy: analysis of 6–oxo-prostaglandin F_1 in maternal plasma. In: *Prostacyclin in Pregnancy*, pp. 79–85, Lewis, P.J., Moncada, S. and O'Grady, J. (eds.) Raven Press, New York.

Benedetto, C., Barbero, M., Rey, L., Zonca, M., Massobrio, M. and Slater, T.F. (1987) Production of prostacyclin, 6–keto-PGF_1 and thromboxane B_2 by human umbilical vessels increases from the placenta towards the fetus. *British Journal of Obstetrics and Gynaecology*, **94**, 1165–1169.

Bobadilla, R.A., Henkel, C.C., Henkel, E.C., Escalente, B. and Hong, E. (1997). Possible involvement of endothelium-derived hyperpolarizing factor in vascular responses of abdominal aorta from pregnant rats. *Hypertension*, **30**, 596–602.

Boccardo, P., Soregaroli, M., Aiello, S., Noris, M., Donadelli, R. and Lojacono, A. (1996) Systemic and fetal-maternal nitric oxide synthesis in normal pregnancy and pre-eclampsia. *British Journal of Obstetrics and Gynaecology*, **103**, 879–886.

Bodelssson, G., Marsal, K. and Stjernquist, M. (1995) Reduced contractile effect of endothelin-1 and noradrenalin in human umbilical artery from pregnancies with abnormal umbilical artery flow velocity waveforms. *Early Human Development*, **42**, 15–28.

Bonnar, J. (1987) Haemostasis and coagulation disorders in pregnancy. In *Haemostasis and Thrombosis*, edited by A.L. Bloom and D.P. Thomas, pp 570–584. Edinburgh, Churchill Livingstone.

Bower, S., Bewley, S. and Campbell, S. (1993) Improved prediction of preeclampsia by two-stage screening of uterine arteries using the early diastolic notch and color doppler imaging. *Obstetrics and Gynecology*, **82**, 78–83.

Branch, D.W. (1994) Pre-eclampsia and serum antibodies to oxidized low-density lipoprotein. *Lancet*, **343**, 645–646.

Bremme, K., Ostlund, E., Almquist, I., Heinonen, K. and Blomback, M. (1992) Enhanced thrombin generation and fibrinolytic activity in normal pregnancy and the puerperium. *Obstetrics and Gynecology*, **80**, 132–137.

Brosens, I., Robertson, W.B. and Dixon, H.G. (1967) The physiological response of the vessels of the placental bed to normal pregnancy. *Journal of Pathology and Bacteriology*, **93**, 569–579.

Brosens, I.A., Robertson, W.B. and Dixon, H.G. (1972) The role of the spiral arteries in the pathogenesis of pre-eclampsia. *Obstetrical and Gynecological Annuals*, **1**, 177–191.

Brown, M.A. and Gallery, E.D.M. (1994) Volume homeostasis in normal pregnancy and pre-eclampsia: physiology and clinical implications. *Baillieres Clinical Obstetrics and Gynaecology*, **8**, 287–310.

Broughton-Pipkin, F., Crowther, C., de Swiet, M., Duley, L., Judd, A., Lilford, R.J. *et al.* (1996) Where next for prophylaxis against pre-eclampsia? *British Journal of Obstetrics and Gynaecology*, **103**, 603–607.

Buttery, L.D., McCarthy, A., Springall, D.R., Sullivan, M.H., Elder, M.G. and Polak, J.M. (1994) Endothelial nitric oxide synthase in the human placenta; regional distribution and proposed regulatory role at the feto-maternal interface. *Placenta*, **15**, 257–265.

Calver A., J. Collier, D. Green, and Vallance, P. (1992) Effect of acute plasma volume expansion on peripheral arteriolar tone in healthy subjects. *Clinical Science*, **83**, 541–547.

Cameron, I.T., van Papendorp, C.L., Palmer, R.M.J., Smith, S.K. and Moncada, S. (1993) Relationship between nitric oxide synthesis and increase in systolic blood pressure in women with hypertension in pregnancy. *Hypertension in Pregnancy*, **12**, 85–92.

Caritis, S., Sibai, B., Hauth, J., Lindheimer, M.D., Klebanoff, M., Thom, E. *et al.* (1998) Low dose aspirin to prevent preeclampsia in women at high risk. *New England Journal of Medicine*, **338**, 701–705.

Castro, L.C., Hobel, C.J. and Gornbein, J. (1994) Plasma levels of atrial natriuretic peptide in normal and hypertensive pregnancies: a meta-analysis. *American Journal of Obstetrics and Gynecology*, **171**, 1642–1651.

Chait, A., Brazg, R., Tribble, D.L. and Krauss, R.M. (1993) Susceptibility of small, low dense low density lipoproteins to oxiative modification in subjects with the atherogenic lipoprotein phenotype pattern B. *American Journal of Medicine*, **94**, 350–356.

Chapman, A.B., Abraham, W.T., Zamudio, S., Coffin, C., Merouani, A., Young, D. *et al.* (1998) Temporal relationships between hormonal and hemodynamic changes in early human pregnancy. *Kidney International.*, **54**, 2056–63.

Chen, G., Wilson, R., Wang, S.H., Zheng, H.Z., Walker, J.J. and McKillop, J.H. (1996) Tumour necrosis factor alpha (TNF-a) gene polymorphism and expression in pre-eclampsia. *Clinical and Experimental Immunology*, **104**, 154–159.

Chesley, L.C. (1978) *Hypertensive Disorders In Pregnancy.* pp 126–131. New York: Appleton-Century-Crofts.

Chester, A.H., Jiang, C., Borland, J.A., Yacoub, M.H. and Collins, P. (1995) Oestrogen relaxes human epicardial coronary arteries through non-endothelium dependent mechanisms. *Coronary Artery Disease*, **6**, 417–422.

Chirico, S., Smith, C., Merchant, C., Mitchinson, M.J. and Halliwell, B. (1993) Lipid peroxidation in hyperlipidaemic patients: a study of plasma using an HPLC-based thiobarbituric acid test. *Free Radical Research Communication*, **19**, 51–57.

Chu, Z.M. and Beilin, L.J. (1993) Mechanisms of vasodilatation in pregnancy: studies of the role of prostaglandins and nitric oxide in changes of vascular reactivity in the in situ blood perfused mesentry of pregnant rats. *British Journal of Pharmacology*, **109**, 322–329.

Chua, W.S., Wilkins, T., Sargent, I. and Redman, C. (1991) Trophoblast deportation in pre-eclamptic pregnancy. *British Journal of Obstetrics and Gynaecology*, **98**, 973–979.

Clark, B.A., Ludmire, J., Epstein, F.H., Alvarez, J., Tavara, L., Bazul, J. et al. (1997) Urinary cGMP, endothelin, and prostaglandin E2 in normal pregnancy and preeclampsia. *American Journal of Perinatology*, **14**, 559–562.

CLASP (Collaborative Low Dose Aspirin Study in Pregnancy) Collaborative group. (1994) CLASP: a randomised trial of low-dose aspirin for the prevention and treatment of preeclampsia among 9364 pregnant women. *Lancet*, **343**, 619–629.

Clifton, V.L., Read, M.A., Letch, I.M., Giles, W.B., Boura, A.L.A., Robinson, P.J. *et al.* (1995) Corticotropin-releasing hormone induced vasodilatation in the human fetal-placental circulation; involvement of the nitric oxide-cyclic guanosine 3'-5'-monophosphate mediated pathway. *Journal of Clinical and Endocrinological Metabolism*, **80**, 2888–2893.

Cockell, A.P. and Poston, L. (1997a) Flow mediated vasodilatation is enhanced in normal pregnancy but reduced in preeclampsia. *Hypertension*, **30**, 247–251.

Cockell, A.P. and Poston, L. (1997b) 17β estradiol stimulates flow-induced vasodilatation in isolated small mesenteric arteries from prepubertal female rats. *American Journal of Obstetrics and Gynecology*, **177**, 1432–1438.

Cockell, A.P., Learmont, J.G., Smarason, A.L., Redman, C.W., Sargent, I.L. and Poston, L. (1997) Human placental syncytiotrophoblast microvillous membranes impair maternal vascular endothelial cell function. *British Journal of Obstetrics and Gynaecology*, **104**, 235–240.

Conger, J.D., Falk, S.A. and Guggenheim, S.J. (1981) Glomerular dynamics and morphologic changes in the generalised Shwartzman reaction in postpartum rats. *Journal of Clinical Investigation*, **67**, 1334–1346.

Conrad, K.P. and Colpoys, M.C. (1986) Evidence against the hypothesis that prostaglandins are the vasodepressor agents of pregnancy. *Journal of Clinical Investigation*, **77**, 230–245.

Conrad, K.P. and Vernier, V.A. (1989) Plasma levels, urinary excretion and metabolic production of cGMP during gestation in rats. *American Journal of Physiology*, **257**, R847–R853.

Conrad, K.P., Joffe, G.M., Kruszyna, H., Kruszyna, R., Rochelle, L.G., Smith, R.P.et al (1993) Identification of increased nitric oxide biosynthesis during pregnancy in rats. *FASEB Journal*, **7**, 566–571.

Conrad, K.P. and Benyo, D.F. (1997) Placental cytokines and the pathogenesis of preeclampsia. *American Journal of Reproductive Immunology*, **37**, 240–249.

Crandon, A.J. and Isherwood, D.M. (1979) Effect of aspirin on incidence of preeclampsia. *Lancet* **i**, 1356.

Curtis, N.E., Gude, N.M., King, R.G., Marriott, P.J., Rook, T.J. and Brennecke, S.P. (1995) Nitric oxide metabolites in normal human pregnancy and preeclampsia. *Hypertension in Pregnancy*, **23**, 1096–1105.

Danielson, L.A. and Conrad, K.P. (1995) Acute blockade of nitric oxide synthase inhibits renal vasodilation and hyperfiltration during pregnancy in chronically instrumented conscious rats. *Journal of Clinical Investigation*, **96**, 482–490.

D'Anna, R., Scilipoti, A., Leonardi, J., Scuderi, M., Jasonni, V.M. and Leonardi, R. (1997) Anticordiolipin antibodies in pre-eclampsia and intrauterine growth retardation. *Clinical and Experimental Obstetrics and Gynecology*, **24**, 135–137.

Davison, J.M. (1984) Renal haemodynamics and volume homeostasis in pregnancy. *Scandinavian Journal of Clinical and Laboratory Investigation*, Supplement **169**, 15–27.

Davison, J.M. and Baylis, C. (1995) Renal Disease. In *Medical Disorders In Obstetric Practice*, edited by M. de Swiet, pp 226–305. Oxford: Blackwell Scientific.

de Boer, K., ten-Cate, J.W., Sturk, A., Borm, J.J., Treffers, P.E. (1989) Enhanced thrombin generation in normal and hypertensive pregnancy. *American Journal of Obstetrics and Gynecology*, **160**, 95–100.

de Groot, C.J.M., Taylor, R.N. (1993) New insights into the etiology of pre-eclampsia. *Annuals of Medicine*, **25**, 243–249.

Dekker, G.A. and Sibai, B.M. (1993) Low-dose aspirin in the prevention of preeclampsia and fetal growth retardation: Rationale, mechanisms and clinical trials. *American Journal of Obstetrics and Gynecology*, **1168**, 214–227.

Dekker, G.A., de Vries, J.I.P., Doelitzsch, P.M., Huijgens, P.C., von Blomberg, B.M.E., Jakobs, C. *et al.* (1995) Underlying disorders associated with severe early onset preeclampsia. *American Journal of Obstetrics and Gynecology*, **173**, 1042–1048.

Department of Health. Why mothers Die. Report on Confidential Enquiries into maternal deaths in the United Kingdom, 1994–1996. (1998). London: HMSO.

Dixon, W.D., Tribe, R.M., Palmer, A.M., Linton, E.A. and Poston, L. (1996) Corticotrophin releasing hormone factor does not relax isolated human fetoplacental resistance arteries. *Journal of the Society of Gynecologie Investigation*, **3**, 225a.

Dizon-Towson, D.S., Nelson, L.M., Easton, K. and Ward, K. (1996) The factor V Leiden mutation may predispose women to severe preeclampsia. *American Journal of Obstetrics and Gynecology*, **175**, 902–905.

Dunlop, W. (1981) Serial changes in renal haemodynamics during normal human pregnancy. *British Journal of Obstetrics and Gynecology*, **88**, 1–9.

Edouard, D.A., Pannier, B.M., London, G.M., Cuche, J.L. and Safar, M.E. (1998) Venous and arterial behaviour during normal pregnancy. *American Journal of Physiology*, **274**, H1605–1612.

Endresen, M.J., Lorentzen, B. and Henriksen, T. (1992) Increased lipolytic activity and high ratio of free fatty acids to albumin in sera from women with preeclampsia leads to triglyceride accumulation in cultured endothelial cells. *American Journal of Obstetrics and Gynecology*, **167**, 440–447.

Endresen, M.J., Tosti, E., Lorentzen, B. and Henriksen, T. (1995) Sera of preeclamptic women is not cytotoxic to endothelial cells in culture. *American Journal of Obstetrics and Gynecology*, **172**, 196–201.

Fickling, S.A., Williams, D., Vallance, P., Nussey, S.S. and Whitley, G.S. (1993) Plasma concentrations of endogenous inhibitor of nitric oxide synthesis in normal pregnancy and pre-eclampsia. *Lancet*, **342**, 242–243.

Fitzgerald, D.J., Entman, S.S., Mulloy, K. and FitzGerald, G.A. (1987a) Decreased prostacyclin biosynthesis precedes the clinical manifestation of pregnancy induced hypertension. *Circulation*, **75**, 956–963.

Fitzgerald, D.J., Mayo, G., Catella, F., Entman, S.S. and FitzGerald, G.A. (1987b) Increased thromboxane biosynthesis in normal pregnancy is mainly derived from platelets. American Journal of Obstetrics and Gynecology, **157**, 325–330.

Fitzgerald, D.J., Rocki, W., Murray, R., Mayo, G. and FitzGerald, G.A. (1990) Thromboxane A_2 synthesis in pregnancy-induced hypertension. *Lancet*, **335**, 751–754.

Flavahan, N.A. and Vanhoutte, P.M. (1995) Endothelial cell signalling and endothelial cell dysfunction. *American Journal of Hypertension*, **8**, 28S-41S.

Ford, G.A., Robson, S.C. and Mahdy, Z.A. (1996) Superficial hand vein responses to NG-monomethyl-L-arginine in post-partum and non-pregnant women. *Clinical Science*, **90**, 493–497.

Fried, G. and Samuelson, U. (1991) Endothelin and neuropetide Y are vasoconstrictors in human uterine blood vessels. *American Journal of Obstetrics and Gynecology*, **164**, 1330–1336.

Gallery, E.D.M., Hunyor, S.N. and Gyory, A.Z. (1979) Plasma volume contraction: a significant factor in both pregnancy associated hypertension (pre-eclampsia) and chronic hypertension in pregnancy. *Quarterly Journal of Medicine*, **48**, 593–602.

Gallery, E.D., Rowe, J., Campbell, S. and Hawkins, T. (1995a) Secretion of prostaglandins and endothelin-1 by decidual endothelial cells from normal and preeclamptic pregnancies: comparison with human umbilical vein endothelial cells. *American Journal of Obstetrics and Gynecology*, **173**, 1557–1562.

Gallery, E.D., Rowe, J., Campbell, S. and Hawkins, T. (1995b) Effect of serum on secretion of prostacyclin and endothelin-1 by decidual endothelial cells from normal and preeclamptic pregnancies. *American Journal of Obstetrics and Gynecology*, **173**, 918–923.

Gant, N.F., Daley, G.L., Chand, S., Whalley, P.J. and MacDonald, P.C. (1973) A study of angiotensin II pressor response throughout primigravid pregnancy. *Journal of Clinical Investigation*, **52**, 2682–2689.

Garland, C.J., Plane, F., Kemp, B.K. and Cocks, T.M. (1995) Endothelium-dependent hyperpolarization: a role in the control of vascular tone. *Trends in Pharmacological Science*, **16**, 23–30.

Gerber, R.T., Anwar, M.A. and Poston, L. (1998) Enhanced acetylcholine induced relaxation in small mesenteric arteries from pregnant rats: an important role for endothelium-derived hyperpolarizing factor. *British Journal of Pharmacology*, **125**, 1212–1157.

Ghabour, M.S., Eis, A.L.W., Brockman, D.E., Pollock, J.S. and Myatt, L. (1995) Immunohistochemical characterisation of placental nitric oxide synthase expression in preeclampsia. *American Journal of Obstetrics and Gynecology*, **173**, 687–694.

Gilligan, D.M., Quyyumi, A.A. and Cannon, R.O. (1994a) Effects of physiological levels of estrogen on coronary vascular vasomotor function in postmenopausal women. *Circulation*, **89**, 2545–2551.

Gilligan, D.M., Badar, D.M., Panza, J.A., Quyyumi, A.A. and Cannon, R.O. (1994b) Acute vascular effects of estrogen in postmenopausal women. *Circulation*, **90**, 786–791.

Goodman, R.P., Killam, A.P., Brash, A.R. and Branch, R.A. (1982) Prostacyclin production during pregnancy and pregnancy complicated by hypertension. *American Journal of Obstetrics and Gynecology*, **142**, 817–822.

Goodrum, L., Saade, G., Jahoor, F., Belfort, M. and Moise, K. (1996) Nitric oxide production in normal human pregnancy. *Journal of the Society of Gynecological Investigation*, **3**, 97A

Greer, I.A., Haddad, N.G., Dawes, J., Johnstone, F.D. and Calder, A.A. (1989) Neutrophil activation in pregnancy-induced hypertension. *British Journal of Obstetrics and Gynaecology*, **96**, 978–982.

Gude, N.M., King, R.G. and Brennecke, S.P. (1990) Role of endothelium-derived nitric oxide in maintenance of low fetal resistance in the placenta Lancet, **336**, 1589–1590.

Gustaffson, J-A. Estrogen receptor ß-Getting in on the action? (1997) *Nature Medicine*, **3**, 493–494.

Haegerstrand, A., Hemsen, A., Gillis, C., Larsson, O. and Lundberg, J.M. (1989) Endothelin; presence in human umbilical vessels, high levels in fetal blood and potent constrictor effect. *Acta Physiologica Scandinavica*, **137**, 541–542.

Hakkinen, L.M., Vuolteenaho, O.J., Leppaluoto, J.P. and Laatikainen, T.J. (1992) Endothelin in maternal and umbilical cord blood in spontaneous labor and at elective cesarean delivery. *Obstetrics and Gynecology*, **80**, 72–75.

Halpern, W., Osol, G., Coy, S. (1984) Mechanical behaviour of pressurized in vitro prearteriolar vessels determined with a video system. *Annals of Biomedical Engineering*, **12**, 463–479.

Hayashi, T., Yamada, K., Esaki, T., Kuzuya, M., Satake, S., Ishikawa, T.et al (1995) Estrogen increases endothelial nitric oxide by a receptor mediated mechansim. *Biochemical, Biophysical Research Communications*, **214**, 847–855.

Hishikawa, K, Nakaki, T., Marumo, T., Suzuki, H., Kato, R. and Saryuta, T. (1995) Upregulation of nitric oxide synthase by estradiol in human aortic endothelial cells. *FEBS letters*, **360**, 291–293.

Holden, D. P., Fickling, S.A., StJ Whitley, G. and Nussey, S.S. (1998) Plasma concentrations of asymmetric dimethylarginine, a natural inhibitor of nitric oxide synthase, in normal pregnancy and pre-eclampsia. *American Journal of Obstetrics and Gynecology*, **178**, 551–556.

Howarth, S.R., Vallance, P. and Wilson, C.A. (1995) Role of thromboxane A2 in the vasoconstrictor response to enodthelin-1, angiotensin II and 5–hydroxytryptamine in human placental vessels. *Placenta*, **16**, 679–689.

Hubel, C.A., McLaughlin, M.K., Evans, R.W., Hauth, B.A., Sims, C.J. and Roberts, J.M. (1996) Fasting serum triglycerides, free fatty acids and malondiadlehyde are increased in preeclampsia, are positively correlated, and decrease within 48 hours post partum. *American Journal of Obstetrics and Gynecology*, **174**, 975–982.

Hubel, C.A., Gandley, R.E., Shakir, Y., Gallaher, M. and Roberts, J.M. (1997) Prevalence of small low-density lipoproteins is increased in preeclampsia. *Journal of the Society of Gynecologic Investigation*, **4**, 95A.

Izumi, H., Makino, Y., Mohti, H., Shirakawa, K. and Garfield, R.E. (1996). Comparison of nitric oxide and prostacyclin in endothelium-dependent vasorelaxation of human umbilical artery at midgestation. *American Journal of Obstetrics and Gynecology*, **175**, 375–381.

Janowiak, M.A., Magness, R.R., Habermehl, D.A. and Bird, I.M. (1998) Pregnancy increases ovine uterine artery endothelial cyclooxygenase expression. Endocrinology, **139**, 765–771.

Jiang, C., Sarrel, P.M., Lindsay, D.C., Poole-Wilson, P.A. and Collins, P. (1991) Endothelium-independent relaxation of rabbit coronary artery by 17ß-oestradiol. *British Journal of Pharmacology*, **104**, 1033–1037.

Jiang, C., Poole-Wilson, P.A., Sarrel, P.M., Mochizuki, S., Collins, P. and MacLeod, K.T. (1992) Effect of 17ß-oestradiol on contraction, Ca^{++} current and intracellular free Ca^{++} in guineapig isolated cardiac myocytes. *British Journal of Pharmacology*, **106**, 739–745.

Jones, C.J.P. and Fox, H. (1980) An ultrastructural and ultrahistochemical study of the human placenta in maternal pre-eclampsia. *Placenta*, **1**, 61–76.

Kaunitz, A.M., Hughes, J.M., Grimes, D.A., Smith, J.C. and Rochat, R.W. (1990) Causes of maternal mortality in the United States, 1979–1986. *American Journal of Obstetrics and Gynecology*, **163**, 460–465.

Khong, T.Y., De Wolf, F., Robertson, W.B. and Brosens, I. (1986) Inadequate maternal vascular response to placentation in pregnancies complicated by pre-eclampsia and by small-for-gestational age infants. *British Journal of Obstetrics and Gynecology*, **93**, 1049–1059.

Khong, T.Y., Sawyer, I.H. and Heryet, A.R. (1992) An immunohistologic study of endothelialization of uteroplacental vessels in human pregnancy – Evidence that endothelium is focally disrupted by trophoblast in preeclampsia. *American Journal of Obstetrics and Gynecology*, **167**, 751–756.

Kniaz, D., Eisenberg, G.M., Elrad, H., Johnson, C.A., Valaitis, J. and Bregman, H. (1992). Postpartum hemolytic uremic syndrome associated with antiphospholipid antibodies. *American Journal of Nephrology*, **12**, 126–133.

Knight, M., Redman, C.W.G., Linton, E.A. and Sargent, I.L. (1998) Shedding of syncytiotrophoblast microvilli into the maternal circulation in pre-eclamptic pregnancies. *British Journal of Obstetrics and Gynaecology*, **105**, 632–640.

Knock, G.A. and Poston, L. (1996) Bradykinin-mediated relaxation of isolated maternal resistance arteries in normal pregnancy and preeclampsia. *American Journal of Obstetrics and Gynecology*, **175**, 1668–1674.

Kopp, L., Paradiz, G. and Tucci, J.R. (1977) Urinary excretion of cyclic 3',5'-adenosine monophosphate and cyclic 3',5'-guanosine monophosphate during and after pregnancy. *Journal of Clinical Endocrinology and Metabolism*, **44**, 590–594.

Kruithof, E.K., Tran-Thang, C., Gudinchet, A., Hauert, J., Nicoloso, G., Genton, C. *et al.* (1987) Fibrinolysis in pregnancy: a study of plasminogen activator inhibitors. *Blood*, **69**, 460–466.

Kublickiene, K.R., Kublickas, M., Lindblom, B., Lunell, N.-O. and Nisell, H. (1997a) A comparsion of myogenic and endothelial properties of myometrial and resistance vessels in late pregnancy. *American Journal of Obstetrics and Gynecology*, **176**, 560–566.

Kublickiene, K.R., Cockell, A.P., Nisell, H. and Poston, L. (1997b) Role of nitric oxide in the regulation of vascular tone in pressurised and perfused resistance myometrial arteries from term pregnant women. *American Journal of Obstetrics and Gynecology*, **177**, 1263–1269.

Kublickiene, K.R., Gruenwald, C., Lindblom, B. and Nisell, H. (1998) Myogenic and endothelial properties of myometrial resistance arteries from women with preeclampsia. *Hypertension in Pregnancy*, **17**, 271–282.

Kupfermine, M., Peaceman, A.M., Wigton, T.R., Rehnberg, K.A. and Socol, M.L. (1994) Tumour necrosis factor a is elevated in plasma and amniotic fluid of patients with severe preeclampsia. *American Journal of Obstetrics and Gynecology*, **170**, 1752–1759.

Kyle, P.M., Buckley, D., Kissane, J., de Swiet, M. and Redman, C.W.G. (1995) The angiotensin sensitivity test and low dose aspirin are ineffective methods to predict and prevent hypertensive disorders in nulliparous pregnancy. *American Journal of Obstetrics and Gynecology*, **173**, 865–872.

Learmont, J.G. and Poston, L. (1996) Nitric oxide is involved in flow induced dilation of isolated human small fetoplacental arteries. *American Journal of Obstetrics and Gynecology*, **174**, 583–588.

Learmont, J.G., Cockell, A.P., Knock, G.A. and Poston, L. (1996) Myogenic and flow mediated responses in isolated mesenteric small arteries from pregnant and non-pregnant rats. *American Journal of Obstetrics and Gynecology*, **174**, 1631–1636.

Letsky, E.A. (1995) Coagulation defects. In Medical Disorders In Obstetric Practice, edited by de Swiet, M. pp 71–115. Oxford, Blackwell Scientific.

Liu, Y.A., Ostlund, E. and Fried, G. (1995) Endothelin-induced contractions in human placental blood vessels are enhanced in intrauterine growth retardation, and modulated by agents that regulate levels of intracellular calcium. *Acta Physiologica Scandinavica*, **155**, 405–414.

Lyall, F., Greer, I.A., Boswell, F., Macara, L.M., Walker, J.J. and Kingdom, J.C.P. (1994) The cell adhesion molecule VCAM-1, is selectively elevated in serum in pre-eclampsia: does this indicate the mechanism of leucocyte activation? *British Journal of Obstetrics and Gynaecology*, **101**, 485–487.

Lyall, F. and Greer, I.A. (1996) The vascular endothelium in normal pregnancy and pre-eclampsia. *Reviews of Reproduction*, **1**, 107–116.

Lyall, F., Robson, S.C. and Bulmer, J.N. (1998) Transformation of spiral arteries in human pregnancy: the role of nitric oxide. *Journal of Obstetrics and Gynaecology*, **18**, S45.

Mabie, W.C. (1992) Acute fatty liver of pregnancy. *Gastroenterological Clinics of North America*, **21**, 951–960.

MacGillivray, I., Rose, G.A. and Rowe, B. (1969) Blood pressure survey in pregnancy. *Clinical Science*, **37**, 395–407.

MacLean, M.R., Templeton, A.G. and McGrath, J.C. (1992) The influence of endothelin-1 on human foeto-placental blood vessels: a comparison with 5–hydroxytryptamine. *British Journal of Pharmacology*, **106**, 937–941.

MacRitchie, A.N., Jun, S.S., Chen, Z., German, Z., Yuhanna, I.S., Sherman, T.S. *et al.* (1997) Estrogen upregulates endothelial nitric oxide gene expression in fetal pulmonary artery endothelium. *Circulation Research*, **81**,

355–362.

Magness, R.R. and Rosenfeld, C.R. (1989) Local and systemic estradiol-17ß: effects on uterine and systemic vasodilatation. *American Journal of Physiology*, **256**, E536–E542.

Magness. R., Rosenfeld, C.R., Faucher, D.J. and Mitchell, M.D. (1992) Uterine prostaglandin production in ovine pregnancy: effects of angiotensin II and indomethacin. American Journal of Physiology, **263**, H188–H197.

Mahdy, Z., Otun, H.A., Dunlop, W. and Gillespie, J.I. (1998). The responsiveness of isolated human hand vein endothelial cells in normal pregnancy and in pre-eclampsia. *Journal of Physiology*, **508.2**, 609–617.

Maigaard, S., Forman, A. and Andersson, K.E. (1986) Relaxant and contractile effects of some amines and prostanoids in myometrial and vascular smooth muscle within the human uteroplacental unit. Acta Physiologica Scandinavica, **128**, 33–40.

Malamitsi-Puchner, A., Antsaklis, A., Economou, E., Mesogitis, S., Papantoniou, N., Koutra, N.et al. (1995) Endothelin 1–21 plasma levels in fetuses at 18–24 weeks of gestation. *Journal of Perinatal Medicine*, **23**, 321–325.

Many, A., Hubel, C.A. and Roberts, J.M. (1996) Hyperuricemia and xanthine oxidase in pre-eclampsia, revisited. *American Journal of Obstetrics and Gynecology*, **174**, 288–291.

Manyonda, I.T., Slater, D.M., Fenske, C., Hole, D., Choy, M.Y. and Wilson, C. (1998) A role for noradrenaline in pre-eclampsia: towards a unifying hypothesis for the pathophysiology. *British Journal of Obstetrics and Gynaecology*, **105**, 641–648.

Matijevic, R., Meekins, J.W., Walkinshaw, S.A., Neilson, J.P. and McFadyen, I.R. (1995) Spiral artery blood flow in the central and peripheral areas of the placental bed in the second trimester. *Obstetrics and Gynecology*, **86**, 289–292.

McCarthy, A.L., Taylor, P., Graves, J., Raju, S.K. and Poston, L. (1994) Endothelium dependent relaxation of human resistance arteries in pregnancy. *American Journal of Obstetrics and Gynecology*, **171**, 1309–1315.

McCarthy, A.L., Woolfson, R.G., Raju, S.K. and Poston, L. (1994) Abnormal endothelial cell function of resistance arteries from women with pre-eclampsia. *American Journal of Obstetrics and Gynecology*, **168**, 1323–1330.

Mikhail, M.S., Anyaegbunam, A., Garfinkel, D., Palan, P.R., Basu, J. and Romney, S.L. (1994) Preeclampsia and antioxidant nutrients: decreased plasma levels of reduced ascorbic acid, alpha-tocopherol, and beta-carotene. *American Journal of Obstetrics and Gynaecology*, 171, 150–157.

Mombouli, J.V., Wasserstrum, N. and Vanhoutte, P.M. (1993) Endothelins 1 and 3 and big endothelin-1 contract isolated human placental veins. *Journal of Cardiovascular Pharmacology*, **22 Supp 8**, S278–281.

Mulvany, M.J. and Halpern, W. (1977) Contractile properties of small arterial resistance vessels in spontaneously hypertensive and normotensive rats. *Circulation Research*; **41**, 19–26.

Myatt, L., Brockman, D.E., Langdon, G. and Pollock, J.S. (1993) Constitutive calcium-dependent isoform of nitric oxide synthase in the human placental villous vascular tree. *Placenta*, **14**, 373–383.

Ness, R.A. and Roberts, J.M. (1996) Heterogeneous causes constituting the single syndrome of preeclampsia: A hypothesis and its implications. *American Journal of Obstetrics and Gynecology*, **175**, 1365–1370.

Nisell, A., Hemsen, A., Lunell, N.-O., Wolff, K. and Lundberg, M.J. (1990) Maternal and fetal levels of a novel polypeptide, endothelin: evidence for release during pregnancy and delivery. *Gynecology and Obstetric Investigation*, **30**, 129–132.

Nobunaga, T., Tokugawa, Y., Hashimoto, K, Kimura, T., Matsuzaki, N., Nitta, Y. *et al.* (1996) Plasma nitric oxide levels in pregnant patients with preeclampsia and essential hypertension. *Gynecologic and Obstetric Investigation*, **41**, 189–193.

Ohno, Y., Mizutani, S., Kurauchi, O., Nishida, Y., Arii, Y. and Tomoda, Y. (1995) Umbilical plasma concentration of endothelin-1 in intrapartum fetal stress: effect of fetal heart rate abnormalities. *Obstetrics and Gynecology*, **86**, 822–825.

Orpana, A.K., Avela, K., Ranta, V., Viinikka, L. and Ylikorkala, O. (1996). The calcium-dependent nitric oxide production of human vascular endothelial cells in preeclampsia. *American Journal of Obstetrics and Gynecology*, **174**, 1056–1060.

Pascoal, I.F. and Umans, J.G. (1996) Effect of pregnancy on mechanisms of relaxation in human omental microvessels. *Hypertension*, **28**, 183–187.

Pascoal, I.F., Lindheimer, M.D., Nalbantian-Brandt, C. and Umans, J.G. (1998) Preclampsia selectively impairs endothelium-dependent relaxation and leads to oscillatory activity in small omental arteries. *Journal of Clinical Investigation*, **101**, 464–470.

Patrignani, P., Filabozzi, P. and Patrono, C. (1982) Selective cumulative inhibition of platelet thromboxane production by low-dose aspirin in healthy subjects. *Journal of Clinical Investigation,* **69**, 366–372.

Pedersen, A.K. and FitzGerald, G.A. (1984) Dose related kinetics of aspirin. *New England Journal of Medicine,* **311**, 1206–1211.

Perkins, A.V. and Linton, E.A. (1995) Identification and isolation of corticotrophin-releasing hormone-positive cells from the human placenta. *Placenta,* **16**, 233–243.

Pijnenborg, R., Dixon, G., Robertson, W.B. and Brosens, I. (1980) Trophoblastic invasion of human decidua from 8 to 18 weeks of pregnancy. *Placenta,* **1**, 3–19.

Poranen AK, Ekbland U, Uotila P and Ahotupa M. (1996) Lipid peroxidation and antioxidants in normal and preeclamptic pregnancies. *Placenta,* **17**, 401–405.

Potter, J.M. and Nestel, P.J. (1979) The hyperlipidaemia of pregnancy in normal and complicated pregnancies. *American Journal of Obstetrics and Gynecology,* **133**, 165–170.

Ramsay, B., Johnson, M.R., Leone, A.M. and Steer, P.J. (1995) The effect of exogenous oestrogen on nitric oxide production in women: a placebo controlled crossover study. *British Journal of Obstetrics and Gynaecology,* **102**, 417–419.

Redman, C.W.G. (1995) Hypertension in pregnancy. In *Medical Disorders In Obstetric Practice,* edited by M. de Swiet, pp. 182–225. Oxford: Blackwell Scientific.

Ritter, J.M., Cockcroft J.R., Doktor H., Beacham J., Barrow S.E. (1989). Differential effect of aspirin on thromboxane and prostaglandin biosynthesis in man. *British Journal of Clinical Pharmacology,* **28**, 573–579.

Roberts, J.M., Taylor, R.N., Musci, T.J., Rodgers, G.M., Hubel, C.A. and McLaughlin, M.K. (1989) Preeclampsia: an endothelial cell disorder. *American Journal of Obstetrics and Gynecology,* **161**, 1200–1204.

Roberts, J.M. and Redman, C.W.G. (1993) Pre-eclampsia: more than pregnancy-induced hypertension. *Lancet,* **341**, 1447–1451.

Roberts, M., Lindheimer, M.D. and Davison, J.M. (1996) Altered glomerular permselectivity to neutral dextrans and heteroporous membrane modeling in human pregnancy. *American Journal of Physiology,* **270**, F338–343.

Robinson, L.J., Weremowicz, S., Morton, C.C. and Michel, T. (1994) Isolation and chromosomal localization of the human eNOS gene. *Genomics,* **19**, 350–357.

Robson, S.C., Hunter, S., Boys, R.J. and Dunlop, W. (1989) Serial study of factors influencing changes in cardiac output during human pregnancy. *American Journal of Physiology,* **256**, H1060–H1065.

Rogers, G.M., Taylor, R.N. and Roberts, J.M. (1988) Pre-eclampsia is associated with a serum factor cytotoxic to human endothelial cells. *American Journal of Obstetrics and Gynecology,* **159**, 908–914.

Rosselli, M., Imthurn, B., Keller, P.J., Jackson, E.K. and Dubey, R.K. (1995) Circulating nitric oxide (nitrite/nitrate) levels in postmenopausal women substituted with 17 beta-estradiol and norethisterone acetate. A two year follow up study. *Hypertension,* **25**, 843–853.

Royal College of General Prctioners. (1967) Oral contraception and thromboembolic disease. *Journal of the Royal College of General Practioners,* **13**, 267–269.

Rubanyi, G.M., Freay, A.D., Kauser, K., Sukovich, D., Burton, G., Lubahn, D.B., *et al.* (1997) Vascular estrogen receptors and endothelium-derived nitric oxide production in the mouse aorta. Gender difference and effect of estrogen receptor gene disruption. *Journal of Clinical Investigation,* **99**, 2429–2437.

Rutherford, R.A., Wharton, J., McCarthy, A., Gordon, L., Sullivan, M.H., Elder, M.G. and Polak, J.M. (1993) Differential localization of endothelin ETA and ETB binding sites in human placenta. *British Journal of Pharmacology,* **109**, 544–552.

Sala, C., Campise, M., Ambroso, G., Motta, T., Zanchetti, A. and Morganti, A. (1995) Atrial natriuretic peptide and hemodynamic changes during normal human pregnancy. *Hypertension Dallas,* **25**, 631–636.

Schiff, E., Friedman, S.A., Stampfer, M., Kao, L., Barrett, P.H. and Sibai, B.M. (1996) Dietary consumption and plasma concentrations of vitamin E in pregnancies complicated by preeclampsia. *American Journal of Obstetrics and Gynecology,* **175**, 1024–1028.

Schneider, F., Lutun, P., Balduf, J.-J., Quirin, L., Dreyfus, M., Ritter, J. et al. (1996) Plasma cyclic GMP concentrations and their relationship with changes of blood pressure levels in pre-eclampsia. *Acta Obstetrica et Gynecologica Scandinvica,* **75**, 40–44.

Schobel, H.P., Fischer, T., Heuszer, K., Geiger, H. and Schmieder, R.E. (1996) Preeclampsia-A state of sympathetic overactivity. *New England Journal of Medicine,* **335**, 1480–1485.

Schrier, R.W. and Briner, V.A. (1991) Peripheral arterial vasodilation hypothesis of sodium and water retention in pregnancy: Implications for pathogenesis of preeclampsia-eclampsia. *Obstetrics and Gynecology,* **77**, 632–639.

Schunkert, H., Danser, A.H.J., Hense, H.W., Derkx, F.H.M., Kurzinger, S. and Riegger, G.A.J. (1997) Effects of estrogen replacement therapy on the renin-angiotensin system in postmenopausal women. *Circulation*, **95**, 39–45.

Seligman, S.P., Buyon, J.P., Clancy, R.M., Young, B.K. and Abramson, S.B. (1994) The role of nitric oxide in the pathogenesis of preeclampsia. *American Journal of Obstetrics and Gynecology*, **171**, 944–948.

Sexton, A.J., Loesch, A., Turmaine, M., Miah, S. and Burnstock, G. (1996) Electron-microsopic immunolabelling of vasoactive substances in human umbilical endothelial cells and their actions in early and late pregnancy. *Cell and Tissue Research*, **284**, 167–175.

Sibai, B.M., Ramadan, M.K., Usta, I., Salama, M., Mercer, B.M., Friedman, S.A. (1993) Maternal morbidity and mortality in 442 pregnancies with hemolysis, elevated liver enzymes and low platelets (HELLP Syndrome). *American Journal of Obstetrics and Gynecology*, **169**, 1000–1006.

Sibai, B.M., Kustermann, L. and Velasco, J. (1994) Current understanding of severe preeclampsia, pregnancy-associated hemolytic uremic syndrome, thrombotic thrombocytopenic purpura, hemolysis, elevated liver enzymes, and low platelet syndrome, and postpartum acute renal failure: different clinical syndromes or just different names? *Current Opinion in Nephrology and Hypertension*, **3**, 436–445.

Silver, R.K., Kupfermine, M.J., Russell, T.L., Adler, L., Mullen, T.A. and Caplan, M.S. (1996) Evaluation of nitric oxide as a mediator of severe preeclampsia. *American Journal of Obstetrics and Gynecology*, **175**, 1013–1017.

Simpkin, J.C., Kermani, F., Palmer, A.M., Campa, J.S., Tribe, R.M., Linton, E.A. et al. Effects of corticotrophin releasing hormone on contractile activity of myomterium from pregnant women. *Br. J. Obstet. Gynecol.*, in press

Sladek, S.M., Magness, R.R. and Conrad, K.P. (1997) Nitric oxide and pregnancy. *American Journal of Physiology*, **272**, R441–R463.

Smarason, A.K., Sargent, I.L., Starkey, P.M. and Redman, C.W.G. (1993) The effect of placental syncytiotrophoblast microvillous membranes from normal and pre-eclamptic women of the growth of endothelial cells in vitro. *British Journal of Obstetrics and Gynecology*, **100**, 943–949.

Smarason, A.K., Sargent, I.L. and Redman, C.W.G. (1996) Endothelial cell proliferation is suppressed by plasma but not serum from women with pre-eclampsia. *American Journal of Obstetrics and Gynecology*, **174**, 787–793.

Smarason, A.K., Allman, K.G., Young, D. and Redman, C.W.G. (1997) Elevated levels of serum nitrate, a stable end product of nitric oxide, in women with pre-eclampsia. *British Journal of Obstetrics and Gynaecology*, **104**, 538–543.

Sorensen, J.D., Secher, N.J. and Jespersen, J. (1995) Perturbed (procoagulant) endothelium and deviations within the fibrinolytic system during the third trimester of normal pregnancy. *Acta Obstetrica Gynecologica Scandinavia*, **74**, 257–261.

Sorensen, T.K., Easterling, T.R., Carlson, K.L., Brateng, D.A. and Benedetti, T.J. (1992). The maternal hemodynamic effect of indomethacin in normal pregnancy. *Obstetrics and Gynecology*, **79**, 661–663.

Thaler-Dao, H., Saintot, M., Baudin, G., Descomps, B. and Crastes de Paulet, A. (1974). Purification of the placental 15–hydroxyprostaglandin dehydrogenase : properties of the purified enzyme. *FEBS letters*, **48**, 204–208.

Tsukimori, K., Hirotaka, M., Shingu, M., Koyanagi, T., Masashi, N., Hitoo, N. (1992) The possible role of endothelial cells in hypertensive disorders during pregnancy. *Obstetrics and Gynecology*, **80**, 229–233.

Uotila, J.T., Tuimala, R.J. and Aarnio, T.M. (1993) Findings on lipid peroxidation and antioxidant function in hypertensive complications of pregnancy. *British Journal of Obstetrics and Gynaecology*, **100**, 270–276.

Vallance, P., Collier, J. and Moncada, S. (1989) Nitric oxide synthesised from L-arginine mediates endothelium dependent dilatation in human veins in vivo. *Cardiovascular Research*, **23**, 1053–1057.

Van Buren, G.A, Yang, D. and Clarke, K.E. (1992) Estrogen-induced uterine vasodilation is antagonised by L-nitroarginine methyl ester, an inhibitor of nitric oxide synthesis. *American Journal of Obstetrics and Gynecology*, **167**, 828–833.

van den Elzen, H.J., Wladimiroff, J.W., Cohen-Overbeek, T.E., de Bruijn, A.J. and Grobbee, D.E. (1996) Serum lipids in early pregnancy and risk of pre-eclampsia. *British Journal of Obstetrics and Gynaecology*, **103**, 117–122.

Vedernikov, Y.P., Belfort, M.A., Saade, G.R. and Mosie, K.J. (1995) Preeclampsia does not alter the response to endothelin-1 in human omental artery. *Journal of Cardiovascular Pharmacology*, **3**, S233–235.

Veille, J., Li, P., Eisenach, J.C., Massman, A.G. and Figueroa, J.P. (1996) Effects of estrogen on nitric oxide synthase biosynthesis and vascular relaxant activity in sheep uterine and renal arteries in vitro. *American Journal of Obstetrics and Gynecology*, **174**, 1043–1049.

Venema, R.C., Nishida, K., Alexander, R.W., Harrison, D.G. and Murphy, T.J. (1994) Organization of the bovine gene encoding the endothelial nitric oxide synthase. *Biochemica Biophysica Acta*, **1218**, 413–420.

Wang, Y. and Walsh, S.W. (1996) Antioxidant activities and mRNA expression of superoxide dismutase, catalase, and glutathione peroxidase in normal and preeclamptic placentas. *Journal of the Society of Gynecologic Investigation*, **3**, 179–184.

Wallenburg, H.C.S., Dekker, G.A., Makovitz, J.W. *et al.* (1986) Low-dose aspirin prevents pregnancy-induced hypertension and pre-eclampsia in angiotensin-sensitive primagravidae. *Lancet* **1**:1–3.

Weiner C.P., Lizasoain, I., S.A. Baylis, R.G. Knowles, I.G. Charles, and S. Moncada (1994). Induction of calcium dependent nitric oxide synthases by sex hormones. *Proceedings of the National Academy of Science*, **91**, 5212–5216.

Weiner, C.P., Knowles, R.G. and Moncada, S. (1994) Induction of nitric oxide synthases early in pregnancy. *American Journal of Obstetrics and Gynecology*, **171**, 838–843.

White, R.E., Darkow, D.J. and Falvo Lang, J.L. (1995) Estrogen relaxes coronary arteries by opening BKCa channels through a cGMP-dependent mechanism. *Circulation*, **77**, 936–942.

Williams, D.J., Vallance, P.J.T., Neild, G.H., Spencer, J.A.D. and Imms, F.J. (1997) Nitric oxide mediated vasodilatation in human pregnancy. *American Journal of Physiology*, **272**, H748–752.

Williams, D.J. and de Swiet, M. (1997) Pathophysiology of Pre-eclampsia.*Intensive Care Medicine*, **23**, 620–629.

Wolfe, C.D.A., Patel, S.P., Linton, E.A., Campbell, Anderson, Dornhurst, et al. (1988) Plasma corticotrophin-releasing factor (CRF) in abnormal pregnancy. *British Journal of Obstetrics and Gynaecology*, **95**, 1003–1006.

Wolff, K., Nisell, H., Modin, A., Lundberg, J.M., Lunell, N.O. and Lindblom, B. (1993) Contractile effects of endothelin 1 and endothelin 3 on myometrium and small intramyometrial arteries of pregnant women at term. *Gynecological and Obstetric Investigation*, **36**, 166–171.

Wolff, K., Nisell, H., Carlstom, K., Kublickiene, K., Lunell, N.O. and Lindblom, B. (1996) Endothelin-1 and big endothelin-1 levels in normal term pregnancy and in preeclampsia. *Regulatory Peptides*, **67**, 211–216.

Wolff, K., Kublickiene K.R., Kublickas, M., Lindblom, B., Lunell, N.-O., Nisell, H. (1996) Effects of endothelin-1 and the ETA receptor antagonist BQ-123 on resistance arteries from normal pregnant and preeclamptic women. *Acta Obstetrica et Gynecologica Scandinavica*, **75**, 432–438.

Wolff, K., Carlstom, K., Fyhrquist, F., Hemsen, A., Lunell, N.O., Nisell, H. (1997) Plasma endothelin in normal and diabetic pregnancy. *Diabetes Care*, **20**, 653–656.

Xu, D., Martin, P., St John, J., Tsai, P., Summer, S.N., Ohara, M., *et al.* (1996). Upregulation of endothelial and constitutive nitric oxide synthase in pregnant rats. *American Journal of Physiology*, **271**, R1739–R1745.

Yallampalli, C. and Garfield, R.E. (1993) Inhibition of nitric oxide synthesis in rats during pregnancy produces signs similar to those of pre-eclampsia. *American Journal of Obstetrics and Gynecology*, **169**, 1316–1320.

Yasuda, M., Takakuwa, K., Tokunaga, A., Tanaka, K. (1995) Prospective studies of the association between anticordiolipin antibody and outcome of pregnancy. *Obstetrics and Gynecology*, **86**, 555–559.

Zammit, V.C., Whitworth, J.A. and Brown, M.A. (1996) Preeclampsia: the effects of serum on endothelial cell. *American Journal of Obstetrics and Gynecology*, **174**, 737–743.

Zemel, M.B., Zemel, P.C., Berry, S., Norman, G., Kowalczyk, C., Sokol, R.J., Standley, P.R., Walsh, M.F. and Sowers, J.R. (1990) Altered platelet calcium metabolism as an early predictor of increased peripheral vascular resistance and pre-eclampsia in urban black women. *New England Journal of Medicine*, **323**, 434–438.

Zhou, Y., Damsky, C.H., Chiu, K., Roberts, J.M. and Fisher, S.J. (1993) Preeclampsia is associated with abnormal expression af adhesion molecules by invasive cytotrophoblasts. *Journal of Clinical Investigation*, **91**, 950–960.

Index